Parasitoids in Pest Management

Pests cause economic damage to crops and stored products, while vectors are responsible for the transmission of disease-causing agents in human beings and livestock. Although the application of synthetic pesticides in agriculture has given immediate relief, it has also caused well-known side effects, leading to a general consensus among entomologists and agriculturists to shift towards other ecofriendly pest management. Insect natural enemies consisting of predators, parasitoids, and pathogens have attracted the attention of scientists across the globe. These natural enemies exist in agroecosystems and suppress the populations of pests. Parasitoids are farmers' friends and the most successful group of natural enemies. Highly specialised in their host choice, the female parasitoid searches for a suitable host to lay eggs and larvae in, on or near the hosts. Exploiting this potency of parasitoids may yield successful results in controlling notorious pests in an ecofriendly way.

This book covers information on the important biocontrol agents effective in pest suppression. It starts with insect parasitic groups followed by specific groups of parasitoids. It is hoped that the book presents a comprehensive account of beneficial parasitoids and will be useful to undergraduate and postgraduate students of Entomology, Biological Control, Plant Protection, Agricultural Zoology, and Zoology, besides those involved in competitive examinations and policy planning.

Features

- Each chapter has been written by very experienced specialists.
- Every chapter includes Learning Objectives and Points to Remember.
- This book offers comprehensive knowledge of parasitoids and their application in pest management in a rational way.

Parasitoids in Pest Management

Edited by
Omkar

CRC Press
Taylor & Francis Group
Boca Raton London New York

CRC Press is an imprint of the
Taylor & Francis Group, an **informa** business

First edition published 2023
by CRC Press
6000 Broken Sound Parkway NW, Suite 300, Boca Raton, FL 33487-2742

and by CRC Press
4 Park Square, Milton Park, Abingdon, Oxon, OX14 4RN

CRC Press is an imprint of Taylor & Francis Group, LLC

Library of Congress Cataloguing-in-Publication Data
Names: Omkar, Omkar, editor.
Title: Parasitoids in pest management / Omkar.
Description: First edition. | Boca Raton, FL : CRC Press, 2023. |
Includes bibliographical references and index.
Identifiers: LCCN 2022043136 | ISBN 9781032406701 (hbk) |
ISBN 9781032406725 (pbk) | ISBN 9781003354239 (ebk)
Subjects: LCSH: Parasitoids. | Pests--Biological control.
Classification: LCC QL496.12 .P38 2023 |
DDC 595.7--dc23/eng/20221018
LC record available at https://lccn.loc.gov/2022043136

ISBN: 978-1-032-40670-1 (hbk)
ISBN: 978-1-032-40672-5 (pbk)
ISBN: 978-1-003-35423-9 (ebk)

DOI: 10.1201/9781003354239

Typeset in Times
by MPS Limited, Dehradun

Contents

Preface

Insects are the most diverse and numerous of all animal groups and occupy almost all habitats except marine habitats. Almost one million species of insects have been reported so far, with more than 6 million species expected to be reported. Extensive research and surveys are resulting in the discovery of new species, which have in some cases led to the change in the classification. The majority of the insects are beneficial and humans could have hardly survived without them as they provide numerous ecosystem services, such as scavenging, crop pollination, medicines, food and yielding fibres (silks), animal resin (lac), honey, propolis, wax, etc. They are also helpful in criminal investigations, as best models for research, and are quite useful in technological innovations.

Yet, insects have often been associated with negative connotations despite only a small fraction (1%) of them being harmful as pests and vectors to human beings. Pests cause economic damage to crop yields and stored products, while vectors are responsible for the transmission of disease-causing agents in human beings as well as livestock. To manage their populations initially various ashes, inorganic chemicals and plant products were used. Gradually, after the discovery of DDT, other organochlorinated, organophosphate and carbamate pesticides were used to suppress the pest and vector populations. While, these provided immediate relief by yielding desired results by minimising the pest populations and reducing the losses caused by them, in the long term they were found to be highly deleterious. Various side effects of pesticides, viz. their persistence and non-degradable nature, accumulation in the environment, eco-degradation, effect on nontarget organisms, especially the beneficial ones, resurgence of new pest species, resistance in target pest species, a threat to biodiversity and human health hazards; have come into the light, which compelled the entomologists and agriculturists to shift towards other ecofriendly pest management strategies.

In search for ecofriendly, safe and alternative strategies, the insect's natural enemies consisting of predators, parasitoids and pathogens have attracted the attention of scientists across the globe. These natural enemies exist in natural agroecosystems and feed upon other insects, which act as pests. Thus, these natural enemies act as natural checks to the pest populations, thereby preventing them from reaching proportions that would cause economic loss. Thus, the conservation and augmentation of natural enemies in nature and insectaries, can be employed safely in pest management programs and they form one of the essential components of Integrated Pest Management.

These natural enemies, such as parasitoids and predators serve as ecofriendly tools in the biological control of insect pests by reducing the use of agrochemicals, leading to self-sustaining agriculture and protecting the environment. Parasitoids are one of the farmers' friends and are the most successful group of natural enemies in the suppression of pest populations. The different parasitoids attack different stages of their hosts. They can be egg, larval, pupal and adult parasitoids. They can also be ecto- or endoparasites. They are usually smaller than their hosts. They are highly specialised in their host choice. The female parasitoid searches for a suitable host to lay eggs. The eggs and larvae are laid in or on or near the hosts. The immatures remain on or in the host and in due course of their development, they always kill the hosts. The parasitoids act slowly and in order to achieve the desired goal they take more time. But, once they establish and colonize in the agroecosystems, the process becomes cost-effective and self-perpetuating.

This book on "Parasitoids in Pest Management" covers information on one of the most important biocontrol agents, which are most effective in the suppression of insect pests of crops. It is so designed that each chapter has been written by senior specialists for better and more valuable information. The book starts with a topic, which deals with important insect parasitic groups followed by specific groups of parasitoids. It is hoped that the proposed book presents a comprehensive account of beneficial insect parasitoids and highlights the positive aspects of insects for the readership. It is hoped that the book will be useful to undergraduate and postgraduate students of Entomology, Biological Control, Plant Protection, Agricultural Zoology and Zoology, to those appearing for specific competitive examinations and to those involved in policy planning.

At the very outset, I am extremely thankful to the contributors of this book for accepting my invitation, sparing their invaluable time to complete the work in time to the best of their ability, and making the revisions of respective chapters as per suggestions. I am also thankful to Dr. Geetanjali Mishra for valuable inputs, Dr. Priya Singh for kind assistance in the cross-checking of references and the members of our research team Dr. Swati Saxena, Apoorva Shandilya, Chandni Verma, Tripti Yadav, Shriza Rai, Lata Verma, Deepali Gupta, Gauranvita Singh, Priya Yadav and Sanjeev Dalal for helping me in various ways during this work. I am thankful to my wife, Mrs. Kusum Upadhyay, for managing the home front and sparing me to complete the task on time.

Prof. Omkar
FNASc, Editor

Editor Biography

Professor Omkar is a former professor and head of the Department of Zoology at the University of Lucknow, India. For his unique contribution to science, Professor Omkar has been conferred with a Fellowship by The National Academy of Sciences, India (FNASc) along with other prestigious fellowships. With 25 PhDs supervised, Professor Omkar has completed 11 research projects funded by State and Central agencies; published more than 300 papers/chapters (including one each in Nature Biotechnology and Conservation Biology), besides having two third research papers in international journals; written and edited more than 20 books published by well recognised national and international publishers. He has also served as a subject expert for Swiss National Science Foundation and many indigenous grant applications. In view of his academic contributions, Professor Omkar has been felicitated for his excellence in academics and awarded the prestigious Saraswati Samman by the Department of Higher Education, Govt. of UP (2017), Professor T N Ananthakrishnan Award, Chennai (2012), Rescholar Award of Excellence in Agricultural Entomology, Patiala (2014), Prof. G. S. Shukla Gold Medal (2014) and Dr. S. Pradhan Memorial Lecture Award (2019). Known globally for identifying and harnessing the potential of beneficial insects, he has worked in collaboration with multiple international scientists and visited the School of Biological Sciences, University of East Anglia (UK) in 2006 for three months and the Department of Zoology, University of South Bohemia, Ceske Budejovice and Institute of Entomology, Czech Academy of Sciences, Czech Republic in 2016 for four Weeks. Professor Omkar is among the Top 2% Scientists of the World in the database of Stanford University published by Elsevier Inc., USA (2021). He has RGScore- 3563, H index-30 & I10-97.

Contributors

Meena Agnihotri
Department of Entomology
G B Pant University of Agriculture and Technology
Pantnagar, Uttarakhand, India

Dhriti Banerjee
Director
Zoological Survey of India
New Alipore, Kolkata

Samiran Chakrabarti
Formerly Department of Zoology
University of Kalyani
Kalyani, India

Kailash Chandra
Former Director Zoological Survey of India, M-Block
New Alipore, Kolkata

Mani Chellappan
AINPVPM, College of Agriculture
Kerala Agriculture University
Thrissur, Kerala

Bidhan Chandra Das
Former Professor of Zoology
Rajshahi University
Rajshahi, Bangladesh

Ankita Gupta
ICAR-National Bureau of Agricultural Insect Resources (NBAIR)
Bengaluru, Karnataka

Devanshu Gupta
Zoological Survey of India, M-Block
New Alipore, Kolkata

Sushil Kumar Jalali
ICAR-National Bureau of Agricultural Insect Resources
Bengaluru, India

Prasun Karmakar
Rain Forest Research Institute
Jorhat, Assam, India

Arvind Kumar
Forest Protection Division
Forest Research Institute
Dehradun, Uttarakhand, India

P. Girish Kumar
Western Ghat Regional Centre
Zoological Survey of India
Kozhikode, Kerala, India

Y. Lalitha
ICAR-National Bureau of Agricultural Insect Resources
Bengaluru, India

Atanu Naskar
Diptera Section, Zoological Survey of India
New Alipore, Kolkata

Omprakash Navik
ICAR-National Bureau of Agricultural Insect Resources
Bengaluru, India

Sunita Patra
Zoological Survey of India, M-Block
New Alipore, West Bengal, India

K Rajmohana
Zoological Survey of India, M-Block
New Alipore, West Bengal, India

M.T. Ranjith
AINPVPM, College of Agriculture
Kerala Agriculture University
Thrissur, Kerala

Aishwarya Ray
Forest Protection Division
Forest Research Institute
Dehradun (Uttarakhand), India

J. M. Samraj
Department of Entomology
G B Pant University of Agriculture and Technology
Pantnagar, Uttarakhand

Jayita Sengupta
Diptera Section
Zoological Survey of India
New Alipore, Kolkata

Sudhendu Sharma
Department of Entomology
Punjab Agricultural University
Ludhiana, Punjab, India

P.S. Shera
Department of Entomology
Punjab Agricultural University
Ludhiana, Punjab, India

Longjam Roni Kumar Singh
Zoological Survey of India, M-Block
New Alipore, Kolkata

Sudhir Singh
Forest Entomology Discipline, Forest Protection Division
Forest Research Institute
Dehradun, Uttarakhand, India

P.M. Sureshan
Western Ghat Regional Centre
Zoological Survey of India
Kozhikode, Kerala, India

Richa Varshney
ICAR-National Bureau of Agricultural Insect Resources
Bengaluru, India

1 Insect Parasitoids

Mani Chellappan and M.T. Ranjith
AINPVPM, College of Agriculture, Kerala Agriculture University, Thrissur, Kerala

CONTENTS

DOI: 10.1201/9781003354239-1

LEARNING OBJECTIVES

i. Insect parasites also called "parasitoids" are the most effective natural enemies and it develops in or on the host insect and eventually kills them during the process of development.
ii. Insect parasitoids are classified based on the stages of the host on which it develops, where it feeds and its developmental strategies.
iii. The Hymenopteran parasitic wasps account for more than 78% of total insect parasitoids and their unique feature is attributed to their remarkable ecological and evolutionary success.
iv. A large number of under-exploited species of parasitoids belong to the endopterygotan insect orders, *viz.*, Diptera, Coleoptera, Lepiodoptera and Neuroptera.
v. The insect parasitic species have been majorly reported in eight families in Diptera, nine families in Coleoptera, two families in Lepidoptera and one family in Neuroptera.
vi. They show high specificity towards their host insects, and capable of suppressing their host insect population and have a potential to become an inevitable component in biocontrol progarmmes.

1.1 INTRODUCTION

Biological control is an inevitable component in the integrated pest management programme and it is defined as the reduction of pest populations by natural enemies and typically involves an active human role (Brodeur et al., 2018). In many instances, the natural enemies associated with an insect pest may be inadequate to suppress the insect population. Under these circumstances, three approaches of biological control, *viz.*, importation, augmentation and conservation of natural enemies, can be followed either alone or in combination (Van Driesche et al., 2008; Nafiu et al., 2014) to reduce the insect pest menace in various agroecosystems.

In general, the natural enemies associated with insect pests fall into three categories, *viz.*, predatory insects, insect parasites and insect pathogens (fungi, bacteria, viruses, or nematodes)

(Bale et al., 2008). Among them, insect parasites also called "parasitoids" are the most effective natural enemies and many of them belong to the non-stinging wasps under the order, Hymenoptera. The term 'parasitoid' has been defined in several ways and according to Eggleton and Gaston (1990), a parasitoid is an organism, which develops on or in another single (host) organism that extracts nourishment from it and kills it as a direct or indirect result of their development.

Due to the high level of species diversity, the parasitoids are usually categorised based on the host they parasitise and wherever their offspring develops (Godfray, 1993). Usually, the adult female locates the host insect and oviposits directly on or into the host. For instance, most of parasitic wasps have specialised ovipositors that aid them simply to easily penetrate the cuticle of host insects or to drill through the plant tissues to reach the hosts hidden within stems or leaves (Strand and Obrycki, 1996). However, some parasitic flies also oviposit directly on or into their hosts, but others deposit their progeny near hosts. For these latter species, hosts become parasitised by consuming the fly's eggs or when the eggs hatch and mobile larvae enter the host (Feener and Brown, 1997). Some parasitoids that oviposit and complete their development in the egg stage of the host are called egg parasitoids, whereas parasitoids that attack other life stages, *viz.*, larva, pupa and adult, are referred to as larval, pupal, or adult parasitoids. However, parasitoids that oviposit in one host stage but whose progeny completes their development at the expense of subsequent stages are referred to as, for example, egg-larval or larval-pupal parasitoids (Godfray, 2007).

Parasitoids can also be classified based on where their progeny feeds. Those parasitoid species that develop within their host body are called endoparasitoids, whereas those that feed externally are called ectoparasitoids. If only a single individual parasitoid develops per the host insect is referred to as solitary parasitoid, whereas species in which more than one individual develops per host are known as gregarious parasitoids (Strand and Obrycki, 1996; Johnson, 2013). Another group of parasitoids is called Hyperparasitoids, where the parasitoid species parasitise another species of parasitoid already present in a host. Even more unusual are heteronomous hyperparasitoids, which are found in the hymenopteran family Aphelinidae in which female wasps are known to develop as primary parasitoids of Homopteran insects, whereas male wasps develop as hyperparasitoids in females of their own species or another (Walter, 1983). Parasitoids can be classified as either koinobiont or idiobiont based on their developmental strategies. Parasitoids whose hosts continue to grow after parasitism are called koinobionts, whereas those parasitoids whose hosts do not develop further after parasitism are called idiobionts (Askew and Shaw, 1986; Gullan and Cranston, 2014).

It is quite fascinating that hymenopteran parasitoids account for nearly 78% of the estimated number of species and consequently have served as models of choice for nearly all recent research on insect parasitoids (Godfray, 1993; Hawkins and Sheehan, 1994; Waage and Greathead, 1986). The suborder Apocrita in Hymenoptera is highly evolved and turned into a huge group called parasitica or parasitoid wasps, which are almost exclusively parasitoids, and the smaller group Aculeata in which the ovipositor has become stung and includes all the species of ants, bees and wasps that are eusocial in nature (Godfray, 2007). Some of the unique features that have contributed to the remarkable ecological and evolutionary success of the parasitic Hymenoptera are the parasitoids in Hymenoptera represent a single evolutionary lineage in contrast to the dozens, perhaps hundreds, of parasitoid lineages in the Diptera, Coleoptera and some other orders (Eggleton and Belshaw, 1992). Similarly, Hymenoptera is alone among the Holometabola insects, which retain the primitive lepsimatid form of ovipositor and associated accessory glands (Gauld and Bolton, 1988). The female hymenopteran parasitoids possess ovipositors that give direct access to concealed hosts and small hosts, such as eggs or early-stage larvae that are not directly accessible to parasitoids in other groups. Moreover, the venom produced in the modified accessory glands allows hymenopteran parasitoids to subdue large active hosts and manipulate the behavior and physiology of hosts in favour of their progeny. Finally, only hymenopteran parasitoids are haplodiploid and such a sex-determining system gives females control over the sex ratio of their progeny, permitting them both to match the sex of their progeny to the size of the host and to reduce the intensity of local mate competition in mixed-sex clutches (Charnov, 1982; Godfray, 1993).

The parasitic hymenopterans play a significant role in many biological control programmes since they suppress the population densities of their host insects. It is estimated that there are about 2,50,000 species of parasitic hymenopterans existing in the world, of which only about 50,000 species have been described so far (Gupta, 2004). In classical biological control programmes, the majority of the success has been achieved through the importation of parasitoids from three families, *viz* Aphilindae, Encyrtidae and Braconidae, whereas for mass culturing and release programmes, the Trichogrammatidae, Braconidae and Pteromalidea have been relied on (Frank et al., 2008). The hymenopteran parasitoids introduced into various parts of India to manage successfully different insect pests infesting economically important crops are given in Table 1.4.

The second major group of parasitoids belongs to the order Diptera and it includes an estimated 16,000 described species of parasitoids or about 20% of the known species with this lifestyle. The parasitoid habit has evolved on many occasions in this order, with the most important taxon being the family Tachinidae, these are almost exclusively parasitoids, with the adults frequently resembling houseflies, Mucidae. The parasitoid habit has also evolved in some of the insect groups, *viz.* beetles (Coleoptera), moths (lepidoptera), Mantispids (Neuroptera), etc. (Eggleton and Belshaw, 1992). Though hymenopteran parasitoids are successfully employed to manage insect pests, the parasitods in other insect order with the potential in suppressing their host densities are also existing in nature. This chapter exclusively deals with such under-exploited groups of parasitoids belonging to the order Diptera, Coleoptera, Lepidoptera and Neuroptera and their prospects in managing the insect pests and vectors in various crop ecosystems.

1.2 ORDER: DIPTERA

The parasitoids belonging to the order Diptera are highly underestimated and often unexplored with respect to their role in biocontrol of insect pests due to their relatively low numbers as compared to the major group of hymenopteran parasitoids. The biological, physiological and behavioural characteristics and the various mode of parasitisation allow the dipteran parasitoids to take possession of the various host insects under different environments and subsequently reduce their populations. Some representatives under the families Tachinidae and Bombyliidae are unique because of their ability to seek host insects thriving hidden in vegetables or in soil (Mellini, 1990). However, the features, like diverse feeding habits and multiple evolutionary origins, make the dipteran parasitoids an underutilised category, but a highly suitable group for quantitative studies of character convergence and adaptive radiation. A brief description of parasitoids belonging to various families under Diptera is given hereunder and the list of important species of parasitoids under various families in Diptera and given in Table 1.1.

1.2.1 Family: Cecidomyiidae

It is the family of dipteran flies known as gall midges or gall gnats. As the name implies, the maggots of the most gall midges feed within plant tissue during the course of development and form characteristic abnormal plant growths called "galls" in economically important crop plants. These flies are very small, fragile insects with body lengths varying from 2 to 3 mm; many of them are less than 1 mm long. They possess hairy wings and long antennae and are well known for the peculiar phenomenon known as "paedogenesis" where the larval stages are capable of reproducing without reaching maturity. Apart from phytophagous species, at least 20 genera and 300 species of entomophagous (predaceous and parasitic form) cecidomyiids have been reported to occur worldwide (Gagne, 2004). But, only a few of them (predaceous cecidomyids), i.e. *Aphidoletes aphidimyaza* (Rondani) and *Feltiella acarisuga* (Vallot), were exploited as biocontrol agents against aphids and spider mites, respectively (Solarska, 2004; Blindeman and van Labeke, 2003). It is quite interesting

TABLE 1.1
Parasitoid Species in Various Families Under the Order Diptera and Their Host Insects

Sl. no.	Family	Parasitoid species	Host insect
1	Cecidomyiidae	*Endaphis perfidus*	Aphids- *Drepanosiphum platanoides*, *Macrosiphum jaceae Aphis punicae*
		Endpahis fugitiva	Banana aphid -*Pentalonia nigronervosa*
		Endaphis muraii sp. nov	Aphids- *Macrosiphum euphorbiae, Aphis glycines*
		Endpahis aphidomyza	Aphids-*Uroleucon sonchi*, *Uromelan compositae compositae*, *Uromelan gobonis*
		Pseudanaphis maculans	Citrus Aphid -*Toxoptera aurantii*
		Occuloxenium compitale	Aphids in different crops
		Endopsylla agilis	Adult Psyllids- *Psylla foersteri, Psylla pyricola, Psylla melanoneura, Psylla mali.*
		Endopsylla endogena	Larval Tingids- *Stephanitis pyri*
2	Nemestrinidae	*Trichopsidea ostracea*	Australian plague locust, *Chortoicetes terminifera*
		sulphurous, N. sackeni	Grasshoppers- *Dichroplus arrogans, D. elongates*
		Neorhynchocephalus vitripennis	Grasshoppers- *Melanoplus mexicanus,*
		Trichopsidea claussa	Grasshoppers- *Camnula pellucida*
		Symmictus costatus	Locust- *Locustana pardalina*
		Symmictus flavopilosus	Locust-*Schistocerca gregaria*
		Hirmoneura sp	Scarabaeid beetles
		Atriadops vespertilio	*Mantodea*
3	Bombyliidae	*Anastoechus mylabricida*	Pupa of *Zonabris* sp
		Anthrax oophagous	Eggs of locust
		Systropus conopoides	Eucleidae larvae
		Heterostylum robustum	Alkali bee, *Nomia melanderi*
		Spogostylum spp.	Solitary bees and wasps, Meloidae, immature stages of Pyralidae
4	Asilidae	*Mallophora ruficauda*	White grub- *Cyclocephala signaticollis*
		Hyperechia bomboides	Bees -*Xylocopa* sp.
5	Phoridae	*Megapropodiphora arnoldi*	Social insects
		Megaselia nigrifinis sp. n.	Termite- *Odontotermes formosanus*
		Mehgaselia setidifferitatis sp. n.	Termite- *Odontotermes formosanus*
		Pseudacteon sp.	South American fire ants - *Solenopsis saevissima*
6	Pipenculidae	*Chalarus* sp.	Potato leaf hopper- *Empoasca fabae*
		Cephalops curtlfrow	Plant hopper, *Stenocranus minutus*
		Eudorylas schreiteri	Corn Leafhopper- *Dalbulus maidis*
7	Tachinidae	*Ceromasia sphenophori*	Sugarcane beetle borer
		Ptychomyia remota	Japanese beetle
		Centeter cinerea	Japanese beetle
		Winthemia lucanae	Armyworm
		Eucelotoria bryani	*Helicoverpa armigera*
		Diatraeophaga striatalis	sugarcane stem borer, *Chilo sacchariphagus indicus*
		Stomatomyia bezziana	Coconut leaf eating caterpillar, *Opisina arenosella*
8	Sarcophagidae	*Myiapis* sp.	Worker honeybees
		Senotainia sp.	Worker honeybees
		Sphixapate sp.	*Metopia* sp.
		Hilarella sp.	Honey bee
		Miltogramma sp.	Honey bee

that some species of cecidomyids belonging to the genus *Endaphis, Occuloxenium, Psuedanaphis* and *Endopsylla* are known to be the endoparasitoids of sucking insect pests, such as aphids and psyllids (Eggleton and Belshaw, 1992).

The first recorded instance of endoparasitism by a member of the dipteran family is that of *Endaphis perfidus* Kieffer on apterous adult aphid *Drepanosiphum platanoides* (Schr.) (Aphididae) from Lorraine and England (Kieffer, 1896; Kieffer,1901). The adult lays eggs on the body surface of aphids and the emerged maggot enters inside the body of aphids through intersegmantal membrane, subsequently becoming bright red in colour and thus it is easily visible in the green aphid by transparency, possessed a well-developed sternal spatula and apparently spent the whole of its feeding period inside the abdomen of its host. The full-grown larva leaves the body of the host through the anal aperture to the soil for pupation. Though it has been found parasitising different species of aphids in England, Germany and the Netherlands, the only one instance of rearing the adult midge *Endaphis* (cf. *perfidus* Kieffer) on aphid *Macrosiphum jaceae* (L.) colonising on carduus have been reported from Germany (Barnes, 1954).

A faunistic survey of natural enemies of pomegranate aphid (*Aphis punicae* Pass.) in Kashmar regions revealed the presence of dipteran endoparasitoid species *Endaphis perfidus* (Namaghi et al., 2012). Muratori et al. (2009) recorded a new species of cecidomyidendo parasitoid from Hawaii, *Endpahis fugitiva* (Plates 1a and 1b), which is parasitic on the banana aphid, *Pentalonia nigronervosa,* an important vector of Banana Buchy Top Virus (BBTV). Another cecidomyid endoparasitiod species, *Endaphis muraii* sp.nov. has been found parasitising the aphid species, such as *Macrosiphum euphorbiae* and *Aphis glycines* (Hemiptera: Aphididae) in Japan (Abe et al., 2011). Similarly, a new species, *Endpahis aphidomyza* has been reported from Maharashtra, India, which is found parasitising the aphid species, *Uroleucon sonchi*, *Uromelan compositae compositae* and *Uromelan gobonis* infesting the crops, *viz.*, mustard, sunflower and niger (Kumar and Chandra, 2015).

The endoparasitiod belongs to the genus *Psedanaphis and Pseudanaphis maculans*, and is the most common and widely distributed endoparasitoid of aphid species, *Toxoptera aurantii* (Boy.) infesting citrusinTrinidad (Kirkpatrick, 1954). Similarly, another genus of endoparasitic gall-midges, *Occuloxenium compitale,* has been reported to parasitise the aphids in different crop ecosystems at Uzbekistan (Mamaev, 1973).

Two species have been described in the genus *Endopsylla*. *Endopsylla agilis* is an endoparasitoid of the adult Psyllidae, whereas *Endopsylla endogena* has been parasitising the larval Tingidae. *Endopsylla agilis* is an endoparasitoid of winged *Psylla foersteri* in the Netherlands and it is also found parasitising the *Psylla pyricola/Psylla melanoneura*on hawthorns and *Psylla mali* on apple in ScotlandandGermany, whereas *Endopsylla endogena* is found parasitising the larva of *Stephanitis pyri* in Portugal (Barnes, 1954). Contrary to all the above-mentioned species of cecidomyid endoparasitoid, the *Endopsylla endogena* pupate inside the body of their host insects.

1.2.2 Family: Nemestrinidae

The Nemestrinidae, also known as tangle-veined flies, comprises of a group of Brachycerous flies with a worldwide distribution (Bernardi, 1973). This family consists of around 250 described species, distributed in 23 genera (Woodley, 2009). The Nemestrinid flies are medium-sized (4–16 mm), and their body is usually covered with dense pilose hairs (Woodley, 2009). The Head is wide, sometimes wider than the thorax (Bernardi, 1973), and large eyes are large, bare to densely pilose, ranging from clearly holoptic to widely dicoptic in males. Their mouthparts are varying from vestigial to very long (Bernardi, 1973; Woodley, 2009). Their thorax is unmodified with long and slender legs (Woodley, 2009). The wings are usually longer than the body, which ranges from hyaline to opaque, with a complex and characteristic venation. The wing has a characteristic diagonal vein formed by many fusions of veins in the radial and medial areas (Bernardi, 1973; Woodley, 2009). Their unique wing venation pattern is usually considered as an autapormophy of the group (Yeates, 1994). The adults are free-living, which frequently visit flowers of many different plants and feed upon the nectar; however, their larvae are known to be the endoparasites of grasshoppers and beetles (Evenhuis, 1994) and all species undergo hypermetamorphosis, in which they have different larval types in different instars (Woodley, 2009).

Plate 1a. *Endaphis fugitiva* – Adult

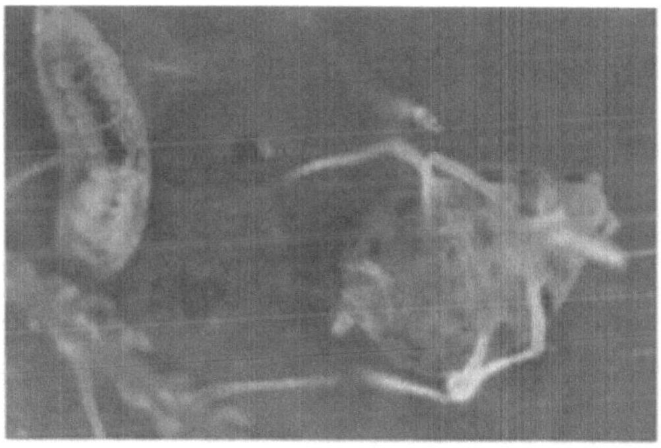

Plate 1b. *Endaphis fugitiva* maggot (Red)

Plate 1c. *Neorhynchocephalus sulphureus*

Plate 1d. *Hirmoneura* sp

Plate 1e. *Atriadops vespertilio*

Plate 1f. *Anastoechus mylabricida* Zack

PLATE 1 Parasitods in the order Diptera.

Plate 1g. *Heterostylum robustum*

Plate 1h. *Spogostylum* sp

Plate 1i. *Mallophora ruficauda*

Plate 1j. *Hyperechia bomboides*

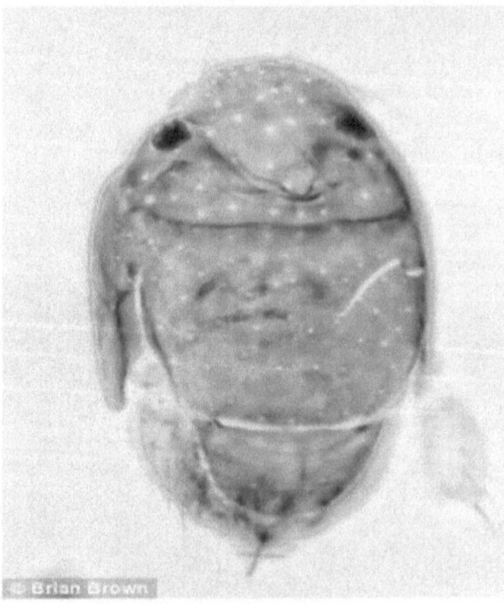

Plate 1k. *Megapropodiphora arnoldi*

Plate 1l. *Pseudacteon* sp

PLATE 1 (*continued*)

Plate 1m. *Eudorylas schreiteri*

Plate 1n. *Ceromasia sphenophori*

Plate 1o. *Metopia* sp

Plate 1p. *Brachicoma* sp

PLATE 1 (*continued*)

The first record of nemestrinid parasitic on Acridids is that of *Trichopsidea ostracea* the nymphs of Australian plague locust, *Chortoicetes terminifera* (Walker) (Olliff, 1891). The other species of nemestrinid parasitising different species of grasshoppers are *Neorhynchocephalus sulphurous* (Plate 1c), *N. sackeni,* (parasitic on grasshoppers: *Dichroplus arrogans and D. elongates*), *N.vitripennis* and *Trichopsidea claussa* (parasitic on grasshoppers: *Melanoplus mexicanus* and *Camnula pellucida*) (Spencer, 1945; Greathead, 1958). Erstwhile, Potgieter (1929) reported the parasitoid, *Symmictus costatus* Loew from *Locustana pardalina* Walker in South Africa; however, the first record of a nemestrinid parasitsing the desert locust, *Schistocerca gregaria* is that of Shulov (1948) who dissected out a young larva of an undetermined species of parasitoid from an immature male locust in Palestine and later confirmed the species as *Symmictus flavopilosus.* The same parasitoid has also been recorded from the field population of the desert locust, *Schistocerca gregaria* from Kenya (Greathead, 1958).

Richter (1997) recorded the nemestrinid endoparasitod under the genus *Hirmoneura* sp.Meigen (Plate 1d) from the larvae of scarabaeidbeetles. The female flies lay eggs in the host's habitat and the emerging first-instar larvae actively search for their host. The larval instars take about 40 days, but the prepupae undergo long diapause (inactive period) before 20 to 30 days of pupation take place. The free-living adults live for about 25 to 45 days (Richter, 1997); whereas, Haehnniand Borer (2007) have reported the species *Atriadops vespertilio* (Plate 1e) parasitising *Mantodea,* in Gambia.

1.2.3 Family: Bombyliidae

Bombyliidae (bee flies) is one of the largest families of Diptera, with over 5000 species described worldwide. Their high diversity may be due to the parasitic nature of the majority of their larvae (Price, 1980; Yeates and Greathead, 1997). The adults are free-living and feed on nectar and pollen from various flowering plants. They occur in all continents except Antarctica; whereas their highest diversities have been observed in semi-arid and arid environments (Hull, 1973). This structural diversity of this bees leads to a higher classification of the family, with a total of 31 subfamilies proposed to date. The family includes a wide variety of morphological forms; the body colour varies among species from gray, yellow, and iridescent green-blue to a striking black. Wings are hyaline with a spanning of 45 mm; they have a tiny, delicate, humpbacked body with a length of about 1 mm.

Most of the Bombyliidae are ectoparasitic on their host larvae, and a shrivelled host larval integument with mouth-hook sized holes, a shed bee flies' larval integument and a fully formed bee fly pupa can be visible after the emergence of the parasitoid. The endoparasitic Bombyliids consume their host after it has pupated and undergo the larval-pupal ecdysis inside the host's pupal case. Under these circumstances, two pupal cases are present in the parasitised cocoon or host shelter – one host pupal case with an exit hole, and the last larval exuvium of the bombyliid larva inside and the Bombyliidae pupal case split along the dorsum of the thorax where the bee fly emerged.

Riley (1880) observed the predaceous bombylids on the egg capsules of Rock Mountain locust. However, many species are external parasitoids of the larvae, prepupae and pupae of Hymenoptera in their cells or cocoons (Apoidea, Sphecoidea, Vespiodea, Tenthredinoidea and Ichneumonoidea) and some species are ectoparasitic on larvae and pupae of soil-inhabiting coleopterans (Meloidea, Cicindelloidea, and Scaraboidea). Zackvatkine (1934) found that the bombyliid species, *Anastoechus mylabricida* Zack (Plate 1f) parasitises the prepupae and pupae of the *Zonabris* sp. at the time when larvae of Carabidae and other insects are present in the soil. Anotherspecies, *Anthrax oophagous* Par. found parasitising the locust eggs, larvae and pupae of *Zonabris* sp., as well as occasionally developing on other bombyliids as secondary parasitoids. Similarly, *Systropus conopoides* and *Systropus* spp. are parasitic on Eucleidae larvae in their egg-like cocoons (Clausen, 1962a). However, some other bomyliid species develop as endoparasitoids in larvae and pupae of other Lepidoptera (Pyralidae, Noctuidae, Totricidae, Tineidae), Coleoptera (Tenebrionidae) and in pupae of Diptera (Muscidae, Tachnidae and Asilidae). Consequently, they are thought to have potential as biocontrol agents of insect pests in agroecosystems (Yeates and Greathead 1997; Dils and Ozbek, 2006; El-Hawagry, 2015 and Evenhuis and Greathead, 2015).

Bohart et al. (1960) recorded a major ectoparasitoid, *Heterostylum robustum* (Plate 1g) from the Alkali bee, *Nomia melanderi,* one of the important pollinators of alfalfa in USA whose rate of parasitism reached up to 90% in some area. The alkali bee usually nests in the ground and the range of *H. robusturn* greatly exceeds that of the alkali bee, and sometimes it is found parasitic on other ground-nesting bees in other parts of its range. Species of the genus *Spogostylum* (Plate 1h) are ectoparasitoids of solitary bees and wasps (order: Hymenoptera), Meloidae (order: Coleoptera), and immature stages of Pyralidae (order: Lepidoptera). Some species were also recorded as predators of egg pods of Acrididae (order: Orthoptera) (Greathead,1963; Yeates and Greathead, 1997).

1.2.4 Family: Asilidae

Asilidae, the dipteran family, consists of a number of voracious predatory flies commonly known as robber flies. However, a few species of all belonging to the *Mallophora* genus, are regarded as parasitoids, because during their larval stages, they attack white grubs and consumed them slowly (Castelo and Lazzari, 2004). *Mallophora ruficauda* (Plate 1i) is endemic of the Pampas region of Argentina inhabiting open grasslands near bee farms (Clements and Bennett, 1969). As adults, *M. ruficauda* feeds mainly on foraging honeybees and other flying insects, and as larvae, they are parasitoids of white grubs (Coleoptera: Scarabaeidae), mainly third instar larvae of *Cyclocephala signaticollis* Burmeister (Castelo and Lazzari, 2004; Castelo and Corley, 2010). Castelo and Crespo (2011) and Grobaand Castelo (2011) studied in detail the host location of *Mallophora ruficauda* larvae. They found that the first-instar larva cannot orient towards their preferred host larvae, *Cyclocephala signaticollis* Burmeister (Coleoptera: Scarabaeidae), because the maxillary palpi that bear the sensory organs are not well developed. Second-instar larvae up to 50 days old have well-developed maxillary palpi and can orient to and locate the host. Third through fifth instars are less active, and pupation occurs during the fifth stadium. Robber fly larvae orient to the host Scarabaeidae larvae using chemical cues generated in the host larva's proctodeum or fermentation chamber. *Mallophora ruficauda* larvae can detect and discriminate the odors of different potential hosts and can specifically orient to *C. signaticollis* larvae. Also, *M. ruficauda* larvae can determine the quality of the potential hosts with respect to parasitism status and can orient to healthy hosts. The chemicals that the robber fly larvae orient to are to be determined. However, *Hyperechia bomboides* (Plate 1j) is found parasitising the bee species, *Xylocopa*(Hymenoptera) larvae thriving in dead wood.

1.2.5 Family: Phoridae

This Dipteran family, also known as coffin flies, humpbacked flies and scuttle flies and consists of 4,000 species, is another group that exploits a wide range of habitats and exhibits diverse feeding habits. India shares only 1.59% of global phorid fauna and among them, the states Maharashtra (20.89%) and West Bengal (19.40%) share the maximum reported species followed by Karnataka (8.95%), Assam (7.46%), Meghalaya (5.97%,), Bihar (4.47%) and Kerala (2.98%). The other states and Union territories (Tamil Nadu, Himachal Pradesh, Uttarakhand, Chhattisgarh, Chandigarh, and Andaman & Nicobar Islands) share the minimum number of species (1.49%). There are also 1.49% of species, which are cosmopolitan in distribution (Mitra et al., 2016). The humpbacked appearance and reduced venation are the major characters help to identify the species under Phoridae. Many species are consumers of decaying organic matter and can infest household garbage cans; the females are strongly attracted to the odor of decay. Among the best-known parasitoid scuttle flies are those that specialize on adult ants, including leaf-cutter and fire ant workers; larvae develop in the host's head, often leading to decapitation (Feener and Brown, 1997). Some species are currently targeted as potential biocontrol agents of fire ants, a serious pest in the southern United States. The endoparasitoids of soil arthropods lay eggs either in soil or (more rarely) insert into the host. Most of them attack the larval or adult ants, provisioning aculeate Hymenoptera and termites, pupae of Coccinellidae and adult Lampyridae (Coleoptera), Sciaridae, Tipulidae and larvae of Bibionidae (Diptera); earthworms; millipedes; and snails. Some phorid females oviposit into adult bees in flight. Many other host records may be of facultative parasitoids (Eggleton and Belshaw, 1992).

Brown (2018) described the smallest phorid parasitoid, *Megapropodiphora arnoldi* (0.395 mm in body length) (Plate 1k) on social insects, which is slightly smaller than the currently recognised smallest fly, *Euryplatea nanaknihali* from Thailand. Further, Liu and Chen (2019) identified two scuttle flies, *Megaselianigrifinis* sp. nov. and *M. setidifferitatis* sp. nov. reared from dead termite, *Odontotermes formosanus* (Isoptera: Termitidae) in Nanjing, China. Chen and Porter (2020) found that *Pseudacteon* flies (Diptera: Phoridae) (Plate 1l) parasitise individual workers of South American fire ants in the *Solenopsis saevissima* (Smith) and causing decapitation of the hosts during pupation.

1.2.6 Family: Pipunculidae

Pipunculidae is a family of flies (Diptera) commonly termed big-headed flies, and possess large compound (holoptic) eyes, which cover nearly the entire head. The family is found all over the globe and more than 1,300 species have been described so far. Adults have small (body length 2.0–11.5 mm), dark in colour, head is semiglobose with either narrow occiput (Chalarinae) or globose with broadened occiput (Nephrocerinae and Pipunculinae). Wings are long and narrow, usually hyaline, iridescent in direct light, pterostigma present or absent, and the cell radius (r $_{4+5}$) is open.

Most pipunculids that have been studied are endoparasitoids of several Hemiptera (Auchenorrhyncha) families, including Cicadellidae, Cercopidae, Delphacidae, Membracidae, Issidae, Cixiidae, and Flatidae. The only known exception occurs in the genus *Nephrocerns* and these unusual flies attack the adult craneflies (Tipulidae) (Rothschild, 1964; Koenig and Young, 2007). The potential value of Pipunculidae for the biocontrol of insect pests has stimulated some research work on the bionomics of this family. The biological control of the potato leafi1opper, *Empoasca fabae* (Harris), a major pest of alfalfa in midwestern and eastern USA and Canada, involved the exploration in Europe for natural enemies to be introduced into the USA (Jervis, 1980) and included rearing of *Chalarus* for release (Skevington and Marshall, 1997). Despite the economic importance of the group, little data on host relationships exist, and some genera have no known hosts (e.g. *Allomethus, Basileunculus, Elmohardyia,* and *Protonephrocerns*).

The pipunculid adults are frequently seen hovering among vegetation, and their flight is similar to that of Syrphidae. Adults utilize all terrestrial habitats, but diversity and numbers are greatest in forest openings and along forest edges. In semiarid regions, adults are commonly found at seeps or along small creeks. Adult flies feed on honeydew and can occasionally be found in large, mixed feeding groups on secretions of Auchenorrhynchaninsects (Skevington and Marshall, 1998). Females of most pipunculid species deposit their eggs in early instar nymphs or adults of their hosts. After visually locating a host, some species remove the host from the food plant and oviposit one egg inside each host while in flight (Huq, 1982; May, 1979; Williams, 1919), whereas others leave the hosts *in-situ* (Benton, 1975).

May (2008) studied the biology of *Cephalops curtlfrow* Coe, an internal parasite of plant hopper, *Stenocranus minutus* (Fabricius), and found causing mortality in the adult stage. It is univoltine and host-specific. The female flies lay eggs in the fourth or fifth instar of *S. rninulus* and they hatch within 38 days. The first larval instars overwinter within their hosts, for a period of 24–25 weeks. After diapause, they moult to the second instar. This stage is relatively short and 30–35 days later the larvae emerge from the host, to pupate in the soil or in leaf-litter. The adult flies emerge after 18–30 days, which coincides with the presence in the field of the fourth and fifth instars of the host. Virla et al. (2009) recorded the *Eudorylas schreiteri* (Shannon) (Diptera: Pipunculidae) (Plate 1m) as a Parasitoid of the Corn Leafhopper, *Dalbulus maidis* (DeLong & Wolcott) (Hemiptera: Cicadellidae) in Argentina and pipunculid-host associations in the Neotropical Region.

1.2.7 Family: Tachinidae

It is the largest and most important group of insect parasitic flies. All species are parasitic in the larval stages and many are important natural enemies of major pests, especially caterpillars, also beetles, grasshoppers, and sawflies. Many species have been introduced into several countries from their native to suppress the population of alien pests. They differ in colour, size, shape and many have a resemblance with houseflies. Adult tachinids range from 2 mm to over 20 mm in length and vary widely in shape and color. They can be distinguished from most other flies by their well-developed subscutellum (postscutellum), and a row of setae on the meron (hypopleuron) (O'Hara, 2008). They usually are either gray-black, or striped and often have many distinct abdominal bristles. This is a very important family in the natural biocontrol of many pests and many have been introduced and successfully established in biocontrol programmes (Dindo, 2011).

Most tachinids attack caterpillars, and larval and adult beetles. Other species kill sawfly larvae, various types of true bugs, grasshoppers, or other types of insects. Egg laying varies considerably. In some species, eggs are deposited on foliage near the host insect. After the eggs hatch, the maggots are ingested during feeding by the host, such as a caterpillar and then develop inside the host. In other species, the adult fly glues her eggs to the body of a caterpillar or other host. After the eggs hatch, the maggot penetrates into the host's body. Some adult female tachinids possess a piercing ovipositor and actually insert the egg inside the host body (Dindo and Nakamura, 2018). Many tachinids exhibit an unusual trait in that the eggs mature within the mother fly, which then lays eggs that immediately hatch. In some species, egg hatching actually occurs within the mother fly and she gives birth to young ones (vivipary). Most, if not all, tachinids are internal parasites within their hosts. Most species are solitary, but sometimes are gregarious, with anywhere from 2–3 up to a dozen or more capable of developing within a single host (Meier et al., 1999).

Tachinids are of great importance in the control of destructive insect pests, particularly caterpillars and beetle larvae and mainly used in biological control programmes. The sugarcane beetle borer population in Hawaii has been reduced considerably by the tachinid, *Ceromasias phenophori* (Plate 1n) from New Guinea (Illingworth, 1919); the coconut moth in Fiji has been controlled by the Malayan tachinids, *Ptychomyia remota;* and *Centeter cinerea,* were imported to the United States to check the destructive Japanese beetle. The caterpillars of the armyworm were controlled by infestation with larvae of the red-tailed tachinids (*Winthemia lucanae*) (Molina-Ochoa et al., 2003).

1.2.8 Family: Sarcophagidae

Sarcophagidae is a species-rich family in the order Diptera, with some 2,500 described species worldwide. Their diversity is markedly concentrated in warmer climates and the family is poorly represented at high latitudes (Pape, 1996). The adult sarcophagids are large, mostly grey, with the abdomen generally chequered grey and black. The eyes are often bright red. These flies feed on flowers for nectar. The males are said to have "stations", where they wait for passing females. Many are scavengers of animal litter and the larvae feed on carrion, manure or dead insects, but a good proportion of the family are parasitoids of insects, such as beetles, grasshoppers, termites or caterpillars (Coupland and Barker, 2004). The adult flies are large, droning, active fliers and are attracted to sugars and they cause some considerable nuisance. A host group frequently attacked by species of the Sarcophagidae are the social wasps and bees. The relationship in some cases is strictly parasitic, while in others it is commensal. *Myiapis* and *Senotainia* are internal parasitoids of worker honeybees, and *Sphixapate* develops within the larvae. *Metopia* (Plate 1o) and *Brachicoma* (Plate 1p) are external parasitoids or predators of the brood of wild bees; the latter genus attacks mainly bumblebees (Pape, 1996). *Hilarella* and *Miltogramma* develop on various insects, which are stored in cells of hunting wasps or on the material with which the cell of bees is provisioned. Several genera have widely different host preferences (Krčmar et al., 2019).

1.3 ORDER: COLEOPTERA

The largest order in the class Insecta is Coleoptera, which consists of both beetles and weevils. The adult beetles/weevils have a hard, dense exoskeleton that covers and protects most of their body surface. The front pair of wings, known as elytra, are just as hard as the rest of the exoskeleton. They fold down elytra over the abdomen and serve as protective covers for the second pair of membranous hindwings. At rest, both elytra meet along the mid-dorsal area, forming a straight line that is probably the most distinctive characteristic of this order. During flight time, the elytra are held out to the sides of the body region where they provide a certain amount of aerodynamic stability (Arnett, 1960; Krinsky, 2002; McHugh and Leibherr, 2009).

Both larvae and adults of the beetles/weevils have mandibulate type of mouthparts. In general, they are clients of kinds of diets, thriving in all terrestrial and freshwater environments, and they do exhibit a variety of feeding habits. Several species are herbivores – feeding on the roots, stems, leaves, or reproductive structures of their host plants. Some species tend to live on fungi, others burrow into hard plant tissues, still, others excavate tunnels in wood or underneath the bark. Several beetles are predaceous in nature. They live in the soil or on vegetation and attack a large

TABLE 1.2
Parasitoid Species in Various Families Under the Order Coleoptera and Their Host Insects

Sl. no.	Family	Parasitoid species	Host insect
1	Carabidae	*Brachinus* sp.	Beetles- Hydrophilidae, Dytiscidae, and Gyrinidae
		Pelecium sulcatum	Chrysomelid pupae
		Lebia scapularis	Elm leaf beetle, *Xanthogaleruca luteola*
		Lebia grandis	Colorado potato beetle, *Leptinotarsa decemlineata*
		*Lebistina*sp.	Chrysomelid – *Diamphidia* sp.
2	Rhipiceridae	*Sandalus niger*	Cicada nymphs
		Polymerius chilensis	Cicada nymphs
		*Rhipicera*sp	Cicada- *Cyclochila australasiae*
3	Cleridae	*Hydnocera verticalis*	Apanteles sp.
		Hydnocera pubescens	Cotton boll weevil, *Anthonomus grandis*
4	Rhipiphoridae	*Rhipiphorus suhdipterus*	Solitary bee- *Halictus sexcinctus*
		Rhipidius pectinicornis	German cockroach -*Blattella germanica*
		Pelecotoma fennica	Wood boring Anobiidae
		Metoecus paradoxus	Yellow jacket wasp- *Vespula vulgaris*
		Macrosiagon ferrugineum	Wasps- *Euodynerus variegatus*, *E. disconotatus*, *Symmorphus murarius* and *Rhynchium oculatum*
5	Meloidae	Tribe- Epicautini and Mylabrini	Egg-pods of grasshoppers
		Nemonagthinae	Honey bee
		*Meloe*sp.	Honey bee
		Physomeloe sp.	Honey bee
		Epicauta spp	Eggs of acridid grasshoppers
		Nemognatha punctulata	Bee- *Megachile* sp.
6	Passandridae	*Aulonosoma insignis*	Larva of *Sinoxylon anale*
7	Bothrideridae	*Dastarcus helophoroides*	Longhorn beetles - *Monochamus alternatus*

kind of invertebrate hosts. Some beetles are scavengers, feeding on fecal matter, carrion, decaying wood, or alternative dead organic matter. There are even some parasitic beetles; some are internal parasites of other insects, some invade the nests of social insects like ants or termites, and a few as external parasites of mammals (Weiss, 1922; Eggleton and Belshaw, 1992). The important species of parasitoids belonging to various families under Coleoptera are enlisted in Table 1.2.

1.3.1 Family: Carabidae

Several genera of carabid beetles are ectoparasitoids of arthropods in soil. Generally, the parasitoid habit is uncommon in beetles; however, only eleven families of beetle include parasitoid species, compared to a much wider diversity of parasitoids as in the Diptera and Hymenoptera (Eggleton and Belshaw, 1993). The evolution and ecology of these parasitoid beetles are fascinating, but their host associations are poorly known (Saska and Honek, 2004). Carabid beetles have been commonly stereotyped as ground-dwelling generalist predators, yet in recent years many counter-examples have shown the Carabidae to be more diverse in form, habit, and trophic association. Many carabids, especially tropical species, are arboreal. Granivory, herbivory, and specialized predatory habits are widespread. Note that three of the 76 recognized tribes are known to have parasitoid species: Brachinini, Peleciini, and Lebiini. All of these are ectoparasitoids on pupae of other beetles or, in one Peleciine genus, on immature millipedes (Weber et al., 2008). In all known parasitoid carabids, the larva passes through three distinct development phases: (i) The free-living first larval instar emerges from an egg laid in the host's habitat and locates a host. (ii) The larva feeds on a single pupal or pre-pupal host, while it molts zero to four times. (iii) The host is consumed; the larva undergoes a nonfeeding larval stage ("pre-pupa") with zero to two molts; it then pupates next to the remains of the host. The total number of larval instars often deviates from the three molts typical for Carabidae, ranging from one (*Pelecium*) to five instars (some *Brachinus*). The adults live in the host habitat and may have a narrow or broad range of prey, including the immature stages of the host (Erwin, 1979; Saska and Honek, 2004).

The best-known genera of parasitoid carabids are *Brachinus*, *Lebia*, and *Lebistina*. *Brachinus*, (Plates 2a, 2b, and 2c) the celebrated bombardier beetle, emits a directed, explosive spray of

Plate 2a. *Brachinus* sp

Plate 2b. *Lebia* sp

Plate 2c. *Lebistina* sp

Plate 2d. *Pelecium sulcatum*

Plate2e. *Rhipicera* sp

Plate2f. *Trichodes* sp

PLATE 2 Parasitoids in the order Coleoptera.

Plate 2g. *Pelecotoma fennica*

Plate 2h. *Epicaucta* sp

Plate 2i. Passandrind species

Plate 2j. *Dastarcus helophoroides*

PLATE 2 (*continued*)

boiling-hot quinone solution, which is considered the most highly evolved defensive secretion of the many types documented in the Carabidae (Weber et al., 2006; Saska and Honek, 2004). Eisner (2003) has shown the elaborate mechanisms, which allow the orchestration of this exothermic reaction while protecting the emitter and instantly repelling potential predators. They have also shown the chain of evolutionary developments leading to this impressive set of defensive organs. North American *Brachinus* are found in littoral habitats near freshwater, where the known beetle hosts in families Hydrophilidae, Dytiscidae, and Gyrinidae emerge to pupate from their larval aquatic habitats. Recently, dryland European *Brachinus* has been associated with carabid hosts of the genus *Amara*, broadening the known hosts to 11 species, for only nine of the approximately 300 *Brachinus* species described. On the basis of fragmentary observation, it appears that *Pelecium sulcatum*(Pelecinii) (Plate 2d) develop as parasitoids on chrysomelid pupae and immature millipedes, and have only one larval instar (Saska and Honek, 2004). *Lebia* species number over 450 and the genus is cosmopolitan, with 47 species in North America. Adults typically seek prey in plant canopies, and all known larvae are ectoparasitoids of chrysomelid beetle pupae, yet only four species' hosts have been documented. Many additional *Lebia* species are reported to be associated (often with adult mimicry) with specific chrysomelids, particularly flea beetles (Alticinae) and casebearers (Cryptocephalinae), implying a host-parasitoid relationship. Two species parasitise economically important hosts: *L. scapularis* on elm leaf beetle, *Xanthogaleruca luteola* in Europe, and *L. grandis* on Colorado potato beetle, *Leptinotarsa decemlineata* in North America. Although elm leaf beetle is a significant invasive pest of ornamental elms in North America and elsewhere, *L. scapularis* apparently has not been considered for classical biocontrol. In contrast, *L. grandis* was introduced to France in the 1930s, and its parasitoid life history was discovered, as part of a USA-France classical biocontrol program (Weber et al., 2006).

Lebia adults are typically found in close association with their host species, and females oviposit in close proximity to the host pupal habitat; in the case of *L. grandis*, this takes place in the soil below infested host plants. *Lebistina*, an African genus closely related to *Lebia*, shows adult mimicry of its chrysomelid hosts, a pattern shared with some *Lebia species*. *Lebistina* is one part of a complex anthro-ecological story involving the San indigenous tribe of Southern Africa. San tribe members dig underground for the pupae of chrysomelids and their carabid parasitoids, both associated with the aromatic shrub *Commiphora* in the incense tree family, Burseraceae. Pupae of both the chrysomelid *Diamphidia*, and especially its parasitoid *Lebistina*, are collected for their potent neurotoxic arrow-poisons, which allow San hunters to fell large prey, such as giraffes, with small bows and arrows, but usually only after several days of tracking the injured animal. Parasitoid carabids present some fascinating evolutionary questions, not the least of which is why both the impressive arrow-poisons and the explosive exocrine toxins are associated with these genera. Yet, at most, 1% of their hosts are known. In addition, the possible management of predator/parasitoid beetles may offer an interesting opportunity for "double control" of chrysomelid pest species (Robertson, 2004).

1.3.2 Family: Scarabidae

Members of this, the sixth largest family in the order Coleoptera, are popular beetles due to their large size, bright colors, and interesting natural histories. Species of scarabs are important agricultural pests, used in the biocontrol of dung and dung flies, are pollinators and have been used as bioindicators of high-quality forest habitat.

Scarab beetles live in a diverse array of microhabitats and consume a wide range of food. Adults and larvae of many species feed on dung, carrion, hides, feathers, or decaying plant material, but several species have evolved the ability to utilize unique resources. A great number of species feeding on dung are specialists, feeding only on one type of vertebrate dung (e.g. deer dung, elephant dung, sloth dung), whereas many other species are generalist dung feeders. Some species will feed on both dung and carrion (Bai et al., 2015). No single species of parasitoids suppressing insect pest have been reported in Scarabidae, however some species acts as provision-directed cleptoparasitoids of other scarabaeid and geotrupid (Coleoptera) brood dung balls (Genus *Aphodeus* and *Onthophagous*). The adult penetrates into the brood ball after they have been made and lay an egg inside them. The host egg is eaten and the brood ball is consumed by cleptoparasitoid (Eggleton and Belshaw, 1992).

1.3.3 Family: Rhipiceridae

The Rhipiceridae, also known as cedar beetles or cicada parasites or is a small family in the order Coleoptera with more than 100 species in seven genera (Lawrence, 2005). The family is represented in the Nearctic region by *Sandalus*, a genus proposed by Knoch (1801) to include two North American species *S. petrophya* and *S. niger*. *Sandalus* also occurs in the New World, Africa, Southeast Asia, China, India, and Japan (Katovich, 2002).

The biology and immature stages of rhipicerids are very poorly known (Lawrence, 2005). In North America the first instar larvae, triungulin of *Sandalus niger* Knoch, have been found attached to cicada nymphs before they enter the soil and transform into ectoparasitic and more grub-like larvae. The final instar larva of *Polymerius chilensis* was described by Solervicens (2005) based on the cast skin of the larva found with remnants of cicada nymph and a newly emerged adult in the subterranean pupal chamber. In Australia, the putative larva of *Rhipicera* (Plate 2e) was found on the cicads nymphs of *Cyclochila australasiae* (Donovan), and it is highly probable that other rhipicerids are also cicada parasites (Elzinga, 1977; Lawrence et al., 1999; Katovich, 2002). Adults of *Sandalus* Knoch

(Katovich, 2002), *Rhipicera* and *Oligorhipis* (collection records) are very often attracted to lights. *Rhipicera femorata* (Kirby) has been collected on sandy swamplands and immediate environs, with sedges, grasses and other swampy land trees (Hawkeswood, 2000). Young and Katovich (2002) recorded the cedar beetle, *Sandalus niger* from Wisconsin for the first time.

1.3.4 Family: Cleridae

Cleridae, commonly known as checkered beetles, has a worldwide distribution and a variety of habitats and feeding preferences. They have elongated bodies with bristly hairs, are usually brightly colored, and have variable antennae. Checkered beetles range in length between 3mm and 24 mm. Cleridae can be identified based on their 5–5–5 tarsal formula, division of sternites, and the absence of a special type of vesicle. Female Cleridae lay between 28 and 42 eggs at a time predominately under the bark of trees. Larvae are predaceous and feed vigorously before pupation and subsequently emerge as adults (McNamara, 1991). Although most species are predaceous, some may develop at times as external parasitoids. Such species are principally in the genus *Hydnocera*, *Hydnocera verticalis* Say has been reared from the cocoons of *Apanteles. Hydnocera pubescens* Lec.seems to be parasitic on the larva of the cotton boll weevil, *Anthonomus grandis* Boh., in its cell in the boll, and the parasitoid finally spins its cocoon and pupates in the host cell (Chapin, 2014). Some are host-directed cleptoparasitoids of Apidae (Hymenoptera) cells in soil and dead plant materials (*Trichodes* sp.) (Plate 2f), feed only short time on provisioning, while others are parasitoids of gall-forming Lepidoptera larvae on aerial parts of the plants (*Pyllobaenus* sp.) (Eggleton and Belshaw, 1992).

1.3.5 Family: Rhipiphoridae

Rhipiphoridae also called "wedge-shaped beetles" is a cosmopolitan family of some 450 described species. Characters include a serrated female antenna, male antenna pectinate or flabellate, 11-segmented in both sexes; humpbacked, wedge-shaped beetles; pronotum large, distinct, narrowed anteriorly; tarsal formula 5-5-4; elytra entire; abdomen with five visible sternites, blunt at apex. The maxillary palps are four-segmented; labial palps three-segmented; legs slender; trochantin absent. In some species females are apterous and larviform (Selander, 1957; Lawrence et al., 2010).

Rhipiphorus suhdipterus is a parasitoid of solitary and subsocial bees, e.g. *Halictus sexcinctus* (Fabricius) (Chobaut, 1906; Bétis, 1912), whereas *Rhipidius pectinicornis* is a parasitoid of the German cockroach (*Blattella germanica* (Linnaeus)), a household pest of tropical origin and it has been recorded from ships infested with German cockroaches in Calcutta (Sundevall, 1831), in Denmark (Stamm, 1936) and in Egypt (Barbier, 1947).

Pelecotoma fennica (Plate 2g) is a parasitoid of wood-boring Anobiidae and is rarely found in woodlands in Germany, Poland, Austria and further into North-Eastern Europe (Horion, 1956; Lucht, 1987). However, *Metoecus paradoxus* is chiefly a Rhipiphorid parasitoid of the larvae of the common yellow jacket wasp,*Vespula vulgaris* (Linnaeus), but it may occasionally occur in the nests of *V. germanica* (Fabricius) and *V. rufa* (Linnaeus) and is commonly found in almost all European countries and further into eastern Europe and Russia (Lucht, 1987; Drees, 1994). *Macrosiagon ferrugineum* is a parasitoid of a number of different hollow stem and twig-nesting eumenid wasp species including *Euodynerus variegatus* (Fabricius), *E. disconotatus* (Lichtenstein), *Symmorphus murarius* (Linnaeus) and *Rhynchium oculatum* (Fabricius) (De Rond et al., 1994), but it is also reared from a nest of a *Eumenes* species (Chobaut, 1891; Vecht& Fischer, 1972).

1.3.6 Family: Meloidae

Meloidae, commonly known as "blister beetles" is a family in the order Coleoptera, which includes 3,000 described species (Bologna et al., 2008). Its vernacular name is related to the capacity to synthesize cantharidin, a sesquiterpenoid toxin (Percino-Daniel et al., 2013). The adult beetles are soft-bodied, long-legged beetles with the head deflexed, fully exposed, and abruptly constricted behind to form an unusually narrow neck, the pronotum is much narrower at the anterior end than the posterior and not carinate (keeled) laterally, the forecoxal cavities open behind, and each of the tarsal claw's cleft into two blades. Body length generally ranges between 3/4 and 2 cm. A few adults are nocturnal, but most are diurnal or show no distinct diel cycle. Since adults are gregarious and often highly coloured, they tend to be conspicuous. However, except for first instar larvae (triungulins) frequenting flowers or clinging to adult bees, larval blister beetles are seldom seen (Bologna and Pinto, 2001; Saha, 1979; Seala).

The four tribes within subfamily, Nemognathinae and six of the eight tribes included in subfamily, Meloinae are parasitoids of many different species of solitary or subsocial bees (superfamily Apoidea) feeding on all resources available at the nest: eggs, bee larvae, and provisions. However, the two exceptions are the tribes Epicautini and Mylabrini, in which the first instar larvae feed on eggs within the egg-pods of grasshoppers of the family Acrididae (Bologna, 1991; Bologna and Di Giulio, 2011). All species of Nemonagthinae are phoretic parasitoids of bees with the probable exception of *Stenodera* (Bologna et al., 2002; Bologna and Di Giulio, 2011), whereas some tribes of Meloinae that are bee-parasitoids (e.g. Meloini) include phoretic (*Meloe*) and non-phoretic (*Physomeloe*) genera. The larvae of some Meloinae, including most *Epicauta* spp., (Plate 2h) prey on the eggs of acridid grasshoppers. A few larvae evidently prey on the eggs of blister beetles (Selander 1981). Of the Florida species, *Nemognatha punctulata* LeConte (misidentified as *Zonitis vittigera* (LeConte)) has been found in a nest of a *Megachile* sp. in Cuba (Scaramuzza 1938) and several members of the genus *Epicauta* have been associated with the egg pods of *Melanoplus* spp.

1.3.7 Family: Passandridae

The family Passandridae "the parasitic flat bark beetles" (Plate 2i) comprises a small group of beetles containing nine genera and about 150 extant species distributed worldwide except in the west Palearctic Region and in New Zealand (Thomas, 2002). They have elongated parallel-side bodies ranging from 6 to 25 mm in length. Most are brown coloured, while others are black, reddish or yellow. Heads are triangular in shape, with filiform antennae of 11 antennomeres, and large mandibles. The pronotum is narrower than the head. The adults live under bark and in the tunnels of wood-boring insects and may be attracted to light at night (Thomas and Chaboo, 2015). Larvae of Passandridae seem to be exclusively ectoparasites on wood-inhabiting insect larvae or pupae, such as longhorn beetles, bark and ambrosia beetles, weevils, and Hymenoptera (Thomas, 1993; Burckhardt and Ślipiński, 2003). In India, during 2007–2008, the larvae of the passandrid, *Aulonosoma insignis* were observed to be ectoparasitic on the larvae of *Sinoxylon anale* infesting wooden logs of subabul (*Leucaena leucocephala* (Lam.) (Deepthi and Ramadevi, 2012).

1.3.8 Family: Bothrideridae

The family Bothrideridae contains about 400 species in 38 genera worldwide (Ślipiński et al., 2010), including 148 Palaearctic species placed in 18 genera (Lee et al., 2017). Members of the Bothrideridae are characterized by a body with oblong to distinctly elongate shape, cylindrical or slightly flattened; surface glabrous, with fine hairs, rarely with scales; head slightly declined; antenna 9–11-segmented, usually with compact club composed of 1–3 segments; eyes large and distinctly prominent laterally; antennal insertions exposed, distinct antennal club; mandible bidentate; mola well-developed and transversely ridged in most, rarely reduced; tarsal formula 4-4-4 (3-3-3 in *Annomatus*); trochantins not visible externally; pretarsal claws simple, empodium reduced (Philips and Ivie, 2002; Ślipiński et al., 2010; Lee et al., 2017).

Bothriderid species except for Anommatinae are usually associated with wood-boring insects and found on or under the bark or trees. They are ectoparasites of larvae and pupae of them (Philips and Ivie, 2002; Ślipiński et al., 2010). Especially, *Dastarcus helophoroides* (Fairmaire) (Plate 2j) is known to be an important natural enemy of longhorn beetles (Ogura et al., 1999). The adult females deposit egg clusters on the gallery walls bored into the stems of trees by host larvae (Gao and Qin, 1992). The hatchlings seek out and paralyze their hosts. Larvae that successfully parasitise hosts feed externally on the host and grow rapidly. Mature larvae spin cocoons and pupate, emerge in August and September (Urano et al., 2004), and overwinter as adults. *D. helophoroides* parasitises the larvae, pupae, and adults of *Monochamus alternatus* Hope (Lim et al., 2012; Okamoto, 1999), the primary vector of the pine wood nematode, *Bursaphelenchus xylophilus* (Steiner et Buhrer) Nickle in Japan.

1.3.9 Family: Curculionidae

Weevil, (family Curculionidae), also called snout beetle, true weevil of the insect order Coleoptera. Curculionidae is one of the largest coleopteran families (about 40,000 species). Most weevils have long, distinctly elbowed antennae that may fold into special grooves on the snout. Many have no wings, whereas others are excellent fliers. Most are less than 6 mm (0.25 inch) in length, although the largest exceed 80 mm (3 inches).

The majority of weevils feed exclusively on plants. The fleshy, legless larvae of most species feed only on a certain part of a plant, i.e. the flower head, seeds, fleshy fruits, stems or roots. Many larvae feed either on a single plant species or on closely related ones. Adult weevils tend to be less-specialized in their feeding habits. However, some curculionid weevils act as provision-directed cleptoparasitoids of curculionid larvae within galls (Eggleton and Belshaw, 1992).

1.4 ORDER: LEPIDOPTERA

The Insect order, Lepidoptera includes both butterflies and moths and about 180,000 species of the Lepidoptera are described in 126 families (Capinera, 2008). and 46 superfamilies(Jim, 2007). Adults are distinctive for their large wings (relative to body size), which are covered with minute overlapping scales. Most entomologists believe that these scales are structurally related to the hair (setae) covering adult caddisflies. Lepidopteran wing scales often produce distinctive colour patterns that play an important role in courtship and intraspecific recognition. The active immature stages are called larvae/caterpillars. They have a well-developed heads with chewing mouthparts. In addition to three pairs of legs on the thorax, they have two to eight pairs of fleshy abdominal prolegs that are structurally different from the thoracic legs. Most lepidopteran larvae are herbivores; some species eat foliage, some burrow into stems or roots, and some are leafminers. The only entirely carnivorous families in the order Lepidoptera are the Epipyropidae and the Cyclotornidae and the list of parasitic species are given in Table 1.3.

TABLE 1.3
Parasitoid Species in Various Families Under the Order Lepidoptera and Neuroptera and Their Host Insects

Sl. no.	Family	Parasitoid species	Host insect
Order: Lepidoptera			
1	Epipyropidae	*Fulgoraecia melanoleuca*	sugarcane *Pyrilla*
2	Cyclotornidae	*Cyclotorna* sp	Leafhoppers
Oder: Neuroptera			
1	Mantispidae	Symphrasinae sp	Larva of bee, wasp, scarab beetles

1.4.1 Family: Epipyropidae

The Epipyropidae is a small family of perhaps as many as 40 species in 11 genera (Davis, 1987). The larvae are parasitic on Homoptera, primarily leafhoppers and also cicadas, and even on other Lepidoptera (Common, 1990). The adults are minute to small (4 to 35 mm wingspan), with head scaling mostly roughened; haustellum absent; labial palpi minute; maxillary palpi absent; antennae bipectinate in males and rather conspicuous. Body robust. Wings quadratic and broadly rounded. Maculation is mostly black or dark brown; sometimes with some spots and iridescence. Adults are crepuscular and nocturnal; females are sedentary. Larvae slug-like with rounded dorsum; parasitic on fulgorids and planthoppers (Hemiptera) (Heppner, 2004).

Fulgoraecia melanoleuca (Plate 3a) has been reported from India, Sri Lanka, Pakistan, and Bangladesh (Kumarasinghe and Wratten, 1996). It has played a major role in the management of the sugarcane *Pyrilla* epidemics (Gangwar et al., 2008; Kumar et al., 2015). Although it has been recorded in India in 1939 (Fletcher, 1939), its biocontrol potential was recognized only during the *Pyrilla* epidemics in Uttar Pradesh and Bihar (Banerjee, 1973). In India, the incidence of *Fulgoraecia* has been recorded in sugarcane growing areas in Maharashtra, Gujarat, Rajasthan, Odisha, Haryana Uttar Pradesh, Punjab, Uttarakhand, Chhattisgarh, Karnataka and Andhra Pradesh (Sankararaman et al., 2020). It has been considered as a potential biocontrol agent against *Pyrilla* and is extensively used in its management (Pawar et al.,2002; Pandey et al., 2008). *Fulgoraecia melanoleuca* has proved its merit in *in situ* parasitisation due to its high multiplication rate, comparatively shorter life cycle, survival under varied agroclimatic conditions and good searching ability of its host by larvae (Rajak, 2007).

1.4.2 Family: Cyclotornidae

Cyclotornidae is a family containing five described species and at least seven undescribed species in the genus *Cyclotorna* (Plate 3b) that is endemic to Australia (Common, 1990). This family and the closely related Epipyropidae are unique among the Lepidoptera in that the larvae are ecto-parasites, the hosts in this case typically being leafhoppers, sometimes scale insects. The larvae of cyclotornids, however, leave the hemipteran host and become predatory on the brood in ant nests, apparently using chemical cues to induce the ants to carry the larvae into the ant nest (Dodd, 1902).

The female moths lay their eggs in large numbers on the twigs in the vicinity of leafhopper colonies. On hatching, young larvae move about until a host is found, after which they attach themselves and begin feeding. They change position on the host body somewhat but later are found primarily on the abdomen. If wing pads are developed on the host, feeding is usually beneath one,

Plate 3a. *Fulgoraecia melanoleuca*

Plate 3b. *Cyclotorna* sp

PLATE 3 Parasitoids in the order Lepidoptera.

TABLE 1.4
Hymenopteran Parasitoids Introduced in India

Sl. no.	Insect pest	Natural enemy	Source/country/year of introduction	Remarks
1. Recovered and Established				
1	Apple woolly aphid, *Eriosoma lanigerum*	*Aphelinus mali* (Hymenoptera: Aphelinidae)	USA via UK, 1936–1937	Introduced at Saharanpur (Uttar Pradesh) for the control of woolly aphid, *Eriosoma lanigerum* on apple. Subsequently established in Kullu valley (Himachal Pradesh), Kashmir valley (Jammu and Kashmir), Coonoor (Tamil Nadu), Shillong (Meghalaya), and all the apple-growing areas of India. More effective in valleys than on mountain slopes
2	San Jose scale, *Quadraspidiotus perniciosus*	*Encarsia perniciosi* (Hymenoptera: Aphelinidae)	California, USA, 1958, 1959; Illinois, USA, 1960; China via Switzerland, 1960; Russia via France, 1960	All strains introduced for the biological suppression of San Jose scale, *Quadraspidiotus perniciosus* performed well, providing 89–95% parasitism in Himachal Pradesh, Jammu & Kashmir and Uttaranchal. Presently established in many areas where the scale population is lo
3	Citrus mealy bug *Planococcus citri*	*Leptomastix dactylopii* (Hymenoptera: Encyrtidae)	Trinidad, West Indies, 1983	Introduced for the control of *Planococcus citri* Established in Karnataka, Tamil Nadu and Kerala.
4	Plassey borer, *Chilo tumidicostalis*	*Telenomus alecto* (Hymenoptera: Scelionidae)	Colombia, 1966;Trinidad, West Indies, 1982	Released in Plassey (West Bengal) and recovered from Plassey borer, *Chilo tumidicostalis*. Also released in Motipur (Bihar) and Golagokarnath (Uttar Pradesh), recovered from Motipur and established at Golagokarnath.
5	Pink bollworm, *Pectinophora gossypiella*	*Bracon kirkpatricki* (Hymenoptera: Braconidae)	Africa via USA, 1969	Recovered from pink bollworm, *Pectinophora gossypiella* on cotton, okra and hollyhock in Gujarat seven years after release indicating its establishment.
6	Coffee Berry Borer, *Hypothenemus hampeii*	*Cephalonomia stephanoderis* (Hymenoptera: Bethylidae)	Mexico, 1995	Released in the field from 1998 to 2001. Established in several areas of Kodagu district (Karnataka), Wyanad (Kerala) and Lower Palanis (Tamil Nadu).
7	Spiralling white fly, *Aleurodicus dispersus*	*Encarsia guadeloupae* (Hymenoptera: Aphelinidae)	Caribbean region & Central America, serendipitously introduced via Lakshadweep Island, 2000	Brought from Minicoy Island of Lakshadweep to main land for the control of spiralling white fly, *Aleurodicus dispersus*. Established in Kerala, Karnataka, Tamil Nadu, and Andhra Pradesh.
8	Spiralling white fly, *Aleurodicus dispersus*	*Encarsia sp. nr. meritoria* (Hymenoptera: Aphelinidae)	Caribbean region & Central America, serendipitously introduced via Lakshadweep Island, 2000	A parasitoid of spiralling whitefly, fortuitously introduced. Present in several parts of south India, but displaced by *E. guadeloupae* in several areas and population remains low.

2. Recovered				
9	San Jose scale, *Quadraspidiotus perniciosus*	*Aphytis sp. nr. diaspidis* (Hymenoptera: Aphelinidae)	Japan via USA, 1966	Colonized on San Jose scale, *Quadraspidiotus perniciosus* on apple and several other deciduous fruits and recovered from Kashmir (Jammu & Kashmir).
10	Sugar cane borer *Chilo auricilius*	*Cotesia flavipes* (Hymenoptera: Braconidae)	Indonesia, 1991	Released in sugarcane fields and recovered from *Chilo auricilius* larvae in Uttar Pradesh.
11	Sugarcane stem borer, *Chilo sacchariphagus indicus*	*Trichogramma australicum* (Hymenoptera: Trichogrammatidae)	Taiwan, 1963; Trinidad, West Indies, 1981	Released in sugarcane fields in Pugalur (Tamil Nadu) and recovered from eggs of sugarcane stem borer, *Chilo sacchariphagus indicus* and other *Chilo* spp.
12	*Chilo* sp. *Helicoverpa armigera, Pectinophora gossypiella*	*Trichogramma brasiliense* (Hymenoptera: Trichogrammatidae)	South America via California, USA, 1968	Released and recovered from sugarcane tissue borers; Helicoverpa armigera on tomatoes, cotton and several other hosts; pink bollworm, Pectinophora gossypiella on cotton, okra and hollyhock in Punjab.
13	Gurdaspur borer, *Acigona steniellus; Chilo infuscatellus, Chilo spp.* and *Scirpophaga spp.*	*Trichogramma japonicum* (Hymenoptera: Trichogrammatidae)	Trinidad, West Indies, 1979	Released in sugarcane fields and recovered from eggs of Gurdaspur borer, *Acigona steniellus; Chilo infuscatellus, Chilo spp.* and *Scirpophaga spp.*
14	Diamond back moth, *Plutella xylostella.*	*Trichogramma- toidea bactrae* (Hymenoptera: Trichogrammatidae)	Taiwan, 1992	Released and recovered from diamond back moth, *Plutella xylostella.*
15	Codling moth, *Cydia pomonella*	*Trichogramma embryophagum* (Hymenoptera: Trichogrammatidae)	Rumania, 1978; Germany, 1988	Released and recovered from codling moth, *Cydia pomonella* on apple in Ladakh region of Jammu & Kashmir state.
16	Potato tuber moth *Phthoremoea operculella*	*Apanteles subandinus* (Hymenoptera: Braconidae)	South America, 1944–45	Released and recovered from potato tuber moth on potato in Rajgurunagar (Maharashtra) and Chickballapur and Hassan (Karnataka).
17	Potato tuber moth *Phthoremoea operculella*	*Copidosoma koehleri* (Hymenoptera: Braconidae)	South America via California, USA, 1965; Peru, 1990	Released and recovered from potato tuber moth on potato in Rajgurunagar (Maharashtra) and Chickballapur and Hassan (Karnataka).
18	Castor semilooper, *Achaea Janata* Tobacco cutworm *Spodoptera litura*	*Telenomus remus* (Hymenoptera: Scelionidae)	New Guinea, 1963	Released and recovered from castor semilooper, *Achaea janata* on castor in Andhra Pradesh; *Spodoptera litura* on papaya, cabbage and tobacco nursery in Rajahmundry (Andhra Pradesh) and on lucerne and cauliflower from Gujarat.
19	Coffee berry borer	*Phymastichus coffea* (Hymenoptera: Eulophidae)	Colombia, 1999–2001	Field released against coffee berry borer and recovered in south India.
20	Coffee berry borer	*Prorops nasuta* (Hymenoptera: Bethylidae)	Mexico, 1995; Colombia, 1999	Introduced against coffee berry borer. Recovered in field cages.

which is as a result forced out of its normal position. One to eight larvae may be found on a single leafhopper, and a silken web, extended at one side to form a delicate wall, is formed underneath the host. A portion of hosts probably dies without reaching the adult stage. The cyclotorinid larvae sometimes move from one host to another and thus they are best-considered predators with considerable advancement toward obligate parasitism (Clausen, 1940; 1962b).

1.5 ORDER: HYMENOPTERA

The order Hymenoptera (*Hymen-membrane*, *Ptera-* Wing) includes ants, wasps, bees, sawflies and hornets. They are well known for their involvement in the largest numbers of interactions with other organisms, including plants in an ecosystem (La Salle and Gauld, 1993) and comprise a significant proportion of arthropod diversity in the terrestrial habitat with more than 1,15,000 species across the globe (Stevens et al., 2007). Generally, the layman used the term wasp to denote the groups of Hymenopterans, such as hornets and yellow jackets that can sting and have aggressive nature. But, according to the entomologists, the term has much larger scope and the vast majority in this group are incapable of stinging. Many of the stingless wasps have an external stinger, which is used for oviposition. However, in stinging bees and wasps, the organ is no longer used as an ovipositor and is used for defence purpose.

Parasitoid wasps are stingless wasps and play key roles in the functioning of natural and agricultural ecosystems by regulating arthropod populations (La Salle and Gauld, 1993). They lay their eggs into or onto other insects (most commonly targeting the egg or juvenile stages) as well as other arthropods, such as spiders and less frequently, ticks. Once the egg laid by a parasitoid wasp hatches, the wasp larva then feeds on the host to complete development, resulting in the death of the host. The high trophic level of parasitoid Hymenoptera as specialised parasitoids within the arthropod community, often with high levels of host specificity, has made them ideal biocontrol agents of a large range of agricultural and horticultural pest insects (La Salle and Gauld, 1993).

The parasitic wasps are very diverse in their size, biology, life cycle and type of host attacked. The smallest known insect are the fairyflies in the family Mymaridae, who have a body size of less than 0.5mm and spend their entire parasitic life at the expense of the eggs stage of another insect (Annecke and Doutt, 1961). Conversely, some of the wasps parasitise the large grubs or wood borers and have sizes of more than 3 inch long. The hymenopteran parasitoids introduced into various parts of India to manage successfully various insect pests infesting economically important crops are given in Table 1.4. The salient features of the important family of parasitic wasps are briefly given below.

1.5.1 FAMILY: BRACONIDAE

The family Braconidae is one of the largest in the Hymenoptera, containing more than 15,000 valid species (Quicke and Achterberg, 1990). Together with the Ichneumonidae, it forms a distinctive superfamily among the assemblage of hymenopterans known as the parasitic wasps (Gauld and Bolton, 1988). Adult braconids oviposit almost exclusively in, on or near other insects, with the immature stages completing their development at the expense of their host insect.

Braconid wasps are small to moderately large sized ranging from 2–3 mm to 15 mm in length. They are generally black or brown in colour, but some have yellow, orange or red markings. They have long antennae with more than 16 segments. The ovipositor may also be long and readily apparent. They resemble ichneumonids but differ in wing vein characters. They do not have more than one **m-cu** cross vein or none in forewing and in the hindwing, cross-vein **rs-m** joins **Sc+R+Rs** before the separation of **Rs.** Usually, their second and third abdominal tergites are fused together. This is one of the richest families of insects between 50,000 to 1,50,000 species existing worldwide. The species are grouped into 45 subfamilies and 1,000 genera. Some important genera are: *Ademon, Apanteles, Aphanta, Asobara, Bracon, Cotesia, Microgaster, Opius, Parapanateles,* etc. The great majority of braconids are primary parasitoids of the other insects and is not unusual that whole subfamilies and tribes are to be associated with single host orders. Aphidiinae parasitises only aphids, whereas Opiinae only on dipterous larvae (Mills, 1992).

One of the important subfamilies of Braconidae is Aphidiinae (Sometimes to be considered as a family: Aphidiidae) comprises very small wasps parasitising the aphids. The genus *Aphidius* (Plate 4a) consists of numerous species attacking the aphids and providing gnatural biocontrol of

Plate 4a. *Aphidius* sp

Plate 4b. *Apanteles glomeratus*

Plate 4c. *Campoletis chloridae*

Plate 4d. *Eriborus argenteopilosus*

Plate 4e. *Isotomia javensis*

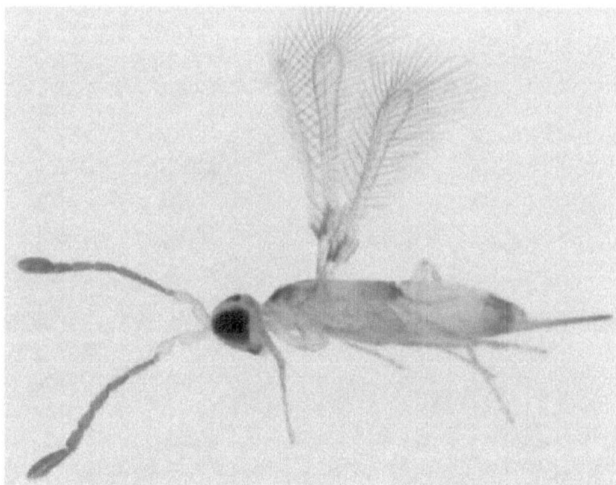

Plate 4f. *Anagrus* sp

PLATE 4 Parasitoids in the order Hymenoptera.

Plate 4g. *Gonatocerus* sp

Plate 4h. *Trichogramma* sp

Plate 4i. *Eulophus* sp

Plate 4j. *Elasmus nephantidis*

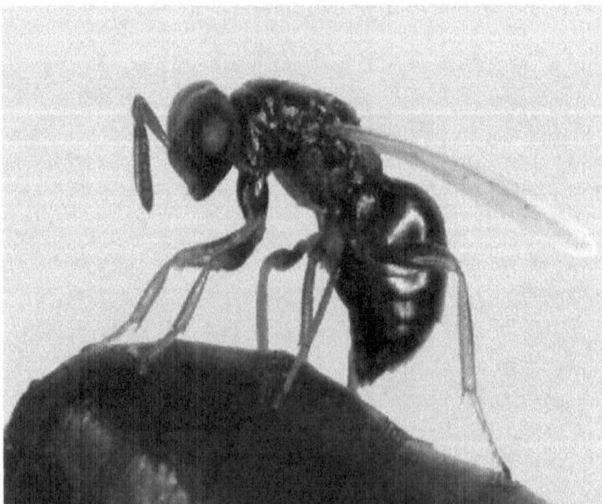

Plate 4k. *Nasonia* sp

Plate 4l. *Acerophaguous papaya*

PLATE 4 (*continued*)

Plate 4m. *Aphelinus mali*

Plate 4n . *Anastatus* sp

Plate 4o. *Brachimeria* spp.

Plate 4p. *Eurytoma* sp

Plate 4q. Torymid wasp

Plate 4r. *Ormyrus orientalis*

PLATE 4 *(continued)*

Plate 4s. Leucopsid wasp

Plate 4t. Euchartid wasp

Plate 4u. Perilampid wasp

Plate 4v. Signiphorid sp

Plate 4w. Eucoilid wasp

Plate 4x. Figit wasp

PLATE 4 (*continued*)

Plate 4y. Platygaster oryzae

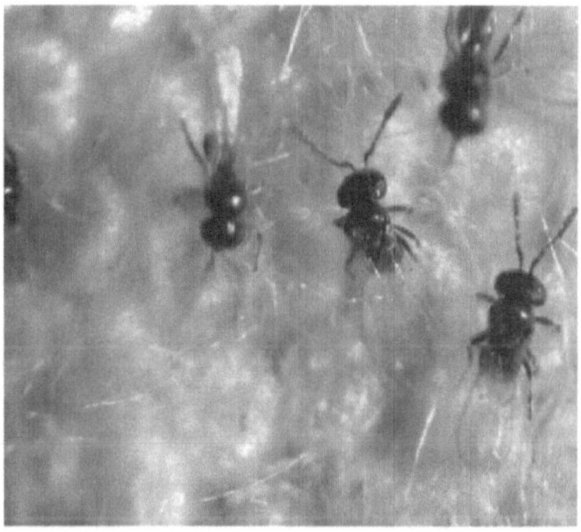

Plate 4z. *Telenomus remus*

Plate 4aa. *Goniozus nephantidis* on *Opisina arenosella*

PLATE 4 (*continued*)

aphids in greenhouse cultivation in Europe and USA. The female wasp lays eggs singly in aphid nymphs and larvae consume the internal organs of aphids during the course of its development. Eventually, the aphids are killed and turn into smooth, shiny and light brown to silvery-gold mummies. The larvae pupate and the adult wasps emerge out through an exit hole cut in the mummy. Apart from killing the larvae, it also makes mechanical disturbance in the colony leads to fall of the aphids from the plant surface and die (Stary, 1973).

Many species of braconids are valuable in the biocontrol of insect pests. *Apanteles glomeratus,* (Plate 4b), for example, parasitises the larvae of the cabbage butterfly (*Pieris rapae*) (Karnavar, 1983) and the cabbage looper (*Trichoplusia ni*) (Hochberg, 1991). *Apanteles congregatus* parasitises the tobacco hornworm (*Manduca sexta*) and the tomato hornworm (*Manduca quinquemaculata*) (Beckage and Riddiford, 1982). Sureshan et al. (2013) reported that *Doryctus* sp. (Braconidae) is found parasitising the wood boring beetle, *Clytocera chinospila* Gahan (Coleoptera: Cerambycidae) from Chinnar Wildlife Sanctuary, southern Western Ghats, Kerala. The braconid, *Chremilus rubiginosus,* is found parasitising the storage pest, granary weevil (*Sitophilus granarius*) in China (Chang et al., 2017). In the Mediterranean region the *Opius concolor* parasitises the olive fly (*Dacusoleae*), the most destructive pest on commercial olives (Farouk et al., 2014).

1.5.2 Family: Ichneumonidae

It is one of the most abundant families in the order Hymenoptera. These wasps are very small to very large in size, vary a lot in colour and form, generally larger than braconids, but possess many unique characteristics that allow them to stand out. The antenna is unusually long with 13 or more segments. They have a characteristic forewing venation, with first discoidal cell and submarginal cells confluent and with two m-cu veins. Hindwings with cross-vein rs-m joining Rs from shortly to very far after that vein has diverged from Sc+R+Rs. The ovipositor is often quite long and larger than the body. The second and third abdominal segments are separated and overlapping. The family is currently subdivided into 42 recognized families (Townes, 1958; Kopylov, 2009).

Ichneumonid wasps attack relatively a narrow range of hosts, which include Lepidoptera, Diptera, Coleoptera, Hymenoptera (Sawflies), Neuroptera, Spiders and Spider egg sacs. Many species are hyperparasiticon other Ichneumonids, Braconids and Tachinids (Gauld, 1988). The larval stages of ichneumonids last from ten days to several weeks and many species have single generation; however, some species may have two, three or more generations per year. Though ichneumonids are well known for the natural biocontrol of many pest species, a few have been widely used in biological control programmes. Some of the important species are *Campoletis chloridae* (Plate 4c), *Eriborus argenteopilosus* (Plate 4d) on *Helicoverpa armigera* and *Isotomia javensis* (Plate 4e) on sugarcane borers (Singh, 2004).

1.5.3 Family: Mymaridae

All species of Mymaridae are solitary or gregarious endoparasitoids of the eggs of other insects. They are known to parasitise eggs in concealed situations, such as those embedded in plant tissue, placed under scales bracts or in soil (Huber, 1986). They do not seem to be particularly host-specific, and mymarid species within one genus may parasitise eggs of insects belonging to several families. The most common hosts (parasitism to the tune of 45%) are the eggs of auchenorrhynchousHomoptera, but eggs of other Hemiptera (Coccoidea, and less frequently Tingidae and Miridae), together with Coleoptera (Curculionidae and Dytiscidae) and Psocoptera are commonly parasitised (Annecke and Doutt, 1961).

Some are the tiniest of all insects, being 0.5–1 mm in length. The smallest one known is about 0.2 mm. The wings are with a fringe of long hairs on the border and the forewing has a narrow base. They can be easily distinguished by a series of unique sutures or lines on the head, and the antennal sockets, which are closer to the eyes than to each other (Yoshimoto, 1990). The common hosts include Lepidoptera, Coleoptera, Diptera, Psocoptera and Hemiptera, especially the plant and leafhoppers. The species of *Anagrus* (Plate 4f) and *Gonatocerus* (Plate 4g) are the most important parasitoids of rice leaf hoppers and plant hoppers (Sahad, 1982).

1.5.4 Family: Trichogrammatidae

Like Mymarids, Trichogrammatids are egg parasitoids, parasitising the eggs of several insects belonging to more than eight orders in aquatic and terrestrial habitats. They are very minute, measuring from 0.3 to 1.2 mm. They are the only chalcidid family with three-segmented tarsi, which readily distinguish them. Their colour varies from yellow or orange to dark brown, but never metallic in colour. Antennae are 5–9 segmented; antennae of male insects are normally with whorls of long setae. The Fore wing lacks a post-marginal vein and the setae are arranged in radiating lines. Their gaster is sessile (Doutt and Viggiani, 1968; Pinto, 1998).

Among the 230 species that have been described worldwide, *Trichogramma* (Plate 4h) is the largest genus in the family constituting over a quarter of the known genera in this family and widely used in augmentative biocontrol programmes. Almost all trichogrammatids are primary, solitary or gregarious endoparasitoids of the eggs of other insects in the order Lepidoptera, Hemiptera, Coleoptera,

Thysanoptera, Hymenoptera, Diptera and Neuroptera (Walter, 1983a). One species of *Trichogramma* is known to occasionally develop as a facultative hyperparasitoidon *Telenomus* parasitising the lepidopteran egg (Strand and Vinson, 1984). A species of *Lathromeris* and one of *Oligosita* have been recognized as larval parasitoids of cecidomyiids (Diptera) (Viggiani and Laudonia, 1994).

Many species oviposit directly into more or less exposed host insect eggs and some maybe even attempt to oviposit into anything that has the same size and shape as an egg such as dried globules of sap. A few trichogrammatids (Plate 4h) parasitise the eggs of aquatic hosts, such as Dytiscidae, Notonectidae or Odonata, whilst the egg is beneath the surface of the water (Henriksen, 1922). Many species of trichogrammatids are of focus of interest because of their widespread use in the biocontrol of various insect pests, especially Lepidoptera (Walter, 1983a & b). In the field, rates of parasitism may often reach 90%. Most species that are used in biocontrol programme are mass-reared under laboratory conditions and then released early in the season to try to control the pest species before their numbers can rise to cause economically significant damage.

1.5.5 Family: Eulophidae

This is one of the largest families of chalcidid wasp, with about 3,400 species worldwide. They are small to minute (1–3 mm) and often have brilliant metallic colour (metallic blue or green). Forewing with long marginal vein, post marginal and stigma veins not usually long, occasionally very short. Eulophidae can be easily recognized by the four–segmented tarsi, axillae extending forward beyond the tegulae, straight spur of foretibia and the non-pedunculate wings (Askew and Ruse, 1974a). The majority of Eulophids are primary parasitoids of concealed larvae, especially those residing in leaf mines. The best-known species attack Lepidoptera, but many species parasitise larvae of other insects living in similar concealed situations (such as, Agromyzidae, Tenthredinidae and Curculionidae) (Gauthier et al., 2000). Other eulophids attack various gall-forming species of insects, eriophyid mites (Boucek and Askew, 1968) and also gall-forming nematodes (Berg et al., 1990).

Eulophidsare also solitary or gregarious, idiobiont ectoparasitoids of the larvae, or sometimes the pupae, of leaf-miners, or similarly concealed hosts, such as leaf- folders, case-bearers, gall-makers and stem-borers. Many species are facultative or obligate hyperparasitoids through other chalcids, braconids and ichneumonids (Gauthier et al., 2000). Some species (*Eulophus* (Plate 4i) and *Euplectrus*) are gregarious ectoparasitoids of the exposed leaf-feeding larvae of large Lepidoptera (Gradwell, 1957). A few species of *Elasmus* have been recorded as primary parasitoids of the larvae of *Polistes* species (Vespidae) (Graham, 1995).

1.5.6 Family: Elasmidae

This family is also with four segmented tarsi as that of Eulophidae, but can be separated by the large, flattened, disc-like hindcoxae and the peculiar dark bristles on their hindtibia, which form a characteristic pattern. The propodeum is medially elongated in shape. The general appearance is like a bristly fly (Burks, 1971; Verma and Hayat, 1988). This family consists of both the primary- and hyperparasites on Lepidoptera and their hymenopteran parasites (Braconidae and Ichneumonidae); e.g.: *Elasmus nephantidis* (Plate 4j) on *Opisina arenosella* (Nasser and Abdurahiman, 2001) and *Elasmus zehntneri* on sugarcane top shoot borer (Singh and Sinha, 1977).

1.5.7 Family: Pteromalidae

This is another large family and consists of a diverse group of insects. In general, the tarsi are five-segmented, the antennal funicle has five or more segments and the pronotum is anteriorly constricted in dorsal view (thus giving it in the shape of a bell) (Boucek and Rasplus, 1991; Stringer et al., 2012). Many develop as solitary or gregarious ectoparasitoids of larvae and pupae of Diptera, Coleoptera, Hymenoptera, Lepidoptera and Siphonaptera. Large numbers of species attack hosts concealed in plant tissue, such as wood-borers, stem- and leaf-miners, gall-formers, etc. Several types of pteromalids parasitise the pupae of filth-breeding flies, such as fleshflies, blowflies and stableflies and storage pests (Flanders, 1959). The most common genera are *Nasonia* (Watt and Shuker, 2010) (Plate 4k), *Musidifurax* (Lysyk, 1999) and *Spalangia* (Boucek, 1963). Some species are commercially available in some countries for the management of flies' population.

1.5.8 Family: Encyrtidae

This family currently includes over 3,000 species under 460 genera and is one of the important chalcidoid families for the biocontrol of insect pests (Noyes, 2012). They are usually 1–2 mm in length and can be differentiated from other chalcidoids by the broad, convex, mesopleura, five-segmented tarsi and large apical spur of mesotibia. The axillae are medially contiguous. The middle coxa with the thorax in lateral view is placed closer to the forecoxa than the hindcoxa (Peck et al., 1964).

About half of the species of Encyrtidae are associated with scale-insects, mealybugs, aphids, psyllids and whiteflies generally as endoparasitoids of immatures or less commonly adults, but with egg, predation practised by some species of *Microterys*. Many other encyrtids (e.g. *Copidosoma* spp.) are polyembryonic parasitoids of larvae of Lepidoptera (Alam, 1959). Some species are hyperparasitoids via other Encyrtidae, or Aphelinidae, Pteromalidae, Braconidae, Dryinidae, etc. Many of the very successful cases of classical biocontrol of scale insects and mealybug pests of fruit trees have involved the encyrtid wasps (Tachikawa, 1981).

In order to manage the papaya mealybug *Paracoccus marginatus* Williams and Granara de Willink (Hemiptera: Pseudococcidae) menace, which was reported from 2008 onwards in south Indian states on various host plants, a classical biological control program was initiated with the importation of three parasitoids, *Acerophagus papayae* Noyes and Schauff, *Pseudleptomastix mexicana* Noyes and Schauff, and *Anagyrus loecki* Noyes and Menezes (Hymenoptera: Encyrtidae) from Puerto Rico in July 2010, and *A. papayae* (Plate 4l) was multiplied and released. Excellent control of the papaya mealybug was obtained within five months, pesticide usage was reduced, and production and income were increased (Myrick et al., 2014).

1.5.9 Family: Aphelindae

They are very small to minute and are usually light yellow to brown in colour. This group of parasitoids can be distinguished by five segmented tarsi, the abdomen broadly attached to propodeum, the elongated marginal veins and the much-reduced post-marginal and submarginal veins. The number of antennal segments is reduced (eight or fewer). The axillae are medially separated and mostly extended forward up to tegulae (Hayat, 1983). The majority of aphelinidsoccur worldwide are parasitoids of sternorrhynchousHomoptera. While a few attack species of Aphidoidea, Aleyrodoidea or Psylloidea, most develop by consuming scale-insects (Coccoidea), either as internal parasitoids, as ectoparasitoids (under the scale) or as predators of their eggs (Viggiani, 1984). The males of a number of species develop as hyperparasitoids of coccoids via their eulophid, aphelinid or encyrtid primary parasitoids (Cendana, 1937). A number of aphelinids are internal parasitoids of the eggs of various Auchenorrhyncha, Reduvioidea, Lepidoptera or Orthoptera whilst very few develop on the larvae or pupae of Dryinidae (Hymenoptera), Cecidomyiidae or Chamaemyiidae (Viggiani, 1984). Members of genera, *Encarsia and Eretmocerus* are whitefly parasitoids, whereas various aphelinid species are aphid parasites, such as *Aphelinus mali* (Plate 4m), a parasite of apple woolly aphid (Long et al., 1960; Hoddle et al., 1998).

1.5.10 Family: Eupelmidae

These are almost similar to encyrtids, but can be distinguished by two characters; marginal vein of the forewing longer and the middle coxa with the thorax in the profile is placed nearer the hind coxa than the front coxa. The mesonotum is flatter and the notular lines are more distinct than in encyrtids (Gibson, 1995). The vast majority of species of Eupelmidae are parasitic and facultative hyperparasitic on the immature stages of other insects, with hosts recorded in the orders Lepidoptera, Homoptera, Hymenoptera, Coleoptera, Neuroptera and Orthoptera. Many eupelmids attack hosts within plant tissues, such as gall-makers or the larvae of wood-boring beetles (Al Khatib et al., 2016). Some are egg parasites. *Anastatus* spp. (Plate 4n) on gypsy moth and *Neanastatus* spp. on paddy gall midge (Mathur, 2009).

1.5.11 Family: Chalcididae

Chalcidids are predominantly solitary, primary endoparasitoids of Lepidoptera and Diptera, though a few species attack Hymenoptera, Coleoptera or Neuroptera; some tropical species are ectoparasitoids, and a few may be gregarious. Some are hyperparasitic mainly on lepidopterans (Narendran, 1989). Some chalcidids are of interest as parasitoids of insect pests. For example, *Brachymeria intermedia* is a parasitoid of *Lymantria dispar*, an introduced lepidopterous pest of a variety of trees in North America, but it has proved to be of little use for biocontrol purposes. Species of the predominantly tropical genus, *Dirhinus,* may be of some economic importance as parasitoids of synanthropic Diptera (Calliphoridae, Muscidae and Sarcophagidae) and tephritid soft-fruit pests (Boucek and Narendran, 1981a). The *Brachimeria* spp. (Plate 4o) and *Antrocephalus* spp. parasitise the coconut black-headed caterpillar, *Opisina arenosella* (Nasser and Abdurahiman, 2001).

1.5.12 Family: Eurytomidae

Eurytomids are characterized by pronotum without a distinct large and subrectangular collar; an antenna frequently with seven funicular segments; sculpture of the head and thorax mostly shallow (Claridge, 1961). Eurytomidae (Plate 4p) contains species, which exhibit a wide range of biologies, but the majority seem to be endophytic, either as phytophagous or as parasitoids of phytophagous insects found within plant tissues, such as stems, seeds, or galls. Most of these entomophagous species are idiobiontecto parasitoids of insect larvae feeding within plant tissue. Hosts attacked include Coleoptera, gall-forming Hymenoptera (mostly Cynipinae), Diptera (especially Tephritidae) and Lepidoptera (Claridge, 1959; Claridge and Askew, 1960)

1.5.13 Family: Torymidae

Torymid wasps are often brightly coloured with either metallic blue or green. The hindcoxae are very large and the notaular lines are distinct (Narendran, 1994). Torymids (Plate 4q) oviposit through plant tissue, usually into galls or developing seeds, but some oviposit into pupae concealed in silk or plant tissue. Torymid larvae may be entomophagous or, less commonly, phytophagous; some may even be both, feeding in turn on gall maker and gall tissue as in some Eurytomidae (Askew, 1961). Many species are parasitic on gall insects, while some are associated with the fruits of *Ficus*. Hosts include the eggs and larvae of moths, beetles, flies and eggs of praying mantis (Askew and Ruse, 1974b).

1.5.14 Family: Ormyridae

It is a very small family with three genera and 265 known species, they are bright metallic green or blue, highly sclerotized, with conspicuous, large, pit-like punctures on abdominal tergites. Gaster of female insects is elongated and pointed at the tip, but ovipoisitor is not protruding (Lotfalizadeh et al., 2012). Many species of ormyrid are parasitoids of various gall-forming insects. They are known to attack cecidogeniccynipids, chalcids and Diptera, and a few may also parasitise phytophagouseurytomids in seeds (Kazmi et al., 2017). *Ormyrus orientalis* (Plate 4r) is frequently associated with the pupae of *Melanagromyza obtuse* on pigeonpea (Patange et al., 2017).

1.5.15 Family: Leucopsidae

It is a small family closely related to Chalcididae, but the species are larger with a very long ovipositor, which is curved upward and forward over the abdomen, reaching beyond the posterior end (Boucek and Narendran, 1981b). Leucospids (Plate 4s) develop as parasitoids of aculeate Hymenoptera. Their hosts are mainly solitary bees, less frequently solitary wasps, e.g. Vespidae and Sphecidae nesting in a similar way to bees (Ye et al., 2017). Occasionally, parasitic bees have also been recorded as hosts and are probably attacked by the leucospid when occupying the cell of a solitary bee after killing its original owner (Cooperband et al., 1999).

1.5.16 Family: Euchartidae

They are a very distinctive-looking insects with black or metallic blue or green in colour, with petiolate abdomen and spinedscutellum (Plate 4t). Thoracic region appears to be humpbacked in profile and all members of the family are parasitic on larvae and pupae of ants (Torrens, 2013).

1.5.17 Family: Perilampidae

Perilampids (Plate 4u) are stout-bodied insects, which are often brilliantly metallic in colour. The thoracic region is high and convex, large and conversely punctate. The abdomen is small, petiolate, shining and triangular (Ferrière and Kerrich, 1958). Many are hyper-parasitoids, parasitising the larvae of flies, parsitoids of lepidopteran and grasshoppers. Few are primary parasites of beetle, lacewing larvae and symphytan hymenopterans (Laing and Heraty, 1981).

1.5.18 Family: Signiphoridae

Signiphorids (Plate 4v) have been reared from scale-insects (Homoptera; Coccoidea), whiteflies (Homoptera: Aleyrodidae) and the puparia of Diptera (Hansen, 2000). All but a few species are endoparasitoids and most are known to be hyperparasitic via other chalcidoids (Ferrière and Kerrich, 1958). Some species are reported to be gregarious hyperparasitoids through encyrtids. These are small, characterised by the broad attachment of the abdomen, antennae of both sexes with an elongated, unsegmented club and 3–4 ring segments/anelli, ribbon like scutellum, middle tibia with long dorsal spines and a pectinate apical spur and the propodeum with a triangular impressed area in the middle (Woolley and Hanson, 2006).

1.5.19 Family: Eucoilidae

Eucoilidae (Plate 4w) are endoparasites of schizophorous Diptera larvae. The family was considered by subfamily under Cynipidae in early classifications. Female wasps oviposit in young host larvae and development proceeds through three to five instars and the pupation takes place within a host puparium. They can be recognized by the rounded, cup-like, elevation on the scutellum or the scutellum may be extended into the spine (Beardsley, 1989). They are mostly parasites of pupae of lies including leafminers. Some species are relatively common parasites of fly maggots that can occur in manure and carrion. Hypermetamorphosis is probably universal with the first instar eucoiliform, with thoracic lobes and caudal projection for emergence from the egg. The respiratory system gradually develops, with the first instar lacking spiracles, while the fifth instar exhibits eight pairs (Beatriz et al., 1996).

1.5.20 Family: Figitidae

They are solitary, primary internal parasitoids (Plate 4x) of Neuroptera and Diptera. The adult lays egg inside the hemocoel of immature neuropteran and the emerging larva feeds internally until the host spins its cocoon and then it migrates outside and feeds externally. Figitids overwinter as diapausing larvae within the host's body. Among dipteran hosts, Syrphidae and saprophagous cyclorrhaphans seem preferred (Buffington et al., 2007; Pujade-Villar et al., 2013).

1.5.21 Family: Platygatseridae

They are small to minute, shiny black insects with reduced wing venation, sometimes completely veinless. The antennae are usually ten-segmented and are attached very low on the face. The second abdominal segment is distinctly longer than other segments and several times longer than the third segment (Johnson, 1992). They are endoparasitiods of the Diptera (gall midges) and on mealybugs and whiteflies (Iranipour et al., 2020); Buffington et al., 2007 e.g. *Platygaster oryzae* (Plate 4y) is a very important parasitoid of rice gall midge (Ogah et al., 2010); *Amitus hesperidum* on whiteflies; some are egg parasitoids of coleopterans (White et al., 2005).

1.5.22 Family: Scelonidae

They are egg parasitoids of other insects and spiders. They are small to minute insects. The head is generally transverse. Antennae are geniculate, 11–12 segmented. Pronotum is arched and extend backwards to reach tegulae, sometimes median portion is not clearly visible. Forewing is always with submarginal and stigma veins, often postmarginal vein also. Gagster is noticeably depressed and strongly sclerotized, with six or less visible segments. The second abdominal tergite is often the largest or sometimes as long as the third tergite. Most Scelonids have abdominal segments divided into large median sclerites and narrow laterotergites or laterostemites. These later structures interlock to form a sharp-angled margin on the abdomen (Kivan and Kilic, 2002; Austin et al., 2005). The eggs of lepidopteran, grasshoppers, mantids, neuropterans, Embidiina, heteropterans, flies, beetles and spiders are commonly parasitised; e.g: -*Telenomus remus* (Plate 4z) on eggs of tobaocco caterpillar, *Spodoptera litura* (Mani and Krishnamoorthy, 1986).

1.5.23 Family: Bethylidae

Bethylids have a somewhat squarish head, which is much longer than broad and depressed. They are ant-like in appearance. Antennae are 12–13 segmented, with the same number of segments in both sexes and situated closer to clypeus. The males are almost always winged, rarely brachypterous; females are winged, brachyperous or wingless. Forewings have reduced venation and hindwings lack closed cells and have conspicuous claval lobe. Seven to eight abdominal segments are visible (Azevedo, 1999; Terayama, 2003).

These are primary parasitoids of larvae of Lepidoptera and beetles found in concealed conditions, such as within rolled leaves, under bark, within pieces of rotten wood or in earthen cells. Carpenter ants are also parasitised by bethylids; e.g: *Goniozus nephantidis* (Plate 4aa) on Coconut leaf-eating caterpillar, *Opisina arenosella* (Murthy et al., 2005).

1.6 ORDER: NEUROPTERA

Lacewings or net-winged insects (order: Neuroptera) are soft-bodied insects, most commonly medium-sized, comprising about 6,000 species in 18 families (Stange, 2004). Lacewings include mantispids (mantidflies), green lacewings, owl flies, antlions and their relatives. The adults of this order possess four membranous wings and as their name suggests their wings appear lace-like with

Plate.5a Symphrasinae, mantispid flies

PLATE 5 Parasitoids in the order Neuroptera.

many intricate veins. Neuropterans are predatory as both adults and larvae feed on other insects, although some species of adults may supplement their diet with honeydew or pollen. However, some neuropterans are parasitic (Table 1.3) in their larval stages (Brooks and Bernard, 1990; Oswald and Machado, 2018).

1.6.1 Family: Mantispidae

Adult mantispids are predators and have front legs strongly modified for catching prey. But, the larvae of most of them are all parasitic to the eggs sac of spiders (Haug et al., 2018), however the members of the subfamily Symphrasinae (Plate 5a)are sedentary parasitoids on bee, wasp or scarab beetle larvae (Eggleton and Belshaw, 1992).

1.7 CONCLUSIONS

- Insect parasites also called "parasitoids" are the most effective natural enemies, which develop on or in another insect (host), extract nourishment from it, and eventually kill it as a direct or indirect result of their development.
- Some parasitoids that oviposit and complete their development in the egg stage of the host insects are called egg parasitoids, whereas parasitoids that attack other life stages, viz., larva, pupa and adult, are referred to as larval, pupal, or adult parasitoids.
- Parasitoids can also be classified based on where their progeny feeds. Those parasitoid species that develop within their host body are called endoparasitoids, whereas those that feed externally are called ectoparasitoids.
- Parasitoids can be classified as either koinobiont or idiobiont based on their developmental strategies. Parasitoids whose hosts continue to grow after parasitism are called koinobionts, whereas, parasitoids whose hosts do not develop further after parasitism are called idiobionts.
- Hymenopteran parasitoids account for nearly 78% of the estimated number of species of total parasitoids and unique features, viz. single evolutionary lineage, possession of ovipositor and associated venomous glands, host-seeking ability and sex-determining system attributed to their remarkable ecological and evolutionary success.
- However, a large number of under-exploited species of parasitoids belong to the insect orders, viz. Diptera, Coleoptera, Lepidoptera and Neuroptera. The eight families in

Diptera (Cecidomyiidae, Nemestrinidae, Bombyliidae, Asilidae, Phoridae, Pipenculidae, Tachinidae and Sacrcophagidae), nine families in Coleoptera (Carabidae, Scarabidae, Rhipiceridae, Cleridae, Rhipiphoridae, Meloidae, Passandridae, Bothrediridae and Curculionidae), two families in Lepiopdtera (Epipyropidae and Cyclotornidae) and one family in Neuroptera (Manitispidae) comprise some of the parasitic species, which bring about the suppression of insect pest and vectors.
- Though parasitoids belonging to the order Diptera, Coleoptera, Lepiodptera and Neuroptera are highly host-specific and capable of suppressing their host insect populations, more focused research works needs to be carried out to exploit them as successful biocontrol agents under integrated pest management programmes.

1.8 POINTS TO REMEMBER

1. Insect parasitoids are the most effective natural enemies, they are capable of surviving at the expense of their host insects and become inevitable components in integrated pest management progarmmes.
2. Insect parasitoids are classified based on the stages of the insect on which it feeds (i.e. egg, larval, pupal, egg-larval, larval-pupal parasitoids), where it feeds (ectoparasitoids, endoparasitoids, hyperparasitoids) and developmental strategies (koinobionts and idiobionts).
3. Hymenopteran parasitic wasps account for 78% of the estimated number of species of total insect parasitoids and consequently have served as models of choice for nearly all recent research on insect parasitoids. They are mass-reared and well-exploited as a component in success full biocontrol programme.
4. However, potential parasitoids exist in the different endopterygote insect orders, which need to be exploited. An overview on different under-exploited parasitoid species in the order Diptera (eight families), Coleoptera (nine families), Lepidoptera (two families) and Neuroptera (one family) and their role in the suppression of insect pest/vectors across the globe has been deduced from the available scientific literature.
5. The unique features of important families of Hymenopteran parasitoids used in successful biocontrol programme are also discussed in this chapter.
6. These parasitoids show high specificity towards their host insects are capable of suppressing the population of insect pests and vectors and have the potential to become an inevitable component in biocontrol progarmmes.

REFERENCES

Abe, J., Sato, S. & Yukawa, J. (2011). Descriptions of two new endoparasitic cecidomyiids (Diptera: Cecidomyiidae) from Japan. *Applied Entomology and Zoology*, 46: 15–25.

Al Khatib, F., Cruaud, A., Fusu, L., Genson, G., Rasplus, J.Y., Ris, N. &Delvare, G. (2016). Multilocus phylogeny and ecological differentiation of the "*Eupelmus urozonus* species group" (Hymenoptera, Eupelmidae) in the West-Palaearctic. *BMC Evolutionary Biology*, 16: 13.

Alam, S.M. (1959). The life-history and the larval anatomy of Euaphycusvariolosus Alam (Hymenoptera, Encyrtidae) - an endoparasite of Asterolecaniumvariolosum Ratzb. (Hemiptera, Coccidae). *Proceedings of the Zoological Society of Calcutta*, 12(1): 35–40.

Arnett, J.R.H. (1960). *The Beetles of the United States*. Catholic University Press, Washington, DC, p. 1112.

Annecke, D.P. & Doutt, R.L. (1961). The genera of the Mymaridae. Hymenoptera: Chalcidoidea. *Entomology Memoir of the Department of Agriculture of the Union of South Africa*, 5: 1–71.

Askew, R.R. (1961). On the biology of the inhabitants of oak galls of Cynipidae (Hym.) in Britain. *Transactions of the Society for British Entomology*, 14(11): 237–268.

Askew, R.R. & Ruse, M. (1974a). Biology and taxonomy of the genus Enaysma Delucchi (Hym., Eulophidae, Entedontinae) with special reference to the British fauna. *Transactions of the Royal Entomological Society of London*, 125(3): 257–294.

Askew, R.R. & Ruse, M. (1974b). The biology of some Cecidomyiidae (Diptera) galling the leaves of birch (Betula) with special reference to their chalcidoid (Hymenoptera) parasites. *Transactions of the Royal Entomological Society of London*, 126(2): 129–167.

Askew, R.R. & Shaw, M.R. (1986). Parasitoid communities: Their size, structure and development. In *Insect Parasitoids*, Ed. J. Waage & D. Greathead, pp. 225–264. Academic Press, London.

Austin A.D., Johnson N.F. & Dowton M. (2005). Systematics, evolution, and biology of scelionid and platygastrid wasps. *Annual Review of Entomology*, 50: 553–582.

Azevedo, C.O. (1999). FamíliaBethylidae; In *Biodiversidade do estado de São Paulo, Brasil: síntese do conhecimentoao final do século XX. Vol. 5: InvertebradosTerrestres*, Ed. C.R. Brandão & E.M. Cancelo, pp. 169–181. FAPESP, São Paulo.

Bai, M., Li, S., Lu, Y., Yang, H., Tong, Y. & Yang, X. (2015). Mandible evolution in the Scarabaeinae (Coleoptera: Scarabaeidae) and adaptations to coprophagous habits. *Front Zool*, 12: 30.

Bale, J., Van Lenteren, J.C. & Bigler, F. (2008). Biological control and sustainable food production.*Philosophical transactions of the Royal Society of London.Series B, Biological Sciences*, 363: 761–776.

Banerjee, S.N. (1973). Integrated control of sugarcane *Pyrilla* in India. *Pesticides: Annual* 73–74.

Barbier, J. (1947). Observations sur les moeurs de *Rhipipidius pectinicomis* Thunbg. et descriptions de la larveprimaire. *Entomologiste*, 3: 163–180.

Barnes, H. (1954). Gall-midge Larvae as Endoparasites, including the description of a species parasitising Aphids in Trinidad, B.W.I. *Bulletin of Entomological Research*, 45(4): 769–775.

Beardsley, J.W. (1989). Hawaiian Eucoilidae (Hymenoptera: Cynipoidea), key to genera and taxonomic notes on apparently non-endemic species. *Proc Hawaiian Entomol Soc.*, 29: 165–193.

Beatriz, D., Gallardo, F., Cabrera, W. & Guillermo. (1996). *Paraganaspis egeria*, a new genus and species of Eucoilidae (Hymenoptera: Cynipoidea). *Annals of the Entomological Society of America*, 89: 497–500.

Beckage, N.E. & Riddiford, L.M. (1982). Effects of parasitism by *Apanteles congregatus* on the endocrine physiology of the tobacco hornworm *Manduca sexta. Gen Comp Endocrinol*. 47(3): 308–322.

Benton, F.P. (1975). Latval taxonomy and bionomics of some British Pipunculidae. Ph.D. Thesis, University of London, p. 208.

Berg, E. van den, Prinsloo, G.L. & Neser, S. (1990). An unusual host association: *Aprostocetus* sp. (Eulophidae), a hymenopterous predator of the nematode *Subanguina mobilis* (Chit; Fisher, 1975) Brzeski, 1981 (Anguinidae). *Phytophylactica*, 22: 125–127.

Bernardi, N. (1973). The genera of the family Nemestrinidae (Diptera: Brachycera). *Arquivos de Zoologia*, 24: 211–318.

Bétis, L. (1912). Les rhipiphorides Gallo-Rhénans. *Annis Soc. Hist. nat. Toulon*, 3: 125–155.

Blindeman, L. & van Labeke, M.C. (2003). Control of the two-spotted spider mite (Tetranychusurticae Koch) in glasshouse roses. *Commun. Agric Appl. Biol, Sci.*, 68: 249–254.

Bohart, G.E., Stephen, W.P. & Eppley, R.K. (1960). The biology of *Heterostylum robustum* (Diptera: Bombyliidae), a parasite of the alkali bee. *Annals of the Entomological Society of America*, 53: 423–435.

Bologna, M.A. (1991). Fauna de Italia. XXVIII. Coleoptera Meloidae. *Calderini, Bologna*. i–xiv, p. 541.

Bologna, M.A. & Di Giulio, A. (2011). Biological and morphological adaptations in the pre-imaginal phases of the beetle family Meloidae. *Atti Accad. Naz. Ital. Ent.* 59: 141–152.

Bologna, M.A., Di Giulio, A. & Pinto J.D. (2002). Review of the genus Stenodera with a description of the first instar larva of S. puncticollis (Coleoptera: Meloidae). *Eur. J. Entomol.*, 99(3): 299–314.

Boucek, Z. (1963). A taxonomic study in Spalangia Latr. (Hymenoptera, Chalcidoidea). Sb. ent. Odd. n r. Mus. Praze., Acta Entomologica Musei Nationalis Pragae, 35: 429–512.

Bologna, M.A., Oliverio, M., Pitzalis, M. & Mariottini, P. (2008). Phylogeny and evolutionary history of the blister beetles (Coleoptera, Meloidae). *Mol. Phylogenet. Evol.*, 48(2): 679–693.

Bologna, M.A. & Pinto, J.D. (2001). Phylogenetic studies of Meloidae (Coleoptera), with emphasis on the evolution of phorsey. *Systematic Entomology*, 26: 33–72.

Boucek, Z. & Askew, R.R. (1968) a. PalaearcticEulophidae sine Tetrastichinae. In *Index of Entomophagous Insects*, Ed. V. Delucchi & G. RemaudièreLe François, Paris, pp. 3–260.

Boucek, Z. & Narendran, T.C. (1981a). Indian chalcid wasps (Hymenoptera) of the genus *Dirhinus* parasitic on synanthropic and other Diptera. *Systematic Entomology*, 6: 229–251.

Bouček, Z. & Narendarn, T.C. (1981b). The Leucospis species of India and adjacent countries (Hymenoptera: Leucospidae). *Oriental Insects*, 15: 1–15.

Boucek, Z. & Rasplus, J.Y. (1991). *Illustrated key to West-Palaearctic genera of Pteromalidae (Hymenoptera: Chalcidoidea)*. Institut National de la Recherche Agronomique, Paris, p. 140.

Brodeur, J., Abram, P.K., Heimpel, G.E. & Messing, R.H. (2018). Trends in biological control: Public interest, international networking and research direction. *BioControl*, 63: 11–26.

Brooks, S.J. & Bernard, P.C. (1990). The green lacewings of the world: A generic review (Neuroptera: Chrysopidae). *Bulletin of the British Museum of Natural History (Entomology)* 59: 117–286.

Brown, B. (2018). A second contender for "world's smallest fly" (Diptera: Phoridae). *Biodiversity Data Journal*, 6: e22396.

Buffington, M.L., Nylander, J.A.A. & Heraty, J. (2007). The phylogeny and evolution of Figitidae (Hymenoptera: Cynipoidea). *Cladistics* 23(1): 29.

Burckhardt, D. & Ślipiński, S.A. (2003). Phylogeny and taxonomy of the world Passandridae (Coleoptera). In *Systematics of the Coleoptera: Papers Celebrating the retirement of Ivan Löbl. Cucordoro. Memoirs on Entomology, International 17*, Ed. G. Cucordoro & R.B. Leschen, pp. 753–883. Associated Publishers, Gainsville.

Burks, B.D. (1971). A North American *Elasmus* parasitic on *Polistes*. *J. Wash. Acad. Sci.*, 61: 194–196.

Capinera, J.L. (2008). Butterflies and moths. *Encyclopedia of Entomology*. 4 (2nd ed.). Capinera J.L.Springer. Haldberg, pp. 626–672.

Castelo, M.K. & Corley, J.C. (2010). Spatial density dependent parasitism and specificity in the robber fly *Mallophoraruficauda* (Diptera: Asilidae). *Austral. Ecol.*, 35: 72–81.

Castelo, M.K. & Crespo, J.E. (2011). Host location in parasitoids with active searching larvae: The case of *Mallophoraruficauda*. In *The Biology of Odors: Sources, Olfaction and Response*, Ed. L.E. Weiss & J.M. Atwood, pp. 185–206. Nova Science Publishers, Hauppage, NY.

Castelo, M.K. & Lazzari, C.R. (2004). Host seeking behavior in larvae of the robber fly *Mallophora ruficauda* (Diptera: Asilidae). *J. Insect Physiol.*, 50(4): 331–336.

Cendana, S.M. (1937). Studies on the biology of Coccophagus (Hymenoptera) a genus parasitic on nondiaspineCoccidae. *University of California Publications in Entomology*, 6(14): 337–399.

Chang, Y., Lee, S.H., Na, J.H., Chang, P.S. & Han, J. (2017). Protection of Grain Products from *Sitophilus oryzae*(L.) Contamination by Anti-Insect Pest Repellent Sachet Containing Allyl Mercaptan Microcapsule. *J Food Sci.*, 82(11): 2634–2642.

Chapin, E.A. (2014). Studies in the genus *Hydnocera* Newman: The species of the New England states, Masters Theses, p. 237.

Charnov, E.L. (1982). *The Theory of Sex Allocation*. Princeton University Press, Princeton, NJ, p. 335.

Chen, L. & Porter, S.D. (2020). Biology of Pseudacteon Decapitating Flies (Diptera: Phoridae) That parasitize ants of the *Solenopsissaevissima* complex (Hymenoptera: Formicidae) in South America. *Insects*, 11 (2): 107.

Chobaut, A. (1891). Moeurs et métamorphoses de EmenadiaHabe 11 ata F., insecteColéoptère de la familie des Rhipiphorides. *Annls Soc. ent. Fr.*, 8: 447–456.

Chobaut, A. (1906). Le triongulinide du *Myiodessubdipterus*Bosc (Col.). *Annls Soc. ent. Fr.* 19: 238–244.

Claridge, M.F. (1959). A contribution of the biology and taxonomy of the British species of the genus Eudecatoma Ashmead (Hym., Eurytomidae). *Transactions for the Society of British Entomology*, 13: 149–168.

Claridge, M.F. (1961). An advance towards a natural classification of eurytomid genera (Hym., Chalcidoidea) with particular reference to British forms. *Transactions of the Society for British Entomology*, 14: 167–185.

Claridge, M.F. & Askew, R.R. (1960). Sibling species in the Eurytomarosae group (Hym., Eurytomidae). *Entomophaga*, 5: 141–153.

Clausen, C.P. (1940). *Entomophagous Insects*. McGraw Hill, New York, p. 688.

Clausen, C.P. (1962a). *Entomophagous Insects*. Hafner Publishing, New York, p. 688.

Clausen, C.P. (1962b). *Entomophagous Insects*. 2nd ed. Hafner Publishing, New York, [2nd Printing], p. 688.

Clements, A.N. & Bennett, F.D. (1969). The structure and biology of a new species of *Mallophora*Macq. (Diptera: Asilidae) from Trinidad. *Bull Entomol Res*, 58(3): 455–463.

Common, I.F.B. (1990). *Moths of Australia*. E. J. Brill, Leiden/ New York/ Copenhagen/ Koeln, p. 535.

Cooperband, M.F., Wharton, R.A., Frankie, G.W. & Vinson, S.B. (1999). New host and distribution records for Leucospis (Hymenopera: Leucospidae) associated primarily with nest of Centris (Hymenoptera: Anthophoridae) in the dry forests of Costa Rica. *Journal of Hymenoptera Research*, 8(2): 154–164.

Coupland, J. & Barker, G. (2004). Flies as Predators and Parasitoids of Terrestrial Gastropods, with Emphasis on Phoridae, Calliphoridae, Sarcophagidae, Muscidae and Fanniidae (Diptera, Brachycera, Cyclorrhapha). In *Natural enemies of terrestrial molluscs*, Ed. G. Barker, pp. 85–158. CABI Publishers, Wallingford.

Davis, D.R. (1987). Epipyropidae. In *Immature insects*, Vol. 1, Ed. F.W. Stehr, pp. 456–460. Kendall/Hunt, Dubuque, Iowa.

De Rond, J., Smit, J., Heitmans, W. & Peeters, T. (1994). A survey of the Western European Rhipiphoridae including the first record of a *Macrosiagon* species in The Netherlands (Coleoptera). *EntomologischeBerichten*, 54: 201–211.

Deepthi, T.R. & Ramadevi, O.K. (2012). New report of *Aulonosoma insignis*Grouvelle (Coleoptera: Passandridae) as a larval ectoparasitoid of *Sinoxylonanale*Lesne (Coleoptera: Bostrichidae). *Journal of Biological Control*, 26(1): 75–76.

Dils, J. & Ozbek, H. (2006). Contribution to the knowledge of the Bombyliidae of Turkey (Diptera). *Linz Biol Beitr*, 38(1): 455–504.

Dindo, M.L. (2011). Tachinid parasitoids: Are they to be considered as koinobionts? *BioControl*, 56, 249–255.

Dindo, M.L. & Nakamura, S. (2018). Oviposition strategies of Tachinid parasitoids: Two *Exorista* species as case studies. *International Journal of Insect Science*, 10: 1–6.

Dodd, F.P. (1902). Contribution to the life history of *Liphyrabrassolis*Westw. *Entomology*, 35: 153–156.

Doutt, R.L. & Viggiani, G. (1968). The classification of the Trichogrammatidae (Hymenoptera: Chalcidoidea). *Proceedings Calif. Acad. Sci.*, 35: 477–586.

Drees, M. (1994). EineGebäudebrut von *Metoecus paradoxus* (L.) (Insecta, Coleoptera, Rhipiphoridae). *Ent. Bl. Biol. Syst. Käfer*, 90: 117–121.

Eggleton, P. & Belshaw, R. (1992). Insect parasitoids: An evolutionary overview. *Trans. R. Soc. Lond.*, 337: 1–20.

Eggleton, P. & Belshaw, R. (1993). Comparisons of dipteran, hymenopteran and coleopteran parasitoids: Provisional phylogenetic explanations. *Biol J Linn Soc*, 48: 213–226.

Eggleton, P. & Gaston, K. (1990). "Parasitoid" species and assemblages: Convenient definitions or misleading compromises? *Oikos*, 59(3): 417–421.

Eisner, T. (2003). *For Love of Insects*. Belknap Press, Harvard University, Cambridge, MA, p. 464.

El-Hawagry, M.S. (2015). *Catalogue of Superfamily Asiloidea*. Lambert Academic Publishing, Saarbrücken, p. 196.

Elzinga, R.J. (1977). Observations on *Sandalusniger*Knoch (Coleoptera: Sandalidae) with a Description of the triungulin Larva. *Journal of the Kansas Entomological Society*, 50: 324–328.

Erwin, T.L. (1979). A review of the natural history and evolution of ectoparasitoid relationships in carabid beetles. In *Carabid Beetles: Their Evolution, Natural History, and Classification*, Ed. T.L. Erwin, G.E. Ball, D.R. Whitehead & A.L. Halpern, pp. 479–484. The Hague, The Netherlands.

Evenhuis, N.L. (1994). Family Nemestrinidae. In *Fossil flies of the world (Insecta: Diptera)*, Ed. N.L. Evenhuis, pp. 313–315. Backhuys, Leiden.

Evenhuis, N.L. & Greathead, D.J. (2015). World catalog of bee flies (Diptera: Bombyliidae). Revised September 2015. Available from: http://hbs.bishopmuseum.org/bombcat. Accessed 1 June 2018.

Farouk, A. Abdel-Galil, Moustafa, M.A., Rizk, Sobhy, A.H., Temerak Dalia, Y.A. & Darwish (2014). Studies on the parasitoid *Opius concolor* Sźepligeti (Hymenoptera: Braconidae), associated with zizyphus fruit fly, *Carpomyiaincompleta* Becker (Diptera: Tephritidae). *Archives of Phytopathology and Plant Protection*, 47(6): 665–674.

Feener, D.H. & Brown, B.V. (1997). Diptera as parasitoids. *Annu Rev. Entomol.*, 42: 73–97.

Ferrière, C. & Kerrich, G.J. (1958). Hymenoptera 2. Chalcidoidea. Section (a) Agaontidae, Leucospidae, Chalcididae, Eucharitidae, Perilampidae, Cleonymidae and Thysanidae. *Handbooks for the Identification of British Insects*, 8(2)(a): 40.

Flanders, S.E. (1959). Differential host relations of the sexes in parasitic Hymenoptera. *EntomologiaExperimentalis et Applicata*, 2: 125–142.

Fletcher, T.B. (1939). A new *Epipyrops* from India (Lepidoptera: Epipyropidae). *Bulletin of Entomological Research*, 30: 293–294.

Frank, J.H., Thomas, M.C., Yousten, A.A., Howard, F.W., Giblin-davis, R.M. & Luna, M.G. (2008). *Parasitic Hymenoptera* (Parasitica). *Encyclopedia of Entomology*, 2: 2730–2736.

Gagne, R.J. (2004). A catalog of the Cecidomyiidae (Diptera) of the world. *Mem Entomol Soc* Wash, 25: 1–408.

Gangwar, S.K., Srivastava, D.C., Tewari, R.K., Singh, M.R. & Rajak, D.C. (2008). Management of *Pyrilla perpusilla* Walker in sugarcane with ectoparasitoid, Epiricania *melanoleuca* (Fletcher), during epidemics in sub-tropical India. *Sugar Tech*, 10(2): 162–165.

Gao, R. & Qin, X. (1992). *Colydiidae*. In *Forest Insects of China*, Ed. G. Xiao, pp. 445–446. Forestry Publishing House, Beijing, China.

Gauld, I.D. (1988). Evolutionary patterns of host utilization by ichneumonoid parasitoids (Hymenoptera: Ichneumonidae and Braconidae), *Biological Journal of the Linnean Society*, 35(4), 351–377.

Gauld, I.D. & Bolton, B. (1988). *The Hymenoptera*. Oxford University Press, New York, p. 332.

Gauthier, N., LaSalle, J., Quicke, D.L.J. & Godfray, H.C.J. (2000). Phylogeny of Eulophidae (Hymenoptera: Chalcidoidea), with a reclassification of Eulophinae and the recognition that Elasmidae are derived eulophids. *Systematic Entomology*, 25: 521–539.

Gibson, G.A.P. (1995). Parasitic wasps of the subfamily Eupelminae: Classification and revision of world genera (Hymenoptera: Chalcidoidea: Eupelmidae). *Memoirs on Entomology, International*, 5: 421.

Godfray, H.C.J. (1993). Parasitoids: *Behavioral and Evolutionary Ecology*. Princeton University Press, Princeton, NJ, p. 473.

Godfray, H.C.J. (2007). Parasitoids. *Encyclopedia of Biodiversity*Ed: Simon A L. Academic Press,London.674-682.

Gradwell, G.R. (1957). Hosts of three species of Eulophus Geoffroy (Hym., Chalcidoidea), one new to science and another new to Britain. *Entomologist's Monthly Magazine*, 93: 140–142.

Graham, M.W.R. (1995). European Elasmus (Hymenoptera: Chalcidoidea, Elasmidae) with a key and descriptions of five new species. *Entomologist's Monthly Magazine*, 131: 1–23.

Greathead, D.J. (1958). Notes on life history of *Symmictusflavopilosus* Bigot (Diptera: Nemestrinidae) as a parasite of *Schistocerca gregaria* (Forskal) (Orthoptera: Acrididae). *Proc. R. Ent. Soc. Lond. Ser. A*, 33: 107–119.

Greathead, D.J. (1963). A review of the insect enemies of Acridoidea (Orthoptera). *Trans Royal Ent. Soc.*, London, 114: 437–517.

Groba, H.F. & Castelo, M.K. (2011). Chemical interaction between the larva of a dipteran parasitoid and its coleopteran host: A case of exploitation of the communication system during searching behavior. *Bulletin of Entomological Research*, 102: 1–9.

Gullan, P.J. & Cranston, P.S. (2014). *The Insects: An Outline of Entomology* (5th ed.). Wiley,UK. p. 362–370.

Gupta, V.K. (2004). Wasps, ants, bees and sawflies (Hymenoptera). *Encyclopedia of Entomology*.In. Ed: Capinera J. L., pp. Springer 2486–2499. Dordrecht.

Haehnni, J.P. & Borer, M. (2007). First report of a Nemestrinidae (Diptera) parasitoid of the Mantodea. *Studia Dipterologica*, 14(1): 61–65.

Hansen, L. (2000). The family Signiphoridae (Hymenoptera, Chalcidoidea) in Norway. *Norwegian Journal of Entomology*, 47: 76.

Haug, J.T., Müller, P. & Haug, C. (2018). The ride of the parasite: A 100-millionyearold mantis lacewing larva captured while mounting its spider host. *Zoological Lett.*, 4, 31.

Hawkeswood, T. (2000). Some notes on the occurrence of the Australian beetle, *Rhipicera femorata* (Kirby) (Coleoptera: Rhipiceridae). *Mauritiana*, 17: 417–419.

Hawkins, B.A. & Sheehan, W. (1994). *Parasitoid Community Ecology*. Oxford University Press, Oxford, p. 516.

Hayat, M. (1983). The genera of Aphelinidae (Hymenoptera) of the world. *Systematic Entomology*, 8: 63–102.

Henriksen, K.L. (1922). Notes upon some aquatic Hymenoptera. *Ann. Biol. Lacustre*, 11: 19–37.

Heppner, J.B. (2004). Planthopper Parasite Moths (Lepidoptera: Epipyropidae)*Encyclopedia of Entomology*. In Ed: Capinera J L, pp. 2921–2922. Springer, Dordrecht.

Hochberg, M. (1991). Intra-Host Interactions between a Braconid Endoparasitoid, *Apanteles glomeratus*, and a Baculovirus for Larvae of *Pieris brassicae*. *Journal of Animal Ecology*, 60(1): 51–63.

Hoddle, M., Van Driesche, R. & Sanderson, J. (1998). Biology and use of the whitefly parasitoid *Encarsiaformosa*. *Annual Review of Entomology*, 43: 645–669.

Horion, A. (1956). Faunistik der mitteleuropäischenKäfer, 5: *Heteromera*: G. Frey, Tutzing. pp. 1–336.

Huber, J.T. (1986). Systematics, biology, and hosts of the Mymaridae and Mymarommatidae (Insecta: Hymenoptera). *Entomography*, 4: 185–243.

Hull, F.M. (1962). Robber flies of the world. *Bull US Natl Mus*, 224(1,2): 1–907

Hull, F.M. (1973). Bee flies of the world.The genera of the family Bombyliidae. In *Bulletin of the United States National Museum No. 286*. Smithsonian Institution Press, Washington, DC, 286:1-687.

Huq, S.B. (1982). *A contribution to the biology of pipencullidflies (Pipunculidae: Diptera)*. PhD thesis, Free University, Berlin, p. 128.

Illingworth, J.F. (1919). The sugarcane beetle borer parasite (*Ceromasia sphenophori*) in Queensland. *J. Econ. Ent.*, 12: 457–459.

Iranipour, S., BenaMoleai, P., Asgari, S. & Michaud, J.P. (2020). Foraging egg parasitoids, *Trissolcusvassilievi*(Hymenoptera: Platygastridae), respond to host density and conspecific competitors in a patchy laboratory environment. *J Econ Entomol.*, 113(2): 760–769.

Jervis, M.A. (1980). Studies on oviposition behaviour and larval development in species of *Chalants*(Diptera, Pipunculidae), parasites of typhlocybine leafhoppers (Homoptera, Cicadellidae). *Journal of Natural History*, 14: 759- 768.

Jim, M. (2007). Taxonomy of Lepidoptera: The scale of the problem. In *The Lepidoptera Taxome Project*. University College, London. Archived from the original on 5 June 2011. Retrieved 8 February 2011.

Johnson, N.F. (1992). Catalog of world species of Proctotrupoidea, exclusive of Platygastridae (Hymenoptera). *Mem.of the American Entomol. Institute*, 51: 825.

Johnson, N.F. (2013). Hymenoptera. *Encyclopedia of Biodiversity 2nd Edition*.in: Ed. Samuel M. S. pp.177–184. Academic Press.

Karnavar, G.K. (1983). Studies on the population control of *Pieris brassicae* L. BY *Apanteles glomeratus* L. *Int J Trop Insect Sci.*, 4: 397–399.

Katovich, K. (2002). 39. RhipceridaeLatreille, 1834. In *American Beetles Volume 2, Polyphaga: Scarabaeoidea through Curculionoidea*, Ed. R.H. Arnett, M.C. Thomas, P.E. Skelley & J.H. Frank, pp. 92–94. CRC Press, Washington DC.

Kazmi, S., Sheela, S., Rameshkumar, A. & Mazumdar, P. (2017). Contribution to the knowledge of the *Ormyrus* (Hymenoptera: Chalcidoidea: Ormyridae) with two new records from India. *Journal of Entomology and Zoology Studies*, 5: 223–227.

Kieffer, J.J. (1896). Observations sur les *Diplosis,* et diagnoses de cinq especesnouvelles (Dipt.). *Bulletin of Entomological Society* France, p. 382–384.

Kieffer, J.J. (1901). Monographie des Ce'cidomyidesd'Europe et d'Alge'rie. Annals of the Entomological Society of France ,p. 181–472.

Kirkpatrick, T. (1954). Notes on *Pseudendaphismaculans* Barnes, a Cecidomyid Endoparasite of Aphids in Trinidad, B. W. I. *Bulletin of Entomological Research*, 45(4): 777–781.

Kivan, M. & Kilic, N. (2002). Host preference: Parasitism, emergence and development of *Trissolcussemistriatus* (Hym., Scelonidae) in various host eggs. *Journal of Applied Entomology*, 126: 395–399.

Knoch, A.W. (1801). *Neue BeyträgezurInsektenkunde. Theil. 1.* Schwickertsche Verlage, Leipzig. p. 208.

Koenig, D.P. & Young, C.W. (2007). First observation of parasitic relations between big-headed flies, *Nephrocerus*Zetterstedt (Diptera: Pipunculidae) and crane flies, *Tipula* Linnaeus (Diptera: Tupulidae: Tipulinae), with larval and puparial descriptions for the genus *Nephrocerus*. *Proceedings of the Entomological Society of Washington*, 109: 52–65.

Kopylov, D.S. (2009). A new subfamily of ichneumonids from the Lower Cretaceous of Transbaikalia and Mongolia (Insecta: Hymenoptera: Ichneumonidae). *Paleontologischeskiizhurnal* (11): 76–85. *Translation in Paleontological Journal*, 43 (1): 83–93.

Krčmar, S., Whitmore, D., Pape, T. & Buenaventura, E. (2019). Checklist of the Sarcophagidae (Diptera) of Croatia, with new records from Croatia and other Mediterranean countries. *Zoo Keys*, 831: 95–155.

Krinsky, W.L. (2002). Beetles *(Coleoptera)*. *Medical and Veterinary Entomology-3rd Edition*. In: Ed. Mullen, G. R. and Durdan, L. A. 'pp. 87–101. Academic Press.

Kumar, M. & Chandra, R. (2015). A brief note on the aphidiphagous *Endaphis aphidimyza* (Diptera Cecidomyiidae) in Chitrakoot Dham region and Parbhani district (India). *Biodiversity Journal*, 6(4): 827–830.

Kumar, R., Mittal, V., Chutia, P. & Ramamurthy, V.V. (2015). "Taxonomy of *Fulgoraecia melanoleuca* (Fletcher, 1939), (Lepidoptera: Epipyropidae) in India, a biological control agent of *Pyrilla perpusilla* (Walker) (Hemiptera: Lophopidae)" (PDF). *Zootaxa*, 3974(3): 431–439.

Kumarasinghe, N.C. & Wratten, S.D. (1996). The sugarcane leaf hopper, *Pyrilla perpusilla* (Homoptera: Lophopidae): A review of its biology, pest status and control. *Bulletin of Entomological Research*, 86: 485–498.

La Salle, J. & Gauld, I.D. (1993). Hymenoptera: Their diversity, and their impact on the diversity of other organisms. In *Hymenoptera and Biodiversity*, Ed. J. La Salle & I.D. Gauld, pp. 1–26. CAB International, London.

Laing, J.E. & Heraty, J.M. (1981). The parasite complex of the overwintering population of Epiblemascudderiana (Lepidoptera: Olethreutidae) in southern Ontario. Proceedings of the Entomological Society of Ontario, 112: 59–67.

Lawrence, J.F. (2005). 16.2. RhipiceridaeLatreille, In *Coleoptera, Beetles. Volume 1: Morphology and Systematics (Archostemata, Adephaga, Myxophaga, Polyphagapartim*, Ed. R.G. Beutel & R.A.B. Leschen, pp. 456–460. *Handbook of Zoology,* Part 38. Walter de Gruyter, Berlin and New York.

Lawrence, J.F., Falin, Z.H. & Ślipiński, A. (2010). RipiphoridaeGemminger and Harold, 1870 (Gerstaecker, 1855). In *Coleoptera, beetles. Volume 2: Morphology and systematics (Elateroidea, Bostrichiformia, Cucujiformiapartim)*, Ed. R.A.B. Leschen, R.G. Beutel & J.F. Lawrence, pp. 538–548. Walter de Gruyter, New York.

Lawrence, J.F., Hastings, A.M., Dallwitz, M.J., Paine, T.A. & Zurcher, E.J. (1999). Beetles of the world: A key and Information System, for Families and MS-Windows. CSIRO Publishing, Melbourne. *European Journal of Entomology*, 98(2): 165–166.

Lee, S.G., Nam, J.W. & Lim, J. (2017). A taxonomic review of the family Bothrideridae Erichson (Coleoptera: Coccinelloidea) in Korea represented by ectoparasites of wood-boring insects. *Journal of Asia-Pacific Biodiversity*, 10: 208–211.

Lim, J., Oh, H., Park, S. & Koh,S. (2012). First record of the family Bothrideridae (Coleoptera) in Korea represented by the wood-boring beetle ectoparasite, *Dastarcus helophoroides*. *Journal of Asia-Pacific Entomology*, 15: 273–275.

Liu, G.C., & Chen, G. (2019). New Scuttle flies (Diptera: Phoridae) reared from dead Termite, *Odontotermesformosanus* (Shiraki, 1909) (Isoptera: Termitidae) in China. *Zootaxa*, 4551(2), 237–243.

Long, C.D., Wang, Y.P. & Tang, P.Z. (1960). Biological characteristics and its utilization of woolly apple aphid parasitoid (Aphelinusmali Haldeman). *Acta EntomologicaSinica*, 10(1): 1–3.

Lotfalizadeh, H., Richard, A., Pablo, F.U. & Majid, T. (2012). The species of *Ormyrus* Westwood (Hymenoptera: Ormyridae) in Iran with description of an unusual new species. *Zootaxa*, 3300: 34–44.

Lucht, W.H. (1987). *Die KäferMitteleuropas*, Katalog: 1–342. Goecke & Evers, Krefeld.

Lysyk, T.J. (1999). Relationships between temperature and life history parameters of *Muscidifurax raptor* (Hymenoptera: Pteromalidae). *Environmental Entomology*, 29: 596–605.

Mamaev, B.M. (1973). A new genus of endoparasitic gall midges, *Occuloxenium compitale* Mamaev gen. et sp. n., developing on aphids (Diptera, Cecidomyiidae). *Uzb. Biol. Zh.*, 17: 52–54.

Mani, M. & Krishnamoorthy, A. (1986). Susceptibility of *Telenomusremus* Nixon, an exotic parasitoid of *Spodoptera litura* (F.) to some pesticides. *Trop Pest Manage.*, 32: 49–51.

Mathur, R. (2009). *A New Species of Anastatusmotschousky (Chacididae: Eupelmidae) from Kashmir.* Proceedings of The Royal Entomological Society of London. Series B, Taxonomy. 25: 93–97.

May, Y.Y. (1979). The biology of *Cephalopscurtifrons (*Diptera: Pipunculidae), an endoparasite of *Stenocranusminutus*(Hemiptera: Delphacidae). *Zoological Journal of the Linnean Society*, 66(1): 15–29.

May, Y.Y. (2008). The biology of *Cephalopscurtifrons* (Diptera: Pipunculidae), an endoparasite of *Stenocranusminutus* (Hemiptera: Delphacidae). *Zoological Journal of the Linnean Society*, 66: 15–29.

McHugh, J.V. & Leibherr, J.K. (2009). Coleoptera.In *Encyclopedia of Insects*, Ed. R.T. Cardé & V.H. Resh, pp. 183–201. Academic Press, Burlingtonp.

McNamara, J. (1991). Family Cleridae: Checkered beetles. In *Checklist of the Beetles of Canada and Alaska*, Ed. Y. Bousquet, pp. 208–211. Canada communication group, Ottawah, canada.

Meier, R., Kotraba, M. & Ferar, P. (1999). Ovoviviparity and viviparity in the Diptera. *Biological Reviews*, 74(3): 199–258.

Mellini, E. (1990). Sinossi di biologiadeiDitteriLarvevoridi.Boll. *Inst. Entomol.*, 45: 1–38.

Mills, N. (1992). Classification and biology of braconid wasps. By M. R. Shaw & T. Huddleston. (London: British Museum (Natural History), 1991). 126 pp. *Bulletin of Entomological Research*, 82(3): 435–436.

Mitra, B., Roy, S., Sumana, H. & Mukhopadhyay, S. (2016). Diversity and distribution pattern of scuttle flies (Diptera: Phoridae) in India. *International Journal of Entomological Research*, 3(1): 33–38.

Molina-Ochoa, J., Carpenter, J., Heinrichs, E. & Foster, J. (2003). Parasitoids and Parasites of *Spodoptera frugiperda* (Lepidoptera: Noctuidae) in the Americas and Caribbean Basin: An Inventory. *The Florida Entomologist*, 86(3): 254–289.

Muratori, F., Gagne, R. & Russel, M. (2009). Ecological traits of a new aphid parasitoid, *Endaphisfugitiva* (Diptera: Cecidomyiidae), and its potential for biological control of the banana aphid, *Pentalonianigronervosa* (Hemiptera: Aphididae). *Biological Control*, 50(2): 185–193.

Murthy, K., Jalali, S. & Venkatesan, T. (2005). Performance of *Goniozusnephantidis* (Bethylidae: Hymenoptera) multiplied on artificial diet-reared *Opisinaarenosella* at variable high temperature. *Indian Journal of Agricultural Sciences*, 75: 529–531.

Musso, J.J. (1983). Nutritive and ecological requirements of robber flies (Diptera: Brachycera: Asilidae). *Entomol. Generalis*, 9(1/2): 35–50.

Myrick, S., George, W., Norton, K.N. Selvaraj, Natarajan, K. & Muniappan, R. (2014). Economic impact of classical biological control of papaya mealy bug in India. *Crop Protection*, 56: 82–86.

Nafiu, B.S., Dong, H. & Cong, B. (2014). Principles of biological control in integrated pest management. *International Journal of Applied Research and Technology*, 3(11): 104–116.

Namaghi, H.S., Hosseini, M. & Jalali, N.M. (2012). First Report of *Endaphisperfidus*Keiffer (Diptera: Cecidomyiidae), as an Endoparasitoid of Pomegranate Aphid (*Aphis punicae* Pass.) in Iran. *Journal of Plant Protection* 26(2): 1–14.

Narendran, T.C. (1989). *Oriental Chalcididae (Hymenoptera: Chalcidoidea).* University of Calicut, Kerala, India, p. 149.

Narendran, T.C. (1994). *Torymidae and Eurytomidae of Indian Subcontinent.* University of Calicut, Kerala, India, p. 500.

Nasser, M. & Abdurahiman, U. (2001). Biological control of the coconut caterpillar *Opisinaarenosella* (Lepidoptera: Xylorictidae): Achievements and prospects. In *Biocontrol Potential and Its Exploitation in Sustainable Agriculture*, Ed. R.K. Upadhyay, K.G. Mukerji & B.P. Chamola, pp. 285–305. Kluwer Academic/Plenum Publishers, New York.

Noyes, J.S. (2012). Universal chalcidoidea database. http://www.nhm.ac.uk/chalcidoids.

Ogah, E.O., Odebiyi, J.A., Omoloye, A.A. & Nwilene, F.E. (2010). Parasitism and development of *Platygasterdiplosisae* (Hymenoptera: Platygastridae) on the African rice gall midge *Orseoliaoryzivora* (Diptera: Cecidomyiidae). *Int J Trop Insect Sci*, 30: 93–100.

Ogura, N., Tabata, K. & Wang, W. (1999). Rearing of the colydiid beetle predator, *Dastarcushelophoroides*, on artificial diet. *Biological Control*, 44:291–299.

O'Hara, J.E. (2008). General information about tachinid flies, http://www.nadsdiptera.org/tach/gerdtachintr.htm.

Okamoto, Y. (1999). Field parasitism and some information on bionomy of *Dastarcushelophoroides* (Fairmaire) parasitizing the Japanese pine sawyer, *Monochamusalternatus* Hope. *Appl. For. Sci.*, 8: 229–232.

Oliff, A.S. (1891). The fly parasite of the plague locust. *Agric.Gaz. New South Wales*, 2: 255–257.

Oswald, J.D. & Machado, R.J.P. (2018). Biodiversity of the Neuropterida (Insecta: Neuroptera, Megaloptera, and Raphidioptera. In *Insect Biodiversity: Science and Society*, Volume 1, Second Edition, Ed. R.G. Foottit & P.H. Adler, pp. 627–672. John Wiley & Sons, New Jersey.

Pandey, K.P., Pandey, M.N. Mishra, V.K.,Singh, S.,Singh,D.N. & Singh,S.B. (2008). Studies on the effect of ecofriendly bioagent *Epiricania melanoleuca* for the control of Sugarcane Pyrilla *(Pyrilla perpusilla)* in eastern U.P. *Indian Agricultural Research*, 23(2): 91–95.

Pape, T. (1996). Catalogue of the Sarcophagidae of the world (Insecta: Diptera). In *Memoirs on Entomology*, International Volume 8. Associated Publishers, Gainesville. p. 558.

Patange, N.R., Sharma, O.P. & Chiranjeevi, B. (2017). New record of parasitoid, associated with pigeonpea pod fly, *Melanagromyzaobtusa* (Malloch) (Diptera: Agromyzidae). *Journal of Biological Control*, 31: 13.

Pawar, A.D., Misra, M.P. & Singh, J. (2002). Role of *Epiricania melanoleuca* Fletcher in biological control of sugarcane pyrilla, *Pyrilla perpusilla* Walker outbreak in 1999 in Western Uttar Pradesh. *Journal of Biological Control*, 16(1): 71–76.

Peck, O., Boucek, Z. & Hoffer, A. (1964). Keys to the Chalcidoidea of Czechoslovakia (Insecta: Hymenoptera). *Memoirs of the Entomological Society of Canada*, 34: 170.

Percino-Daniel, N., Buckley, D. & García-París, M. (2013). Pharmacological properties of blister beetles (Coleoptera: Meloidae) promoted their integration into the cultural heritage of native rural Spain as inferred by vernacular names diversity, traditions, and mitochondrial DNA. *J. Ethnopharmacol.*, 147(3): 570–583.

Philips, T.K. & Ivie, M.A. (2002). Bothrideridae Erichson 1845. In *American Beetles, vol. 2: Polyphaga: Scarabaeoidea through Curculionoidea*, Ed. R.H. Arnett, M.C. Thomas, P.E. Skelley & J.H. Frank, pp. 358–362. CRC Press LLC, Boca Raton.

Pinto, J.D. (1998). Systematics of the North American species of Trichogramma Westwood (Hymenoptera: Trichogrammatidae). *Memoirs of the Entomological Society of Washington*, 22: 1–287.

Potgieter, J.T. (1929). A contribution to the biology of the Brown Swarm Locust *Locustanapardalina* (Wlk.) and its natural enemies. *Pan Afr. Agric. Vet. Conf., Pretoria*, (1929.) Papers Agric. Sect. pp. 265–308, 1 map, 7 pls.; Rep. Proc. pp. 113–127.

Price, P.W. (1980). Evolutionary biology of parasites. *Monogr. Popul. Biol.*, 15: 1–237.

Pujade-Villar, J., Petersen-Silva, R. & Paretas-Martínez, J. (2013). Description of a New Genus and Three New Species of Figitinae (Hymenoptera: Cynipoidea: Figitidae) from Colombia. *Neotrop.Entomol*, 42: 63–71.

Quicke, D.L. & Achterberg, V.C. (1990). Phylogeny of the subfamilies of the family Braconidae (Hymenoptera: Ichneumonoidea). *Zool. Verh.*, 258: 1–95.

Rajak, D.C. (2007). Colonization and redistribution of *Epiricania melanoleuca* Fletcher against *Pyrilla perpusilla* Walker. *Annals of Plant Protection Sciences*, 15: 83–86.

Richter, V.A. (1997). Family Nemestrinidae. In *Contributions to a manual of Palearctic Diptera*, Ed. L. Papp & B. Darvas, pp. 459–468. Science Herald, Budapest.

Riley, C.V. (1880). The rocky mountain locust. *U. S. Entomology communication.* 2nd Report, p. 322.

Robertson, H. (2004). How San hunters use beetles to poison their arrows. Iziko museum of Cape Town. Available at www.biodiversityexplorer.org/beetles/chrysomelidae/alticinae/arrows.htm (accessed online 24 April 2021).

Rothschild, G. (1964). The biology of *Pipunculussemifumosus* (Kowarz) (Diptera: Pipunculidae), a parasite of Delphacididae (Homoptera), with observations on the effects of parasitism on the host. *Parasitology*, 54(4): 763–769.

Saha, G.N. (1979). Revision of Indian blister beetles (Coleoptera: Meloidae: Meloinae). *Records of the Zoological Survey of India*, 74: 1–146.

Sahad, K.A. (1982). Biology and morphology of Gonatocerus sp. (Hymenoptera, Mymaridae), an egg parsasitoid of the green rice leafhopper, Nephotettixcincticeps Uhler (Homoptera, Deltocephalidae). *II.Morphology. Kontyû*, 50(3): 467–476.

Sankararaman, H., Naveenadevi, G. & Manickavasagam, S. (2020). Occurrence of *Fulgoraecia* (= *Epiricania*) *melanoleuca* (Lepidoptera: Epipyropidae) as a parasitoid of sugarcane lophopid planthopper *Pyrilla perpusilla* in Tamil Nadu (India) with brief notes on its life stages. *Journal of Threatened Taxa*, 12(8): 15927–15931.

Saska, P. & Honek, A. (2004). Development of the beetle parasitoids, *Brachinusexplodens* and *B.crepitans* (Coleoptera: Carabidae). *J. Zool.* (Lond), 262: 29–36.

Scaramuzza, L.C. (1938). Breve nota acerca de dos parásitos de "*Megachile* sp." (Hymenoptera, Apoidea, Megachilidae). *Mem. Soc. Cubana Hist. Nat.*, 12: 87–88.

Selander, R.B. (1957). The systematic position of the genus *Nephrites* and the phylogenetic relationships of the higher groups of Rhipiphoridae (Coleoptera). *Ann. Ent. Soc. Am.*, 50: 88–103.

Selander, R.B. (1981). Evidence for a third type of larval prey in blister beetles (Coleoptera: Meloidae). *Journal of the Kansas Entomological Society*, 54: 757–783.

Shulov, A. (1948). A Nemestrinid Parasite of *Schistocerca gregaria* (Forsk.). *Nature*, 162: 255.

Singh, R. & Sinha, T.B. (1977). Bionomics of ElasmuszehntneriFerr. (Elasmidae: Hymenoptera), an ectolarval parasitoid of *Tryporizanivella* Fabr., a pest of sugarcane. 1. Description of developmental stages. *Bulletin of Entomology*, 18: 46–52.

Singh, S.P. (2004). *Some Success Stories in Classical Biological Control of Agriculture Pest in India.* APAARI, Bangkok Thailand, pp. 121–154.

Skevington, J. & Marshall, S.A. (1997). First record of a big-headed fly, *Eudorylasalternatus* (Cresson) (Diptera: Pipunculidae), reared from the subfamily Cicadellinae (Homoptera: Cicadellidae), with an overview of pipunculid-host associations in the Nearctic region. *The Canadian Entomologist*, 129: 387–398.

Skevington, I. & Marshall, S.A. (1998). Systematics of New World *Pipunculus* (Diptera: Pipunculidae). Thomas Say Publications in Entomology: Monographs. *Entomological Society of America*, Lanham, Maryland, 103–108.

Ślipiński, S.A., Lord, N.P. & Lawrence, J.F. (2010). Bothrideridae Erichson, 1984. In *Handbook of Zoology, Coleoptera*, Vol. II. , Ed. R.G. Beutel, J.F. Lawrence & R.A.B. Leschen, pp. 411–422. Walter de Gruyter, Berlin Press, Boca Raton.

Solarska, E. (2004). The use of *Aphidiuscolemani* and *Aphidoletesaphidimyza* to control damson-hop aphid (PhorodonhumuliSchrank) on hop. *J. Plant Prot. Res.*, 44: 85–90.

Solervicens, J. (2005). Descripción de estados juveniles de *Polymerius chilensis* (Laporte, 1834) (Coleoptera: Rhipiceridae). *Acta EntomológicaChilena*, 29(1): 71–75.

Spencer, G.J. (1945). X note on the tangle-winged flies of British Columbia (Diptera: Nemestrinidae). *Proc. Entomol. Soc.* Brit. Columbia, 42: 18.

Stamm, R.H. (1936). A new find of *RhipidiuspectinicomisThbg.* (*SymbiusblattarumSund.*). *Ent. Meddr*, 19: 286–297.

Stange L. (2004). Lacewings, Antlions and Mantispids (Neuroptera). In *Encyclopedia of Entomology* Ed. Capinera JL. Springer, Dordrecht, 1244–1249.

Stary, P. (1973). A review of the *Aphidius* species (Hymenoptera: Aphidiidae) of Europe. *Annot. Zool. Bot.*, 84: 1–85.

Stevens, N.B., Stephens, C.J., Iqbal, M., Jennings, J.T., La Salle, J. & Austin, A.D. (2007). What wasp is that? An interactive identification guide to the Australasian families of Hymenoptera. *CDROM.* Australian Biological Resources Study, Environment Australia, Canberra. ISBN-10: 0 642 56851 0.

Strand, M.R. & Obrycki, J.J. (1996). Host specificity of insect parasitoids and predators: Many factors influence the host ranges of insect natural enemies. *BioScience*, 46(6): 422–429.

Strand, M.R. & Vinson, S.B. (1984). Facultative hyperparasitism by the egg parasitoid *Trichogrammapretiosum* (Hymenoptera: Trichogrammatidae). *Annals of the Entomological Society of America*, 77(6):679–686.

Stringer, D., Mantel, S., Jennings, J. & Austin, A. (2012). FamilyPteromalidae. Australian Centre for Evolutionary Biology and Biodiversity, and the School of Earth and Environmental Science, The University of Adelaide, South Australia, Australia, p. 140.

Sundevall, J.C. (1831). BeschreibungeinerneuenColeopteren-Gattung, SymbiusBlattarum. *Isis, Jena*, 11: 1222–1228.

Sureshan, P.M., Narendran, T.C. & Nikhil, K. (2013). Parasitoids (Hymenoptera) of xylophagous beetles (Coleoptera) attacking dead wood in southern Western Ghats, Kerala, India, with descriptions of two new species. *Journal of Threatened Taxa*, 5: 4385.

Tachikawa, T. (1981). Hosts of encyrtid genera in the world (Hymenoptera: Chalcidoidea). *Memoirs of the College of Agriculture, Ehime University*, 25(2): 85–110.

Terayama, M. (2003). Phylogenetic systematics of the family Bethylidae (Insecta: Hymenoptera) Part II. Keys to subfamilies, tribes and genera in the world. *The Academic Reports of the Faculty of Engineering of Tokyo Polytechnic University*, 26: 16–29.

Thomas, M. & Chaboo, C. (2015). Beetles (Coleoptera) of Peru: A survey of the families. Cucujidae, Laemophloeidae, Silvanidae, Passandridae (Cucujoidea). *Journal of the Kansas Entomological Society*, 88(2): 251–257.

Thomas, M.C. (1993). The flat bark beetles of Florida (Coleoptera: Silvanidae, Passandridae, Laemophloeidae). In *Arthropods of Florida and Neighboring Land Areas*. Vol. 15. Florida Department of Agriculture & Consumer Services, Gainesville, p. 93.

Thomas, M.C. (2002). Passandridae Erichson, 1845. In *American Beetles Volume 2, Polyphaga: Scarabaeoidea through Curculionoidea*, Ed. R.H. Arnett, M.C. Thomas, P.E. Skelley & J.H. Frank, pp. 327–328. CRC Press, Washington DC.

Torréns, J. (2013). A review of the biology of Eucharitidae (Hymenoptera: Chalcidoidea) from Argentina. *Psyche*, 3: 1–14.

Townes, H. (1958). Some biological characteristics of the Ichneumonidae (Hymenoptera) in relation to biological control, *Journal of Economic Entomology*, 51(5): 650–652.

Urano, T., Inoue, M., Ishii, S., Ando, Y., Shiomi, S., Jikumaru, S., Fukui, S., Sugimoto, H., Takemoto, M. & Inada, T. (2004). Emergence pattern of *Dastarcushelophoroides* in the Kansai region. *Trans. Mtg. Jpn. For. Soc.*, 115: 245 (In Japanese).

Van Driesche, R.G., Lyon, S., Sanderson, J.P., Bennett, K.C., Stanek, E.J. & Zhang, R.T. (2008). Greenhouse trials of *Aphidiuscolemani* (Hymenoptera: Braconidae) banker plants for control of aphids (Hemiptera: Aphididae) in greenhouse spring floral crops. *Florida Entomologist*, 91, 583–591.

Vecht, J.V. & Fischer, F.C.J. (1972). Palaearctic Eumenidae (Hymenoptera, Vespoidea). *Hym. Cat.* (Nova Edition) 8: 1–199.

Verma, M. & Hayat, M. (1988). Family Elasmidae. In *The Chalcidoidea (Insecta: Hymenoptera) of India and the Adjacent Countries*, Ed. B.R. SubbaRao & M. Hayat, *Oriental Insects*, 19: 329.

Viggiani, G. (1984). Bionomics of the Aphelinidae. *Annual Review of Entomology*, 29: 257–276.

Viggiani, G. & Laudonia, S. (1994). Description of a new species of *Lathromeris* Förster (Hymenoptera: Trichogrammatidae) larval parasitoid of *Lasioptera* sp. (Diptera: cecidomyiidae). *BollettinodelLaboratorio di EntomologiaAgraria 'FilippoSilvestri'*, 49: 169–172.

Virla, E.G., Raygoza, G.M. & Rafael, J.A. (2009). First Record of *Eudorylas schreiteri* (Shannon) (Diptera: Pipunculidae) as a Parasitoid of the Corn Leafhopper (Hemiptera: Cicadellidae) in Argentina, with a Table of Pipunculid-Host Associations in the Neotropical Region. *Neotropical Entomology*, 38: 898–900.

Waage, J.K. & Greathead, D.J. (1986). *Insect Parasitoids*. Academic Press, London, p. 389.

Walter, G.H. (1983). Divergent male ontogenies in Aphelinidae (Hymenoptera: Chalcidoidea): A simplified classification and a suggested evolutionary sequence. *Biological Journal of the Linnean Society*, 19: 63–82.

Walter, S. (1983a). The biology and ecology of egg parasites of the genus Trichogramma Westwood (Hym., Chalc.). *Part I.: Studies in selected forest biocoenoses in the G.D.R. ZoologischeJahrbücher, AbteilungfürSysteamtik, Ökologie und Geographie der Tiere*, 110(3): 271–299.

Walter, S. (1983b). Biological and ecological studies on egg parasites of the genus *Trichogramma* Westwood (Hym.Chalc.). *Part 2. Investigations carried out under laboratory conditions. ZoologischeJahrbücher, AnteilungfürSystematik, Ökologie und Geographie der Tiera*, 110(4): 419–441.

Watt, R. & Shuker, D.M. (2010). Nasonia wasp behavior genetics. In *Encyclopedia of Animal Behavior*, Vol. 2, Ed. M.D. Breed & J. Moore, pp. 513–519. Academic Press, Oxford.

Weber, D.C., Rowley, D.R., Greenstone, M.H. & Athanas, M.M. (2006). Prey preference and host suitability of the predatory and parasitoid carabid beetle, *Lebia grandis*, for several species of *Leptinotarsa* beetles. *J Insect Sci*, 6: 9.

Weber, D.C., Saska, P., Chaboo, C.S. (2008). Carabid beetles (Coleoptera: Carabidae) as Parasitoids. In *Encyclopedia of Entomology*, Ed. J.L. Capinera, p. 492. Springer, Dordrecht.

Weiss, H.B. (1922). A summary of the food habits of North American Coleoptera. *The American Naturalist*, 56(643): 159–165.

White, G., Kairo, M. & Lopez, V. (2005). Classical biological control of the citrus blackfly *Aleurocanthuswoglumi* by *Amitushesperidum* in Trinidad. *Biocontrol*, 50: 751–759.

Williams, F.X. (1919). Some observations of *Pipunculus* flies which parasitize the cane leathopper, at Pahala, Hawaii. *Proceedings of the Hawaiian Entomological Society*, 4: 68–71.

Wood, G.C. (1981). Asilidae. In *Manual of Nnearticdiptera*. volume 1, Ed. J.F. McAlpine, pp. 549–574. Department of Agriculture Research Branch, Ottawap.

Woodley, N.E. (2009). Nemestrinidae. In: *Manual of Central American Diptera*. Vol. 1, Ed. B.V. Brown, A. Borkent, J.M. Cumming, D.M. Wood, N.E. Woodley & M.A. Zumbado, pp. 557–560. NRC Research Press, Ottawa.

Woolley, J.B. & Hanson, P.E. (2006). Familia Signiphoridae. In *Hymenoptera de la Región Neotropical*, Ed. P.E. Hanson & I.D. Gauld. Memoirs of the American Entomological Institute 77, pp. 422–425.

Ye, X.H., van Achterberg, C., Yue, Q. & Xu, Z.F. (2017). Review of the Chinese Leucospidae (Hymenoptera, Chalcidoidea). *ZooKeys*.651(1):107–157.

Yeates, D.K. (1994). The cladistics and classification of the Bombyliidae (Diptera: Asiloidea). *Bulletin of the American Museum of Natural History*, 219: 1–191.

Yeates, D.K. & Greathead, D. (1997). The evolutionary pattern of host use in the Bombyliidae: A diverse family of parasitoid flies. *Biological Journal of the Linnean Society*, 60: 149–186.

Yoshimoto, C.M. (1990). A review of the genera of new world Mymaridae (Hymenoptera: Chalcidoidea). In *Flora & Fauna Handbook*, Sandhill Crane Press, Gainsville, Florida, 7: 1–166.

Young, D.K. & Katovich, K. (2002). First record of Rhipiceridae (Coleoptera: Polyphaga: Dascilloidea) From Wisconsin. *The Great Lakes Entomologist*, 35(1): 33–34.

Zackvatkine, A.A. (1934). The parasites of the Moroccan cricket in Azerbaidjan. *Zashch. Rast. OtVred. (Plant Protect.)*, 9: 52–71.

2 Ichneumonid Parasitoids

Longjam Roni Kumar Singh, Devanshu Gupta, and Kailash Chandra

Zoological Survey of India, M-Block, New Alipore, Kolkata

CONTENTS

LEARNING OBJECTIVES

i. To know the biodiversity, systematics and life cycle of braconids.
ii. To know important braconids for biocontrol of pests.
iii. To learn the raring techniques of braconids for the biocontrol of pests.

2.1 INTRODUCTION

The term 'biological control', first used by Smith (1919), refers to the use of predators, parasitoids, and pathogens for the control of insect pests. Insect parasitoids are those that during a stage of their life-cycle feed and develop in or on a host killing it in the process (Godfray, 1994). Furthermore, natural enemies (parasitoids) exert continuous associations with insect pests, keeping them suppressed and regulating the population densities of pests without any specific human intervention. Biological control is considered as an essential component of IPM and the world is now aware of judiciously using chemical pesticides in food production, and looking towards pest management strategies that use habitat manipulation, i.e. to conserve or augment natural enemies. According to Greathead and Greathead (1992) out of 509 species of biological control agents (predators and parasitoids) that have been established in classical biological control programmes, 384 (75.4%) are parasitic Hymenoptera. Among the Hymenoptera parasitoids, the most successful superfamily is Chalcidoidea and followed by Ichneumonoidea (Greathead, 1986). The family Ichneumonidae are solitary or gregarious, idiobiont ectoparasitoids, koinobiont endoparasitoids or hyperparasitoids, on

DOI: 10.1201/9781003354239-2

a variety of holometabolous insects and spiders, particularly the larvae or pupae of Lepidoptera, Coleoptera, Diptera and other Hymenoptera (Wahl and Gauld, 2014). In the Indian context, Lepidoptera is the host of more than 75% of the ichneumonid genera, while about 6.5% use sawflies, 5% Coleoptera, 9% other Hymenoptera (Braconidae, Chalcididae, Ichneumonidae, Pteromalidae, Vespidae) and the remaining utilise Neuroptera, Trichoptera, etc. (Gupta, 1987).

Although ichneumonids are one of the most important natural regulatory factors in preserving a balance of pest populations and maintaining biodiversity in terrestrial ecosystems, our knowledge of their taxonomy, biology, ecology, distribution, and host associations is poorly known. No extensive survey has been made to explore these natural enemies of major ecosystems for conservation and augmentation of biological control agents; and only a few species have been of practical use in biological control. And also due to a lack of suitable manuals, it is often not possible to identify the species correctly, thus hampering the progress of the biological control programme.

2.1.1 Systematic Account

The Ichneumonidae is a cosmopolitan and morphologically a most diverse family in the insect order Hymenoptera, having an estimated number of 100,000 species worldwide (Gauld, 2002), of which 24,281 are described under 39 subfamilies (Wahl and Gauld, 2014). The Indian ichneumonid fauna comprises about 1,200 species under 24 subfamilies (Jonathan, 2005; Chandra, 2011), accounting for about 5% of the total diversity. Ichneumonids can be separated from its sister group Braconidae by having two recurrent veins in the forewing; abdomen petiolate or subpetiolate; tergites 2 and 3 are usually separated.

2.1.2 Life History

Ichneumonid parasitoids occur in several orders of holometabolous insects, particularly the larvae or pupae of Lepidoptera, Coleoptera, Diptera and other Hymenoptera (Symphyta) and spiders. Generally, the development of parasitoid follows one of the two biological strategies: either they are endoparasitoids, which feed and develops within their host's body, and koinobiont, which allows further development of the host, which continue to feed, and moult; or they are ectoparasitoids, that feed on the host from outside, and idiobiont, whose hosts are paralysed and cease further development of the hosts. Most koinobionts are endoparasitoids of the larval stage of insects, although a few are ectoparasitoids (e.g. *Netelia* sp.). While some ichneumonids show some form of hyperparasitism, they use other parasitoids through their primary hosts. For instance, *Mesochorus orientalis* (Viereck) hyperparasitic on *Snellenius maculipennis* (Szépligeti) (Braconidae), a larval parasitoid of castor semilooper, *Achaea janata* (Linnaeus) (Lepidoptera: Erebidae) (Gupta, 1987).

2.2 BIOLOGICAL CONTROL UTILISING ICHNEUMONIDAE

Ichneumonidae hold great potential for controlling insect pest populations. Some successful examples of biocontrol using ichneumonid parasitoids are: *Gambroides javensis* (Rohwer) (= *Isotima javensis*, Campopleginae) a key parasitoid of sugarcane top borer, *Scirpophaga excerptalis* (Walker) (Lepidoptera: Crambidae) has been successfully colonised in Northern and Southern India and plays an important role in the suppression of this pest; control of black-headed caterpillar, *Opisina arenosella* Walker (Lepidoptera: Oecophoridae) by *Eriborus trochanteratus* Morley (Campopleginae) in Southern India. *Xanthopimpla nana nana* Schulz (Pimplinae) is also known to be a potentially important parasitoid of *O. arenosella*. Some of the biocontrol potential species that are recorded from India are *Campoletis chlorideae* Uchida, an important parasitoid for the control of the polyphagous pest, *Helicoverpa armigera* (Hübner) (Nikam and Gaikwad, 1989); *Campoplex haywardi* Blanch (Campopleginae) on potato tuber moth, *Phthorimaea operculella* (Zeller) (Lepidoptera: Gelechiidae) in India and Australia and on the winter moth, *Operophtera brumata* (Linnaeus) (Lepidoptera: Geometridae) in Cyrus; *Bathyplectes curculionia* (Thomson) (Campopleginae) on Alfalfa weevil, *Hypera postica* (Gyllenhal) (Coleoptera: Curculionidae) in the USA, but this parasitoid is recorded from Uttar Pradesh without host information. Therefore, it shows potential for the suppression of this pest. Recently, some species of potential interest were investigated by Sutar and Sathe (2016, 2017) on the genus *Diadegma* Foerster (Campopleginae) from Maharashtra.

2.2.1 Subfamilies of Indian Ichneumonidae

In the account given below, a brief treatment has been given on the subfamilies of Ichneumonidae recorded from India, to show how important they are in nature as well as in biological control programmes undertaken against insect pests. The major contributions to the current knowledge of the Indian Ichneumonidae and most of the host records compiled herein are adapted from Gupta (1987) and Jonathan (2006). As far as possible, names of genera and species mentioned under each subfamily are mostly from India stating their possible potential biological control. Some representative of the Indian Ichneumonidae is given in Plate I. (a) *Isotima rufipleuralis* Jonathan (Cryptinae), (b) *Netelia facialis* Kaur & Jonathan, (c) *N. fumosa* Kaur & Jonathan, and (d) *N. parca* Kaur & Jonathan (Tryphoninae).

i. **Acaenitinae:** This is a moderately small subfamily and acaenitines are koinobiont endoparasitoids of larvae of beetle. Despite their large size, they are generally not common, and little is known about their biology. Most species occur in forests and their host relations are unknown in India.
ii. **Agriotypinae:** This is a small group, aquatic idiobiont ectoparasitoids of caddisfly larvae, pupae and prepupae inhabiting in streams. In India, it is represented by a single species, *Agriotypus himalensis* Mason parasitic on *Neophylax* sp. (Trichoptera) larvae inhabiting in streams.
iii. **Anomaloninae:** This is a moderately small cosmopolitan group and anomalonines are koinobiont endoparasitoids mostly larvae of Lepidoptera and Coleoptera. *Agrypon nox* Morley parasitic on lepidopterous larvae, *Digama hearseyana* Moore (Erebidae); *Aphanistes eupterotes* Cushman parasitic on *Eupterote germinata* (Walker) (Eupterotidae); *Trichomma decorum* (Cameron) parasitic on a serious pest of teak, *Paliga machoeralis* (Walker) and bamboo leaf roller, *Crypsiptya coclesalis* (Walker) (Crambidae); *T. dioryctri* Gauld parasitic on *Dioryctria abietella* (Denis & Schiffermuller) (Pyralidae), *Eucosma pylonitis* Meyrick (Tortricidae); *T. nigricans* Cameron parasitic on *Dichocrocis nigrilinealis* Walker, *P. machoeralis* (Walker) and *Syllepta derogata* (Fabricius) (Crambidae).
iv. **Banchinae:** Blachinae is a large subfamily, koinobiont endoparasitoids of lepidopterous larvae. The majority of Banchinae attack concealed microlepidoptera, while some species of Banchini parasitise exposed larvae. Genus *Lissonota* Gravenhorst is a common parasitoid of Lepidoptera larvae in stems, buds and leaf rolls; *L. castaneae* Chandra & Gupta are internal parasitoids of *Dioryctria castanea* Bradley (Pyralidae) and *Retinia cristata* (Walsingham) (Tortricidae). Some members of the genera, *Banchus* Fabricius and *Exetastes* Gravenhorst attack larvae and pupae of Noctuidae.
v. **Brachycyrtinae:** This is a small, uncommon and has generally been believed to be ectoparasitoids. Members of this attack on cocoons of *Eublemma* (Erebidae), pupae or prepupae of Chrysopidae and spider egg sacs. Only three species are known from India, in a single genus *Brachycyrtus* Kriechbaumer. *B. eublemmae* (Rao) is parasitic on *Eublemma* sp. (Erebidae) and *Chrysopa* sp. (Neuroptera: Chrysopidae).
vi. **Campopleginae:** Campopleginae is a large, cosmopolitan subfamily. Most of the species are solitary koinobiont endoparasitoids mainly larvae of Lepidoptera and

Symphyta. This subfamily has been widely investigated from the biocontrol point of view and most species are fairly host-specific. Species of *Campoletis* Foerster is larval parasitoids of Noctuidae and Tortricidae: *C. chlorideae* Uchida and *C. multicinctus* (Gravenhorst) parasitic on *Adisura* sp., *Helicoverpa armigera* (Hübner) (Noctuidae) and *Cryptophlebia ombrodelta* (Lower) (Tortricidae); *Campoplex haywardi* Blanchard is a potential biological agent on potato tuber moth, *Phthorimaea operculella* (Zeller) (Gelechiidae); *Diadegma semiclausum* (Hellén) and *D. fenestrale* (Holmgren) is an important parasitoids of diamondback moth, *Plutella xylostella* (Linnaeus) (Plutellidae); *D. dominans* parasitic on *H. armigera* (Hübner) and *Spodoptera mauritia* (Boisduval) (Noctuidae); *Eriborus argenteopilosus* (Cameron) is also an important parasitoids against cotton bollworm, *H. armigera* (Hübner). Species of the genera *Casinaria* Holmgren, *Charops* Holmgren, *Dusona* Cameron, *Hyposoter* Foerster, *Venturia* Schrottky and *Xanthocampoplex* Morley are larval parasitoids of the family Crambidae, Erebidae, Geometridae, Hesperiidae and Pyralidae (Lepidoptera). While a small number of species parasites on sawflies.

vii. **Cremastinae:** This is a moderately large subfamily, of solitary koinobiont endoparasitoids of lepidopterous larvae, which are leaf rollers or stem borers and some are parasitics on coleopterous larvae. Species of the genus *Pristomerus* Curtis attacks on larvae of Lepidoptera; *P. euzopherae* Viereck parasitic on *Euzophera perticella* Ragonot (Pyralidae); *P. marginicollis* (Cameron) parasitic on *Dichomeris metatoxa* (Meyrick) (Gelechiidae); *P. microdon* Cushman parasitic on *Paliga machoeralis* (Walker) (Crambidae), *Hyblaea puera* (Cramer) (Hyblaeidae) and hyperparasitic on *Microchelonus* sp., *Apanteles* sp., *Diolcogaster indica* (Wilkinson) (Braconidae). Species of the genera *Trathala* Cameron and *Temelucha* Foerster are parasitoids of Lepidoptera larvae and occasionally on Coleoptera and Hymenoptera.

viii. **Cryptinae:** This is a very large, cosmopolitan subfamily, idiobiont ectoparasitoids and some species appear to be koinobiont endoparasitoids. Some members oviposit in cocoons of Lepidoptera, Symphyta, Neuroptera, and other Hymenoptera (hyperparasitoids on Braconidae and Ichneumonidae). Species of the genus *Amauromorpha* Ashmead are parasitoids of lepidopterous stem borers, particularly of paddy pests, *Chilo sacchariphagus* (Bojer), *C. suppressalis* (Walker), *Scirpophaga excerptalis* (Walker), *S. incertulas* (Walker), *S. nivella* (Fabricius), *S. praelata* (Scopoli) (Crambidae), *Sesamia inferens* (Walker), *Spodoptera exigua* (Hübner) (Noctuidae); *Gambroides javensis* (Rohwer) is an important parasitoids of Lepidoptera stem-borers, *S. nivella* (Crambidae) and *Dineutus unidentatus* Aubé (Coleoptera: Gyrinidae). Some of the genera, which are of potential interest found in India, are *Diatora* Foerster, *Goryphus* Holmgren, *Gotra* Cameron, *Isotima* Foerster and *Skeatia* Cameron.

ix. **Ctenopelmatinae:** This is a large group and koinobiont endoparasitoids mainly larvae of sawfly. Only one species of the genus *Perilissus* Holmgren, *P. cingulator* (Morley) is recorded parasitising *Athalia proxima* (Klug) (Tenthredinidae).

x. **Diplazontinae:** This is a relatively small, cosmopolitan subfamily. Diplazontines are solitary koinobiont endoparasitoids of syrphids (Diptera). Species of the genus *Diplazon* Viereck are parasitoids of syrphid larvae in grasses, meadows, fields and in agricultural crops; *D. orientalis* (Cameron) parasitic on *Ischiodon aegyptius* (Wiedemann) and *D. pectoratorius* (Thunberg) on *Eupeodes* (= *Syrphus*) *confrater* (Wiedemann) and *Ischiodon scutellaris* (Fabricius) (Syrphidae).

xi. **Eucerotinae:** Members of this subfamily are hyperparasitoids of Anomaloninae, Banchinae, Ctenopelmatinae, Porizontinae, and Phygadeuontinae (Ichneumonidae), through their sawfly or lepidopterous hosts. This subfamily contains a single species *Euceros latitarsus* Barron from India and their host relation is unknown.

xii. **Ichneumoninae:** This is a very large, cosmopolitan subfamily and ichneumonines are pupal or larva-pupal koinobiont endoparasitoids of Lepidoptera. Several species of the subfamily are of biocontrol potential of insect pests; however, so far, only a few species have host records to date from India. *Diadromus collaris* (Gravenhorst) parasitic on pupa of diamondback moth, *Plutella xylostella* (Linnaeus) (Plutellidae); *Holcojoppa coelopyga* (Morley) parasitic on *Papilio helenus* Linnaeus (Papilionidae); *Ischnojoppa luteator* (Fabricius) parasitic on *Scirpophaga incertulas* (Walker) (Crambidae); *Lissosculpta alecto* (Morley) parasitic on *Pelopidas mathias* (Fabricius) (Hesperiidae).

xiii. **Mesochorinae:** Mesochorinae is a moderately large, cosmopolitan subfamily and members of this subfamily are koinobiont endoparasitoids (as endophagous obligatory hyperparasitoids) that attack larvae of Lepidoptera, sawflies, adult Coleoptera, Diptera and Hymenoptera (Braconidae and Ichneumonidae). *Mesochorus discitergus* (Say) parasitic on castor semilooper, *Achaea janata* (Linnaeus), *Lymantria concolor* Walker (Erebidae), *Bocchoris artificalis* (Lederer), *Pycnarmon cribrata* Fabricius (Crambidae), hyperparasitic on *Apanteles* sp. (Braconidae); *M. orientalis* (Viereck) hyperparasitic on *Snellenius maculipennis* (Szépligeti) (= *Microplitis ophiusae*) (Braconidae) through *A. janata* (Linnaeus) (Erebidae); *M. pilicornis* (Cameron) hyperparasitic on *S. maculipennis* (Szépligeti) (= *Microplitis maculipennis*) through *A. janata* (Linnaeus) (Erebidae); *M. plusiaephilus* Viereck hyperparasitic on *Apanteles taragamae* Viereck(= *Apanteles plusiae* Viereck), *A. machaeralis* Wilkinson through *P. machoeralis* (Walker) (Crambidae), *Hyblaea puera* (Cramer) (Hyblaeidae); *M. srinaraini* Gupta hyperparasitic on *Cotesia ruficrus* (Haliday). Hyperparasitoids have generally been regarded as enemies of biocontrol programmes.

xiv. **Metopiinae:** This is a cosmopolitan subfamily and metopiines are koinobiont endoparasitoids of lepidopterous larvae. Very little detail is known about their biology and host insects. *Triclistus aitkini* (Cameron) parasitic on rice leaf roller, *Cnaphalocrocis medinalis* (Guenée) (Crambidae) from Taiwan and *Colpotrochia pilosa pilosa* (Cameron) parasitic on *Deltaes uriens* (Fabricius) (Vespidae) recorded from India.

xv. **Nesomesochorinae:** This subfamily is rather small and is represented by five species and two genera in India and their host relations are unknown.

xvi. **Ophioninae:** Ophionines are solitary koinobiont endoparasitoids on lepidopterous larvae belonging to Crambidae, Erebidae, Eupterotidae, Geometridae, Lasiocampidae, Lymantriidae, Noctuidae, Saturniidae and Sphingidae. Several species of the genus *Enicospilus* Stephens are common parasitoids of a variety of agricultural pests; *Enicospilus capensis* (Thunberg) parasitic on larvae of *Euproctis fraterna* Moore (Erebidae); *E. dasychirae* Cameron parasitic on *Psalis pennatula* (Fabricius) (Lymantriidae); *E. dolosus* (Tosquinet) parasitic on *Syllepta derogata* (Fabricius) (Crambidae); *E. grandis* (Cameron) parasitic on *Trabala vishnou* (Lefèbvre) (Lasiocampidae); *E. heliothidis* Viereck parasitic on *Olene mendosa* Hübner (Lymentriidae); *E. lineolatus* (Roman) parasitic on *Orgyia postica* (Walker) (Erebidae); *E. melanocarpus* Cameron parasitic on *Streblota siva* (Lefèbvre) (Lasiocampidae) and *E. fraterna* Moore (Erebidae); *E. signativentris* (Tosquinet) parasitic on *Helicoverpa armigera* (Hübner), *Trichoplusia ni* (Hübner) (Noctuidae); *E. xanthocephalus* Cameron parasitic on *E. fraterna* Moore (Erebidae); *Dicamptus flavoplagiatus* (Cushman) parasitic on *Eupterote* sp. (Eupterotidae); *Ophion areolatus* Cameron parasitic on *Acronicta major* Bremer and *A. maxima* Moore (Noctuidae). Despite their abundance and having a considerable number of host records, only a few species were released in biocontrol programmes.

xvii. **Orthocentrinae:** Orthocentrines are koinobiont endoparasitoids of Mycetophilidae and Stratiomyiidae (Diptera). It is represented by three species and two genera in India and their host relations are unknown.

xviii. **Pimplinae:** This is a large cosmopolitan subfamily and members of this are idiobiont ecto- and endoparasitoids, koinobiont ectoparasitoids of pupae of certain Lepidoptera. Some of the genera which are of potential interest found in India are *Acropimpla* Townes, *Afrephialtes* Benoit, *Camptotypus* Kriechbaumer, *Dolichomitus* Smith, *Exeristes* Foerster, *Flavopimpla* Betrem, *Gregopimpla* Momoi, *Scambus* Hartig and *Sericopimpla* Kriechbaumer (parasitic on larvae or cocoon of Coleoptera and Lepidoptera); *Clistopyga* Gravenhorst and *Zaglyptus* Foerster (parasitic on spider egg sacs); *Echthromorpha* Holmgren, *Itoplectis* Foerster, *Lissopimpla* Kreichbaumer, *Theronia* Holmgren and *Xanthopimpla* Saussure (parasitic on pupae and prepupae of Lepidoptera, including those of Noctuidae, Hesperiidae, Agaristidae, Anthelidae, Bombycidae, Lycaenidae, Lymantriidae, Nymphalidae, Psychidae, Xyloryctidae, Papilionidae, Tortricidae, and Saturniidae).

xix. **Poemeniinae:** Poemeniines appear to be idiobiont ectoparasitoids of Coleoptera larvae and solitary aculeate Hymenoptera (Sphecidae). In India, it is represented by 14 species and four genera and host relations of this subfamily are unknown.

xx. **Rhyssinae:** This is a small cosmopolitan subfamily and rhyssines are apparently idiobiont ectoparasitoids of wood-boring holometabolous insects, particularly of Siricidae, Xiphydriidae and Cerambycidae. Rhyssine species are important in the regulation of wood wasp pests of conifers; *Rhyssa persuasoria himalayensis* Wilkinson parasitic on wood-borers *Sirex imperialis* Kirby, *S. juvencus* (Linnaeus), *Urocerus xanthus* (Cameron) and *Xeris himalayensis* Bradley (Siricidae).

xxi. **Sisyrostolinae:** This is a small subfamily and it is represented by a single species, *Brachyscleroma assamensis* in India. Members of this subfamily are endoparasitoids of Coleoptera larvae (Curculionidae).

xxii. **Tersilochinae:** This is a cosmopolitan subfamily and tersilochines are koinobiont endoparasitoids of larvae or pupae of Coleoptera. Six species under five genera are reported from India and their host relations are unknown.

xxiii. **Tryphoninae:** Tryphoninae is a cosmopolitan subfamily and members of this are koinobiont ectoparasitoids of larvae of Lepidoptera or Symphyta. *Netelia Gray are* large parasitoids of lepidopterous larvae and comprised of 60 species from India. *N. kashmirensis* (Cameron) parasitic on lepidopterous larvae *Adisura* sp. (Noctuidae); *N. lineata* ((Brullé) and *N. ocellaris* (Thomson) parasitic on castor semilooper, *Achaea janata* (Linnaeus) (Erebidae); *N. testacea* (Gravenhorst) parasitic on *Spodoptera litura* (Fabricius) (Noctuidae); *N. vinulae* (Scopoli) parasitic on *Cerura himalayana* Moore (Notodontidae), and species of *Polyblastus* Hartig and *Kristotomus* Mason are larval parasitoids of sawflies.

xxiv. **Xoridinae:** Xoridinae is a small widely distributed, idiobiont ectoparasitoids of the larvae and pupae of wood-boring Coleoptera and Symphyta. The larval parasitoid, *Xorides caerulescens* (Morley) is an important parasitic hymenopteran of coffee white stem-borer, *Xylotrechus javanicus* (Castelnau & Gory) (= *X. quadripes* Chevrolat) (Coleoptera: Cerambycidae), one of the serious pest of coffee in Southern India.

2.3 MASS REARING

Most of the ichneumonid species collected by using net sweeping, pan traps, light traps, Malaise traps, etc. are deployed for the studies of taxonomic and faunal diversity. But this hardly gets the biological information of the host-parasitoid relationships. Most of the rearing is done in the greenhouse and laboratory; however, the efficient and cost-effective methods of rearing parasitoids are done in near-natural conditions; which are likely to have more success where it provides all the necessary environmental conditions. Once collected the potential host larvae, the individual should be isolated and placed with the respective or alternative host plant in a rearing cage or jar and covered with suitable mesh-size cloths; a moderate level of humidity should be maintained to check for both desiccation and fungal growth. Care should be taken to check the introduction of other hyperparasitoids into the rearing cage. The adults' emergence should be noted along with the details of the host, stage, date of collection and emergence, place of collection, habitat, and brood size should be counted in the case of gregarious parasitoids and the host remains and cocoons should be preserved. Mass-rearing *(in-vitro)* system of parasitoids for biological control has yet to be developed, particularly only nine species of Hymenoptera (four species of *Trichogramma* and five ichneumonoids have been done (King et al., 1984).

2.4 CONCLUSIONS

India is very rich in terms of the diversity of arthropods, with many species of natural enemies of economic insect pests. A total of about 1,200 species under 24 subfamilies of the family Ichneumonidae are known from India. Ichneumonid parasitoids are important in nature as well as in biological control programmes undertaken against insect pests. Several species have been investigated from a biological control point of view and a few have been introduced into different parts of the world. Members of the subfamilies Campopleginae, Cryptinae, Ichneumoninae, and Pimplinae, have been utilised for successful biological control, mostly larvae and pupae of Lepidoptera, Coleoptera, Diptera, and other Hymenoptera (Symphyta). In the Indian context, the Lepidoptera are the hosts of more than 75% of the ichneumonid genera, while about 15.5% utilise other Hymenoptera (including sawflies) and 5% Coleoptera and the remaining Neuroptera, Trichoptera, etc. However, hyperparasitoids have generally been regarded as enemies of biocontrol programmes.

Despite their importance, our knowledge of their taxonomy, biology, ecology, distribution, and host associations is poorly known; no extensive survey has been made to explore these natural enemies. The presence of natural enemies has potential value for the core of pest management techniques; however, many indigenous natural enemies have decreased in abundance mainly due to the side effects of chemical pesticides and the destruction of natural habitats. Thus, any addition to our knowledge of them has potential value, thus there is a need to increase efforts towards studies on these natural enemies to explore them in major ecosystems for conservation and augmentation of biological control.

2.5 POINTS TO REMEMBER

i. Ichneumonidae is one of the most diverse families in order Hymenoptera (Insecta).
ii. Indian ichneumonid fauna comprises about 1,200 species under 24 subfamilies.
iii. Ichneumonids are one of the most important natural regulatory factors in preserving a balance of pest populations and maintaining biodiversity in terrestrial ecosystems.
iv. Ichneumonid parasitoids are important in nature as well as in biocontrol programmes undertaken against insect pests.
v. Ichneumonid parasitoids occur in several orders of holometabolous insects and spiders.
vi. Many species hold great potential for controlling insect pest populations.
vii. Lepidoptera is the host of more than 75% of the Indian ichneumonid genera.
viii. Hyperparasitoids have generally been regarded as enemies of biocontrol programmes.
ix. There is a need for biosystematics research on these natural enemies to explore them in major ecosystems for conservation and augmentation of biocontrol.

REFERENCES

Chandra, K. (2011). Insect diversity of Sikkim, India. In M.L. Arrawatia & S. Tambe (Eds.), *Biodiversity of Sikkim-Exploring and Conserving a Hotspot.* Information and Public Relations Dept. Govt. of Sikkim, Gangtok, pp. 186–206.

Gauld, I.D. (2002). The family Ichneumonidae. In *The Ichneumonidae of Costa Rica,* 4. I. Gauld, C. Godoy, J. Ugalde & R. Sithole (Eds.), *Memoirs of the American Entomological Institute*, 66: 1–768.

Godfray, H.C.J. (1994). *Parasitoids: Behavioural and Evolutionary Ecology*. Princeton University Press, Princeton, NJ, pp. 1–473.

Greathead, D.J. (1986). Parasitoids in classical biological control. In J. Waage & D. Greathead (Eds.), *Insect Parasitoids.* Academic Press, London, U.K., pp. 289–318.

Greathead, D.J. & Greathead, A.H. (1992). Biological control of insect pests by insect parasitoids and predators: The BIOCAT database. *Biocontrol News and Information*, 13(4): 61N–68N.

Gupta, V.K. (1987). The Ichneumonidae of the Indo-Australian area (Hymenoptera). *Memoirs of the American Entomological Institute*, 41(1–2): 1–1210.

Jonathan, J.K. (2005). Insecta: Hymenoptera: Ichneumonidae. In *Fauna of Western Himalaya* (*Part-2*): *Himachal Pradesh.* 1-359 + 16 plates, The Director, Zool. Surv. India, Kolkata, pp. 191–220.

Jonathan, J.K. (2006). *Ichneumonologia Indica*, Part-1 Hymenoptera: Ichneumonidae. The Director, Zool. Surv. India, Kolkata, pp. 1–680..

King, E.G., Ridgway, R.L. & Hartstack, A.L. (1984). Propagation and release of entomophagous arthropods for control by augmentation. In P.L. Adkisson & Ma. Shih-chün (Eds.), *Proceedings of the Chinese Academy of Science – U.S. National Academy of Science Joint Symposium on Biological Control of Insects*. Beijing, China: Science Press. pp. 99–121.

Nikam, P.K. & Gaikwad, A.M. (1989). Role of hymenopterous parasitoids in the biological control of *Heliothis armigera* (Hubner) (Lepidoptera: Noctuidae) with special reference to *Campoletis chlorideae* Uchida (Hymenoptera: Ichneumonidae) in India. *Journal of Entomological Research*, 13: 6–20.

Smith, H.S. (1919). On some phases of insect control by the biological method. *Journal of Economic Entomology*, 12: 288–292.

Sutar, M.V. & Sathe, T.V. (2016). Diversity and bio-control potential of the genus *Diadegma* Forster (Hymenoptera: Ichneumonidae) from Western Maharashtra. *Biolife*, 4(1): 202–208.

Sutar, M.V. & Sathe, T.V. (2017). Biocontrol potential of *Diadegma trichoptilus* (Cameron) (Hymenoptera: Ichneumonidae) against Tur Plume Caterpillar *Exelastis atomosa* Walsingham (Lepidoptera: Pterophoridae). *Indian Journal of Applied Research*, 7(3): 410–411.

Wahl, D.B. & Gauld, I.D. (2014). *Genera Ichneumonorum Nearcticae*. The American Entomological Institute. Version: 2014.09.22. http://www.amentinst.org/GIN/. (Accessed on 30 July 2020).

3 Braconid Parasitoids

Ankita Gupta
ICAR-National Bureau of Agricultural Insect Resources (NBAIR), Bengaluru, Karnataka

CONTENTS

LEARNING OBJECTIVES

i. To understand the taxonomic diversity of braconid parasitoids.
ii. To learn the phylogenetic relationship between different Indian species.
iii. To know the host specificity of the group.
iv. To resolve taxonomic ambiguity or description of new taxa using morphological and molecular analysis.

DOI: 10.1201/9781003354239-3

3.1 INTRODUCTION

Family Braconidae Nees, the second largest family of the order Hymenoptera, belongs to the superfamily Ichneumonoidea. Braconids are ecto- or endoparasitoids of immature stages of Coleoptera, Diptera, Lepidoptera, Hemiptera (Quicke, 2015), and also form plant galls (Zaldivar-Riveron et al., 2014). Braconids have a key role in natural pest control. Even though of enormous economic importance in agriculture, this group is taxonomically extremely challenging as they are minute, morphologically homogeneous, and highly speciose. The global diversity estimates of the total valid species described to date are 18,000. Perhaps barely 3000 species are known from Asia (tropical), and approximately 1083 species from India, as compared to 17,000 species from Africa and 11,000 species from the Neotropics. Indian species account for a diminutive 6% of the total diversity under 27 subfamilies and 59 tribes.

3.2 DIAGNOSIS AND KEY CHARACTERS

A. Family Diagnosis

The family Braconidae can be identified with the following key characteristics: Forewing cross vein 2m-cu almost always absent, but present in Apozyginae and rarely in Rhyssalinae and Helconinae. Hindwing cross-vein r-m is almost always antefurcal but interstitial in Trachypetinae. Hindwing vein C+SC+R fused along all its length except for Trachypetinae. The joint between terga of metasomal segments 2 and 3 is inflexible with suture, if present, sclerotised except in Aphidinae. The key identifying characters are illustrated in Figs A–K (van Achterberg, 1993)

3.2.1 Classification – Subfamily and Tribes

The present classification includes 43 subfamilies under the family Braconidae across the globe-Acampsohelconinae, Agathidinae, Alysiinae, Amicrocentrinae, Aphidiinae, Apozyginae, Betylobraconinae, Blacinae, Braconinae Cardiochilinae, Cenocoeliinae, Charmontiinae, Cheloninae, Dirrhopinae, Doryctinae, Euphorinae, Exothecinae, Gnamptodontinae, Helconinae, Homolobinae, Hormiinae, Ichneutinae, Khoikhoiinae, Lysiterminae, Macrocentrinae, Masoninae, Maxfischeriinae, Mendesellinae, Mesostoinae, Meteorideinae, Microgastrinae, Microtypinae, Miracinae, Opiinae, Orgilinae, Pambolinae, Rhysipolinae, Rhyssalinae, Rogadinae, Sigalphinae, Telengaiinae, Trachypetinae, and Xiphozelinae (Quicke, 2015).

3.2.2 List of Subfamilies and Tribes Reported from India

The subfamilies and tribes reported from India are mentioned below:

FAMILY BRACONIDAE Nees, 1811
- **AGATHIDINAE** Haliday
  - **AGATHIDINI** Haliday, 1833
  - **CREMNOPTINI** Sharkey
  - **DISOPHRINI** Sharkey, 1992
- **ALYSIINAE** Leach, 1815
  - **ALYSIINI** Leach, 1815
  - **DACNUSINI** Förster
- **APHIDIINAE** Haliday, 1833
  - **APHIDIINI** Haliday, 1833
  - **EPHEDRINI** Mackauer
  - **PRAINI** Mackauer, 1961
  - **BLACINI** Förster, 1863

BRACONINAE Nees, 1811
ADESHINI van Achterberg
APHRASTOBRACONINI Ashmead
BATHYAULACINI Quicke
BRACONINI Nees, 1811
COELOIDINI Tobias, 1957
EUUROBRACONINI Ashmead, 1900
GLYPTOMORPHINI Tobias, 1957
PHYSARAIINI van Achterberg, 1984
CARDIOCHILINAE Ashmead, 1900
CARDIOCHILINI Ashmead, 1900
CENOCOELIINAE Szépligeti, 1901
CENOCOELIINI Szépligeti
CHARMONTINAE van Achterberg
CHARMONTINI van Achterberg, 1979
CHELONINAE Förster, 1863
CHELONINI Förster, 1863
PHANEROTOMINI Baker
DORYCTINAE Förster, 1863
DORYCTINI Förster, 1863
ECPHYLINI Hellén
HECABOLINI Förster, 1863
HOLCOBRACONINI Cameron, 1905
RHACONOTINI Fahringer, 1928
SPATHIINI Marshall, 1872
EUPHORINAE Förster, 1863
CENTISTINI Capek, 1970
DINOCAMPINI Shaw, 1985
EUPHORINI Förster, 1863
HELORIMORPHINI Schmiedeknecht, 1907
METEORINI Cresson, 1887
PERILITINI Förster, 1863
SYNTRETINI Shaw, 1985
EXOTHECINAE Förster, 1863
EXOTHECINI Förster, 1863
GNAMPTODONTINAE Fischer, 1970
GNAMPTODONTINI Fischer, 1970
HELCONINAE Förster, 1863
BRACHISTINI Förster, 1863 **(earlier in** BRACHISTINAE Förster, 1863)
HELCONINI Förster, 1863
HOMOLOBINAE van Achterberg, 1979
HOMOLOBINI van Achterberg, 1979
HORMIINAE Förster, 1863
AVGINI Belokobylskij, 1993
HORMIINI Förster, 1863
LYSITERMINAE Tobias, 1968
CEDRIINI Belokobylskij, 1993
LYSITERMINI Tobias
PENTATERMINI Belokobylskij
TETRATERMINI van Achterberg
MACROCENTRINAE Förster, 1863
METEORIDEINAE Tobias, 1967
MICROGASTRINAE Förster, 1863
COTESIINI Mason, 1981
FORNICIINI Mason, 1981

MICROGASTRINI Förster, 1863
MICROPLITINI Mason, 1981
SEMIONINI Tobias, 1989
MICROTYPINAE Szépligeti, 1908
MIRACINAE Viereck, 1918
OPIINAE Blanchard, 1845
BIOSTERINI Fischer, 1970
OPIINI Blanchard, 1845
ORGILINAE Ashmead, 1900
MIMAGATHIDINI Enderlein, 1905
PAMBOLINAE Marshall, 1885
PAMBOLINI Marshall, 1885
RHYSSALINAE Förster, 1863
RHYSSALINI Förster, 1863
ROGADINAE Förster, 1863
ALEIODINI Muesebeck
CLINOCENTRINI van Achterberg
ROGADINI Förster, 1863
YELICONINI van Achterberg, 1991
XIPHOZELINAE van Achterberg, 1979

3.3 MOLECULAR CHARACTERISATION AND PHYLOGENY

Morpho-phylogenetic hypothesis for Braconidae: Source: (Quicke, 2015, page 203) (a) tree redrawn by Quicke (2015) from Tobias (1989); (b) tree redrawn by Quicke (2015) from Capek (1970); (c) tree proposed by van Achterberg (1984) and used in Quicke (2015).

Capek (1970) presented relationships based on the characters of the final instar larval head capsule as in Fig. 12.1b (Quicke, 2015), which has cyclostomes branching off the main stem at a number of separate basal nodes. van Achterberg in 1984 (Fig. 12.1c in Quicke, 2015) illustrated bifurcations leading to the subfamilies (recognised by him at that point of time). As per the hypothesis by Sharanowski et al. (2011) there are three competing hypotheses – first, the extant cyclostomes and endoparasitic non-cyclostomes are sister groups (presently this hypothesis is most accepted); second, the extant cyclostomes are paraphyletic grade giving rise to the non-cyclostomes (this hypothesis is now rejected) and lastly, the non-cyclostomes are a paraphyletic grade giving rise to the cyclostomes (Quicke, 2015).

3.3.1 Morphological and Molecular Phylogenetic Analysis of the Genus *Glyptapanteles*

The study by Gupta et al. (2016d) was based on 60 populations reared from 35 host species, 100+ individual caterpillars, from 12 different geographical locations of the country (11 states and one Union territory) that represent 26 provisional *Glyptapanteles* species within eight species groups. Maximum likelihood (Figure 3.1) and Bayesian inference (Figure 3.2) methods displayed three and four major discrete *Glyptapanteles* clades, respectively. In clade of a few Indian species were grouped along with Neotropical and Thailand species. The other clades B and C grouped the majority of the Indian species and showed considerable host specificity in both trees. All parasitic wasp species were gregarious in nature, except for two populations. Three different sets of data (morphology, host records, and COI) were integrated in order to generate accurate boundaries between species/species-groups. The present study, perhaps the most comprehensive done to date in India, suggests the presence of several additional *Glyptapanteles* species, which were previously unrecognised.

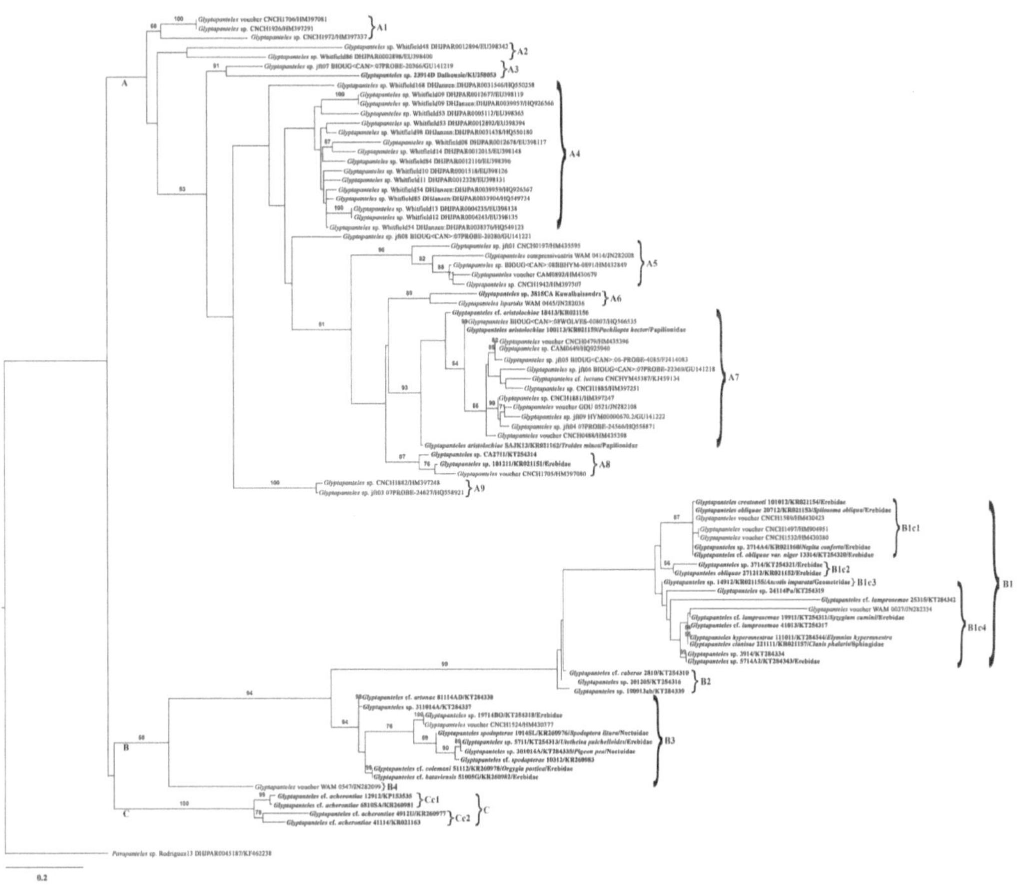

FIGURE 3.1 Maximum likelihood phylogenetic tree of partial COI gene sequences of *Glyptapanteles* spp. The scale bar indicates the number of substitutions per nucleotide position. Numbers at nodes represent bootstrap values with 1000 replicates (only values >50 are shown) (Source Gupta et al., 2016d).

3.3.2 Biodiversity and Species Richness

The Old World tropics, Africa and South-East Asia, in particular, are far richer with more than 170 valid genera of Ichneumonoidea (Yu et al., 2012). Roughly 3,000 species are known from Asia (tropical), as compared to 17,000 species from Africa and 11,000 species from the Neotropics (Jones et al., 2009). The total number of Indian species accounts for approximately 1083, which is about 6% of the total world braconid diversity. It is better not to shy away from the fact that we do not have a real figure from India as the present data is based on information extracted from Yu et al. (2012).

Jones et al. (2009) provided the total number of described braconid species and estimated numbers for seven biogeographic regions (±SE), *viz.* Africa, Neotropics, Asia temperate, Europe, Asia tropical, Nearctic and Australasia as less than 3000, less than 3000, nearly 4000, less than 2000, nearly 3000, nearly 3000 and nearly 1000 species, respectively for the described ones and for the estimated species nearly 17,000, 11,000, 5000, nearly 4000, less than 3000, less than 2000 and less than 1000 species, respectively.

3.4 SEXUAL DIMORPHISM

Male and female ichneumonoids are generally somewhat similar apart from primary and secondary sexual characteristics; such as body and metasomal size, sex-related glands and numbers of

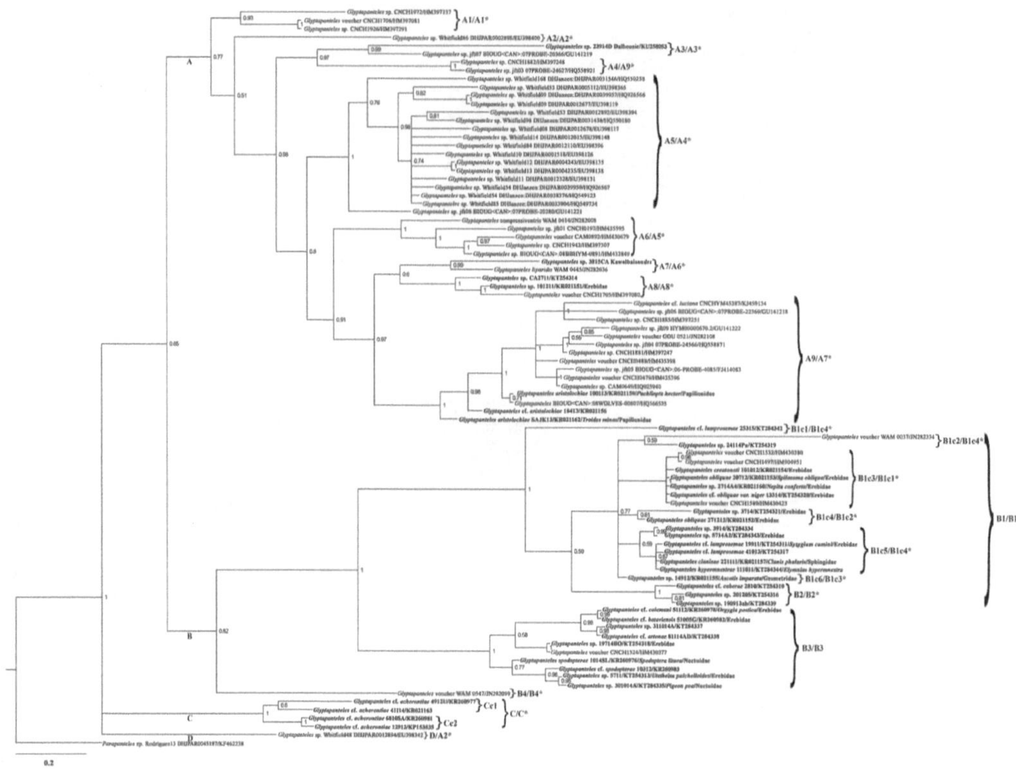

FIGURE 3.2 Bayesian inference phylogenetic tree of partial COI gene sequences of *Glyptapanteles* spp. The scale bar indicates the number of substitutions per nucleotide position. Posterior probabilities supporting nodes (> 0.50) are shown. The sequences of isolates obtained in the present study are shown in bold and the closest sequences are shown with their GenBank accession number. * Symbol represents the similar ML tree clade number (Source Gupta et al., 2016d).

antennal segments and their numbers of various classes of sensilla. In the Braconinae, there is a tendency for males to have darker heads and a darker apex to the metasoma (Figure 3.3 and Figure 3.4) (Quicke, 2015).

3.5 BIOLOGY

Braconid wasp genus/species have set preferences for their respective host family/genus/species as well as for the stage of the host they parasitize. Cheloninae are ovo-larval parasitoids of Lepidoptera; Doryctinae are parasitoids of wood-dwelling Coleoptera larvae; Braconinae parasitize larvae as well as pupae, for instance, *Habrobracon* are in general larval parasitoids of many members of Lepidoptera while *Crinibracon* parasitizes pupae of *Hasora chromus* (Cramer) (Hesperiidae); Lysiterminae more specifically genus *Acanthormius* parasitizes bagworm larvae (Psychidae) (Gupta and Quicke, 2018); and Opiinae in general are fruitfly parasitoids.

Microgastrinae are most frequently encountered in agroecosystems. The most common species of the Indian fauna reported by Gupta and Fernández-Triana (2014) are mentioned in Table 3.1. In general *Apanteles* species are predominant parasitoids of crambid, hesperiid and lycaenid hosts, while the members of families Plutellidae, Pieridae and Lasiocampidae are mostly parasitized by *Cotesia*. Additionally, *Apanteles sauros* Nixon was reported from *Cajanus cajan* leaf webber *Cydia critica* Meyrick (PhD thesis). Some exclusive relationships recorded are between Limacodidae with *Fornicia* and Noctuidae with *Microplitis* (Gupta and Fernández-Triana, 2014).

FIGURE 3.3 *Phanerotoma andamanensis* Gupta & van Achterberg (female).

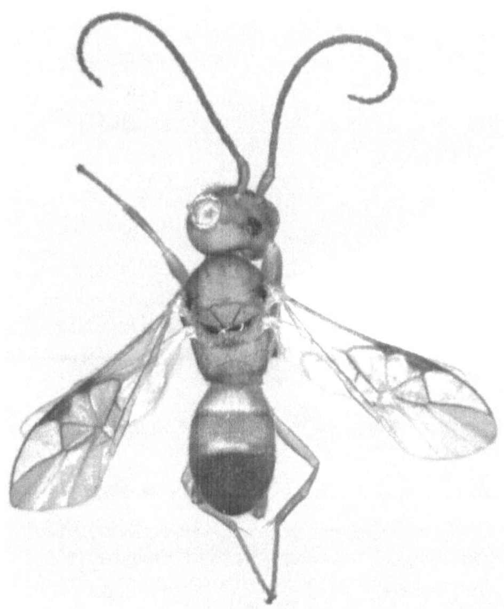

FIGURE 3.4 *Phanerotoma andamanensis* Gupta & van Achterberg (male).

3.5.1 Braconid Parasitoids of Recent Invasive Pest

In 2018, the most notorious invasive pest the fall armyworm (FAW) *Spodoptera frugiperda* (J. E. Smith) (Lepidoptera: Noctuidae) was first reported in India, an extensive search was made for the presence of indigenous natural enemies. Shylesha et al. (2018) have documented natural parasitism by braconid parasitoids – *Glyptapanteles creatonoti* (Viereck) (Braconidae). Later Gupta et al. (2019a) reported *Cotesia ruficrus* (Haliday) as a larval parasitoid followed by Gupta et al. (2019b) and Gupta et al. (2020) where additional reports of *Chelonus formosanus* Sonan and *Coccygidium transcaspicum* (Kokujev), respectively, were made from FAW larvae. Also *Phanerotoma* sp. and *Meteorus* sp. were collected by NBAIR team during various surveys which were identified by the author.

TABLE 3.1
Host – Parasitoid Association

S. no.	Species	Host	Host family
1	*Apanteles folia* Nixon	*Arhopala amantes* (Hewitson), *Tajuria cippus* (Fabricius) and *Spindasis vulcanus* (Fabricius)	Lycaenidae
2	*Apanteles galleriae* Wilkinson	*Galleria mellonella* L.	Pyralidae
3	*Apanteles javensis* Rohwer	*Pelopidas conjuncta* (Herrich-Schaffer)	Hesperiidae
4	*Apanteles javensis* Rohwer	*Pelopidas mathias* (Fabricius)	Hesperiidae
5	*Apanteles machaeralis* Wilkinson	*Eutectona machaeralis* (Walker)	Crambidae
6	*Apanteles mamitus* Nixon	*Spoladea recurvalis* (Fabricius)	Crambidae
7	*Apanteles mamitus* Nixon	*Maruca vitrata* (Fabricius)	Crambidae
8	*Apanteles phycodis* Viereck	*Argina syringa* Cramer	Erebidae: Arctiinae
9	*Apanteles phycodis* Viereck	*Phycodes radiata* Ochsenheimer	Brachodidae
10	*Apanteles stantoni* (Ashmead)	*Parotis marginata* (Hampson)	Crambidae
11	*Apanteles tachardiae* Cameron	*Holcocera pulverea* (Meyrick)	Blastobasidae
12	*Apanteles taragamae* Viereck	*Diaphania indica* Saunders	Crambidae
13	*Apanteles taragamae* Viereck	*Opisina arenosella* Walker	Oecophoridae
14	*Cotesia erionotae* (Wilkinson)	*Udaspes folus* (Cramer)	Hesperiidae
15	*Cotesia erionotae* (Wilkinson)	*Udaspes folus* (Cramer)	Hesperiidae
16	*Cotesia flavipes* Cameron	*Chilo partellus* (Swinhoe)	Crambidae
17	*Cotesia glomerata* (Linnaeus)	*Pieris brassicae* L.	Pieridae
18	*Cotesia vestalis* (Haliday)	*Plutella xylostella* (L.)	Plutellidae
19	*Cotesia ruidus* Wilkinson	*Eutectona machaeralis* (Walker)	Crambidae
20	*Cotesia ruficrus* (Haliday)	*Spodoptera* sp.	Noctuidae
21	*Distatrix papilionis* (Viereck)	*Papilio demoleus* L.	Papilionidae
22	*Dolichogenidea kunhi* Gupta & Kalesh	*Thoressa evershedi* (Evans)	Hesperiidae
23	*Fornicia ceylonica* Wilkinson	*Spatulifimbria castaneiceps* Hampson	Limacodidae
24	*Glyptapanteles aristolochiae* (Wilkinson)	*Pachliopta hector* (Linnaeus)	Papilionidae
25	*Glyptapanteles hypermnestrae* Gupta & Pereira	*Elymnias hypermnestra* (Linnaeus)	Nymphalidae
26	*Glyptapanteles spodopterae* Ahmad	*Spodoptera litura* Fab.	Noctuidae
27	*Microgaster carinicollis* Cameron	*Psalis pennatula* (Fab.)	Erebidae
28	*Microplitis demolitor* Wilkinson	*Helicoverpa armigera* (Hubner)	Noctuidae
29	*Microplitis indicus* Marsh	*Helicoverpa* sp. and *Helicoverpa assulta* (Guenee)	Noctuidae
30	*Microplitis maculipennis* (Szepligeti)	*Acanthodelta janata* L.	Erebidae
31	*Microplitis manilae* Ashmead	*Spodoptera* sp.	Noctuidae
32	*Microplitis prodeniae* Rao & Kurian	*Spodoptera litura* (Fabricius)	Noctuidae
33	*Microplitis spodopterae* Rao & Kurian	*Spodoptera mauritia* (Boisduval)	Noctuidae
34	*Protapanteles creatonoti* Viereck	*Amsacta albistriga* (Walker)	Erebidae
35	*Protapanteles obliquae* (Wilkinson)	*Spilosoma obliqua* Walker	Erebidae

FIGURE 3.5 Depicts ball like cocoon spun by *Glyptapanteles clanisae* Gupta on host *Clanis phalaris* Cramer (Lepidoptera: Sphingidae).

3.5.2 Immature Taxonomy

Ideally cocoons are not part of the wasps but studies have proved their crucial role in species differentiation. Gupta and Fernández-Triana (2014) showed each microgastrine wasp species has a fixed cocoon spinning/laying pattern, structure and colour, which are quite distinctive and help in ascertaining its identity and irrespective of the host species it attacks (Figures 3.5–3.8).

3.5.3 Integrative Taxonomy

Form India, nothing majorly has been done in integrative taxonomy when it comes to Braconidae. Gupta et al. (2016d) pioneered the first integrative comprehensive study of any Braconidae genus in India to

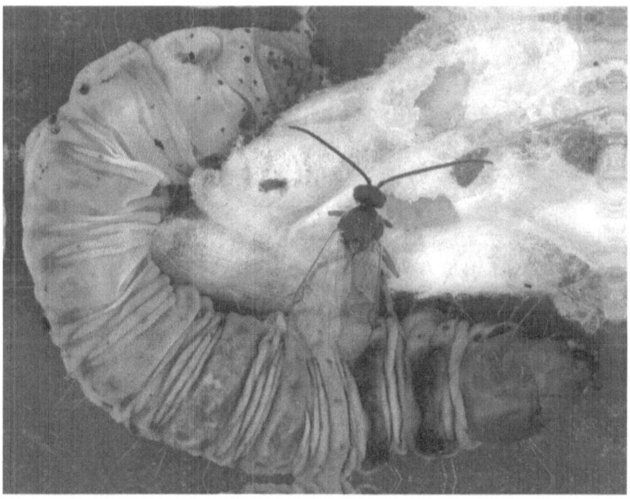

FIGURE 3.6 *Apanteles phycodis* Viereck lays white coloured gregarious cocoons on the lateral side of the host *Phycodes radiata* Ochsenheimer.

FIGURE 3.7 *Cotesia trabalae* Gupta lays cocoons on the dorsal side of its host, *Trabala vishnou* (Lefèbvre).

FIGURE 3.8 *Microplitis maculipennis* Szepligeti lays a single brown cocoon at the rear end of its host, Castor semilooper *Acanthodelta janata* (L.).

resolve a diverse and geographically realistic subset of species within the hyperdiverse genus *Glyptapanteles* to correlate the host specialisation in India. The study was based on 60 populations reared from 35 host species with 100+ individual caterpillar rearings (containing 1100 wasp specimens pinned and 2000 in alcohol) collected from 12 different geographical locations of the country (11 states and 1 Union Territory) that represented 26 provisional *Glyptapanteles* species within eight species-groups.

3.5.4 Taxonomic Problems

Despite their enormous species richness and economic importance in biocontrol programs, the Indian Braconidae still remain a taxonomically challenging group due to inadequate taxonomic expertise in India, missing type/voucher specimens (not to be located in repositories mentioned in the literature) and erroneous identifications. It is extremely difficult to narrow down even to the generic level in highly speciose subfamilies like Braconinae, Doryctinae, Cheloninae, etc. and to the species level in subfamily Microgastrinae. There is an ardent need for a well-identified reference repository as types either are not to be located in India or if present are in museums abroad. The taxonomists working on this group need to have compulsory exposure visits to understand the genus/species diversity.

3.5.5 Economic Importance

The majority of the braconid parasitoids are beneficial in naturally controlling pest populations (Figures 3.9 and 3.10). Here, a few wasp species of high economic importance are listed. *Bassus relativus* (Bhat & Gupta) (Figure 3.9a) is common parasitoid of *Maruca vitrata* (Fabricius) (Lepidoptera: Crambidae). *Chelonus blackburni* Cameron (Figure 3.9b) is an effective egg-larval

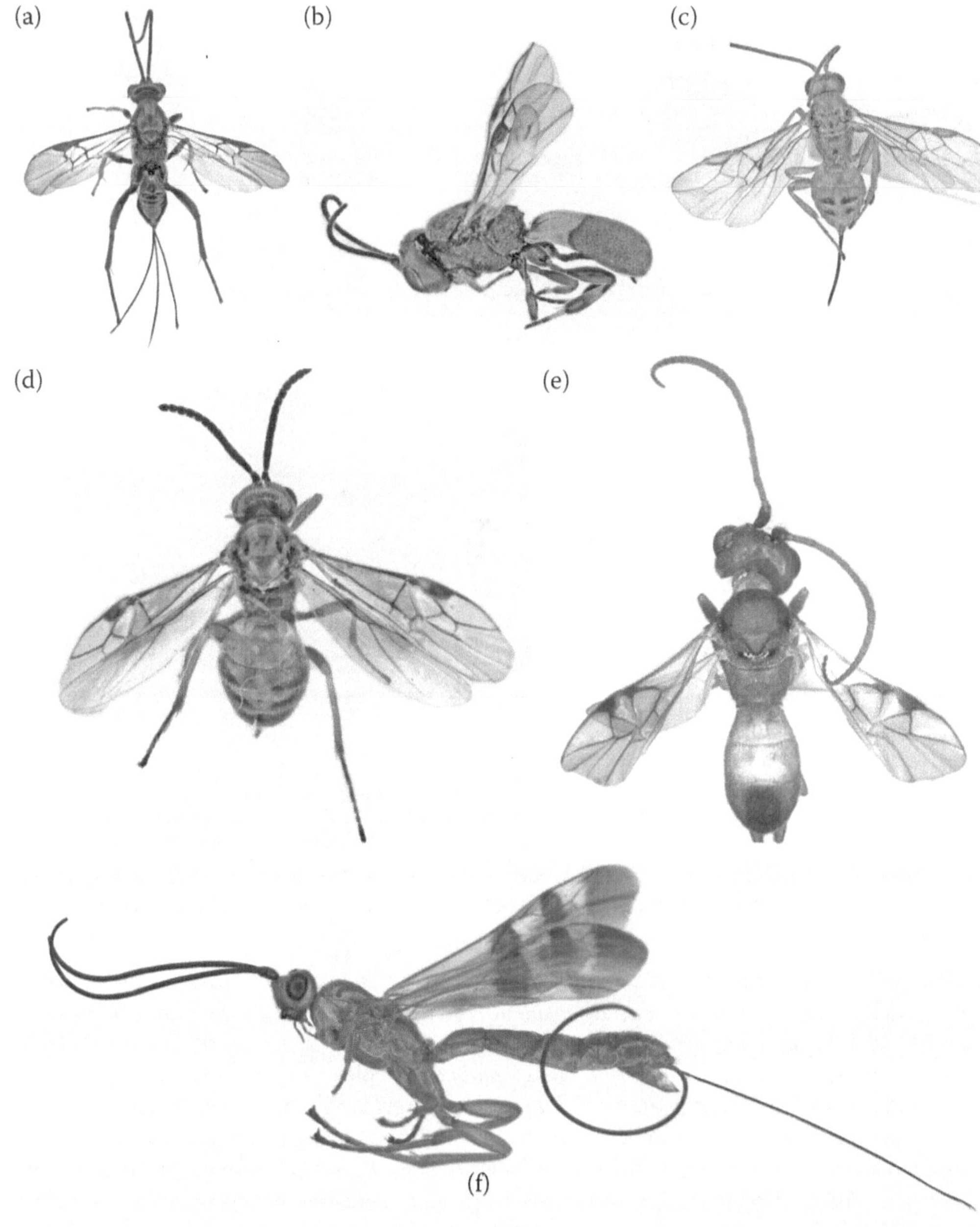

FIGURE 3.9 Common braconid parasitic wasps of agroecosystem: (a) *Bassus relativus* (Bhat & Gupta); (b) *Chelonus blackburni* Cameron; (c) *Diachasmimorpha longicaudata* complex; (d) *Habrobracon brevicornis* (Wesmael); (e) *Phanerotoma* sp.; (f) *Stenobracon* sp.

FIGURE 3.10 Common microgastrine parasitoids and their hosts: (a) *Apanteles galleriae* Wilkinson cocoons along with larvae of *Galleria mellonella* L. and wasp in habitus; (b) Parasitized caterpillars of *Acanthodelta janata* L. with cocoon of *Microplitis maculipennis* (Szepligeti) and wasp in habitus; (c) *Cotesia vestalis* (Haliday) from *Plutella xylostella* (L.) on cabbage and wasp in habitus; (d) *Distatrix papilionis* (Viereck), a gregarious endoparasitoid of *Papilio demoleus* L. on *Citrus* spp. and wasp in habitus.

parasitoid of several lepidopterous pests, like *Phthorimaea operculella* (Zeller), *Hellula undalis* (Fab.), *P. xylostella, Earias vittella, E. insulana, P. gossypiella* (Saunders) and *H. armigera* (Hubner), etc. *Phanerotoma syleptae* Zettel is known to parasitize larvae of *Diaphania indica* (Saunders). Wasps of *Diachasmimorpha longicaudata* complex (Figure 3.9c) are solitary endoparasitoids of tephritid fruitfly larvae. *Habrobracon brevicornis* (Wesmael) (Figure 3.9d), an extremely polyphagous ectoparasitoid, prasitizes many hosts belonging to the family Crambidae and Pyralidae; the common ones being *Chilo partellus* (Swinhoe), *Pectinophora gossypiella* (Saunders); *Corcyra cephalonica* (Stainton), *Opisina arenosella* (Walker), etc. For the recent invasive pest, *Tuta absoluta* (Meyrick), infesting tomatoes in many parts of India *Habrobracon* sp. (Hymenoptera: Braconidae) was found to be associated (Ballal et al., 2016). *Stenobracon* spp. (Figure 3.9f) are solitary larval parasitoids of several graminaceous borers, such as *Chilo partellus* Swinhoe, *C. infuscatellus* Snellen, *C. sacchariphagus indicus* (Kapur), *C. auricilius* Dudgeon, *C. suppressalis*

(Walker), *Scirpophaga excerptalis* (Walker), *S. incertulas* (Walker), *S. innotata* (Walker), *S. nivella* (F.), *Sesamia inferens* (Walker), etc.

3.5.6 Collection and Preservation Techniques

The collection is either by sweeping, setting pan traps, pitfall tarps, light trapping, malaise traps, canopy fogging or the best way is by rearing the hosts. Usually, caterpillars are handpicked from various crop fields, forest trees, and wild plantations throughout the year. Caterpillars from the same host plant are kept separately in individual plastic containers with fine-meshed lids carrying specific code numbers in the laboratory. Unparasitized caterpillars from the same brood are reared to adults for host confirmation or preserved in 95% alcohol for molecular identification. Caterpillar rearings were conducted at 25±2°C and 50±10% relative humidity and were fed with the natural host plant from which the respective caterpillar was collected.

The success of any rearing depends on the trickiest part of the rearing which is providing uninterrupted natural host plants for the caterpillars/immature ones for feed and by following flawless coding methods so that one does not get confused later. A very successful method yet practical method is followed for coding. The caterpillars/cocoons should be coded based on the date of collection, host plant and locality. For example, 61212CN-K—could be interpreted as a caterpillar collected on December 6, 2012 on *Cocus nucifera* from Kerala. If two species of caterpillars are collected on the same host plant on the same date, then they are maintained along with an additional supporting numeral/alphabet (e.g., 61212CN-K-1 and 61212CN-K-2). Occasionally, when multiple species (as in the case of hyperparasitoids) emerged on the same date from the same host caterpillar, the date is supported with alphabets either as a prefix or suffix to avoid confusion (e.g. *Glyptapanteles clanisae* Gupta—Code: 2211CLP for the primary parasitoid and Code: 2211CLH for the associated hyperparasitoid *Eurytoma* sp.).

3.5.7 Workers on Indian Fauna

Contributions to Indian fauna were done by: Ayyar, 1927, 1928; Wilkinson, 1927, 1928, 1929; Bhatnagar, 1948; Rao, 1961; Nixon, 1965, 1967, 1968; Rao et al., 1969, 1971; Veneeta Sharma, 1972 (Ph. D. thesis), 1984; Mason, 1981; Sumodan and Narendran, 1990; Austin and Dangerfield, 1992; Saeed et al., 1999; Rema and Narendran, 1996; Rema and Sheeba, 2004; Ahmad et al., 2009; Valerio et al., 2009; Gupta et al., 2011a, 2011b; Gupta and Kalesh, 2012; Gupta and Pereira, 2012; Gupta, 2013a, 2013b; Gupta and Lokhande, 2013; Gupta et al., 2013a, 2013b; Gupta et al., 2014a, 2014b; Gupta and Fernandez-Triana, 2014, 2015; Gupta and van Achterberg, 2014; Gupta et al., 2016a, 2016b, 2016c, 2016d; and Gupta and Quicke, 2018; Gupta et al., 2019a, 2019b, 2020. Recently a new subfamily Miracinae is added to the Indian fauna of Braconidae by Ranjith et al., 2018.

3.6 CONCLUSIONS

As indicated in this chapter, the Indian braconid fauna is poorly studied. Many species await identification and many more are expected to be reared. Braconidae in India needs more exploratory surveys, systematic host rearings, and molecular characterisation with multiple genes to deal with ambiguous taxa. Simple trap collection methodology does not work in this group, as a lot of identification is dependent on associated host information. Post-collection and transportation of caterpillars from different geographical locations of the country to the laboratory cause considerable in-transit mortality of many caterpillar species. This is a major hindrance in host rearing. Hence good networking with like-minded individuals for rearing in respective geographical regions should be encouraged and for this healthy collaboration with citizen scientists should be promoted. It is worth mentioning that since the sampling involves individuals, there are many chances of inherent sampling biases and recurrent sampling in some parts of the country as compared to others.

Though phylogenetic studies to resolve a diverse and geographically realistic subset of species within the genus *Glyptapanteles* to correlate the host specialisation in India were initiated and attempted, only some relationships could be resolved. Additional gene/genes evolving more slowly than COI, but fast enough to separate closely related groups of species, may provide the extra resolution required to completely resolve the complex species-groups of Indian Braconidae. Furthermore, it is necessary to incorporate these findings into a global context by adding additional geographic, biological and economically important subsets of species and to describe the Indian species delineated through molecular studies.

3.7 POINTS TO REMEMBER

1. Systematic and exhaustive sampling is necessary for thorough diversity documentation.
2. Rearing of the host species is mandatory for parasite biology studies.
3. Details of host insect species are one of the key factors leading to the correct identification of the parasitoid and the study of host-parasitoid relationships.
4. Always consult subject experts for species identity before publishing in haste.

REFERENCES

Ahmad, Z., Hussain, K.Z. & Shanthi, R. (2009). *Glyptapanteles spodopterae* Ahmad sp. nov., (Hymenoptera: Braconidae: Microgastrinae) parasitic on larvae of *Spodoptera litura* in India. *Journal of Entomological Research*, 33: 97–99.

Austin, A.D. & Dangerfield, P.C. (1992). Synopsis of the Australasian Microgastrinae (Hymenoptera: Braconidae), with key to genera and description of new taxa. *Invertebrate Taxonomy*, 6: 45–47.

Ayyar, T.V.R. (1927). The parasitic Hymenoptera of economic importance from south India. *Bulletin of Entomological Research*: 18, 73–75.

Ayyar, T.V.R. (1928). Contribution to our knowledge of south Indian Braconidae-Vipionidae. *India Deptartment of Agriculture Memoirs – Entomological Series*, 10: 29–60.

Ballal, C.R., Gupta, A., Mohan, M., Lalitha, Y. & Verghese, A. (2016). The new invasive pest *Tuta absoluta* (Meyrick) (Lepidoptera: Gelechiidae) in India and its natural enemy complex. *Current Science*, 110(11): 2155–2159.

Bhatnagar, S.P. (1948). Studies on *Apanteles* Förster (Vipionidae: parasitic Hymenoptera) from India. *Indian Journal of Entomology*, 10: 133–203.

Capek, M. (1970). A new classification of the Braconidae (Hymenoptera) based on the cephalic structures of the final instar larva and biological evidence. *Canadian Entomologist*, 102, 846–875.

Gupta, A. (2013a). Revision of the Indian *Microplitis* Foerster (Hymenoptera: Braconidae: Microgastrinae), with description of one new species. *Zootaxa*, 3620: 429–452.

Gupta, A. (2013b). Three new species of reared parasitic wasps (Hymenoptera: Braconidae: Microgastrinae) from India. *Zootaxa*, 3701: 365–380.

Gupta, A., Churi, P.V., Sengupta, A. & Mhatre, S. (2014a). Lycaenidae parasitoids from peninsular India with description of four new species of microgastrine wasps (Hymenoptera: Braconidae) along with new insights on host relationships. *Zootaxa*, 3827(4): 439–470. 10.11646/zootaxa.3827.4.2

Gupta, A., Das A.K., Neog, K. & Verghese A. (2016a). First report of *Cotesia dictyoplocae* (Hymenoptera: Braconidae), a larval parasitoid of *Antheraea assamensis* (Lepidoptera: Saturniidae), from India. *Florida Entomologist*, 99(3): 541–543.

Gupta, A. & Fernández-Triana, J.L. (2014). Diversity, host association, and cocoon variability of reared Indian Microgastrinae (Hymenoptera: Braconidae) *Zootaxa*, 3800(1): 001–101. 10.11646/zootaxa.3800.1.1

Gupta, A. & Fernandez-Triana, J.L. (2015). Four new species of the genus *Diolcogaster* Ashmead, 1900 (Hymenoptera: Braconidae: Microgastrinae) from South East Asia with a key to the Indian species. *Systematic Parasitology*, 90, 3: 285–300. 10.1007/s11230-014-9546-8

Gupta, A., Ghosh, A., Baby, N.L. & Jalali, S.K. (2011a). Morphological and molecular characterization of *Apanteles mohandasi* Sumodan & Narendran (Hymenoptera: Braconidae), a solitary endoparasitoid of *Pammene critica* Meyrick (Lepidoptera: Tortricidae), with notes on biology from India. *Entomological News*, 122: 354–365.

Gupta, A. & Kalesh, S. (2012). Reared parasitic wasps attacking hesperiids from Western Ghats (Kerala, India) with description of a new species of *Dolichogenidea* (Hymenoptera: Braconidae) as a larval parasitoid of *Thoressa evershedi* (Evans) (Lepidoptera: Hesperiidae). *Zootaxa*, 3413: 29–43.

Gupta, A., Khot, R. & Chorge, S. (2014b). A new species of *Parapanteles* Ashmead, 1900 (Hymenoptera: Braconidae: Microgastrinae) parasitic on *Charaxes athamas* (Drury) (Lepidoptera: Nymphalidae) in India. *Systematic Parasitology*, 88: 273–279. 10.1007/s11230-014-9492-5.

Gupta A., Lalitha, Y., Varshney, R., Shylesha, A.N. & Van Achterberg, C. (2019b). *Chelonus formosanus* Sonan (Hymenoptera: Braconidae) an egg-larval parasitoid of the invasive pest *Spodoptera frugiperda* (J. E. Smith) (Lepidoptera: Noctuidae) amenable to laboratory mass production in India. Journal of Entomology and Zoology Studies 2020; 8(1): 1521–1524

Gupta, A. & Lokhande, S.A. (2013). A new host record and a new combination in Cotesia *Cameron* (Hymenoptera: Braconidae) from India. *Journal of threatened taxa*, 5: 3678–3681.

Gupta, A., Lokhande, S.A. & Soman, A. (2013a). Parasitoids of Hesperiidae from peninsular India with description of a new species of *Dolichogenidea* (Hymenoptera: Braconidae) parasitic on caterpillar of *Borbo cinnara* (Wallace) (Lepidoptera: Hesperiidae). *Zootaxa*, 3701: 277–290.

Gupta, A., Mark, S., Cardinal, S. & Fernandez-Triana, J.L. (2016b). A review of unusual species of *Cotesia* (Hymenoptera, Braconidae, Microgastrinae) with the first tergite narrowing at midlength, *ZooKeys*, 580: 29–44. 10.3897/zookeys.580.8090

Gupta, A. & Pereira, B. (2012). A new species of *Glyptapanteles* (Hymenoptera: Braconidae: Microgastrinae), a larval parasitoid of *Elymnias hypermnestra* (Linnaeus) (Lepidoptera: Nymphalidae), along with some new host records of parasitoids from Peninsular India. *Zootaxa*, 3227: 54–63.

Gupta, A., Pereira, B. & Churi, P.V. (2011b). Illustrated notes on some reared parasitic wasps (Braconidae: Microgastrinae) with new host and distribution records from India along with reassignment of *Glyptapanteles aristolochiae* (Wilkinson) as a new combination. *Entomological News*, 122: 451–468.

Gupta, A., Pereira, B. & Churi, P.V. (2013b). A new species of *Parapanteles* Ashmead (Hymenoptera: Braconidae) from India reared from *Abisara echeria* Stoll (Lepidoptera: Riodinidae) with key to the Indian *Parapanteles* species. *Zootaxa*, 3709, 363–370.

Gupta, A. & Quicke, D.L.J. (2018). A new species of *Acanthormius* (Braconidae: Lysiterminae) reared as a gregarious parasitoid of psychid caterpillar (Lepidoptera: Psychidae) from India. *Zootaxa*, 4388(3): 425–430. 10.11646/zootaxa.4388.3.8

Gupta A., Ramesh Babu, S. & Sampath Kumar, M. (2019a). *Cotesia ruficrus* (Haliday, 1834) (Hymenoptera: Braconidae) emerging as a common natural parasitoid of *Spodoptera frugiperda* (J. E. Smith) (Lepidoptera: Noctuidae) in Indian maize fields. *Journal of Biological Control*, 33(3): 193–196. 10.1 8311/jbc/2019/24118

Gupta A., Soujanya, P.L., van Achterberg, C. & Sekhar, J.C. (2020). *Coccygidium transcaspicum* (Kokujev) (Hymenoptera: Braconidae) parasitizing larvae of invasive pest *Spodoptera frugiperda* (J. E. Smith) (Lepidoptera: Noctuidae) in India. *Zootaxa*, 4750(2): 293–297.

Gupta, A. & van Achterberg, C. (2014). A new species of *Phanerotoma* Wesmael (Hymenoptera: Braconidae: Cheloninae) from the Andaman Islands, India. *Zootaxa* 3856(4): 595–600. 10.11646/zootaxa.3856.4.8

Gupta, A., van Achterberg, C. & Malathi C. (2016c). A new species of *Crinibracon* Quicke (Hymenoptera: Braconidae) parasitic on pupae of *Hasora chromus* (Cramer) (Lepidoptera: Hesperiidae) from India. *Zootaxa*, 4158(2): 281–291. 10.11646/zootaxa.4158.2.9

Gupta, A., Venkatesan, T. & More, R.P. (2016d). Morphological and Molecular Characterization of Reared Parasitoid Wasps of the Genus *Glyptapanteles* Ashmead 1904 (Insecta: Hymenoptera: Braconidae: Microgastrinae) Associated with Lepidoptera in India. *PLoS ONE*, 11(3): e0150765. 10.1371/ journal.pone.0150765

Jones, O.R., Purvis, A., Baumgart, E. & Quicke, D.L.J. (2009). Using taxonomic revision data to estimate the geographic and taxonomic distribution of undescribed species richness in the Braconidae (Hymenoptera: Ichneumonoidea). *Insect Conservation and Diversity*, 2: 204–212.

Mason, W.R.M. (1981). The polyphyletic nature of *Apanteles* Förster (Hymenoptera: Braconidae): a phylogeny and reclassification of Microgastrinae. *Memoirs of the Entomological Society of Canada*, 115: 1–147.

Nixon, G.E.J. (1965). A reclassification of the tribe Microgasterini (Hymenoptera: Braconidae). *Bulletin of the British Museum (Natural History), Entomology series, Supplement* 2, 1–284.

Nixon, G.E.J. (1967). The Indo-Australian species of the *Ultor* group of *Apanteles* Förster Hymenoptera Braconidae. *Bulletin of the British Museum Natural History Entomology*, 21: 1–34.

Nixon, G.E.J. (1968). A revision of the genus *Microgaster* Latreille (Hymenoptera: Braconidae). *Bulletin of the British Museum (Natural History)-Entomology series*, 22: 33–72.

Quicke, D.L.J. (2015). The *Braconid* and Ichneumonid Parasitoid Wasps: Biology, Systematics, Evolution and Ecology. John Wiley & Sons, Sussex, pp. 681.

Ranjith, A.P., van Achterberg, C., Priyadarsanan, D.R., Kim, I.-K., Keloth, R., Mukundan, S. & M. Nasser, M. (2018). First Indian record of *Centistidea* Rohwer (Hymenoptera: Braconidae, Miracinae) with description of eight new species. *Insect Systematics & Evolution*, 1–36. 10.1163/1876312X-00002194

Rao, S.N. (1961). A key to the Oriental species of *Apanteles* Foerster (Hymenoptera). Proceedings of the National Academy of Sciences India (B), 31, 32–46.

Rao, V.P., Chako, M.J., Phalak, V.R. & Dinesh Rao, H. (1969). Leaf feeding caterpillars of paddy and their natural enemies in India. *Journal of Bombay National History Society*, 66, 455–477.

Rao, V.P., Ghani, A.M., Sankaran, T. & Mathur, K.C. (1971). *A review of the biological control of insects and other pests in South East Asia and the Pacific region*. Commonwealth Institute of Biological Control Technical Communication, (6), CAB, Slough, pp. 149.

Rema, C.G. & Narendran, T.C. (1996). A remarkable new genus of Braconidae (Hymenoptera) from India. *Journal of the Bombay Natural History Society*, 93: 264–267.

Rema, C.G. & Sheeba, M. (2004). A new species of *Diolcogaster* Ashmead (Hymenoptera: Braconidae: Microgastrinae) from Kerala, India with a key to Indian species. In. *Perspectives on Biosystematics and Biodiversity* Eds K. Rajmohan, K. Sudheer, P. Girish Kumar, S. Santhosh, pp. 509–513. Calicut, India: Harvest Media Services.

Saeed, A., Austin, A.D. & Dangerfield, P.C. (1999). Systematics and host relationships of Australasian *Diolcogaster* (Hymenoptera: Braconidae: Microgastrinae). *Invertebrate Taxonomy*, 13: 117–178.

Sharma, V. (1972). Taxonomic studies on *Apanteles* Foerster (Hymenoptera: Braconidae: Microgasterinae) from India III. The vitripennis group. *Oriental Insects*, 6: 553–560.

Sharma, V. (1984). A new species of genus *Fornicia* (Hymenoptera, Braconidae, Microgasterinae). *Reichenbachia*, 22: 209–211.

Sharanowski, B.J., Dowling, A.P.G. & Sharkey, M.J. (2011). Molecular phylogenetics of Braconidae (Hymenoptera: Ichneumonoidea), based on multiple nuclear genes, and implications for classification. *Systematic Entomology*, 36: 549–572.

Shylesha, A. N. S. K. Jalali, Gupta, A., Varshney, R., Venkatesan, T., Shetty, P., Ojha, R., Ganiger, P.C., Navik, O., Subaharan, K., Bakthavatsalam, N. & Ballal, C.R. (2018). Studies on new invasive pest *Spodoptera frugiperda* (J. E. Smith) (Lepidoptera: Noctuidae) and its natural enemies. *Journal of Biological Control*, 32(3): 1–7, 2018, 10.18311/jbc/2018/21707

Sumodan, P.K. & Narendran, T.C. (1990). Five new species of *Apanteles* Foerster (Hymenoptera: Braconidae) from Kerala, India. *Journal of Ecobiology*, 2: 239–248.

Tobias, V.I. (1989). Application of W. Hennig's method of phylogenetic analysis for the construction of the phylogenetic tree of the family Braconidae (Hymenoptera). *Trudy Zoologicheskogo Instituta, Leningrad*, 202: 67–83 (in Russian).

Valerio, A.A., Whitfield, J.B. & Janzen, D.H. (2009). Review of world *Parapanteles* Ashmead (Hymenoptera: Braconidae: Microgastrinae), with description of fourteen new Neotropical species and the first description of the final instar larvae. *Zootaxa*, 2084: 1–49.

van Achterberg, C. (1984). Essay on the phylogeny of Braconidae (Hymenoptera: Ichneumonoidea). *Entomologisk Tidskrift*, 105: 41–58.

van Achterberg C. (1993). Illustrated key to the subfamilies of the Braconidae (Hymenoptera: Ichneumonoidea). *Zoologisches Verhandelingen, Leiden*, 283: 1–189.

Wilkinson, D.S. (1927). On the Indo-Malayan species of the genus *Microgaster* (Hymenoptera: Braconidae). *Bulletin of Entomological Research*, 18: 171–178.

Wilkinson, D.S. (1928). A revision of the Indo-Australian species of the genus *Apanteles* (Hym. Bracon.). Part II. *Bulletin of Entomological Research*, 19: 109–146.

Wilkinson, D.S. (1929). A revision of the Indo-Australian and Ethiopian species of the genus *Microgaster* (Hym. Bracon.). *Transactions of the Royal Entomological Society of London*, 77: 99–123.

Yu, D.S.K., van Achterberg, C. & Horstmann, K. (2012). Taxapad 2012, Ichneumonoidea 2011. Database on flash-drive. www.taxapad.com, Ottawa, Ontario, Canada.

Zaldivar-Riveron, A., Martinez, J.J., Belokobylskij, S.A., Pedraza-Lara, C., Shaw, S.R., Hanson P.E. & Varela-Hernandez, F. (2014). Systematics and evolution of gall formation in the plant-associated genera of the wasp subfamily Doryctinae (Hymenoptera: Braconidae). *Systematic Entomology*, in press, 10.1111/syen.12078.

4 Aphidiinae Parasitoids (Braconidae: Hymenoptera)

Bidhan Chandra Das
Department of Zoology, Rajshahi University, Rajshahi, Bangladesh

Samiran Chakrabarti
Formerly Department of Zoology, University of Kalyani, Kalyani, India

CONTENTS

DOI: 10.1201/9781003354239-4

LEARNING OBJECTIVES

i. To learn the diversity and zoogeographical distribution of aphidiine parasitoids.
ii. To know the collection and identification methods of aphidiines.
iii. To get an idea of the biology and ecology of aphiidine parasitoids.
iv. To learn their potential utilisation in biocontrol programme, and
v. To know their success stories, mass rearing and constraints of using in biocontrol.

4.1 INTRODUCTION

Aphid parasitoids are a well-known group of insects, exclusively associated with aphids (Rakhshani and Starý, 2021). All members of aphid parasitoids belong to the subfamily Aphidiinae, a monophyletic subfamily of Braconidae (Hymenoptera) that parasitizes only aphids (Hemiptera: Aphidomorpha: Aphididae) (Kambhampati et al., 2000).

Till the early 1990s, the aphidiines were treated as a separate family, the Aphidiidae, based on their behavioural and morphological traits, but later phylogenetic studies established them as a lineage within the family Braconidae (van Achterberg and Quicke, 1992; Wharton et al., 1992).

All members of the Aphidiinae are solitary endoparasitoids and only one individual completes its development inside a single individual host aphid (Starý, 1970; Mackauer and Chow, 1986). They are small insects with adults usually ranging from less than 1 mm to 3.5 mm (Starý, 1988). This group is koinobiont (Haeselbarth, 1979), as the parasitoid larva develops in a living host that continues to feed and grow (Boivin et al., 2012). Except for some thelytokous species, in which males are rare or absent (Nemec and Starý, 1985), the majority of aphidiine wasps are bisexual (Völkl et al., 2007). These parasitoids comprise numerous specialist species but some generalists exist that may parasitize dozens of aphid species (Benelli et al., 2014).

DNA analysis, however, reveals that most generalists actually consist of many cryptic species (Derocles et al., 2015; Loxdale et al., 2019). However, this hypothesis is opposed by the argument that generalist species actually exist in nature and that trophic generalism has its own benefits. One of the main benefits is the reduction of the risk of its extinction, which is significantly reduced if a species is able to maintain itself on several resources in an environment, where exploitable resources are hardly available or already exploited by other competitors (Dennis et al., 2011).

Most of the aphidiine species have several generations in a year. Exceptions are reported in *Monoctonia pistaciaecola* Starý and *Pseudopauesia prunicola* Halme, which are apparently obligatory monovoltine (Starý, 1988). They appear in the field in the late spring after the appearance of their hosts aphids (Campbell et al., 1974). The combination of high fecundity, short generation time, and female-biased sex ratio result in a high intrinsic rate of increase (r_m), ranging between 0.2 and 0.6 females per female per day (Force and Messenger, 1964; Völkl, 1990).

Certain members of this subfamily are important biocontrol agents for many aphid pest species. In fact, Aphidiinae is one of the most important and best-investigated group because of their significant role in the biocontrol of aphid species (Quicke, 2015).

4.2 DIVERSITY AND ZOOGEOGRAPHICAL DISTRIBUTION

Žikić et al. (2017) reported altogether 505 world species of aphidiines under 38 genera, excluding synonyms (Table 4.1). Earlier, Mackauer and Finlayson (2012) estimated ~~the number of~~ 700 described species of Aphidiinae worldwide. The subfamily has generally been subdivided into four tribes, *viz.* Aclitini Mackauer, Aphidiini Haliday, Ephedrini Mackauer and Praini Mackauer (Mackauer and Starý, 1967; Starý, 1988; Smith and Kambhampati, 2000). The Aphidiini is the largest tribe and includes a majority of known genera and species. It has three subtribes, Aphidiina, Monoctonina and Trioxina. Aphidiines are characterised by a number of synapomorphies (e.g. host specialisation and the presence of a flexible suture between the second and third tergites of the gaster). However, significant differences exist in morphology, biology and behaviour among tribes, genera and species (Smith and Kambhampati, 2000).

The aphid parasitoids can be found both in natural and managed agricultural ecosystems, in association with their preferred host aphids. The majority of species are found in the temperate to subtropical zones of the Northern hemisphere but some species are also purposefully or accidentally introduced to other parts of the world, particularly for the biocontrol of aphid pests (Mackauer and Starý, 1967; Starý, 1970; Powell and Pell, 2007).

Genus *Aphidius* Nees consists of the highest number (104 species) of species that have been reported from 119 countries and territories. However, the genus *Diaeretiella* Starý consists of single species (monotype) but has been recorded from 87 countries and territories. Genera *Binodoxys* Mackauer (60 species), *Lysiphlebus* Förster (22 species), *Ephedrus* Haliday (37 species), *Praon* Haliday (59 species), *Trioxys* Haliday (59 species) and *Pauesia* Quilis (75 species) are reported from 85, 85, 83, 76, 66 and 51 countries/territories, respectively. Genus *Parephedrus* Starý & Carver is reported only from Australia (Yu et al., 2016). Žikić et al. (2017) did not include *Eoaphidius* in their list of aphidiine genera. *Indoephedrus* and *Neoephedrus* were wrongly described as aphid parasitoids (personal communications, emails from Dr. Ehsan Rakhshani, of 11 August 2022; and Dr. Xue-xin Chen, 15 August 2022). Thus, the total number of valid aphidiine species would be 501 under 36 genera.

Twenty-two aphidiine genera, i.e. more than sixty percent (61.11%) of the genera, are present at least in the two biogeographical regions. Rest of the fourteen genera (38.89%), *viz. Acanthocaudus* Smith, *Aclitus* Förster, *Archaphidus* Starý & Schlinger, *Calaphidius* Mackauer, *Cristicaudus* Starý & Remaudiere, *Indaphidius* Starý, *Kashmiria* Starý & Bhagat, *Metaphidius* Starý & Sedlag, *Monoctonia* Starý, *Paralipsis* Förster, *Parephedrus* Starý & Carver, *Protaphidius* Ashmead, *Pseudephedrus* Starý and *Pseudopauesia* Halme have restricted distribution, only in the single biogeographical region. Genera *Binodoxys, Ephedrus, Pauesia* and *Praon* are known from five zoogeographical regions, whereas *Aphidius, Diaeretiella, Lysiphlebus* and *Trioxys* are distributed in all six zoogeographical regions (Table 4.1).

4.2.1 Diversity in India and Adjacent Countries

India and its adjacent countries, *viz.* Bangladesh, Bhutan, Nepal, China (south), Pakistan and Sri Lanka, represent a great part of the Oriental region. Studies on aphids in those areas have been conducted satisfactorily, but explorations for aphid parasitoids have not been done elaborately. From India and adjoining areas following aphidiine species are so far known: India (157 species in 23 genera from the 3 reviews: Ghosh and Starý, 1983, Raychaudhuri, 1990 and Akhtar et al., 2011, and subsequent publications of Rakhshani et al., 2012; Lopes et al., 2016; Yu et al., 2016), Bangladesh (4 species in 3 genera: Das, 1994); Bhutan (0 species), Nepal (14 species in 7 genera: Tamrakar and Singh, 2000; Todorov, 2009; Lopes et al., 2016; Yu et al., 2016), China (131 species in 23 genera: Ji et al., 1992; Jiahua and Quanxiu, 2001; Starý et al., 2010; Petrović et al., 2013; Lopes et al., 2016; Yu et al., 2016; Tian et al., 2017; Davidian, 2020), Pakistan (54 species in

TABLE 4.1
Name of the Genera and the Corresponding Number of Valid Species of World Aphidiinae with their Zoogeographical Distribution (Yu et al., 2016; Žikić et al., 2017)

Genus	Number of Species	Zoogeographical Distribution
Acanthocaudus Smith	3	Neotropical
Aclitus Förster	2	Palearctic
Adialytus Förster	6	Neotropical, Palearctic
Aphidius Nees	104	Nearctic, Afrotropical, Palearctic, Neotropical, Oriental, Australian
Archaphidus Starý & Schlinger	1	Oriental
Areopraon Mackauer	4	Palearctic, Oriental
Betuloxys Mackauer	5	Nearctic, Palearctic, Oriental
Binodoxys Mackauer	60	Nearctic, Afrotropical, Palearctic, Neotropical, Oriental
Calaphidius Mackauer	1	Palearctic
Cristicaudus Starý & Remaudiere	4	Oriental
Diaeretellus Starý	4	Nearctic, Palearctic, Oriental
Diaeretiella Starý	1	Nearctic, Afrotropical, Palearctic, Neotropical, Oriental, Australian
Diaeretus Förster	2	Neotropical, Palearctic, Oriental,
Dyscritulus Hincks	1	Palearctic
Ephedrus Haliday	37	Nearctic, Afrotropical, Palearctic, Neotropical, Oriental
Falciconus Mackauer	1	Palearctic, Oriental
Harkeria Cameron	2	Nearctic, Palearctic
Indaphidius Starý	1	Oriental
**Indoephedrus* Samanta, Pramanik & Raychaudhuri	2	Oriental
Kashmiria Starý & Bhagat	1	Oriental
Lipolexis Förster	4	Nearctic, Neotropical, Palearctic, Oriental
Lysiphlebia Stary & Schlinger	3	Palearctic, Oriental
Lysiphlebus Förster	22	Nearctic, Afrotropical, Palearctic, Neotropical, Oriental, Australian
Metaphidius Starý & Sedlag	1	Palearctic
Monoctonia Starý	2	Palearctic
Monoctonus Haliday	18	Nearctic, Palearctic, Neotropical, Oriental,
**Neoephedrus* Samanta, Tamili, Saha & Raychaudhuri	2	Oriental
Paralipsis Förster	2	Palearctic
Parephedrus Starý & Carver	1	Australian
Pauesia Quilis	75	Nearctic, Afrotropical, Palearctic, Neotropical, Oriental
Praon Haliday	59	Nearctic, Afrotropical, Palearctic, Neotropical, Oriental
Protaphidius Ashmead	2	Palearctic
Pseudephedrus Starý	5	Neotropical
Pseudopauesia Halme	1	Palearctic

TABLE 4.1 (Continued)
Name of the Genera and the Corresponding Number of Valid Species of World Aphidiinae with their Zoogeographical Distribution (Yu et al., 2016; Žikić et al., 2017)

Genus	Number of Species	Zoogeographical Distribution
Pseudopraon Starý	2	Nearctic, Palearctic, Neotropical, Oriental
Toxares Haliday	4	Nearctic, Palearctic, Oriental
Trioxys Haliday	59	Nearctic, Afrotropical, Palearctic, Neotropical, Oriental, Australian
Xenostigmus Smith	1	Nearctic, Palearctic, Neotropical
38	505	

* *Notes*: *Indoephedrus* Samanta, Pramanik & Raychaudhuri and *Neoephedrus* Samanta, Tamili, Saha & Raychaudhuri were wrongly described as aphidiid parasitoids (personal communications, emails from Dr. Ehsan Rakhshani, 11 August 2022 and Dr. Xue-xin Chen, 15 August 2022). So, the actual world genera will be 36, and the world species will be 501.

16 genera: Starý et al.,1998; Bodlah, 2010; Bodlah et al., 2012; Ullah et al., 2015; Lopes et al., 2016; Yu et al., 2016), and Sri Lanka (3 species in 3 genera: Wickremasinghe, 2007; Lopes et al., 2016; Yu et al., 2016).

In view of the preponderance and diversity of aphids in the Himalaya and its foothills, most of the faunistic explorations for aphidiine parasitoids in India were also conducted in that region (Chakrabarti, 2021). Of the above mentioned 157 aphidiines species recorded from India so far, the highest number, 75 (16.82%) species are from Jammu and Kashmir: followed by Meghalaya 70 (15.70%). Other States and Territories represent aphidiine species as (alphabetically): Arunachal Pradesh 12 (2.69%), Assam 1 (0.22%), Bihar 1 (0.22%), Delhi 8 (1.79%), Himachal Pradesh 23 (5.16%), Karnataka 15 (3.36%), Kerala 1 (0.22%), Madhya Pradesh 1 (0.22%), Manipur 33 (7.40%), Mizoram 4 (0.90%), Nagaland 30 (6.73%), Punjab 6 (1.35%), Sikkim 57 (12.78%), Tamil Nadu 1 (0.22%), Tripura 4 (0.90%), Uttar Pradesh 19 (4.26%), Uttarakhand 36 (8.07%), West Bengal 49 (10.99%)(Figure 4.1). From Nepal 60 and Bhutan 91 odd aphid species respectively, are known (Chakrabarti, Personal Data bank). But so far only 14 aphidiines species in 7 genera from Nepal and none from Bhutan respectively, are recorded, which may be a poor estimation from these above two countries. The above results suggest that serious explorative taxonomic studies on aphidiines need to be conducted properly in those countries.

4.3 FOSSIL APHIDIINES

The aphid parasitoids co-evolved with the divergence of their respective host aphids (Starý, 1981). The fossil fauna of Aphidiinae consists of 14 genera and 26 species. The oldest fossil genus and species, *Archephedrus stolamissus* Ortega-Blanco, Bennett, Delclòs & Engel was described from the late Albian (Early Cretaceous) Álava amber of Spain (Ortega-Blanco et al., 2009). The majority of fossil aphidiines have been described from the early Rupelian (Oligocene) Anna pit in the Alsace potash field (Quilis, 1940; Starý 1973; Berger et al., 2005; Ortega-Blanco et al., 2009). Aphidiinae are also found in the late Eocene Baltic (Starý, 1970, 1973) and in the middle Eocene Sakhalinian amber (Rasnitsyn, 1980). The first aphidiine wasp species, *Ephedrus rasnitsyni* Davidian and Kaliuzhna from Sakhalinian amber (middle Eocene) has been described recently, which is the first fossil aphidiine reported from Asia (Davidian et al., 2021).

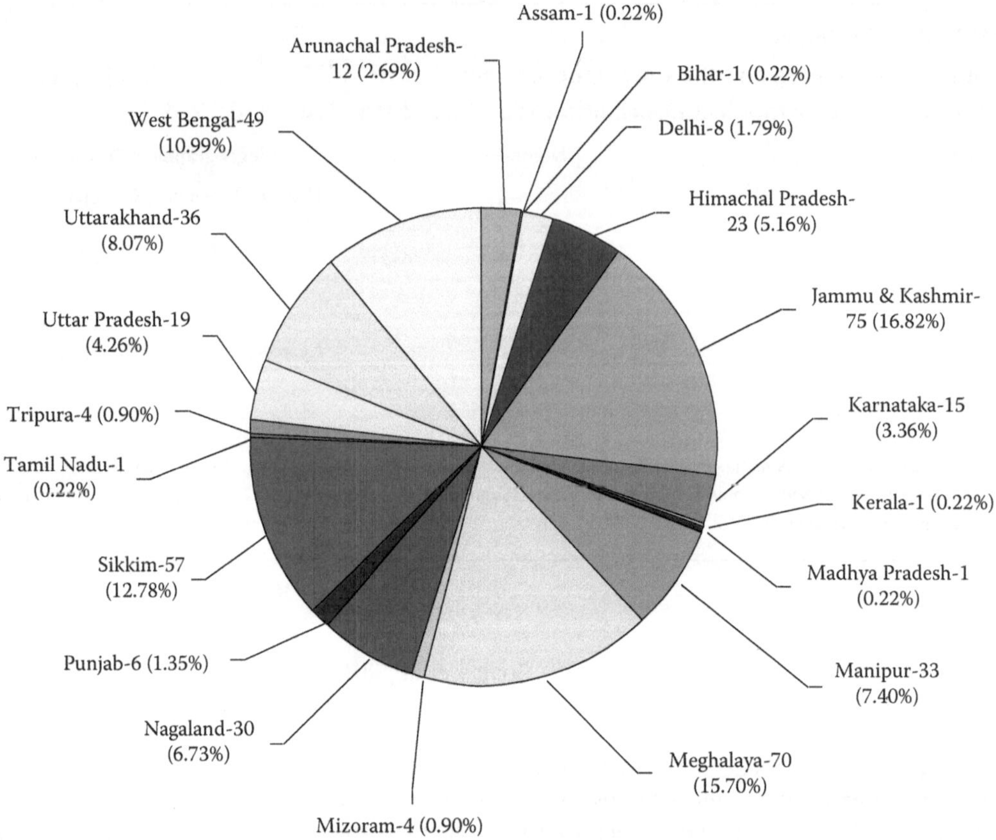

FIGURE 4.1 Aphidiinae species reported from different States and Union Territories of India (based on 157 species).

4.4 COLLECTION, REARING AND IDENTIFICATION

The most useful method for the collection of aphidiine parasitoids is the rearing of aphids and mummies (Figures. 4.2–4.4). Though symptoms of parasitization of aphids are not always apparent by naked eyes, yet it may be assumed that aphid colonies of moderate size may contain an appreciable number of parasitized aphids. Hence colonies of aphids with infested plant parts are to be collected and reared in spite of not having parasitization detectable by the naked eye. Observations can be made for up to 25 days to ascertain the emergence of parasitoids as well as hyperparasitoids within the containers. As soon as the adult parasitoids emerge, they should be immediately captured using an aspirator or camel hair brush and transferred into vials containing 70% ethyl alcohol. A camel hair brush soaked in alcohol may also be used to entangle adult parasitoids and put them in the vials. Sweeping is another common method that can be used for the collection of aphidiines. However, the names of the aphid and the plants infested by them are not revealed for the parasitoids when collected by sweeping.

Identification of the primary parasitoids of aphids is highly important for the successful biocontrol of aphid pests (Pons et al., 2011). Both dry mounts and slide mounts (different body parts) are required to study specific characters for the identification of the parasitoids. These are necessary for the conventional method of identification. Different types of microscopes and accessories are used to study, draw and capture photographs.

FIGURE 4.2 Aphidiine mummies in *Pterocomma populeum* (Kaltenbach) colony (Photo: Courtesy of Prof. Jan Havalka, Biology Centre CAS, Czech Republic).

Members of this subfamily can be distinguished structurally by their medium body size usually about 1.5–3.5 mm long; females with short ovipositor; metasomal tergum weakly sclerotised; flexible suture present between the second and third metasomal segments; smooth scutellar sulcus present; occipital carina present (at least laterally); antenna short with 10 to 28 segments; ovipositor sheath or hypopygium in females often modified; with a single, large median cell in the forewing in most species; cross vein cu-a absent in the hindwing (Starý, 1970; van Achterberg, 1997).

In the conventional method of identification, the following morphological characters and measurements are generally considered for the identification and construction of the keys: number of antennal segments; number of longitudinal placodes on first and second flagellomere; number of palpomeres of maxillary palpus; number of cells in forewing; length of forewing stigma; length of forewing R1 (metacarpus), 3RSa, 3RSb, RS, 2RS and r and RS veins; width of forewing stigma; existence of forewing M and m-cu, RS þM, and r-m veins; length of setae on fringe of forewing; existence of notaulices on mesonotum; setation of lateral lobes of mesonotum; sculpture of propodeum; shape, sculpture, setation, width and length of petiole; existence of secondary tubercles on petiole; distance between primary and secondary tubercles on petiole; width of spiracles on petiole, existence, setation, shape, and curvature of prongs; shape, direction of curvature and setation of ovipositor sheath; and color of flagellomere one, petiole, distal tergites and mummy (Kavallieratos et al., 2016).

Nowadays, many researchers apply the molecular method to identify species and determine the percentage of parasitism (Ye et al., 2017a,b; Tomanović et al., 2018; Zhu et al., 2019; Hwang and Lee, 2021). But, the morphotaxonomy is a more holistic approach than molecular taxonomy, as

FIGURE 4.3 Aphidiine mummy (*Praon* sp.) of *Drepanosiphum platanoides* (Schr.) (Photo: Courtesy of Prof. Jan Havalka, Biology Centre CAS, Czech Republic).

this is based on only a few markers. In this connection, integrative taxonomy may be a reliable tool because it goes beyond the naming of species and gives priority to species delineation and the processes underlying it (Schlick-Steiner et al., 2010).

4.5 BIOLOGY

Life Cycle: The aphidiine life cycle can be simplified into a few key stages (Figure. 4.5): (1) The adult female inserts an egg into the body of the aphid. (2) The egg develops into a larva. (3) The larva consumes the aphid from within, eventually killing the aphid. (4) The dead aphid forms a puparia (mummy) for the transformation of the larvae into an adult. (5) An adult parasitoid emerges from the rear end of the mummy. (6) The newly emerged adult female parasitoid either: (a) mates with a male and infects aphids, or (b) infects aphids without mating. The life cycle of aphidiine is haplodiploidy, meaning that males are haploid (1n), which is the result of offspring from unfertilised eggs and females are diploid (2n), which is the result of offspring from fertilised eggs (Purkiss, 2018).

Immature Stages: After eclosion from the egg, the larva starts to feed on the aphid's haemolymph. Later, it consumes other tissues, which leads to the death of the host (van Emden and Harrington, 2007). Once a parasitoid reaches second larval instar, the reproduction of the aphid ceases due to the consumption of its reproductive organs by the developing larva (Miller, 2018).

There is a controversy regarding the number of larval instars in aphidiines (Quicke, 2015). O'Donnel (1987) on the basis of quantitative studies on 22 species belonging to 10 different genera

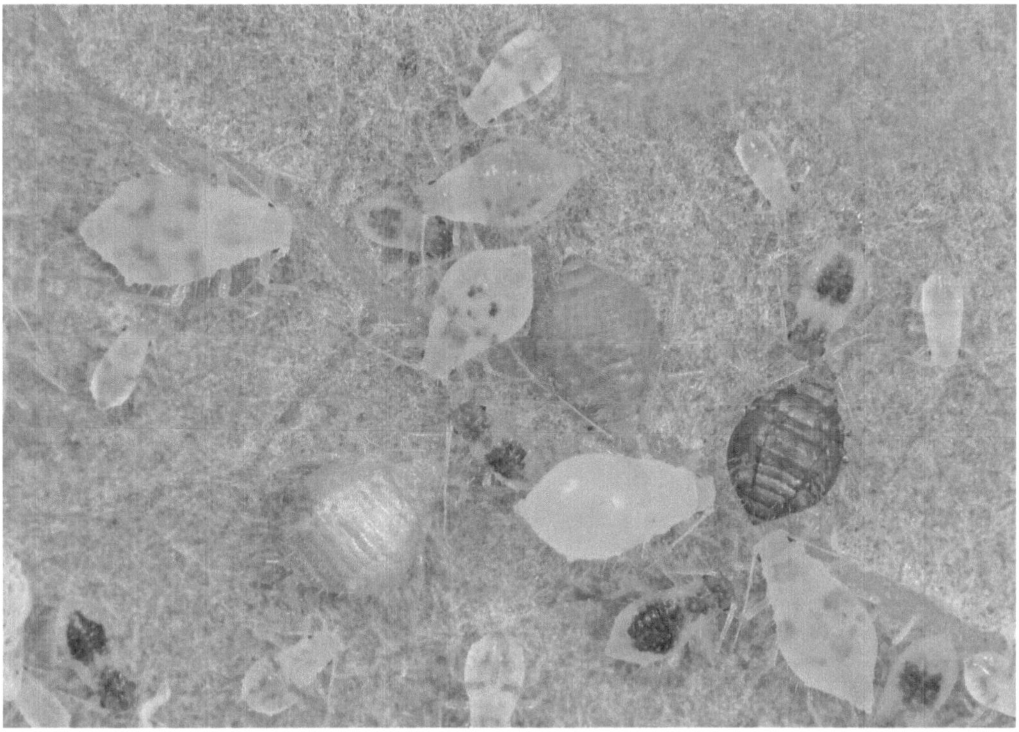

FIGURE 4.4 Aphidiine mummies in *Chaitophorus populialbae* (BdF) colony (Photo: Courtesy of Prof. Jan Havalka, Biology Centre CAS, Czech Republic).

of Aphidiinae and also many other workers (Schlinger and Hall, 1960; Starý, 1966; O'Donnell, 1987, 1989; Pennacchio and Digilio, 1990; Muratori et al., 2004) proposed three larval instars while others (Starý, 1970; Chow and Sullivan, 1984; Das and Chakrabarti, 1988; Hughes, 2010; Jamhour, 2017; Zhang et al., 2018) suggested four larval instars in aphidiine parasitoids. On the contrary, some authors (Vevai, 1942; Starý, 1962; van der Hoek, 1971) suggested five larval instars for these parasitoids. In fact, more works on the larval development of aphidiines are needed to redress the questions, especially the number of larval instars, morphological characters of each instar, duration of each instar, etc. In this connection, *in vitro* observation may be useful, since this method may unequivocally determine the number of instars through the counting of exuviae (Pennacchio and Digilio, 1990).

The larva consumes all the contents of the aphid and finally kills it and thereafter starts to spin the cocoon. However, prior to this, the mature larva makes a small hole on the ventral side of the aphid skin for fixation of the aphid body on the surface by the secretion from its silk gland. Pupation takes place within the aphid body. The adult parasitoid emerges from the pupa and quits the mummy through a circular hole, which it cuts in the abdomen. The development of aphidiine species is greatly affected by the physical factors of the environment.

Longevity: The longevity of adult aphidiine depends upon the physical factors of the environment and the quantity and quality of food (Starý, 1970). Within the range of 10–30 °C, aphidiine longevity follows a hyperbolic curve (Martinou, 2005). This group of parasitoids lives longest at temperatures range 10–15 °C. Their life span is shortened with the increase of temperature. Generally, they live 6 to 25 days at a temperature range of 20–25 °C. However, some species may live for more than 50 days at low temperatures. Aphidiine longevity is significantly varied in terms of sex, mating condition and ovipositional status.

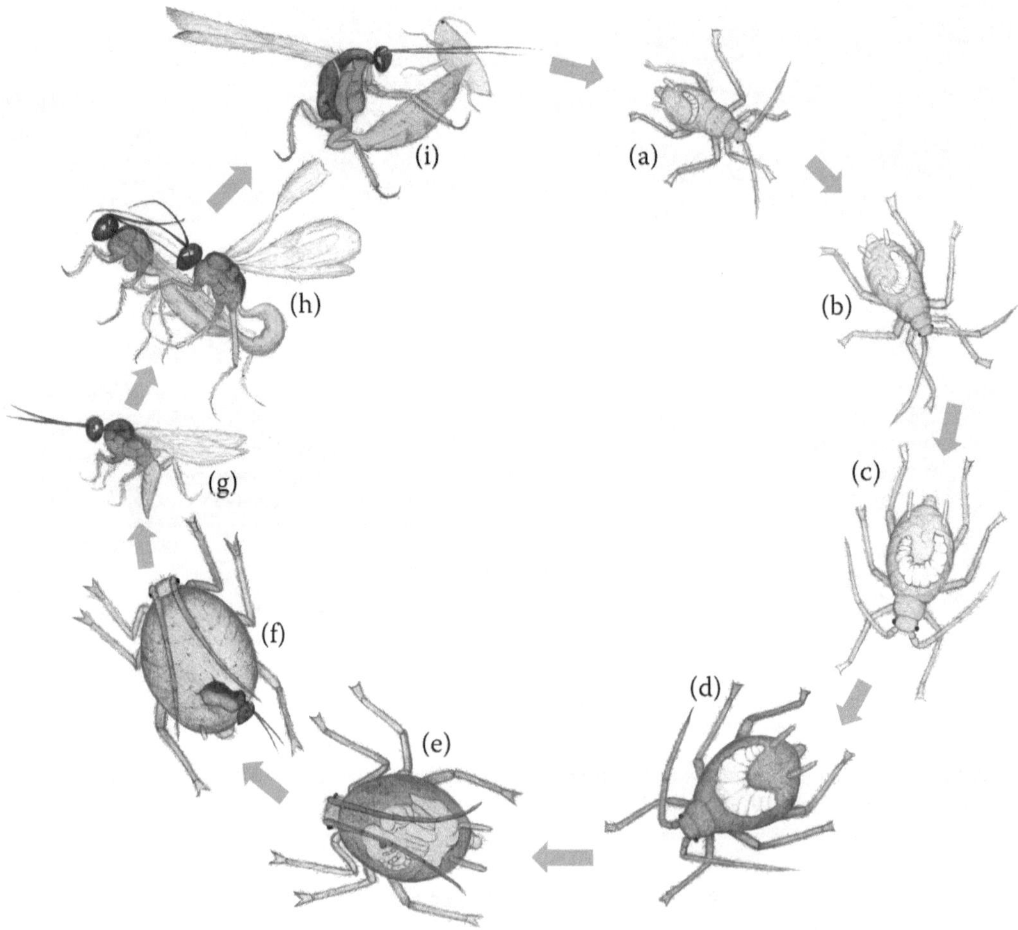

FIGURE 4.5 a-i: Aphidiine Life Cycle. a. First instar larva in aphid body b. Second instar larva in aphid body c. Third instar larva in aphid body d. Fourth instar larva in aphid body e. Pupa in aphid body f. An aphidiine emerging from the aphid body g. Free-living aphidiine h. Mating posture of aphidiines i. Oviposition posture of aphidiine (Painting: Sanjay Kumar Pramanik).

Mating: The aphidiines include mostly biparental species and for this reason mating is normal and a necessary phenomenon of their life cycle (Starý, 1970). The interval between the emergence and mating is different in different species. It may occur immediately after emergence in many species (Schlinger and Hall, 1960; Hafez, 1961; Wiackowski, 1962). Some female parasitoids are not receptive to males immediately after emergence and mating may take place 2 hours after their emergence (Vevai, 1942). Aphidiine females are monoandrous and males are polygynous (Kant et al., 2012). However, report on repeated mating in female is also available (He and Wang, 2008). Partial sib-mating and also local mate competition on the natal patch and off-patch matings among non-sibs are common features of aphidiine biology (Mackauer and Völkl, 2002).

Oviposition: The first relationship between adult parasitoid and host is established through oviposition with the deposition of egg into the host body (Figure 4.6). During oviposition, the parasitoid shows conspicuous differences in their behaviour (Starý, 1970). Females of most aphidiine species oviposit in a broad range of hosts differing in size, including aphid embryos (Mackauer and Kambhampati, 1988). However, females generally exhibit a preference for particular larval instars and morphs and this preference appears to be correlated with the host's nutritional quality (Völkl et al., 2007).

FIGURE 4.6 Oviposition posture of an aphidiine female (Photo: Ed Phillips, UK).

The oviposition behaviour of aphidiine wasps has been classified into six groups (Matin et al., 2005). The first type is unspecialised oviposition behaviour with a long oviposition time of 5 to 10 seconds (e.g. Ephedrini). The second and third types have various morphological and behavioural adaptations, like immobilisation of the potential host for oviposition (e.g. Trioxini). The fourth type is characterised by females with a highly flexible abdomen and very long oviposition time (e.g. *Pauesia*). In the fifth type, females usually avoid direct contact with a potential host by administering a quick sting that ensures a short oviposition time (less than 5 sec), e.g. *Aphidius* species. In the sixth type, aphidiine species (such as *Adialytus* and *Lysiphlebus)* have evolved chemical strategies to counter the guarding behaviour of trophobiotic ants (Jamhour, 2017).

The larvae of Aphidiinae are obligatory parasitoids and they entirely depend on the host for their development, feeding exclusively on its tissues until they reach the stage of their maturity (Jamhour, 2017). The reproductive success of parasitoids is strongly dependent on the fitness of females and their ability to select a suitable aphid host for the development of their progeny (Godfray, 1994). The process of host location and selection is complex and can be classified into five general steps, which sometimes may be overlapping: (1) host habitat location, (2) host location, (3) host acceptance, (4) host suitability and (5) host regulation (Hågvar and Hofsvang, 1991; Rehman and Powell, 2010).

Natural or artificial diets can change the oviposition behavior of parasitoids, modifying either the propensity to oviposit or the number of eggs laid. The supply of an optimal sugar-rich diet to parasitoid females increases fecundity (Benelli et al., 2017).

Fecundity: The number of eggs in the ovaries of aphidiine at emergence is lower than the realised fecundity, indicating maturation of eggs during their life (synovigenic) (Starý, 1962; Shirota et al., 1983). In fact, in aphidiine eggs mature throughout their reproductive life (Völkl et al., 2007). Eggs are relatively small and nutrient-poor (Le Ralec, 1991). A female may have between 200 and 400 mature eggs in her ovaries. However, eggs not laid cannot be resorbed (Völkl and Mackauer, 1990).

Lifetime fecundity is high, ranging from 300 to 1800 eggs per female under optimal conditions of the adequate number of hosts (Hågvar and Hofsvang, 1991), but fecundity is lower in fields where adults do not survive for more than 2–3 days (Mackauer, 1983). A significant positive correlation was found between parasitoid size and mature egg load after emergence (Collins and Dixon, 1986).

Sex Ratio: The sex ratio represents one of the important characteristics for understanding the effectiveness of the aphidiines under various conditions (Starý, 1970). An equilibrium ratio (1:1) of offspring was postulated by Fisher (1930). Hagen and van den Bosch (1968) reported and Hartl and Brown (1970) predicted 1: 1 sex ratio for the members of Aphidiinae. But this ratio has been reported to be variable and often strongly female-biased (Hamilton, 1967). The spermatheca plays a key role in sex-determining mechanism owing to sperm storage (Starý, 1970). The sex of the egg is determined immediately before the oviposition and the discharge of spermatozoa into the oviduct is regulated by external stimuli. Accordingly, the sex ratio is variable as a result of inconsistent physical factors of the environment. Sperms stored in spermatheca need activation to migrate down the sperm duct to fertilise the eggs. The source of the activating agent appears to be the spermathecal gland that responds to external stimuli (Ahmed, 1990). Factors affecting the sex ratio are many, e.g. successive matings of the male (Vevai, 1942), age of the female (Sekhar, 1959), seasons (Hafez, 1961), times of mating of female (Wiackowski, 1962) and temperature extremes (Force and Messenger, 1964).

4.6 FORAGING

The foraging behaviour of aphidiine parasitoids includes host habitat selection, host location and host acceptance (Vinson, 1976). They use a variety of host-plants or host-borne cues to locate an aphid colony (Mackauer et al., 1996). The relative importance of these cues varies considerably depending on the spatial scale (Völkl, 2000). When foraging between habitats or host plants (i.e. situated at relatively long distances so that hosts cannot be found by walking), females respond to olfactory cues. Volatile host-plant compounds (Volatile organic compounds) provide the parasitoid first information about the potential hosts (aphids) (Micha et al., 2000). Parasitoids show a low patch-leaving tendency when the host aggregation level is high, which influences the foraging behaviour of parasitoids (Louâpre et al., 2018).

To understand the foraging behaviour of parasitoids, it is necessary to study: (1) the ways species use the information to localise and assess resources, (2) the mechanism used in searching and (3) the factors that act as sources of variability (Bell, 1991). Field investigations are important to understand better the scale of a patch for a foraging male of an aphid parasitoid (Delphine, 2012).

The foraging behaviour of aphidiine parasitoids has a major effect on the population dynamics of both the parasitoids and aphid hosts. The patchiness of aphid host populations and behaviour of the parasitoids influence the way the patches are exploited (Wellings, 1993). It has been reported that virgin females abandoned bean shoots infested with pea aphid earlier than their mated counterparts. Virgin females attack fewer aphids within a patch than mated females, and parasitise fewer hosts in a 2-h period. Females, therefore, tend to produce a smaller brood within a patch when unable to fertilise their eggs and produce daughters (Michaud, 1994).

Foraging behaviour is no longer fixed and predictable, but rather it varies in response to the insect's physiological condition and genetic composition in relation to environmental factors. The quest for factors inducing variability in parasitoid foraging behaviour has largely centered on the influence of learning (Vet et al., 2003).

4.7 HYPERPARASITOIDS

An insect parasite of another parasitoid is called a hyperparasitoid (Coppel and Mertins, 1977). Hyperparasitism is the parasitism of a parasite (Askew, 1971). This phenomenon is highly evolved in insects, especially so among the hymenopterans associated with aphids (Sullivan, 1986). Aphid hyperparasitoid communities are divided into ecto- and endohyperparasitoids, with ectohyperparasitoids being less host-specific than endohyperparasitoids. The reproductive capacity and intrinsic rate of increase of hyperparasitoids are generally lower than those of their primary hosts. Aphid ectohyperparasitoids do not use specific cues for host searching. They search randomly whereas endohyperparasitoids gain information that originates from host plants or aphid hosts for long-range search. Aphid-attending ants typically prevent successful hyperparasitoid foraging (Sullivan, 1999).

The action of hyperparasitoid is generally considered to be harmful as they reduce the number of primary parasitoids considerably and cause a decrease in parasitism (Starý, 1970). Hyperparasitoids have negative effects on the populations of aphidiines, ~~and~~ decrease the density of aphidiines and the parasitism rates of the next generation, which leads to weak control of aphidiine on aphids and a rapid increase of aphid density. Higher hyperparasitism affects the biocontrol efficacy of aphidiines for aphids (Guo'An et al., 2001).

Some opposite views regarding the impact of hyperparasitoids on the primary parasitoids also exist (Hofsvang et al., 2014). In fact, the effect of hyperparasitoids on biocontrol of aphids is much debated (Brodeur and Rosenheim, 2000; Schooler et al., 2011). Many field studies have shown that primary parasitoids are able to regulate aphid population in spite of high levels of native hyperparasitoids (Farrell and Stufkens, 1990; Hougardy and Mills, 2009). However, hyperparasitoid ecology is of basic importance for the success of biocontrol programs (Sullivan, 1999).

The aphid–primary parasitoid–hyperparasitoid food-web has been used as a model system in the community ecology, not only because of the economic importance of aphids as worldwide pests on a variety of agricultural crops and forest trees, but also because of the relatively easier methods of rearing aphids, their primary parasitoids, and hyperparasitoids in the laboratory and/or greenhouse for precise behavioral and ecological studies (Sullivan, 2003). In multitrophic interactions, aphidiine hyperparasitoids stand at the fourth trophic level (Figure 4.7). Major studies on hyperparasitism are known on the primary parasitoids that attack the family Aphididae (Das and Chakrabarti, 2018); because of the importance of aphids that harbor a species-rich and taxonomically diverse hyperparasitoids (Zamora-Mejías and Hanson, 2016).

True aphid hyperparasitoids (endohyperparasitoids) are mainly found in Charipinae (Figitidae). Altogether 143 species under 3 genera (*Alloxysta* Förster having 111 species, *Phaenoglyphis* Förster having 31 species and *Lytoxysta* Kieffer having 1 species) of this subfamily are reported as endoparasitoids of Aphidiinae and Aphelinidae (Ferrer-Suay et al., 2012). Many members from

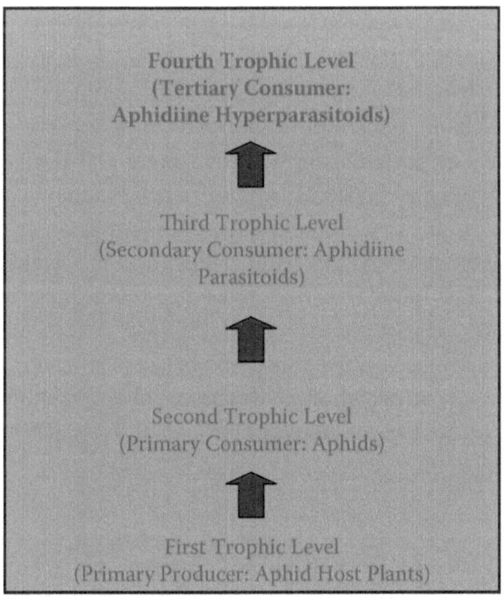

FIGURE 4.7 Tetra-trophic/multi-trophic interactions among plants, aphids, aphidiine parasitoids and hyperparasitoids.

four other families, *viz*. Pteromalidae (*Asaphes* Walker, *Corona* Walker *Pachyneuron* Walker), Encyrtidae (*Syrphophagus* Ashmead), Eulophidae (*Aprostocetus* Westwood) and Megaspilidae (*Dendrocerus* Ratzeburg) are also known as hyperparasitoids of aphids (Sullivan, 1988; 1999). But all species under these genera (*Corona, Pachyneuron, Syrphophagus, Dendrocerus*) are not exclusively hyperparasitoids of aphids. Only species of the genus *Asaphes* are considered to be exclusively hyperparasitoids of aphids (Gibson and Vikberg, 1998). In fact, an authentic world list of true hyperparasitoids of aphidiines is not available to date.

4.8 DIAPAUSE

Diapause is an adaptive strategy of insects to overcome adverse environmental conditions and ensure the continuity of the population (Kostál, 2006). Insects may enter summer diapause (aestivation) to escape from the hot and dry weather or winter diapause (hibernation) to survive during the cold weather and food shortage (Tauber et al., 1986). In aphidiine parasitoids, there is a large plasticity in the occurrence of diapause, only a part of the population enters diapause while the other part remains active or undergoes quiescence during winter (Colinet et al., 2010). They represent a typical group where both a genetically determined diapause and a variable response to environmental cues occur (Starý, 1970).

Aphidiine parasitoids enter diapause as prepupae, which allows measurement of potential trade-offs with adults and studies the costs and benefits of the diapause strategy (Tougeron, 2017). There are reports that in populations of mild winter areas (e.g. Western Europe), the incidence of diapauses has been decreasing for a few years and parasitoids of different species mostly remain active throughout the year (Andrade et al., 2016; Tougeron et al., 2017). This might be the impact of climate change on the life cycle of aphidiines. Parasitoids that do not enter diapause should be cold-resistant enough to withstand the low temperatures during their immature and adult stages (Tougeron et al., 2018).

Aestival diapause and adaptive diapause in mummies in *Trioxys pallidus* (Haliday) during summer and autumn respectively, were reported in Iran (Rakhshani and Starý, 2021) . In cold natural conditions, most Aphidiinae arrest their development and enter hibernal quiescence (dormancy) inside the cocoon in a mummified aphid (Starý, 1970; Jarry and Tremblay, 1989). Aestivation has been reported for only a few species in the Aphidiinae (Polgár and Hardie, 2000).

Aphidius nigripes Ashmead has the capacity to enter winter diapause, and the emergence of this parasitic wasp in spring coincided with the hatching dates of the eggs of its host aphid, *Macrosiphum euphorbiae* (Thomas) (Brodeur and McNeil, 1994). A perfect synchronisation was also reported in spring between the emergence of diapausing *Aphidius ervi* Haliday and its host, *Acyrthosiphon pisum* (Harris) in France (Bournoville, 1973). Non-diapausing parasitoids, associated with anholocyclic aphids, need available hosts during winter. On the other hand, parasitoids associated with holocyclic aphids should enter diapause because of non-availability of hosts during winter (Langer and Hance, 2000). Langer and Hance (2000) found that in the two parasitoid species, *Aphidius ervi* Haliday and *Aphidius rhopalosiphi* (Deste.-Perez.) the diapausing mummies exhibited greater cold hardiness than non-diapausing mummies, resulting in significantly lower supercooling points (SCP) and in a higher survival rate during longterm exposures at 0 and 21°C. The induction of increased cold hardiness in parasitoids was thus associated with the diapause state.

All aspects of diapause have applied interest in parasitic insects since this could be useful for mass-rearing protocols through long-term cold storage. However, information on the practical applications of diapause for cold storage has been reported only in a few parasitoids (Colinet and Boivin, 2011).

4.9 HOST PARASITOID RELATIONSHIP

The host-parasitoid relationship in aphid-parasitoid system is particularly interesting (Ortiz-Martínez et al., 2019). Some aphidiine species (e.g. *Aphidius ervi* Haliday) are considered habitat/

host generalist species, which attack several aphid species on different plant families. In contrast, some other species (e.g. *Aphidius uzbekistanicus* Luzhetzki) exhibit a narrow host range and are considered to be habitat specialists foraging on specific aphids (Starý, 1993; Žikić et al., 2017). In terms of competitive abilities, some parasitoids are known to win the competition against other aphidiine parasitoids, as they often lay two or more eggs in the same aphid in contrast to others (Žikić et al., 2017).

Some parasitoids (e.g. *A. rhopalosiphi* (Deste.-Perez.) are known to be affected by the defensive behaviour of aphids (e.g. *Sitobion avenae*) and as soon as aphids emit defensive signals, the parasitoids leave the host patches partially exploited (Ortiz-Martínez, 2019). Conversely, some group of other parasitoids (e.g. *A. ervi*), which are not affected by the defensive behaviour of aphids (e.g. *S. avenae*), tends to fully exploit host patches (van Baaren et al., 1994; Žikić et al., 2017). Parasitoids are capable of discriminating unparasitized aphids from parasitized one. They more often accept those that are unparasitized (van Baaren et al., 2009). Behavioural experiments have shown that the recognition of unparasitized and parasitized hosts occurs before oviposition through external cues (Le Lann et al., 2011).

Body sizes of host aphids and parasitoids in individual feeding relationships are also interesting (Cohen et al., 2005). The relationship between emerging parasitoid body size and final aphid host body size depends on the type of parasitoid (primary, hyper-, or mummy parasitoid) as well as on the specific parasitoid species (Cohen et al., 2005). The relation between consumer and resource body sizes generates a misleading picture of the real feeding interactions between an individual host and an individual parasitoid. The body lengths of parasitoid are related to the body lengths of aphid hosts by a power law with an exponent close to ¾ (Cohen et al., 2005).

The percentage of parasitism is directly related to the abundance of host aphids (Thies et al., 2005). Thus, parasitoids cannot be found or their population is generally very poor if the aphid population on certain host plants is low, both in the cultivated and non-cultivated habitats (Jaureguiberry et al., 2010). A strong relationship was obtained between early parasitism with the maximum subsequent aphid population (Fougeroux et al., 1988). In the field condition, if 20% parasitism occurs early in the season, the aphid population never exceeds the economic threshold (Glinwood, 1998).

Aphidiine parasitoids are synchronised with their host aphids (Cameron et al., 1981). Host size/stage preference affects host–parasitoid population dynamics. When parasitoids attack only a narrow window of the developmental stages of a host, host–parasitoid population dynamics may be stabilised (Murdoch et al., 1987). Generally, the highest degree of parasitism coincides with the peak population density of their host or shortly afterwards. When prone to mate-finding difficulties, parasitoids with cyclic dynamics or low searching efficiency go perished whereas the parasitoids with high searching efficiency may either persist or perish, depending on host intraspecific competition (Bompard et al., 2013).

Interestingly, a few host plants in the non-cultivated habitat maintain parasitized aphids that do not cause damage to these plants (Blackman and Eastop, 2000). In the crop season, parasitoids emerge from the nearby plants and these are highly suitable for biocontrol of aphids infesting the crop plants (Jaureguiberry et al., 2010).

4.10 POPULATION DYNAMICS

Population dynamics of aphidiine parasitoids have received considerable attention by researchers (Völkl, 1994; Mackauer et al., 1996; Jarosik and Lapchin, 2001). In fact, parasitoids are ideal subjects for developing relatively simple population models. Most of the work on the dynamics of host-parasitoid interactions has taken a mechanistic approach (Hassell, 2000). Parasitoid population dynamics can be uncoupled from the population dynamics of the pest species, leading to high densities of the control agent, even at low pest densities (Hawkins, 1993). Specific knowledge of the population dynamics and feeding range of each organism is necessary for understanding

FIGURE 4.8 Intraguild Predation (Photo: Mark Yokoyama, USA).

multi-guild interactions (Fig. 4.8). This knowledge is likely to provide a powerful tool for biocontrol (Ahmed, 1990).

The physical factors of the environment, *viz.* temperature, relative humidity, light etc., have profound effects on the abundance of aphidiine parasitoids. The influence of environmental conditions on the populations of host and parasitoid in various natural communities is complex in nature and the influence of individual factors cannot be separated in an isolated way (Starý, 1970).

Various methods for the study of aphidiine population are followed. These are: 'mummy counting' (Pimentel, 1961), aphid dissection (Starý, 1968; Cameron et al., 1981; Wilson et al., 1982), simultaneous sweeping, dissection and mummy counting (Cameron *et al.,* 1981; Wilson et al., 1982), leaf and stem counting (Starý, 1978), electrophoresis or isoenzyme analysis or soft X-rays (wavelength > 0.25°A) and fine-grain film (Wool et al., 1978; Unruh et al., 1983) and molecular techniques (Gariepy and Messing, 2012; Zhu et al., 2019).

4.11 EFFECTIVE APHIDIINE BIOCONTROL AGENTS

Aphid parasitoids are commonly used in biocontrol programs to control aphid pests in greenhouses and field situations (Boivin et al., 2012). In fact, the biocontrol of aphids has a long history. In 1734, eggs of 'Aphidovorous Fly' were recommended by De Reaumur to control greenhouse aphids (Simmonds et al., 1976). Carver (1989) made a summary of all such attempts in classical biocontrol throughout the world, where aphids were also included. Joshi et al. (2010) prepared a chronology of the history of classical biocontrol of aphids. They actually covered the years from 1920 to 1986 for different groups of natural enemies of aphids and mentioned that the parasitoids, *viz. Trioxys complanatus* Quilis and *Praon exsoletum* (Nees) were utilised first during 1953 to control *Therioaphis trifolii maculata* (Buckton) in Western Australia (Table 4.2). Here, they included altogether 19 aphidiine species [*Aphidius eadyi* Starý, *Aphidius ervi* Haliday, *Aphidius*

TABLE 4.2
Success and Failure History of Classical Biological Control of Aphids with Aphidiine Parasitoids (Schlinger *et al.*, 1960; Simmonds *et al.*, 1976; Persad *et al.*, 2007; Joshi *et al.*, 2010; Brodeur, 2013; Cross and Poswal, 2013)

Year of Introduction	Aphid Species	Parasitoid Species	Introduced from	Introduced into	Results
1953	*Therioaphis trifolii maculata* (Buckton)	*Trioxys complanatus* Quilis *Praon exsoletum* (Nees)	Middle East	Western Australia	Quickly established but *T. complanatus* played major role
1958	*Acyrthosiphon pisum* (Harris)	*Aphidius smithi* Sharma & Subba Rao	India	California	Readily established
1959	*Chromaphis juglandicola* (Kaltenbach)	*Trioxys pallidus* (Haliday)	France	California	Established temporarily
1962	*Cavariella aegopodii* (Scopoli)	*Aphidius salicis* Haliday	California	Australia	Controlled aphid and carrot motley dwarf virus disease
1963	*Acyrthosiphon pisum* (Harris)	*Aphidius ervi* Haliday	Europe	North America	Very successful but *A. ervi* displaced earlier existing parasitoid, *A. smithi*
1968, '69	*Chromaphis juglandicola* (Kaltenbach)	*Trioxys pallidus* (Haliday)	Iran	California	Established
1970s	*Tinocallis platani* (Kaltenbach)	*Trioxys curvicaudus* Mackauer	Europe	California	Established
1973-74	*Aphis* (*Toxoptera*) *aurantii* BdF	*Lysiphlebus testaceipes* (Cresson)	Cuba	France	Established
1976	*Acyrthosiphon pisum* (Harris)	*Aphidius smithi* Sharma & Subba Rao *Aphidius eadyi* Starý Gonzalez & Hall	India	New Zealand	*A. smithi* could not establish but *A. eadyi* is common now
1976	*Metopolophium dirhodum* (Walker) *Sitobion avenae* (Fabricius)	*Aphidius ervi* Haliday *Aphidius rhopalosiphi* de Stefani-Perez *Aphidius uzbekistanicus* Luzhetzki	Mediterranean France	Southern Chile	Considered as potential regulator
1977	*Therioaphis trifolii maculata* (Buckton)	*Trioxys complanatus* Quilis *Praon exsoletum* (Nees)	North America	Western Australia	Quickly established but *T. complanatus* played major role
1977	*T. trifolii maculata* (Buckton)	*Trioxys complanatus* Quilis	Israel	Western Australia	Established

(*Continued*)

TABLE 4.2 (Continued)
Success and Failure History of Classical Biological Control of Aphids with Aphidiine Parasitoids (Schlinger *et al.*, 1960; Simmonds *et al.*, 1976; Persad *et al.*, 2007; Joshi *et al.*, 2010; Brodeur, 2013; Cross and Poswal, 2013)

Year of Introduction	Aphid Species	Parasitoid Species	Introduced from	Introduced into	Results
1977-'79	*Acyrthosiphon (Acyrthosiphon) kondoi* Shinji	*Aphidius ervi* Haliday	Japan via USA and New Zealand, Italy, France and Greece	New South Wales	Established from Italy, France and Greece but not from Japan
1977-'79	*A. (A.) kondoi* Shinji	*Ephedrus plagiator* (Nees)	Japan via USA and New Zealand,	New South Wales	Not established
1978	*Acyrthosiphon pisum* (Harris)	*Aphidius smithi* Sharma & Subba Rao	India	Hawai, South America Europe	Results not known
1979	*A. pisum* (Harris)	*Aphidius smithi* Sharma & Subba Rao *Aphidius eadyi* Starý Gonzalez & Hall *Aphidius pisivorus* Smith	Europe	Australia	Not established
1981	*Hyperomyzus lactucae* (Linnaeus)	*Aphidius sonchi* Marshall *Praon volucre* (Haliday)	Mediteranean Areas	Australia	Established
1982	*Therioaphis trifolii maculata* (Buckton)	*Trioxys complanatus* Quilis	Australia	New Zealand	Spotted aphid is no longer a pest
1983	*Cinara cronartii* Tissot & Pepper	*Pauesia* spp.	USA	South Africa	Established
1984	*Aphis craccivora* Koch	*Lysiphlebus fabarum* (Marshall)	Mediteraneun areas	Australia	Not recovered
1984	*A. craccivora* Koch	*Lysiphlebus testaceipes* (Cresson)	California	Australia	Established, also taking care of other species of aphid *T. aurantii*
1985	*Metopolophium dirhodum* (Walker)	*Aphidius rhopalosiphi* de Stefani-Perez	England and France	New Zealand	Established
1986	*Aphis craccivora* Koch	*Binodoxys indicus* (Subba Rao & Sharma)	India	Australia	Results not known
1986	*Acyrthosiphon kondoi* Shinji	*Ephedrus plagiator* (Nees) *Praon barbatum* Mackauer	Europe	Australia, New Zealand	Not successful

1986	*A. kondoi* Shinji	*Aphidius ervi* Haliday	Europe	North America, Australia, New Zealand	Now widespread
1987-'88	*Metopolophium dirhodum* (Walker)	*Aphidius rhopalosiphi* de Stefani-Perez	New Zealand	Australia	Established
1990	*Macrosiphum rosae* (Linnaeus)	*Aphidius rosae* Haliday	Italy	South Australia	Established
1995	*Cinara cupressivora* Watson & Voegtlin	*Pauesia juniperorum* (Starý)	UK and France	Malawi and Kenya	Established
1997	*Pterochloroides persicae* (Cholodkovsky)	*Pauesia antennata* (Mukerji)	Pakistan	Yemen	Aphid populations collapsed within two months and it has been established.
1999	*Aphis (Toxoptera) citricidus* (BdF)	*Lipolexis oregmae* (Gahan)	Guam	Florida	Established
2007	*Aphis glycines* Matsumura	*Binodoxys communis* (Gahan)	Asia	USA	Not established

pisivorus Smith, *Aphidius rhopalosiphi* de Stefani-Perez, *Aphidius salicis* Haliday, *Aphidius smithi* Sharma & Subba Rao, *Aphidius salicis* Haliday, *Aphidius uzbekistanicus* Luzhetzki (=*Aphidius rhopalosiphi* de Stefani-Perez), *Ephedrus plagiator* (Nees), *Lysiphlebus fabarum* (Marshall), *Lysiphlebus testaceipes (*Cresson), *Pauesia* sp., *Praon barbatum* Mackauer, *Praon exsoletum* (Nees), *Praon volucre* (Haliday), *Trioxys curvicaudus* Mackauer, *Trioxys complanatus* Quilis, *Binodoxys indicus* (Subba Rao & Sharma) and *Trioxys pallidus* (Haliday)], which were used in classical biocontrol of aphids in different countries from 1953 to 1986 (Table 4.2). Several workers (Persad et al., 2007; Brodeur, 2013; Cross and Poswal, 2013) also mentioned some other species that were used till 2007 in the classical biocontrol of aphids. Quoting Hughes (1989), Joshi et al. (2010) mentioned five aphidiine species, [*viz. Aphidius ervi* Haliday, *Trioxys complanatus* Quilis, *Trioxys tenuicaudus* Starý, *Trioxys curvicaudus* Mackauer and *Trioxys pallidus* (Haliday)], which were substantially successful as the biocontrol agents of aphids in the open fields.

Although there are many parasitoid species that attack aphid species in the open field, only a few of them have shown effectiveness to control aphids in greenhouse environments. However, historically, augmentative biocontrol has been more successful in greenhouse crops than in outdoor crops. This is due to the unique characteristics of greenhouse environments and the experiences of growers that manage these crops. In some modern greenhouses, there is a precisely controlled climate where temperature, ventilation, light intensity, light cycles, and carbon dioxide levels are computer-controlled and monitored (Hale and Elliott, 2002).

A chronology of the success and failure history of classical biocontrol of aphids with aphidiine parasitoids is provided in Table 4.2. Information about the successful utilisation of four aphidiine parasitoids in the greenhouses is also provided in Table 4.3. The lists provided in these two tables are not exhaustive. Actually, these are the examples cited in many works.

4.11.1 Success Story

Many aphidiine species have been successfully used in biocontrol programmes against aphid pests in the field and glasshouses around the world (Carver, 1989; van Steenis and El-Khawass, 1995; Yano 2006). Several reviews on case histories indicated some common attributes of natural enemies leading to their successful use in biocontrol programme (Joshi et al., 2010). Success stories of biocontrol of six aphid species including one in greenhouse, using aphidiine parasitoids from five continents, are provided briefly in the following sections.

4.11.1.1 Biocontrol of Walnut Aphid, *Chromaphis juglandicola* (Kaltenbach) in USA

Chromaphis juglandicola was a serious pest of walnut in California, USA since the early part of the 20th century. This small aphid is an autoecious holocyclic species and monophagous on the Persian walnut, *Juglans regia* L. and its cultivated forms and hybrids (Hougardy and Mills, 2009). For decades, during its annual outbreaks spreading through thousands of acres of land, plants were regularly treated with chemical insecticides and no attention at all was paid to its biocontrol. Initially, nicotine sulfate was the most widely used material to control this aphid. However, since the mid-1940s it has been supplemented by a variety of synthetic organochlorines and organophosphates. But aphids develop resistance against these chemical insecticides, thereby causing resurgences of pests. Besides, such chemicals were causing environmental pollution due to the leakage of toxic materials outside the treated area. Subsequent to the striking success of classical biocontrol against the spotted alfalfa aphid in California in the middle of 1950s (van den Bosch et al., 1964), an extensive search was conducted to find out a suitable biocontrol agent for controlling this aphid. A parasitoid, *Trioxys pallidus* Haliday was found in Southern France in 1959 that showed a high degree of host specificity. It was successfully imported to California and established as an effective agent against this aphid in the coastal plain of southern California. The parasitoid was found spreading rapidly and destroying a high percentage of the aphid population.

TABLE 4.3
Successful utilization of aphidiine parasitoids in the greenhouse

Parasitoid Species	Target	Crop	Country	Effect	Reference
Aphidius matricariae Haliday	*Myzus persicae* (Sulzer)	Sweet pepper	France	Significant impact on aphids	Rabasse *et al.*, 1983
A. matricariae Haliday	*M. persicae* (Sulzer)	Sweet pepper	Former USSR	Good Control	Shijko, 1989
A. matricariae Haliday	*Aphis gossypii* Glover	Cucumber	UK	Good Control	Bennison, 1992
Aphidius colemani Viereck	*A. gossypii* Glover	Cucumber	UK	Good Control	Bennison and Corless, 1993
Lysiphlebus testaceipes (Cresson)	*Myzus persicae* (Sulzer)	Bean	Czech Republic	Good Control	Starý, 1993
Aphidius colemani Viereck	*M. persicae* (Sulzer)	Bean	Czech Republic	Good Control	Starý, 1993
A. colemani Viereck	*Aphis gossypii* Glover	Cucumber	The Netherlands	Significant Effect (Spring Experiment)	van Steenis, 1995
Ephedrus cerasicola Starý	*Aphis gossypii* Glover	Cucumber	Norway	Significant Effect	Hågvar and Hofsvang, 1994
Aphidius colemani Viereck	*A. gossypii* Glover	Cucumber and melon	Italy	Efficient control of aphids on cucumber but not on melon	Burgio *et al.*, 1997
A. colemani Viereck	*A. gossypii* Glover	Melon	France	Satisfactory Control	Schoen and Martin, 1997
A. colemani Viereck	*A. gossypii* Glover	Cucumber	The Netherlands	Good Control	Mulder *et al.*, 1999
A. colemani Viereck	*Myzus persicae* (Sulzer)	Sweet pepper	The Netherlands	Good Control	van Schelt, 1999
A. colemani Viereck	*Aphis gossypii* Glover	Chrysanthemum	USA	Good Control	Vásquez *et al.*, 2006
A. colemani Viereck	*A. gossypii* Glover	Watermelon	Korea	Good Control	Moon *et al.*, 2011
A. colemani Viereck	*A. gossypii* Glover	Wheat	Egypt	Good Control together with *Coccinella septempunctata* Linnaeus	Eid *et al.*, 2018

But the aphid problem was not completely solved because this parasitoid was not well adapted to conditions in northern and especially central California where, in spite of large-scale releases, it never became permanently established (van den Bosch et al., 1985). It was evident that the French strain of *T. pallidus* lacked genetic characteristics enabling it to reproduce and survive in areas of extreme summer heat and low relative humidity (DeBach, 1974).

Further, *T. pallidus* had also been observed in the hot, dry central plateau of Iran during 1960, and subsequently in 1968 some live specimens of this parasitoid from Iran were shipped to California. The parasitoids were cultured and colonised in central California with spectacular results. First, small releases made in June and July gave rise to abundant parasitoids forming large populations, which dispersed significantly by fall. Further releases were made in the Central Valley in 1969 from insectary cultures. Surveys of release plots in October 1969 showed that the parasitoids have been established in all new release sites and had spread to a distance of at least eight miles. In fact, the results were prompt and spectacular. Even though only small colonisations were made at the most unfavorable time of year (midsummer), recoveries were successful at all colonisation sites. Moreover, an autumn survey at one of the colonisation sites showed that the parasite had affected a high level of parasitization and had spread at least a mile from the colonisation focus. The wasp then survived the winter of 1968–1969 and continued to increase in abundance and disperse. Additional colonisations were made in 1969, and the establishment was again invariably affected. By the autumn of 1969, *T. pallidus* was widely established in the Great Central Valley and in several valleys peripheral to San Francisco Bay. Surveys conducted in the spring and early summer of 1970 showed that the parasitoid was having a heavy impact on *C. juglandicola* at a number of places. By the end of that year, it had virtually covered all of the walnut-growing areas of central and northern California, an area of perhaps 50,000 square miles (van den Bosch et al., 1970). By the spring of 1971, *T. pallidus* had a generally crushing impact on the aphid. Surveys indicated that economically injurious infestations of the aphid were virtually nonexistent in the major production areas. In two monitored groves, parasitization of the fundatrix generation (aphids hatching from overwintering eggs) exceeded 90%. This degree of parasitization appeared to be general in commercial groves (van den Bosch et al., 1985).

After ten years of vigorous activity, the Iranian *T. pallidus* strain continues to be a major population determinant of *C. juglandicola* populations. Already the wasp can be credited with almost total commercial control of this aphid (van den Bosch et al., 1985).

4.11.1.2 Biocontrol of Citrus Aphids, *Aphis* (*Toxoptera*) *aurantii* BdF in France

The citrus aphid, *Aphis* (*Toxoptera*) *aurantii* (BdF) (=*Toxoptera aurantii*) was described in 1841 from *Citrus aurantium* L. in France. Since then, this aphid has also been observed in several European and other countries. This aphid is a pest of, among others, *Citrus* and *Coffea*. It damages the leaves and transmits viruses resulting in dwarf trees and shrubs, besides producing fewer and smaller fruits. It has more than a hundred host plants including tea mainly growing in the tropics and subtropics (Piron et al., 2019).

Lysiphlebus testaceipes (Cresson) was selected as a biocontrol agent of citrus aphids based on its available basic information on its host range and field ecology required for the suitability of its culture (Starý et al., 1988). The founder population of *L. testaceipes* consisted of several dozen specimens reared from *Aphis nerii* BdF on *Nerium oleander* L. in Cuba in 1972. This population was shipped to the insectary (Institute of Entomology, Czechoslovak Academy of Sciences, Prague) and reared on *Aphis craccivora* Koch and *Aphis fabae* Scopoli on horse or broad bean, *Vicia faba* L. (=*Faba vulgaris* Moench) (1972–1973). Population number in the insectary varied significantly from several dozens to many hundreds or thousands of specimens (Starý et al., 1988).

Releases were made in the I.N.R.A. (Institut National de la Recherche Agronomique) garden at Antibes, France and in the citrus plantations of the Citrus Experimental Station, San Giuliano in eastern Corsica, France in May-June 1973; in Antibes and in the western Corsica (Porto) in April

1974. Additional small releases were made in glasshouses (Avignon, Perpignan) during the winter 1982–1983. Six different release methods of confined (nylon mesh bags on branches with aphid colonies), semi-confined (the bags with parasitoids being removed after one day and transferred to another aphid-infested tree) and open-type were used. Laboratory-parasitized live aphids, mummies and/or adult parasitoids were released (Starý et al., 1988).

Some 2300 mummies were imported from the then Czechoslovakia for initial release in France in 1973, and another 3000 in 1974. Part of this material was used for laboratory rearing in Antibes, whereas about 5000 mummies and/or adults were released in the field. Both in Antibes and Corsica, 1000 parasitoids each were released in 1973 and 1000 in Antibes and 2000 in Corsica in 1974. The numbers released in each occasion varied from several dozens to 200–300 (Starý et al., 1988).

Sampling at all the release sites as well as in other localities (1975, 1976, 1977, 1978, and 1986) demonstrated that *L. testaceipes* had been established. Extensive sampling was made in 1978 (area of Antibes, Valbonne, Opio, Nice, Menton) and in 1986 (area of Antibes to St. Tropez, Montpellier to Perpignan). A total of 1000 samples were taken in 1973–1986, of them, about 20 % yielded actual or potential hosts of *L. testaceipes*. At least 26 aphid species have been determined as hosts of *L. testaceipes* in southern France. The occurrence of mummies in aphid colonies since the establishment of *L. testaceipes* in southern France has been virtually impressive. A high percent of aphids to almost whole colonies of pests as well as non-pest species have become dense groups of prevalently whitish-yellow mummies, owing to the infection of *L. testaceipes*. In fact, the introduced *L. testaceipes* was very effective bioagent significantly complemented the overall effectiveness of the parasitoid complex associated with the pest and non-pest aphid species in crop and non-crop situations. It was recovered later from Italy. The parasitoid was also sent to eastern Spain where it was established and spread. Ultimately, it became the predominant parasitoid in that region. It also attacks *A. gossypii* on citrus and a number of other aphids on other host plants (Starý et al., 1988).

4.11.1.3 Biocontrol of Rose-grain Aphid, *Metopolophium dirhodum* (Walker) in Australia

Metopolophium dirhodum of European origin arrived in Australia through cut flowers and occurs widely in eastern Australia on barley and oats. This aphid is also known from other plants, *viz.* wheat, sorghum and a number of grasses. It is common in south-eastern Australia on the underside of leaves of barley and oats and less common on wheat, triticale and forage sorghum. It not only occurs on several grass species, especially prairie grass (*Bromus catharticus* Vahl), but also on phalaris (*Phalaris canariensis* L. =*Phalaris aquatica* Delile), cocksfoot (*Dactylis glomerata* L.) and *Lolium* sp. In addition to the damage caused by sap removal and honeydew production, the aphid has been shown to be an effective vector of an Australian isolate of Barley Yellow Dwarf Virus (Waterhouse and Helms, 1985; Waterhouse and Sands, 2001).

In 1984, surveys were carried out to find out the natural enemies of *M. dirhodum.* It was found that this aphid is attacked by *Aphidius ervi* Haliday in southeastern Australia and Tasmania (Carver, 2000). A review of the literature reveals that this parasitoid had been established in Australia at an earlier stage as a result of introductions between 1977 and 1981 against *Acyrthosiphon kondoi* Shinji (Milne, 1986). Another study has also shown that this parasitoid was present in Tasmania before these introductions (Carver, 2000). However, this parasitoid was not very effective against *M. dirhodum* in southeastern Australia. Accordingly, a suggestion came to find out an effective parasitoid to control *M. dirhodum* in southeastern Australia. In this connection, *Aphidius rhopalosiphi* de Stefani-Perez was selected for release and the first introduction of this parasitoid was made from an imported stock from England in March 1985. Although the wasps of this introduction attacked *M. dirhodum* readily in quarantine, the culture was destroyed in favour of obtaining another parasitoid from an area with climate similar to Australia. For this reason, mummies of *M. dirhodum* parasitised by *A. rhopalosiphi* were imported from Chile. In late 1985 and early 1986, some 91,000 laboratory-reared adults were released in the Australian Capital Territory, New South Wales and Victoria, but recoveries were made only for a short period.

Further introductions of *A. rhopalosiphi* from Chile did not alter this picture (Waterhouse and Sands, 2001).

Thereafter, *A. rhopalosiphi* from *Sitobion berchemiae* (Takahashi) [=*Sitobion avenae* (Fabricius)] mummies were imported from the south of France, but they did not respond to *M. dirhodum* in quarantine in Australia. Perhaps this was a parasitoid strain closely adapted to *S. berchemiae*, an aphid that does not occur in Australia. Literature showed that *A. rhopalosiphi* was established at an earlier stage in New Zealand on *M. dirhodum* from stocks obtained from England and France. Thus, mummies of *A. rhopalosiphi* were imported from New Zealand in 1986 and 1988 and mass-produced in *M. dirhodum* developing on barley seedlings. Releases were made in 1987 in the Australian Capital Territory, New South Wales, Victoria and Tasmania and again in August 1988. At the latter time, *M. dirhodum* was present in moderate numbers at release sites. The specimens of *A. rhopalosiphi* cultured from New Zealand stock was found to be widely established on *M. dirhodum* by 2001. This parasitoid now occurs widely on *M. dirhodum.* So, reports are now seldom found regarding aphid populations attaining damage threshold (Waterhouse and Sands, 2001).

4.11.1.4 Biocontrol of Brown Peach Aphid, *Pterochloroides persicae* (Cholodkovsky) in Yemen

In Yemen, an aphid attacking fruit trees, particularly peach and almond, was first reported in 1993 and was identified as the Brown Peach Aphid, *Pterochloroides persicae* (Cholodkovsky) (Harvey and Abdul-Moghni, 1994). The aphid had spread across Yemen by the mid-1990s and caused severe losses to the country's 70,000 farmers who grew fruit tree crops for sale and home consumption. Yields and fruit quality were affected and, in some places, the aphids have also caused tree losses. Chemical pesticides were used first to control this aphid. However, due to the high cost and environmental issues of these chemical pesticides, an emergency response project was set up by the Food and Agriculture Organization of the United Nations (FAO) and General Department of Plant Protection (GDPP), Republic of Yemen, in collaboration with the Yemeni–German Plant Protection Project, to find out suitable parasitoid against this aphid pest (Cock et al., 2015).

Since *P. persicae* was not included before in the biocontrol programme, CABI (as the International Institute of Biological Control) was tasked to find out a potential biocontrol agent. Through literature surveys and assessments, CABI found the name of *Pauesia antennata* (Mukerji) as the most commonly recorded and predominant natural enemy of *P. persicae* in Central Asia. The extensive survey in Pakistan found that it was the only parasitoid of *P. persicae* there. Thereafter, cultures of both the aphid and the parasitoid were established at CABI in Pakistan and the CABI quarantine facility in the UK. Research were conducted on the biology, ecology and rearing, host specificity, other potential risks and possible impact of *P. antennata* on the aphid (*P. persicae*). A dossier was prepared (Cross and Poswal, 2013) and submitted to the Government of Yemen via FAO. Clearance for the introduction of the parasitoid was obtained in December 1996.

P. antennata was imported from Pakistan to the GDPP laboratories located in the city of Sana'a at the end of January 1997 and a rearing colony was established there. By July, the GDPP reared more than 65,000 parasitoids, and over 50,000 of these were released in the field (IIBC, 1998). Releases were concentrated at three main sites around Sana'a. Altogether 350,000 parasitoids were released in the country. Within 2 months and beyond, the aphid populations in these areas were completely collapsed as a result of the parasitoid attack. Parasitism rates reached 40–90% in orchards within 25 km of Sana'a. Since the release of parasitoids, these were found at farms more than 50 km and 120 km away from release sites after 2 and 4 months, respectively. The release programme was extended to other areas of the country, and successful country-wide control was achieved as the parasitoid spread and established in Yemen. Through this programme over 7.6 million trees were saved in Yemen (Cock et al., 2015).

4.11.1.5 Biocontrol of Cypress Aphid, *Cinara cupressivora* Watson & Voegtlin in Africa

The Cypress aphid, *Cinara cupressivora* was first reported from an African country, Malawi in 1986, but, within a short period this aphid spread to many countries in eastern and southern Africa (Mills, 1990; FAO, 1991). This aphid is highly aggregative and exploits a wide range of feeding sites ranging from green branches to woody stems. In Africa, *C. cupressivora* infests a wide range of Cupressaceae, including indigenous species, such as Mulanje cedar, *Widdringtonia cupressoides* (L.) Endl. (The National tree of Malawi) and *Juniperus procera* Hochst. ex Endl., important trees in many water catchment areas in Kenya. In most countries, however, this aphid caused the greatest damage on the exotic plantations or ornamental species, particularly the Mexican Cypress, *Cupressus lusitanica* Mill, an important plantation tree grown in large monocultures. It was estimated that the aphid had killed US$41 million worth of trees in Africa by 1991, and was causing an annual loss of growth increment worth US$13.5 million (Murphy, 1996).

A range of options for controlling *C. cupressivora* was initially considered that included chemical control, the use of resistant varieties of plants and silvicultural techniques. However, it was clear from the beginning that biocontrol should be the most promising option (Mills, 1990). Discussions in several regional workshops held during the early 1990s reached at the same conclusions (FAO, 1991).

A biocontrol programme commenced in 1991, spearheaded by CABI in collaboration with several national and regional agencies. In this connection, four species of parasitoids, all in the genus *Pauesia* were assessed. Of these, *Pauesia juniperorum* (Starý) showed the greatest promise, and this species was imported from United Kingdom and France. However, before bringing them to Africa, *P. juniperorum* was assessed and screened at CABI, UK. (Day et al., 2003).

Before the introduction of *P. juniperorum* in Africa, a dossier was prepared for submitting the same to the National Regulatory Authorities, along with the procedures prescribed by the Code of Conduct for the Import and Release of exotic Biocontrol Agents (FAO, 1996). However, at that time the Code had not been officially published. The first shipment reached Kenya in 1993, and the import permit was granted under the conditions that the insects be held in quarantine, and host-specificity tests are to be conducted to demonstrate that it would not attack coccinellids, predatory mites or other local conifer aphids. Permission for field releases was granted in 1994 following the presentation of a report on quarantine and host-specificity tests, indicating that the parasitoids were cultured for nine generations. In Malawi and Uganda, permissions for direct releases were granted on the basis of the same dossier. The first release of parasitoids was made in Malawi and Kenya in 1994 and in Uganda in 1995. Repeated releases in small numbers (<500 wasps) were made over a period of 1–2 years in all these three countries. Following more than a year of releases in Malawi, there was no evidence of the establishment of the parasitoid, although initial results in field cages were encouraging (Chilima et al., 1995). Further releases were made in 1995 and in the following year, the establishment of the parasitoid was confirmed. The establishment however, was slower in Kenya and Uganda. The last releases were made in late 1996 with no sign of establishment. The parasitoid was found in both countries till early 1999. After two years, the parasitoids were also recorded from the field. *P. juniperorum* is now widespread in Malawi and Kenya, and impact assessment on the activities of these parasitoids has commenced in Kenya. The severity of damage by the aphid has however, declined in both countries (Day et al., 2003).

4.11.1.6 Biocontrol of *Aphis gossypii* Glover on Cucurbits in India

Aphis gossypii has been reported from more than 25 plant species in Gorakhpur, northeastern Uttar Pradesh, India (Ahmad and Singh, 1994). Among these, the highly infested food plants were *Abelmoschus esculentus* (L.) Moench, *Cajanus cajan* (L.) Millsp., *Capsicum frutescens* L., *Solanum melongena* L., *Lagenaria siceraria* (Molina) Standl. (=*Lagenaria vulgaris*), *Luffa cylindrica* (L.)

Roem, *Cucurbita maxima* Duchesne, *Tagetes erecta* L. and *Hibiscus rosa-sinensis* L. *Binodoxys indicus* (Subba Rao and Sharma, 1958) [=*Trioxys* (*Trioxys*) *indicus* Subba Rao and Sharma] was selected for the trial since it is a common but dominant species of parasitoid complex of *A. gossypii* in India (Singh, 2000). Earlier biological studies on *B. indicus* revealed its potential as a promising biocontrol agent against *A. gossypii* on cucurbits (Bhatt and Singh, 1989, 1991a, b, c).

The parasitoids were released into five fields in Gorakhpur district at least 5 km away from each other. Two initial inoculative releases of the parasitoid in 1990–1991 trials produced effective results. The extent of the peak aphid population had rapidly decreased in the fields of *L. siceraria* and *L. cylindrica*. This reduction was 20% within a fortnight after first introduction of the parasitoid in the above fields. A subsequent second introduction of mummies in the field reduced the aphid population between 66% and 76% on *L. siceraria* and *L. cylindrica* respectively, within three weeks (on January 16, 1991). However, on *C. maxima*, 50% reduction in the aphid population was noticed only on February 1, 1991. No significant variation was estimated in the percentage of aphid parasitism in the control and experimental fields of *C. maxima* (Singh and Rao, 1995; Singh, 2000).

In 1991–1992 trial, effective control of *A. gossypii* was also achieved after two inoculative releases of the parasitoids. The peak of the aphid population rapidly reduced in *L. siceraria* and *L. cylindrica* fields. This reduction was 15% and 20% on *L. siceraria* and *L. cylindrica* respectively, within a fortnight after the first parasitoid introduction. After second introduction of the mummies, the aphid population was reduced upto 85%. However, on *C. maxima*, the percentage of reduction of aphid population was only 20%. In these trials, satisfactory control of aphids was achieved in 1990–91 and 1991–92 only on *L. siceraria* and *L. cylindrica* but not on *C. maxima* (Singh and Rao, 1995; Singh, 2000).

4.11.1.7 Biocontrol of Cotton Aphid, *Aphis gossypii* Glover on Chrysanthemums in Greenhouse

Chrysanthemum is one of the major floricultural crops grown in greenhouses throughout the world. USA is among the top five major producers of Chrysanthemum (Parrella et al., 1999). *Aphis gossypii* is the most serious pest of greenhouse-grown chrysanthemum (Storer and van Emden, 1995; Bethke et al., 1998) causing direct damage to the plants through feeding that result in wilting, leaf distortion, and transmission of several viruses. Besides indirect damage, these aphids make physical contamination with their exuviae and honeydew, a nutrient source for sooty mold on plant surfaces reducing the aesthetic value of the plants (Agrios, 1988).

Vásquez et al. (2006) assessed the efficacy of *Aphidius colemani* Viereck controlling *A. gossypii* in greenhouse-grown Chrysanthemum. They compared the performance of *A. colemani* with a pesticide standard, imidacloprid (Marathon 1% G). Plants (Kory chrysanthemum) were grown in a greenhouse in the North Carolina State University Phytotron under controlled conditions (22/18°C day/night, 40–83% RH, and a photoperiod of 16:8 [L:D] h). The plants were infested by *A. gossypii* and two treatments, *viz.* (i) three weekly releases of *A. colemani* (five adults per square meter at sundown), and (ii) one application of imidacloprid (Marathon 1% G) used as a standard insecticide treatment. Untreated plants acted as controls. The results were satisfactory.

Vásquez et al. (2006) concluded, 'Production of chrysanthemums that will meet grower esthetic standards can be achieved by inoculative releases of *A. colemani*, providing an effective *A. gossypii* management alternative to traditional insecticide control'.

4.12 MASS REARING

The rearing of aphidiines starts with the selection of parasitoid species, its common host aphid and the aphid host plant. Mass rearing of aphidiines is a chain of procedures that consists of the mass growing of plants, mass rearing of host aphids, and mass rearing of aphidiines usually maintaining a synchronous tritrophic (plants, aphids and aphidiines) rearing system (Hill and Hoy, 2003; Walker and Hoy, 2003). It is true that if any one of the three becomes unsuccessful, will lead to a complete failure of mass production (Wei et al., 2003). Therefore, the selection of a suitable host plant and suitable aphid species is important in the mass rearing programme of any aphidiine parasitoid species.

These plants, aphids and their parasitoids can be cultured in small or big greenhouses. The entire greenhouse is divided into several chambers so that the plants, aphids and parasitoids can be cultured separately or simultaneously. In case of limited space facility in the greenhouse, cabinets or cages of suitable sizes can be prepared with fine nylon nets and can be used as separate chambers.

Mass culture may be readily contaminated by unwanted organisms, like hyperparasitoids. For this reason, continuous collaboration with a taxonomist or a concerned expert is necessary, so that contamination may be noticed earlier and necessary actions can be taken accordingly (Delucchi et al., 1976).

A number of studies have reported that the suitable temperature for storage of aphidiine parasitoids ranges from 2 to 7°C (Lins et al., 2013; Mahi et al., 2017). The longer duration of storage of aphidiine mummies is technically more advantageous for biocontrol but their performance (emergence, fecundity, mobility, etc.) generally reduces with the increase of storage duration (Amice et al., 2008; Ismail et al., 2010).

A number of factors, *viz.* temperature, light, relative humidity, food, genetic deterioration of the mass-produced organisms, unnatural hosts, microbes, etc. are responsible for the production of low-quality aphidiine parasitoids in the greenhouse. Thus, before mass production, sound knowledge of these factors is necessary. Before selecting a particular strain of a natural enemy, the concerned personnel should ensure that the traits of such strain are appropriately matched with the targeted situations in the greenhouse. Sufficient care (hygiene, temperature, light, R.H., etc.) should be taken to maintain the above conditions in the greenhouse otherwise the parasitoids would be of low quality. Adult parasitoids should be well fed (honey or sugar source available in mass rearing), have had opportunities to mate, and have had time for the pre-oviposition period before releases are made. It is best to rear natural enemies under conditions that are as natural as possible—a conclusion that is supported by a number of researchers with experience in mass production (van Lenteren 2003). In mass rearing, the quality of parasitoids may be improved by selective hybridisation between the population reared on different host plant-host aphid combinations (Tripathi and Singh, 1997) or using semiochemicals (Srivastava and Singh, 1988a, b; Zhua et al., 2006).

4.13 CONSTRAINTS OF USING APHIDIINES IN BIOCONTROL

There are several groups of natural enemies, *viz.* predators, hyperparasitoids, viruses, bacteria and fungi that generally attack aphid parasitoids (Brodeur and Rosenheim, 2000) (Fig. 4.8). All developmental stages of aphid parasitoids are vulnerable to natural enemies (Boivin et al., 2012). Results of surveys as well as experiments conducted under field conditions provide some evidence for the negative influence of predators, hyperparasitoids and fungi on aphid parasitoid populations (Brodeur and Rosenheim, 2000). Coccinellids, spiders, lacewings, anthocorids, nabids, predatory midges, syrphid flies, carabids, staphylinids and ants are major components of the predatory guild associated with aphid colonies (Fig. 4.8) (Sunderland, 1988).

Climate change is another threat, which may not only impact reducing crop production but also cause rapid multiplication of many pests and at the same time change the life cycle pattern of natural enemies. In this connection, an example can be cited where the combined effects of increased temperature and drought had a detrimental effect on the aphid parasitoid, *Diaeretiella rapae* (M'Intosh), which resulted in a less efficient control of aphids (Romo and Tylianakis, 2013).

In order to address constraints more research in relation to the influence of biotic (predators, hyperparasitoids, viruses, bacteria, fungi, plants etc.) and abiotic (temperature, drought, relative humidity, photoperiod, etc.) factors on the life cycle and performance of aphidiine parasitoids should be carried out. In fact, the efficacy of parasitoids as biocontrol agents is influenced by both biotic and abiotic factors (Mkenda et al., 2019).

Some biocontrol critics mentioned that aphidiine parasitoids can evolve, adapt to new hosts, and eventually pose a threat to non-target species of different insect families (Follett et al., 2000). van Driesche and Hoddle (2016) reviewed the nontarget effects of insect biocontrol agents including aphidiine members since 1985 and prepared guidelines for the introduction of biocontrol agents. Thus, the non-target issues should also be taken into consideration in the biocontrol programmes using aphidiine parasitoids.

4.14 CONCLUSIONS

Aphidiinae parasitoids occupy third trophic level in complex food webs. Among the biotic factors, plants (producers), aphids (second trophic level) and hyperparasitoids (fourth trophic level) greatly influence the life cycle of the aphidiine parasitoids. Climate change has profound impact on each and every trophic level; thus, the interactions among the whole complex would be changed. The biological and ecological information on aphidiine parasitoids, especially in relation to its other trophic levels under the changing climate are necessary to utilise them efficiently in biocontrol programme. Accordingly, new research on the biology and ecology under various biological and physical factors of the environment and also the optimisation of multiplication of parasitoids and techniques for their release under laboratory and field conditions should be carried out to develop better and effective ecosystem services.

Research on the exploration of aphidiine fauna from unexplored ecological zones should also be conducted. Actually, there are many potential locations on this planet where efforts have not yet been made to explore this important group of beneficial insects. For example, data for the Indian regions reveals that there are still vast potential areas, where large numbers of aphidiine species are awaiting exploration. Utilising native species has many benefits over importing species from outside in classical biocontrol programmes.

4.15 POINTS TO REMEMBER

i. Aphidiine parasitoids are solitary endoparasitoids that exclusively attack aphids.

ii. To date, there are 501 valid species of aphidiine parasitoids. The majority of them are found in the temperate to subtropical zones of the Northern hemisphere.

iii. Aphidiines can be distinguished by their medium body size, metasomal tergum weakly sclerotised; flexible suture between the second and third metasomal segments; ovipositor sheath or hypopygium in females often modified; a single, large median cell in the forewing in most species; smooth scutellar sulcus; cross vein cu-a absent in the hindwing and occipital carina.

iv. The most successful aphidiine species for the classical biocontrol of aphids are: *Aphidius eadyi* Starý, Gonzalez & Hall, *Aphidius ervi* Haliday, *Aphidius rhopalosiphi* de Stefani-Perez, *Aphidius rosae* Haliday, *Aphidius salicis* Haliday, *Aphidius smithi* Sharma & Subba Rao, *Aphidius sonchi* Marshall, *Aphidius uzbekistanicus* Luzhetzki, *Lipolexis oregmae* (Gahan), *Lysiphlebus testaceipes* (Cresson), *Pauesia antennata* (Mukerji), *Pauesia juniperorum* (Starý), *Praon exsoletum* (Nees), *Praon volucre* (Haliday), *Trioxys complanatus* Quilis, *Trioxys curvicaudus* Mackauer, *Trioxys pallidus* (Haliday) and *Binodoxys indicus* (Subba Rao and Sharma, 1958).

v. In the greenhouses, the aphidiine species, *Aphidius matricariae* Haliday, *Aphidius colemani* Viereck, *Lysiphlebus testaceipes* (Cresson) and *Ephedrus cerasicola* Starý have been used extensively for the control of aphids.

vi. The classical biocontrol of walnut aphid, *Chromaphis juglandicola* (Kaltenbach) by *Trioxys pallidus* Haliday in USA; Citrus aphid, *Aphis* (*Toxoptera*) *aurantii* BdF by *Lysiphlebus testaceipes* (Cresson) in France; Rose-grain aphid, *Metopolophium dirhodum* (Walker) by *Aphidius rhopalosiphi* de Stefani-Perez in Australia; Peach aphid,

Pterochloroides persicae (Cholodkovsky) by *Pauesia antennata* (Mukerji) in Yemen; Cypress aphid, *Cinara cupressivora* Watson & Voegtlin by *Pauesia juniperorum* (Starý) in Africa are the examples from few successful biocontrol programme of aphid pests by aphidiines.

vii. Selection of strain of aphidiine is important for successful classical biocontrol of aphid pests.

viii. Many ecological zones are waiting for aphidiine parasitoids explorations.

ix. New research on the aphidiine biology and ecology are necessary under the changing climate.

ACKNOWLEDGMENTS

We are indebted to Professor Dr. Ehsan Rakhshani, Department of Plant Protection, College of Agriculture, University of Zabol, Zabol, Iran for his sustained support and critical suggestions during the preparation and improvement of this article. We owe great gratitude to him for sending several documents and arranging to send a few photographs from Dr. Jan Havelka, Institute of Entomology, Biology Centre CAS, Czech Republic. Sincere thanks are due to Dr. Jan Havelka for giving us a few photographs for this paper. We would also like to express our sincere thanks to Prof. Rajendra Singh, formerly the Department of Zoology, D.D.U. Gorakhpur University, India for his kind help and cooperation. Thanks are also due to Mr. Brian Valentine (UK), Mr. Ed Phillips (UK) and Mr. Mark Yokoyama (USA) for according permission to use some photographs of aphid parasitoids taken by them in this article. The first author (BCD) is thankful to the Authorities of Rajshahi University, Rajshahi-6205, Bangladesh for providing various facilities.

REFERENCES

Agrios, G.N. (1988). *Plant pathology*, 3rd ed. Academic, San Diego, CA.

Ahmad, M.E. & Singh, R. (1994). *Aphis gossypii* Glover on different food plants and its associations with parasitoids and hyperparasitoids in northeastern Uttar Pradesh. *Annals of Entomology*, 12: 63–67.

Ahmed, S.E.M.K.H. (1990). *Biological Control of Glasshouse Myzus persicae (*Sulzer*) using Aphidius matricariae* Haliday. Ph.D. Dissertation. University of London, pp. 1–341.

Akhtar, M.S., Dey, D. & Usmaniz, M.K. (2011). A catalogue of aphid parasitoids (Hymenoptera: Braconidae: Aphidiinae) from India. *Insecta Mundi*, 151: 1–31.

Amice, G., Vernon, P., Outreman, Y., van Alphen, J.J.M. & van Baaren, J. (2008). Variability in responses to thermal stress in parasitoids. *Ecol Entomol*, 33: 701–708.

Andrade, T.O., Krespi, L., Bonnardot, V., van Baaren, J. & Outreman, Y. (2016). Impact of change in winter strategy of one parasitoid species on the diversity and function of a guild of parasitoids. *Oecologia*, 180: 877–888.

Askew, R.R. (1971). *Parasitic Insects*. Heinemann Educational Books Ltd., London, pp. 1–316.

Bell, W.J. (1991). *Searching Behaviour: The Behavioural Ecology of Finding Resources*. Chapman and Hall, London, pp. 1–370.

Benelli M., Marchetti E. & Dindo M.L. (2017). Effects of storage at suboptimal temperatures on the in vitro-reared parasitoid *Exorista larvarum* (Diptera: Tachinidae). *Journal of Economic Entomology*, 110 (4): 1476–1482.

Benelli, G., Messing, R.H., Wright, M.G., Giunti, G., Kavallieratos N.G. & Canale, A. (2014). Cues triggering mating and host-seeking behavior in the aphid parasitoid *Aphidius colemani* (Hymenoptera: Braconidae: Aphidiinae): implications for biological control. *J. Econ. Entomol.*, 107: 2005–2022.

Berger, J.P., Reichenbacher, B., Becker, D., Grimm, M., Grimm, K., Picot, L., Storni, A., Pirkenseer, C. & Schaefer, A. (2005). Eocene–Pliocene time scale and stratigraphy of the Upper Rhine Graben (URG) and the Swiss Molasse Basin (SMB). *International Journal of Earth Sciences (Geologische Rundschau)*, 94: 711–731.

Bethke, J.A., Redak, R.A. & Schuch, U.K. (1998). Melon aphid performance on chrysanthemum as mediated by cultivar, and differential levels of fertilization and irrigation. *Entomologia Experimentalis et Applicata*, 88: 41–47.

Bhatt, N. & Singh, R. (1989). Bionomics of an aphidiid parasitoid *Trioxys indicus* Subba Rao and Sharma 30, Effect of host plants on reproductive and developmental factors. *Biological Agriculture & Horticulture*, 6: 149–157.

Bhatt, N. & Singh, R. (1991a). Bionomics of an aphidiid parasitoid *Trioxys indicus* Subba Rao and Sharma 35, Influence offood plants on life-table statistics of the parasitoid through its host *Aphis gossypii* Glover. *International Journal of Tropical Insect Science*, 12: 385–389.

Bhatt, N. & Singh, R. (1991b). Bionomics of an aphidiid parasitoid *Trioxys indicus* Subba Rao and Sharma 33, Impact of food plants on the behaviour and sex allocation by the female parasitoid at her varying densities. *Biological Agriculture & Horticulture*, 7: 247–259.

Bhatt, N. & Singh, R. (1991c). Bionomics of an aphidiid parasitoid *Trioxys indicus* Subba Rao and Sharma 32, Effect of food plants on the functional response, area of discovery and sex ratio of FJ offspring at varying densities of its host, Apis gossypii Glover. *Journal of Applied Entomology*, III: 263–269.

Blackman, R.L. & Eastop, V.F. (2000). *Aphids on the World s Crops: An Identification Guide. Coventry*, John Wiley and Sons, pp. 1–475.

Bodlah, I. (2010). *Biosystematics of Aphid Parasitoids from Punjab Province of Pakistan*. Ph.D. Dissertation. Arid Agriculture University, Rawalpindi, pp. 1–189.

Bodlah, I., Naeem, M. & Rakhshani, E. (2012). Record of Genus *Proan* Haliday (Hymenopter, Braconidae, Aphidiinae) from arid tract of Punjab Province of Pakistan. *The Journal of Animal & Plant Sciences*, 22(4): 944–947.

Boivin, G., Hance, T. & Brodeur, J. (2012). Aphid parasitoids in biological control. *Canadian Journal of Plant Science*, 92(1): 1–12.

Bompard, A., Amat, I., Fauvergue, X. & Spataro, T. (2013). Host-Parasitoid dynamics and the success of biological control when parasitoids are prone to allee effects. *PLoS ONE*, 8(10): e76768. 10.1371/journal.pone.0076768

Bournoville, R. (1973). Observations e´cologiques sur l'hivernation du puceron du pois, *Acyrthosiphon misum* (Harris) et de ses parasites dans la re´gion de Versailles. *Annales de Zoologie Ecologie Animale*, 5(1): 13–28.

Brodeur, J. (2013). *Aphis glycines* Matsumura, Soyabin aphid (Hemiptera: Aphididae). In: *Biological Control Programmes in Canada 2001–2012*, eds: P.G. Mason & D.R. Gillespie, pp. 93–97. CAB International.

Brodeur, J. & McNeil, J.N. (1994). Seasonal ecology of *Aphidius nigripes* (Hymenoptera, Aphidiidae), a parasitoid of *Macrosiphum euphorbiae* (Homoptera, Aphididae). *Environmental Entomology*, 23: 292–298.

Brodeur, J. & Rosenheim, J.A. (2000). Intraguild interactions in aphid parasitoids. *Entomologia Experimentalis et Applicata*, 97: 93–108.

Cameron, P.J., Walker, G.P. & Allan, D.J. (1981). Establishment of the introduced parasite *Aphidius eadyi* (Hymenoptera: Aphidiidae) in the north island of New Zealand, and its initial effect on pea aphid. *New Zealand Journal of Zoology*, 8: 105–112.

Campbell, A., Frazer, B.D., Gilbert, N., Gutierrez, A.P. & Mackauer, M. (1974). Temperature requirements of some aphids and their parasites. *Journal of Applied Ecology*, 11: 431–438.

Carver, M. (1989). Biological control of aphids. In: *Aphids: their Biology, Natural Enemies and Control*, eds. A.L. Minks & P. Hourenijn, pp. 141–165, Vol. 2c. New York, USA: Elsvier Publication.

Carver, M. (2000). *Natural enemies and mutualists of aphids of Australia, New Guinea and the South Pacific region*. Division of Entomology, CSIRO, Aphid Database.

Chakrabarti, S. (2021). Aphids (Hemiptera: Aphididae) in the Himalaya: Present Status. *Indian Entomologist*, 2: 6–24.

Chilima, C.Z., Day, R.K. & Meke, G. (1995). *The 1994 Release and Establishment of Pauesia juniperorum, a Natural Biological Control Agent for Cupress Aphids in Malawi*. Report No. 950007, Forestry Research Institute of Malawi, Zomba, Malawi.

Chow, F. & Sullivan, D. (1984). Developmental stages of *Praon pequodorum* Viereck (Hymenoptera: Aphidiidae), a pea aphid parasitoid. *Annals of the Entomological Society of America*, 77: 319–322.

Cock, M.J.W., Day, R.K., Hinz, H.L., Pollard, K.M., Thomas, S.E., Williams, F.E., Witt, A.B.R. & Shaw, R.H. (2015). The impacts of some classical biological control successes. *CAB Reviews*, 10: 1–042.

Cohen, J.E., Jonsson, T., Müller, C.B., Godfray, H.C.J. & Savage, van M. (2005). Body sizes of hosts and parasitoids in individual feeding relationships. *The Proceedings of the National Academy of Sciences (PNAS)*, 2(3): 694–689.

Colinet, H. & Boivin, G. (2011). Insect parasitoids cold storage: A comprehensive review of factors of variability and consequences. *Biological Control*, 58: 83–95.

Colinet, H., Muratori, F. & Hance, T. (2010). Cold-induced expression of diapause in *Praon volucre*: fitness cost and morpho-physiological characterization. *Physiological Entomology*, 35: 301–307.

Collins, M.D. & Dixon, A.F.G. (1986). The effect of egg depletion on the foraging behaviour of an aphid parasitoid. *Journal of Applied Entomology*, 102: 342–352.

Coppel, H.C. & Mertins, J.W. (1977). *Biological Insect Pest Suppression*. Springer-Verlag, Berlin-Heidelberg, pp. 1–314.

Cross, A.E. & Poswal, M.A. (2013). *Dossier on Pauesia antennata (Mukerji): biological control agent for the brown peach aphid, Pterochloroides persicae, in Yemen*. CABI Working Paper 5, pp. 1–24.

Das, B.C. (1994). Aphids (Aphididae: Homoptera) and their natural enemies in Bangladesh. *Environment & Ecology*, 12: 795–801.

Das, B.C. & Chakrabarti, S. (1988). Immature stages of *Kashmiria aphidis* (Hymenoptera: Aphidiidae). *Acta Entomologica Bohemoslovakia*, 85: 16–20.

Das, B.C. & Chakrabarti, S. (2018). Diversity and distribution of hyperparasitoids of Aphidiinae (Insecta: Braconidae: Hymenoptera) in the Garhwal Himalayas. In: *Biodiversität und Naturausstattung im Himalaya* VI (eds: M. Hartman, M.V.L. Barclay and J. Weipert). Biodiversität und Naturausstattung im Himalaya VI.- Verein der Freunde & Foerderer des Naturkundemuseums Erfurt e. V., Erfurt, pp. 565–570.

Davidian, E.M. (2020). New species of the genus *Praon* Haliday, 1833 (Hymenoptera: Aphidiidae) from China. *Caucasian Entomological Bulletin*, 16(2): 207–211.

Davidian E.M., Kaliuzhna M.O. & Perkovsky E.E. (2021). First aphidiine wasp from the Sakhalinian amber. *Acta Palaeontologica Polonica*, 66, (Supplement to x): xxx–xxx. 10.4202/app.00843.2020 (Retrieved on 12 August 2021).

Day, R.K., Kairo, M.T.K., Abraham, Y.J., Kfir, R., Murphy, S.T., Mutitu, K.S. & Chilima, C.Z. (2003). Biological control of homopteran pests of conifers in Africa. In: *Biological Control in IPM Systems in Africa*, eds P. Neuenschwander, C. Borgemeister & J. Langewald, pp. 101–112. UK: CABI Publishing.

DeBach, P. (1974). *Biological Control by Natural Enemies*. Cambridge University Press, London & New York. pp. 1–323.

Delphine, B. (2012). *Inbreeding avoidance mechanism: how premating conditions may influence the mating system in a quasi-gregarious parasitoid of aphids, Aphidius matricariae*. Ph.D. Dissertation. UCL, ELI-B, Belgium, pp. 1–260.

Delucchi, V., Rosen, D. & Schlinger, E.I. (1976). Relationship of systematics to biological control. In: *Theory and Practice of Biological Control*, eds. C.B. Huffaker & P.S. Messenger, pp. 81–91. New York, London: Academic Press.

Dennis, R.L.H., Dapporto, L., Fattorini, S. & Cook, L.M. (2011). The generalism-specialism debate: The role of generalists in the life and death of species. *Biological Journal of the Linnean Society*, 104: 725–737.

Derocles, S.A.P., Plantegenest, M., Rasplus, J.Y., Marie, A., Evans, D.M., Lunt, D.H., Le Ralec, A. (2015). Are generalist Aphidiinae (Hym. Braconidae) mostly cryptic species complexes? *Systematic Entomology*, 41: 379–391.

FAO (Food and Agriculture Organization). (1991). Exotic *Aphid Pests of Conifers: a Crisis in African Forestry*. Proceedings of FAO Workshop, 3–6 June 1991, Muguga, Kenya. FAO, Rome, Italy.

FAO (Food and Agriculture Organization). (1996). *Code of Conduct for the Import and Release of Exotic Biological Control Agents*. FAO, Rome, Italy.

Farrell, J.A. & Stufkens, M.J. (1990). The impact of *Aphidius rhopalosiphum* (Hymenoptera: Aphidiidae) on populations of the rose grain aphid (*Metopolophium dirhodum*) (Hemiptera: Aphididae) on cereals in Canterbury, New Zealand. *Bulletin of Entomological Research*, 80: 377–383.

Ferrer-Suay, M., Paretas-Martínez, J., Selfa, J. & Pujade-Villar J. (2012). Taxonomic and synonymic world catalogue of the Charipinae and notes abouth this subfamily (Hymenoptera: Figitidae: Charipinae). *Zootaxa*, 1376, p. 92.

Fisher, R.A. (1930). *The genetical theory of natural selection*. Oxford: Oxford University Press.

Follett, P.A., Duan, J., Messing, R.H. & Jones, V.P. (2000). Parasitoid drift after biological control introductis: Re-examining Pandora's Box. *American Entomologist*, 46(2): 82–94.

Force, D.C. & Messenger, P.S. (1964). Fecundity, reproductive rate and innate capacity of increase of three parasites of *Therioaphis maculata* (Buckton) reared at various constant temperatures. *Ecology*, 45: 706–715.

Fougeroux, A., Bouchet, C., Reboulet, J.N. & Tisseur, M. (1988). Microhymenoptera for aphid population regulation in French cereal crops. In: *Integrated Crop Protection in Cereals*, eds. R. Cavalloro & K.D. Sunderland, pp. 61–68. Proc. Meeting EC Experts Group, Littlehampton, England, 1986. Rotterdam: A.A. Balkema.

Gariepy, T.D. & Messing, R.H. (2012). Development and use of molecular diagnostic tools to determine trophic links and interspecific interactions in aphid–parasitoid communities in Hawaii. *Biological Control*, 60: 26–38.

Ghosh, A.K. & Starý, P. (1983). *Aphid parasitoids of India and adjacent countries (Hymenoptera: Aphidiidae)*. Technical Monograph No. 7. Zoological Survey of India. pp. 1–97.

Gibson, G.A.P. & Vikberg, V. (1998). The species of *Asaphes* Walker from America north of Mexico, with remarks on extralimital distributions and taxa (Hymenoptera: Calcidoidea, Pteromalidae). *Journal of Hymenoptera Research*, 7: 209–256.

Glinwood, R.T. (1998). *Responses of aphid parasitoids to aphid Sex pheromones: Laboratory and Field Studies*. Ph.D. Dissertation. The University of Nottingham, pp. 1–209.

Godfray, H.C.J. (1994). *Parasitoids: Behavioural and Evolutionary Ecology*. Ph.D. Dissertation. Princeton University Press, Princeton, NJ, pp. 1–473.

Guo'An, Z., Yue, G. & QingHai, W. (2001). Correlation between Aphids, Aphidiinae and Hyperparasitoids in Different Habitats. *Proceedings of National Biological Control & VIII Microbial Insecticides Workshop*, 2000. 11:29.

Haeselbarth, E. (1979). Zur Parasitierung der Puppen der Forleule (*Panolis flammea*) [Schiff.]) und Heidelbeerspanner *(Boarmia bistortana*) [Goeze]) in bayerischen Kiefernwäldern. *Zeitschrift für Angewandte Entomologie*, 87: 186 202, pp. 311 322.

Hafez, M. (1961). Seasonal fluctuations of population density of the cabbage aphid. *Brevicorvne brassicae* (L.), in the Netherlands, and the role of its parasite. *Aphidius* (*Diaeretiella) rapae* (Curtis). *Tijdschrift Plantenziekten*, 67: 445–548.

Hagen, K.S. & van den Bosch., R. (1968). Impact of pathogens, parasites, and predators on aphids. *Annual Review of Entomology*, 13: 325–384.

Hågvar, E.B. & Hofsvang T. (1991) Aphid parasitoids (Hymenoptera, Aphidiidae): biology, host selection and use in biological control. *Biocontrol News and Information*, 12: 13-42.

Hale, M.A. & Elliott, D. (2002). *Successes and challenges in augmentative biological control in outdoor agricultural applications: a producer's perspective*. First International Symposium on Biological Control of Arthropods. USDA Forest Service. pp. 185–188.

Hamilton, V.D. (1967). Extraordinary sex ratios. *Science*, 156: 477- 488.

Hartl, D.L. & Brown, S.V. (1970). The origin of male haploid genetic systems and their expected sex ratio. *Theoretical Population Biology*, 1(2): 165–190.

Harvey, A.W. & Abdul-Moghni, A.A. (1994). *New aphid pest infesting fruit-trees in Yemen*. FAO Plant Protection Bulletin, 42(4): 1–220.

Hassell, M.P. (2000). Host–parasitoid population dynamics. *Journal of Animal Ecology*, 69: 543–566.

Hawkins, B.A. (1993). Refuges, host population dynamics and the genesis of parasitoid diversity. In: *Hymenoptera and Biodiversity*, eds. J. LaSalle & I.D. Gauld, pp. 235–256. Oxford: C.A.B International.

He, X.Z. & Wang, Q. (2008). Reproductive strategies of *Aphidius ervi* Haliday (Hymenoptera Aphidiidae). *Biological Control*, 45: 281–287.

Hill, S.L. & Hoy M.A. (2003). Interactions between the red imported ire ant *Solenopsis invicta* and the parasitoid *Lipolexis scutellaris* potentially affect classical biological control of the aphid *Toxoptera citricida. Biological Control*, 27: 11–19.

Hofsvang, T., Godonou, I., Tepa-Yotto, G.T. & Sæthre, M.G. (2014). The native hyperparasitoid complex of the invasive aphid parasitoid *Lysiphlebus testaceipes* (Hymenoptera: Braconidae: Aphidiinae) in Benin, West Africa. *International Journal of Tropical Insect Science*, 34(1): 9–13.

Hougardy, E. & Mills, N.J. (2009). Factors influencing the abundance of *Trioxys pallidus*, a successful introduced biological control agent of walnut aphid in California. *Biological Control*, 48: 22–29.

Hughes, G.E. (2010). *Thermal biology and establishment potential of two non-native candidate biological control agents, Nesidiocoris tenuis Reuter (Hemiptera: Miridae) and Lysiphlebus testaceipes (Cresson) (Hymenoptera: Braconidae, Aphidiinae), in the U.K.* Ph.D. Dissertation. University of Birmingham, pp. 1–141.

Hughes, R.D. (1989). Biological control in the open field. In: *Aphids: their Biology, Natural Enemies and Control*, eds A.K. Minks & P. Hourenijn, pp. 167–198, Vol. 2C. New York, USA; Elsevier Publication.

Hwang, H.-S. & Lee, K-Y. (2021). Molecular identification of three aphid parasitoids, *Aphidius areolatus, A. absinthii, and A. uzbekistanicus* (Hymenoptera: Braconidae) from Ulleungdo and Dokdo Islands in Korea. *Journal of Asia-Pacific Biodiversity*, https://reader.elsevier.com/reader/sd/pii/S2287884X21000297?token=A6FEF2827C5E867611B70BFF4D20A208C8C734AA36D413C7C05A73F5444827BCBFBFA6202BDE-B732530AE73470D669AC&originRegion=eu-west-1&origin Creation=20210727154749 (Retrieved on 27 July 2021).

IIBC [International Institute of Biological Control]. (1998). *Annual Report 1997.* CAB International, Wallingford, UK, pp. 1–135.

Ismail, M., Vernon, P., Hance, T. & van Baaren, J. (2010). Physiological costs of cold exposure on the parasitoid *Aphidius ervi*, without selection pressure and under constant or fluctuating temperatures. *BioControl*, 55: 729–740.

Jamhour, A.M. (2017). *Molecular characterization and phylogenetic relationships among European Aphidius Nees (Hymenoptera, Braconidae, Aphidiinae).* Ph.D. Dissertation. University of Belgrade, pp. 1–105.

Jarosik, V.C. & Lapchin, L. (2001). An experimental investigation of patterns of parasitism at three spatial scales in an aphid-parasitoid system (Hymenoptera: Aphidiidae). *European Journal of Entomology*, 98: 295–299.

Jarry, I. & Tremblay, E. (1989). Cold storage of *Lysiphlebus fabarum* (Marsh.) mummies (Hymenoptera: Braconidae). *Bollettino del Laboratorio di Entomologia Agraria Filippo Silvestri*, 46: 199–206.

Jaureguiberry, P., Buffa, L.M. & Delfino, M.A. (2010). Cross-habitat usage by crop aphids and their parasitoids in the cropnoncrop interface in an organic vegetable farm. *Revista Brasileira de Agroecologia*, 5(2): 39–49.

Ji, Z.D., Lv, N. & Tian, S.B. (1992). *Protaphidius nawaii* Ashmead, a New Record from China. *Journal of Hebei Forestry College*, 7(4): 273–274.

Jiahua, C. & Quanxiu, S. (2001). *Sestematic studies on aphidiidae of China (Hymenoptera: Aphidiidae).* (In Chinese with English summary). Chinese Edition). Fujian Science and Technology Publishing House, pp. 1–31.

Joshi, S., Rabindra, R.J. & Rajendran, T.P. (2010). Biological control of aphids. *Journal of Biological Control*, 24(3): 185–202.

Kambhampati, S., Völkl, W., Mackauer, M. (2000). Phylogenetic relationships among genera of Aphidiinae (Hymenoptera: Braconidae) based on DNA sequence of the mitochondrial 16S rRNA gene. *Systematic Entomology*, 25: 437–445.

Kant, R., Trewick, S.A., Sandanayaka, W.R.M., Godfrey, A.J.R. & Minor, M.A. (2012). Effects of multiple matings on reproductive fitness of male and female *Diaeretiella rapae*. *Entomologia Experimentalis et Applicata*, 145(3): 215–221.

Kavallieratos, N.G., Tomanovič, Ž., Prtrović, A., Kocić, K., Jankovič, M. & Starý, P. (2016). Parasitoids (Hymenoptera: Braconidae: Aphidiinae) of aphids feeding on ornamental trees in southeastern Europe: Key for identification and tritrophic associations. *Annals of the Entomological Society of America*, 109(3): 473–487.

Kostál, V. (2006). Eco-physiological phases of insect diapause. *Journal of Insect Physiology*, 52: 113–117.

Langer, A. & Hance, T. (2000). Overwintering strategies and coldhardiness of two aphid parasitoid species (Hymenoptera: Braconidae: Aphidiinae). *Journal of Insect Physiology*, 46: 671–676.

Le Lann, C., Outreman, Y., van Alphen, J.J.M. & van Baaren, J. (2011). First in, last out: asymmetric competition influences patch exploitation of a parasitoid. *Behavioral Ecology*, 22: 101–107.

Le Ralec, A. (1991). *Les Hyménoptères parasitoides: adaptions de l'appareil reproducteur femelle. Morphologie et ultrastructure de l'ovaire, de l'oeuf et de l'ovipositeur.* Ph.D. Dissertation. Université de Rennes I, Rennes, France.

Lins, J.C., Bueno, V.H.P., Sidney, L.A., Silva, D.B., Sampaio, M.V., Pereira, J.M., Nomelini, Q.S.S. & van Lenteren, J.C. (2013). Cold storage affects mortality, body mass, lifespan, reproduction and flight capacity of *Praon volucre* (Hymenoptera: Braconidae). *European Journal of Entomology*, 110: 263– 270.

Lopes, T. Libert, P.N. Starý, P. Japoshvili, G. Hatt, S. & Francis, F. (2016). Checklist of Aphidiinae (Hymenoptera: Braconidae) and *Aphelinus* (Hymenoptera: Aphelinidae) species from Belgium with respectively four and three new records. *Zootaxa*, 4092(4): 548–560.

Louâpre, P., Lann, C.L. & Hance, T. (2018). When parasitoids deal with the spatial distribution of their hosts: consequences for both partners. *Insect Science*, 00, 1–9, 10.1111/1744-7917.12583.

Loxdale, H.D., Balog, A. & Harvey, J.A. (2019). Generalism in Nature. The Great Misnomer: Aphids and wasp parasitoids as examples. *Insects*, 10: 314.

Mackauer, M. (1983). Quantitative assessment of *Aphidius smithi* (Hymenoptera, Aphidiidae): fecundity, intrinsic rate of increase and functional response. *Canadian Entomologist* 115: 399–415.

Mackauer, M. & Chow F.J. (1986). Parasites and parasite impact on aphid populations. In: *Plant Virus Epidemics: Monitering, Modeling and Predicting Outbreaks*, eds. G.D. McLean, R.G. Garret & W.G. Ruesink, pp. 95–117. Sydney: Academic Press.

Mackauer, M. & Finlayson, T. (2012). *Choreopraon totarae* (Hymenoptera: Braconidae: Aphidiinae), a new parasitoid of *Neophyllaphis totarae* (Hemiptera: Aphidoidea: Drepanosiphidae) in New Zealand. *New Zealand Journal of Zoology*, 39(1): 77–84.

Mackauer, M. & Kambhampati, S. (1988). Parasitism of aphid embryos by *Aphidius smithi*: some effects of extremely small host size. *Entomologia Experimentalis et Applicata*, 49: 167–173.

Mackauer, M. & Starý, P. (1967). *World Aphidiidae (Hym. Ichneumonoidea).* Index of Entomophagous Insects. Paris, Le François. pp. 1–195.

Mackauer, M. & Völkl, W. (2002). Brood-size and sex-ratio variation in field populations of three species of solitary aphid parasitoids (Hymenoptera: Braconidae, Aphidiinae). *Oecologia*, 131: 296–305.

Mackauer, M., Michaud, J.P. & Völkl, W. (1996). Host choice by aphidiid parasitoids (Hymenoptera: Aphidiidae): host recognition, host quality and host value. *Canadian Entomologist*, 128: 959–980.

Mahi, H., Rasekh, A. & Shishehbor, P. (2017). The impact of cold storage on mating behavior of emerging adults and progeny fitness, in a parasitoid wasp, *Lysiphlebus fabarum*, under constant and fluctuating temperatures. *Applied Entomology and Phytopathology*, 85: 1–10.

Martinou, A. (2005). *Studies on intraguild predation and biological control of the aphid Myzus persicae (Sulzer) on pepper and aubergine*. Ph.D. Dissertation. Imperial College London, pp. 1–179.

Matin, S.B., Sahragard, A. & Rasoolian, G. (2005). Some Behavioural Characteristics of *Lysiphlebus fabarum* (Hym: Aphidiidae) Parasitizing *Aphis fabae* (Hom: Aphididae) under Laboratory Conditions. *Journal of Entomology*, 2: 64–68.

Micha, S.G., Kistenmacher, S., Mölck, G. & Wyss, U. (2000). Tritrophic interactions between cereals, aphids and parasitoids: discrimination of different plant-host complexes by *Aphidius rhopalosiphi* (Hymenoptera: Aphidiidae). *European Journal of Entomology*, 97: 539–543.

Michaud J.P. (1994). Differences in foraging behaviour between virgin and mated aphid parasitoids (Hymenoptera: Aphidiidae). *Canadian Journal of Zoology*, 72(9): 1597–1602.

Miller, T.L.P. (2018). *Evaluation of the Aphidius colemani-Rhopalosiphum padi banker plant system in greenhouse biological control.* Ph.D. Dissertation. Oklahoma State University, pp. 1–120.

Mills, N.J. (1990). Biological control of forest aphid pests in Africa. *Bulletin of Entomological Research*, 80: 31–36.

Milne, W.M. (1986). The release and establishment of *Aphidius ervi* Haliday (Hymenoptera: Ichneumonoidea) in lucerne aphids in Eastern Australia. *Journal of the Australian Entomological Society*, 25: 123–130.

Mkenda, P.A., Ndakidemi, P.A., Stevenson, P.C., Arnold, S.E.J., Belmain, S.R., Chidege, M., Gurr, G. & Woolley, V.C. (2019). Characterization of hymenopteran parasitoids of *Aphis fabae* in an African smallholder bean farming system through sequencing of COI 'Mini-barcodes'. *Insects, Communication*, 10: 331. 10.3390/insects10100331

Muratori, F., Lannic. J.L., Nénon, J.-P. & Hance, T. (2004). Larval morphology and development of *Aphidius rhopalosiphi* (Hymenoptera: Braconidae: Aphidiinae). *The Canadian Entomologist*, 136: 169–180.

Murdoch, W.W., Nisbet, R.M., Gurney, W.S.C. & Reeve, J.D. (1987). An invulnerable age class and stability in delay-differential parasitoid–host models. *American Naturalist*, 129: 262–282.

Murphy, S.T. (1996). Status and impact of invasive conifer aphid pests in Africa. In: *Impact of Diseases and Insect Pests in Tropical Forests*, eds K.S.S. Nair, J.K. Sharma & R.V. Varma. Proceedings of the IUFRO Symposium, 23–26 November 1993, Peechi, India. Kerala Forest Research Institute, Peechi, India, pp. 289–297.

Nemec, V. & Starý, P. (1985) Population diversity in deuterotokous Lysiphlebus species, parasitoids of aphids (Hymenoptera, Aphidiidae). *Acta Entomologica Bohemoslovaca*, 82: 170–174.

O'Donnell, D.J. (1987). Larval development and the determination of the number of instars in aphid parasitoids (Hymenoptera: Aphidiidae). *International Journal of Insect Morphology and Embryology*, 16: 3–15.

O'Donnell, D.J. (1989). A morphological and taxonomic study of the first instar larvae of Aphidiinae (Hymenoptera: Braconidae). *Systematic Entomology*, 14: 197–219.

Ortega-Blanco, J., Bennett, D.J., Delclòs, X. & Engel, M.S. (2009). Primitive aphidiine wasp in Albian amber from Spain and a Northern Hemisphere origin for the subfamily (Hymenoptera: Braconidae: Aphidiinae). *Journal of the Kansas Entomological Society*, 82: 273–282.

Ortiz-Martínez, S., Pierre, J.S., van Baaren, J., Lann, C.L., Zepeda-paulo, F. & Lavandero, B. (2019). Interspecific competition among aphid parasitoids: molecular approaches reveal preferential exploitation of parasitized hosts. *Sci. Rep.*, 9: 19641, 10.1038/s41598-019-56187-3 (Retrieved on 20 July 2021).

Parrella, M.P., Hansen, L.S. & van Lenteren, J.C. (1999). Glasshouse environments. In: *Handbook of biological control*, eds, T.S. Bellows & T.W. Fisher, pp. 819–839. New York: Academic Press.

Pennacchio, F. & Digilio, M.C. (1990). Morphology and development of larval instars of *Aphidius ervi* Haliday (Hymenoptera, Braconidae, Aphidiinae). *Bolletino del Laboratorio di Entomologia Agraria Filippo Silvestri*, 46: 163–174.

Persad, A.N., Hoy, M.A. & Nguyen, R. (2007). Establishment of *Lipolexis oregmae* (Hymenoptera: Aphidiidae) in a classical biological control program directed against the brown citrus aphid (Homoptera: Aphididae) in Florida. *Florida Entomologist*, 90(1): 204–213.

Pimentel, D. (1961). Natural control of aphid populations on cole crops. *Journal of Economic Entomology*, 4: 885–888.

Polgár, L.A. & Hardie, J. (2000). Diapause induction in aphid parasitoids. *Entomologia Experimentalis et Applicata*, 97: 21–27.

Pons, X., Lumbierres, B., Antoni, R. & Starý, P. (2011). Parasitoid complex of alfalfa aphids in an IPM intensive crop system in northern Catalonia. *Journal of Pest Science*, 84: 437–456.

Powell, W. & Pell, J.K. (2007). Biological control. In: *Aphids as crop pests*, eds. H.F. Emden & R. Harrington, pp. 469–513. Wallingford: CAB International.

Petrović, A., Mitrovic, M., Starý, P., Petrović-Obradović, O., Žikić, V., Tomanović, Ž. & Vorburger, C. (2013). *Lysiphlebus orientalis* (Hymenoptera, Braconidae), a new invasive aphid parasitoid in Europe - evidence from molecular markers. *Bulletin of Entomological Research*, 103: 451–457.

Piron, P.G.M., de Haas, M.C. (M.) & Sonnemans, M.A.H.M. (Ad). (2019). The presence of *Aphis* (*Toxoptera*) *aurantii* (Homoptera: Aphididae) in the Netherlands. *Entomologische Berichten*, 79(5): 162–164.

Purkiss, S.A. (2018). *Community genetic and environmental effects in extremely plastic aquatic and terrestrial arthropods*. M.Phil. Dissertation. University of Manchester. pp. 1–166.

Quicke, D.L.G. (2015). *The Braconid and Ichneumonid Parasitoid Wasps: Biology, Systematics, Evolution and Ecology*. John Wiley & Sons, Ltd. pp. 1–681.

Quilis, P.M. (1940). Los Aphidiidae fósiles de Wittenheim (Haut-Rhin, Francia) (Hym. Brac.). *Eos, Revista Española de Entomología*, 14: 23–61.

Rakhshani, E. & Starý, P. (2021). Aphid Parasitoids: Aphidiinae (Hym., Braconidae). In: *Biological Control of Insect and Mite Pests in Iran*. Progress in Biological Control, eds. J. Karimi & H. Madadi, pp. 333–399, vol 18. Springer Nature Switzerland.

Rakhshani, E., Barahoei, H., Ahmed, Z., Tomanovič, Ž., Jankovič, M., Petrović, A., Bodlah, I. & Starý, P. (2012). New species and additional evidence of aphid parasitoids (Hymenoptera: Braconidae. Aphidiinae) from India. *Zootaxa*, 3397: 45–54.

Rasnitsyn, A.P. (1980). Origin and evolution of Hymenoptera [in Russian]. *Trudy Paleontologičeskogo instituta Akademii Nauk SSSR*, 174: 1–191.

Raychaudhuri, D. (1990). *Aphidiids (Hymenoptera) of Northeast India*. Indira Publishing House, MI, USA: 1–155.

Rehman, A. & Powell, W. (2010). Host selection behaviour of aphid parasitoids (Aphidiidae: Hymenoptera). *Journal of Plant Breeding and Crop Science*, 2: 299–311.

Romo, C.M. & Tylianakis, J.M. (2013). Elevated temperature and drought interact to reduce parasitoid effectiveness in suppressing hosts. *PLoS ONE*, 8: e58136. https://journals.plos.org/plosone/article?id=10.1371/journal.pone.0058136 (Retrieved on 2 August 2021).

Schlick-Steiner, B.C., Steiner, F.M., Seifert, B., Stauffer, C., Christian, E. & Crozier, R.H. (2010). Integrative taxonomy: a multisource approach to exploring biodiversity. *Annual Review of Entomology*, 55: 421–438.

Schlinger, E.I., Hagen, K.S. & van den Bosch, R. (1960). Parasite of walnut aphid established in California. *California Agriculture*, 14(11): 3–4

Schlinger, E.I. & Hall, J.C. (1960). The biology, behavior, and morphology of *Praon palitans* Muesebeck, an internal parasite of the spotted alfalfa aphid, *Therioaphis maculata* (Buckton) (Hymenoptera: Braconidae, Aphidiinae). *Annals of the Entomological Society of America*, 53: 144–160.

Schooler S.S., De Barro P. & Ives A.R. (2011). The potential for hyperparasitism to compromise biological control: why don't hyperparasitoids drive their primary parasitoid hosts extinct? *Biological Control*, 58: 167–173.

Sekhar, P.S. (1959). Life history of *Aphidius testaceipes* (Cresson) and *Praon aguti* Smith (Hym., Braconidae) primary parasites of aphids with notes on the effects of parasitism on hosts. *Current Science*, 28: 333–335.

Shirota, Y., Carter, N., Rabbinge, R. & Ankersmit, G.W. (1983). Biology of *Aphidius rhopalosiphi*, a parasitoid of cereal aphids. *Entomologia Experimentalis et Applicata*, 34: 27–34.

Simmonds, F.J., Franz, J.M. & Sailer, R.I. (1976). History of Biological Control. In: *Theory and Practice of Biological Control*, eds. C.B. Huffaker & P.S. Messenger, pp. 17–39. New York, London: Academic Press.

Singh, R. (2000). Biological Control of Aphids by using their Parasitoids In: *Biocontrol potential and its exploitation in sustainable agriculture*, eds. R.K. Upadhyay, K.G. Mukerji & B.P. Chamola, pp. 57–73, Volume 2: Insect Pests. Springer Science+Business Media, LLC.

Singh, R. & Rao, S.N. (1995). Biological control of *Aphis gossypii* Glover (Homoptera: Aphididae) on cucurbits by *Trioxys indicus* Subba Rao & Sharma (Hymenoptera: Aphidiidae). *Biological Agriculture and Horticulture*, 12, pp. 227–236.

Smith, P. & Kambhampati, S. (2000). Evolutionary transitions in Aphidiinae (Hymenoptera: Braconidae). In: *Hymenoptera: Evolution, Biodiversity and Biological Control*, eds. A.D. Austin & M. Dowton, pp. 106–113. Collingwood, Victoria: CSIRO.

Srivastava, M. & Singh, R. (1988a). Bionomics of *Trioxys indicus*, an aphidiid parasitoid of *Aphis craccivora*. 26. Impact of host extract on the ovipostion response of the parasitoid. *Biological Agriculture and Horticulture*, 5: 169–176.

Srivastava, M. & Singh, R. (1988b). Bionomics of *Trioxys (Binodoxys) indicus* Subba Rao and Sharma, an aphidiid parasitoid of *Aphis craccivora* Koch. 29. Impact of host extract on reproductive behaviour of the parasitoid at different host densities. *Journal of Applied Entomology*, 106: 408–412.

Starý, P. (1962). Hymenopterous parasites of the pea aphid *Acyrthosiphon onobrychis* (Boyes) in Czechoslovakia. I. Bionomics and ecology of *Aphidius ervi*. *Folia* Zoologica, 11: 265–278.

Starý, P. (1966). *Aphid parasites of Czechoslovakia. A review of the Czechoslovak Aphidiidae (Hymenoptera)*. Publishing House of the Czechoslovak Academy of Sciences, Prague.

Starý, P. (1968). Impact of an indigenous parasite, *Aphidius ervi* Hal. on pea aphid, *Acyrthosiphon pisum* (Harris) populations on alfalfa in Czechoslovakia. *Bollettino del Laboratorio di Entomologia Agraria Filippo Silvestri*, 26: 293–313.

Starý, P. (1970). *Biology of aphid parasites (Hymenoptera: Aphidiidae)*. Dr. W. Junk Publ., The Hague.

Starý, P. (1973). A revision of the fossil Aphidiidae (Hymenoptera). *Annotationes Zoologicae et Botanicae*, 87: 1–22.

Starý, P. (1978). Seasonal relations between lucerne, red clover, wheat and barley agro-ecosystems through the aphids and parasitoids (Homoptera, Aphldidae; Hymenoptera, Aphidiidae). *Acta entomologica bohemoslovaca*, 75: 296–311.

Starý, P. (1981). On the strategy, tactics and trends of host specificity evolution in aphid parasitoids (Hymenoptera: Aphidiidae). *Acta EntomActa entomologica bohemoslovaca*, 78: 65–75.

Starý, P. (1988). Aphidiidae. In: *Aphids. Their Biology, Natural Enemies and Control*, eds. A.K. Minks & P. Harrewijn, pp. 171–184, Volume 2B. Amsterdam: Elsevier.

Starý, P. (1993). Alternative host and parasitoid in first method in aphid pest management in glasshouses. *Journal of Applied Entomology*, 116: 187–191.

Starý, P., Lyon, J.P. & Leclant, F. (1988). Biocontrol of aphids by the introduced *Lysiphlebus testaceipes* (Cress.) (Hym., Aphidiidae) in Mediterranean France. *Journal of Applied Entomology*, 105: 74–87.

Starý, P., Naumann-Etienne, K. & Remaudiere, G. (1998). A review and tritrophic associations of aphid parasitoids (Hym., Braconidae, Aphidiidae) of Pakistan. *Parasitica.*, 54(1): 3–21.

Starý, P., Rakhshani, E., Tomanović, Ž., Hoelmer, K., Kavallieratos, N.G., Yu, J., Wang, M. & Heimpel, G.E. (2010). A New Species of *Lysiphlebus* Förster 1862 (Hymenoptera: Braconidae, Aphidiinae) Attacking Soybean Aphid, *Aphis glycines* Matsumura (Hemiptera: Aphididae) from China. *Journal of Hymenoptera Research*, 19(1): 179–186.

Storer, J.R. & van Emden, H.F. (1995). Antibiosis and antixenosis of chrysanthemum cultivars to the aphid *Aphis gossypii*. *Entomologia Experimentalis et Applicata*, 77: 307–314.

Subba Rao. B.R. & Sharma A.K.(1958).*Trioxys (Trioxys) indicus,* new species, a parasite of *Aphis gossypii* Glover. *Indian Journal of Entomology*, 20: 199–202.

Sullivan, D.J. (1986). Aphid hyperparasites: Taxonomy and ovipositional behavior (Minireview). In: *Ecology of Aphidophaga*, ed. I. Hodek, pp. 1–562. Praha: Academia.

Sullivan, D.J. (1988). Aphid hyperparasites. In: *Aphids, Their Biology, Natural Enemies and Control*, eds. A.K. Minks & P. Harrewijn, pp. 189–203, Vol. 2B. Amsterdam: Elsevier.

Sullivan, D.J. (1999). Hyperparasitism: Multitrophic Ecology and Behavior. *Annual Review of Entomology*, 44: 291–315.

Sullivan, D.J. (2003). Hyperparasitism. In: *Encyclopedia of Insects*, eds. V.H. Resh & T.T. Cardé, Academic Press.

Sunderland, K.D. (1988). Carabidae and other Invertebrates. In: *World crop pests. Aphids. Their biology, natural enemies and control*, eds. A.K. Minks & P. Harrewijn, pp. 293–310, Vol. B. Amsterdam, the Netherlands: Elsevier.

Tamrakar, A.S. & Singh, R. (2000). Records of aphid parasitoids (Hymenoptera: Braconidae, Aphidiinae) from Madhyanchal of Nepal. *Journal of Aphidology*, 14: 129–132.

Tauber, M.J., Tauber, C.A. & Masaki, S. (1986). *Seasonal Adaptations of Insects*. Oxford University Press, New York.

Thies, C., Roschewitz, I. & Tscharntke, T. (2005). The landscape context of cereal aphid parasitoid interactions. Proceedings of the *Royal Society of London B., London*, 272: 203–210.

Tian, H.-W., van Achterberg, K. & Chen, X.-X. (2017). A new genus of the tribe *Praini Mackauer* (Hymenoptera: Braconidae: Aphidiinae) from China. *Zootaxa*, 4362(1): 131–134.

Todorov, O. (2009). New records of Aphidiidae (Hymenoptera) from Nepal. *Linzer Biologische Beitraege*, 4(1): 885–888.

Tomanović, Ž., Mitrović M., Petrović, A., Kavallieratos, N.G., Žikić, V., Ivanović, A., Rakhshani, E., Starý P. & Vorburger, C. (2018). Revision of the European *Lysiphlebus* species (Hymenoptera: Braconidae: Aphidiinae) on the basis of COI and 28SD2 molecular markers and morphology. *Arthropod Systematics & Phylogeny*, 76(2): 179–213.

Tougeron, K. (2017). Variabilité de la diapause chez les parasitoïdes de pucerons dans le cadre des changements climatiques: implications en lutte biologique. *Sciences agricoles*. Université Rennes.

Tougeron, K., Damien, M., Le Lann, C., Brodeur, J. & van Baaren, J. (2018). Rapid responses of winter aphid-parasitoid communities to climate warming. *Frontiers in Ecology and Evolution*, 180: 877–888.

Tougeron, K., Le Lann, C., Brodeur, J. & van Baaren, J. (2017). Are aphid parasitoids from mild winter climates losing their winter diapause? *Oecologia*, 183: 619–629.

Tripathi, S.K. & Singh, R. (1997). Effect of host-food plant transfer on the life-table of *Lysiphlebia mirzai* Shuja-Uddin, an aphid parasitoid of cereal aphids. *Malayasian Journal of Applied Biology*, 26: 45–55.

Ullah, M., Zaman, M. Ahmed, N., Ali, M. & Shah, J.A. (2015). Description and key to different species of genera of Aphidiinae (Homoptera: Aphididae) of District D.I. Khan, KPK. *Journal of Entomology and Zoology Studies*, 3(5): 221–224.

Unruh, T.R., White, V., González, D., Gordh, G. & Luck, R.F. (1983). Hererozygosity and effective size in laboratory populations of *Aphidius ervi* (Hymenoptera: Aphidiidae). *Entomophaga*, 28: 245–258.

van Achterberg, C. (1997). Subfamily Aphidiinae. In: *Manual of the New World genera of the family Braconidae (Hymenoptera)*, eds. R.A. Wharton, P.M. Marsh & M.J. Sharkey, pp. 119–132. Washington DC: The International Society of Hymenopterists.

van Achterberg, K. & Quicke, D.L.J. (1992). Phylogeny of the subfamilies of the family Braconidae- A reassessment assessed. *Cladistics*, 8: 237–264.

van Baaren, J., Boivin, G. & Nénon, J.P. (1994). Intra- and interspecific host-discrimination in two closely related egg parasitoids. *Oecologia*, 100: 325–330.

van Baaren, J., Lann, C.L., Pichenot, J., Pierre, J.S., L. Krespi, L. & Outreman, Y. (2009). How could host discrimination abilities influence the structure of a parasitoid community? *Bulletin of Entomological Research*, 99: 299–306.

van den Bosch, R., Frazer, B.D., Davis, C.S., Messenger, P.S. & Hom, R. (1970). *Trioxys pallidus*-An effective new walnut aphid parasite from Iran. *California Agriculture*., 28: 8–10.

van den Bosch, Messenger, P.S. & Gutierrez, A.P. (1985). *An introduction to biological control*. Springer Science+Business Media New York.

van den Bosch R., Schlinger, E.I., Hall, J.C. & Puttler, B. (1964). Studies on succession, distribution and phenology of imported parasites of *Therioaphis trifolii* (Monell) in southern California. *Ecology*, 45: 602–621.

van der Hoek, G. (1971). The larval instars of *Aphidius pulcher* Baker (Hymenoptera: Aphidiidae). *Proceedings of the Entomological Society of Ontario*, 101, 55–59.

van Driesche, R. & Hoddle, M. (2016). Non-target effects of insect biocontrol agents and trends in host specificity since 1985. *CAB Reviews*, 11(44): 1–66.

van Emden & Harrington, R. (2007). *Aphids as crop pests*. CAB International, Oxford, UK.

van Lenteren, J.C. (2003). *Quality Control and Production of Biological Control Agents*. Theory and Testing Procedures. Wallingford, CABI Publishing.

van Steenis, M.J. & El-Khawass, K.A.M.H. (1995). Life history of *Aphis gossypii* on cucumber: influence of temperature, host plant and parasitism. *Entomologia Experimentalis et Applicata*, 76: 121–131.

Vásquez, G.M., Orr, D.B. & Baker, J.R. (2006). Efficacy Assessment of *Aphidius colemani* (Hymenoptera: Braconidae) for Suppression of *Aphis gossypii* (Homoptera: Aphididae) in Greenhouse-Grown Chrysanthemum. *Journal of Economic Entomology*, 99(4): 1104–1111.

Vet, L.E.M., Lewis, W.J., Papaj, D.R. & van Lenteren, J.C. (2003). A Variable-response Model for Parasitoid Foraging Behaviour. In: *Quality Control and Production of Biological Control Agents: Theory and Testing Procedures*, ed. J.C. van Lenteren, pp. 25–40. CAB International.

Vevai, E.J. (1942). On the bionomics of *Aphidius matricariae* Hal., a braconid parasite of *Myzus persicae* Sulz. *Parasitology*, 34: 141–151.

Vinson, S.B. (1976). Host selection by insect parasitoids. *Annual Review of Entomology*, 21: 09–133.

Völkl, W. (1990). *Fortpflanzungsstrategien von Blattlausparasitoiden (Hymenoptera, Aphidiidae): Konsequenzen ihrer Interaktionen mit Wirten und Ameisen*. Ph. D. Dissertation, University of Bayreuth, Bayreuth, Germany.

Völkl, W. (1994). Searching at different spatial scales: the foraging behaviour of the aphid parasitoid *Aphidius rosae* in rose bushes. *Oecologia*, 100: 177–183.

Völkl, W. (2000). Foraging behaviour and sequential multisensory orientation in the aphid parasitoid, *Pauesia picta* (Hymenoptera, Aphidiidae) at different spatial scales. *Journal of Applied Entomology*, 124: 307–314.

Völkl, W. & Mackauer, M. (1990). Age-specific pattern of host discrimination by the aphid parasitoid *Ephedrus californicus* Baker (Hymenoptera: Aphidiidae). *Canadian Entomologist*, 122: 349–361.

Völkl, W., Mackauer, M., Pell, J.K. & Brodeur, J. (2007). Predators, Parasitoids and Pathogens. In: *Aphids as Crop Pests*, eds. H.F. van Emden & R. Harrington, pp. 187- 234. CAB International.

Walker, A.M. & Hoy, M.A. (2003). Responses of *Lipolexis oregmae* (Hymenoptera: Aphidiidae) to different instars of *Toxoptera citricida* (Homoptera: Aphididae). *Journal of Economical Entomology*, 96: 1685–1692.

Waterhouse, D.F. & Sands, D.P.A. (2001). *Classical Biological Control of Arthropods in Australia*. CSIRO Entomology (Canberra) and CSIRO Publishing (Melbourne). pp. 1–559.

Waterhouse, P.M. & Helms, K. (1985). *Metopolophium dirhodum* (Walker): a newly arrived vector of barley yellow dwarf virus in Australia. *Australasian Plant Pathology*, 14: 64–66.

Wei J.N., Li T.F., Kuang R.P., Wang Y., Yin, T.S., Wu, X.F., Zou, L., Zhao, W.Y., Cao, J. & Deng, J.H. (2003). Mass rearing of *Aphidius gifuensis* (Hymenoptera: Aphidiidae) for biological control of *Myzus persicae* (Homoptera: Aphididae). *Biocontrol Science and Technology*, 13: 87–97.

Wellings, P.W. (1993). Foraging behaviour in aphid parasitoids: Spatial scale and resource assessment. *European Journal of Entomology*, 90: 377–382.

Wharton, R.A., Shaw, S.R., Sharkey, M.J., Wahl, D.B., Woolley, J.B., Whitfield, J.B., Marsh, P.M. & Johnson, W. (1992). Phylogeny of the subfamilies of the family Braconidae (Hymenoptera, Ichneumonoidea) – A reassessment. *Cladistics*, 8: 199–235.

Wiackowski, S.K. (1962). Studies on the biology and ecology of *Aphidius smithi* Sharma & Subba Ran (Hymenoptera: Braconidac), a parasite of the pea aphid. *Acyrthosiphon pisum* (Harr.) [Homoptera: Aphididae). *Bulletin of Entomology, Pologne*, 32: 253–310.

Wickremasinghe, M.G.V. (2007). Searching behaviour of the aphid parasitoid *Aphidius rhopalosiphi* (Hymenoptera: Aphidiidae) in response to honeydew excreted by aphids on different host plants. *Ruhuna Journal of Science*, 2: 89–95.

Wilson, C.G., Swincer, D.E. & Valden, K.J. (1982). The introduction of *Trioxys complanatus* Quilis (Hymenoptera: Aphidiidae), an internal parasite of the spotted alfala aphid, into South Australia. *Journal of the Australian Entomological Society*, 21: 13–27.

Wool, D., van Emden, H.F. & Bunting, S.W. (1978). Electrophoretic detection of the internal parasite, *Aphidius matricariae* in *Myzus persicae*. *Annals of Applied Biology*, 90: 21–26.

Yano, E. (2006). Ecological considerations for biological control of aphids in protected culture. *Population Ecology*, 48: 333–339.

Ye, Z., Vollhardt, I.M.G., Girtler, S., Wallinger, C., Tomanovic, Z. & Traugott, M. (2017a). An effective molecular approach for assessing cereal aphid-parasitoid-endosymbiont networks. *Scientific Reports*, 7: 3138. 10.1038/s41598-017-02226-w. (Retrieved on 20 July 2021).

Ye, Z., Vollhardt, I.M.G., Tomanović, Ž. & Traugott, M. (2017b). Evaluation of three molecular markers for identification of European primary parasitoids of cereal aphids and their hyperparasitoids. *PLoS ONE*, 12(5): e0177376. 10.1371/journal.pone.0177376 (Retrived on 27 July 2021).

Yu, D.S., van Achterberg, C. & Horstmann, K. (2016). *Taxapad*, World Ichneumonoidea. www.taxapad.com (now only in 'database flash-drive'). Ottawa, Ontario.

Zamora-Mejias, D. & Hasan, P. (2016). Trophic relationships of aphid hyperparasitoids (Hymenoptera) in Costa Rica. *Journal of Entomology Research Society*, 18 (3): 35–44.

Zhang, H.Z., Li, Y.Y., An, F.X.T., Huang, F.X., Wang, M.Q., Liu, C.X., Mao, J.J. & Zhang, L.S. (2018). Comparative transcriptome and iTRAQ proteome analyses reveal the mechanisms of diapause in *Aphidius gifuensis* Ashmead (Hymenoptera: Aphidiidae). *Frontiers in Physiology*, 9(1697): 1–13.

Zhua, J., Zhang, A., Parkc, K., Bakera, T., Lange, B., Jurenkac, R., Obryckic, J.J., Gravesg, W.R., Picketth, J.A., Smileyh, D., Chauhand, K.R. & Klund, J.A. (2006). Sex pheromone of the soybean aphid, *Aphis glycines* Matsumura, and its potential use in semiochemical-based control. *Environmental Entomology*, 35(2): 249–257.

Zhu, Y.L., Yang, F., Yao, Z.W., Wu, Y.-K., Liu, B., Yuan, H-B. & Lu, Y-H. (2019). A molecular detection approach for a cotton aphid-parasitoid complex in northern China. *Sci Rep*., 9: 15836. 10.1038/s41598-019-52266-7 (Retrieved on 22 July 2021).

Žikić, V., Lazarević, M. & Milošević, D. (2017). Host range patterning of parasitoid wasps Aphidiinae (Hymenoptera: Braconidae). *Zoologischer Anzeiger*, 268: 75–83.

5 Pteromalid Parasitoids

Arvind Kumar and Aishwarya Ray
Forest Protection Division, Forest Research Institute, Dehradun
Uttarakhand India

Meena Agnihotri
G.B. Pant University of Agriculture and Technology, Pantnagar
Uttarakhand India

CONTENTS

LEARNING OBJECTIVES

1. To know the diversity of Pteromalidae parasitoids in the ecosystem.
2. To understand the biology and its role in the biocontrol of insect pests and vectors.
3. Come to know about the effective biocontrol agents and their mass multiplication for field release.

5.1 INTRODUCTION

The family Pteromalidae is one of the largest and second earliest available families of super-family Chalcidoidea (Hymenoptera) to define taxonomically and phenotypically. The species of the family Pteromalidae is diversified with their morphological features, biology and behavior. These attributes play a vital role in the diversification of the pteromalid species with the geographical variations, therefore categorisations of species, like desert, tropical climate, semitropical and many of them complete their life cycle in a temperate climate. Subsequently, the pteromalid species are categorised

DOI: 10.1201/9781003354239-5

on the basis of their food requirements. Some of the species prefer to feed on plant-based protein (plant feeders) but the majority of them prefer animal protein (entomophagous). Therefore, species are found to feed on the grasses; some are plant gall formers and a few of them complete their development in the flower buds or seeds. But the majority of pteromalids are primary or secondary parasitoids, endo- and ectoparasitoids attacking a large number of insects in various stages of their development as idiobionts and koinobionts. The pteromalids may attack the egg, larval, prepupal, pupal or adult stages of the insects of major insect orders.

They play an important role in the biocontrol of serious insect pests of crops, *viz.* field insect pest parasitoids- *Catolaccus grandis, Cephaleta australiensis, Cheiropachus quadrum, Monoska dorsiplana, Pteromalus puparum, Scutellista caerulea, Trichomalus perfectus*, etc. Some of the most effective stored-product insect pest parasitoids have been well recognised, *viz. Anisopteromalus, Dibrachys, Lariophagus, Pteromalus and Theocolax species are* registered with Environmental Protection Agency for use as biocontrol agents in the United States (Hagstrum and Subramanyam, 2006). The family also has a good number of parasitoids of medical and veterinary importance insect pests, *viz. Muscidifurax zaraptor, M. raptor, Nasonia vitripennis, Pachycrepoideus vindemmiae, Spalangia nigroaena, Spalangia cameroni, Trichomalopsis sarcophagae, Urolepis rufipes*, etc.

5.1.1 Identification Features

The members of the family Pteromalidae cannot be identified by any unique attribute but need a combination of morphological attributes. In spite of great structural diversity and various morphological variations, Pteromalidae can be recognised on the basis ofcharacteristics like small to very large chalcids wasp, body slender to quite robust, from 1 to 48 mm in length. Body is usually black, metallic black, green or yellow. Head and thorax reticulately punctuate. They possess 5-segmented tarsi, antennae usually 8–13 segmented with 1–3 anelli and 5–7 funicular segments; both mandibles with at least 2 teeth. Forewings with well-developed marginal, post-marginal and stigmal veins, mesopleuron not swollen and convex. Metasoma is usually sessile or petiolated, and the ovipositor is rarely exerted. Pteromalinae with the head orthognathous, the back of the head always possesses a broad channel between the oral fossa and foramen magnum. Other characters, such as the elongated cerci present in a few genera are apparently plesiomorphic (Boucek, 1988; Boucek and Rasplus, 1991; Sureshan, 2003). Some morphological features have been shown in the Figures 5.1 & 5.2.

5.1.2 Brief History and Systematic Position

Pteromalidae is one of the oldest families of the order Hymenoptera. The study of pteromalid fauna o have begun with Carolus Linnaeus with his publication of *Sytema Naturae* in 1758 and later Walker has systematically classified the group in his publication "*Monographia Chalciditum*" (1833–1839) followed by Thomson's *Hymenoptera Scandinaviae* (1876–1878). Dalman (1820) first isolated the family Pteromalini, the second earliest named group available in the superfamily Chalcidoidea. The subfamilies Miscogasterinae, Ormocerinae and Cleonyminae have been separated as a new family from Pteromalidae in 1933. Later it again resumed and subsequently, 'Spalangiides' of Torymidae was placed under 'Pteromalidae' (Walker, 1833; Westwood, 1839). During this period another group called Tribus Pirenianii was separated from Pteromalidae by Haliday, (1844). Subsequently Pirenini was placed under the subfamily Miscogasterinae, Pteromalidae (Graham, 1969). Forster (1856) attempted an elaborate classification system and treated all the previously proposed groups as different families; which was changed by Thomson (1876, 1878), who replaced the Ormocerides by Tridymina, and treated it as 'tribes', (now re-added as subfamilies) Pirenina, Spalangiina and Pteromalina. Pteromalina has been subdivided into several subtribes including 'Mischogastrides' and 'Cleonymides', which were later elevated as

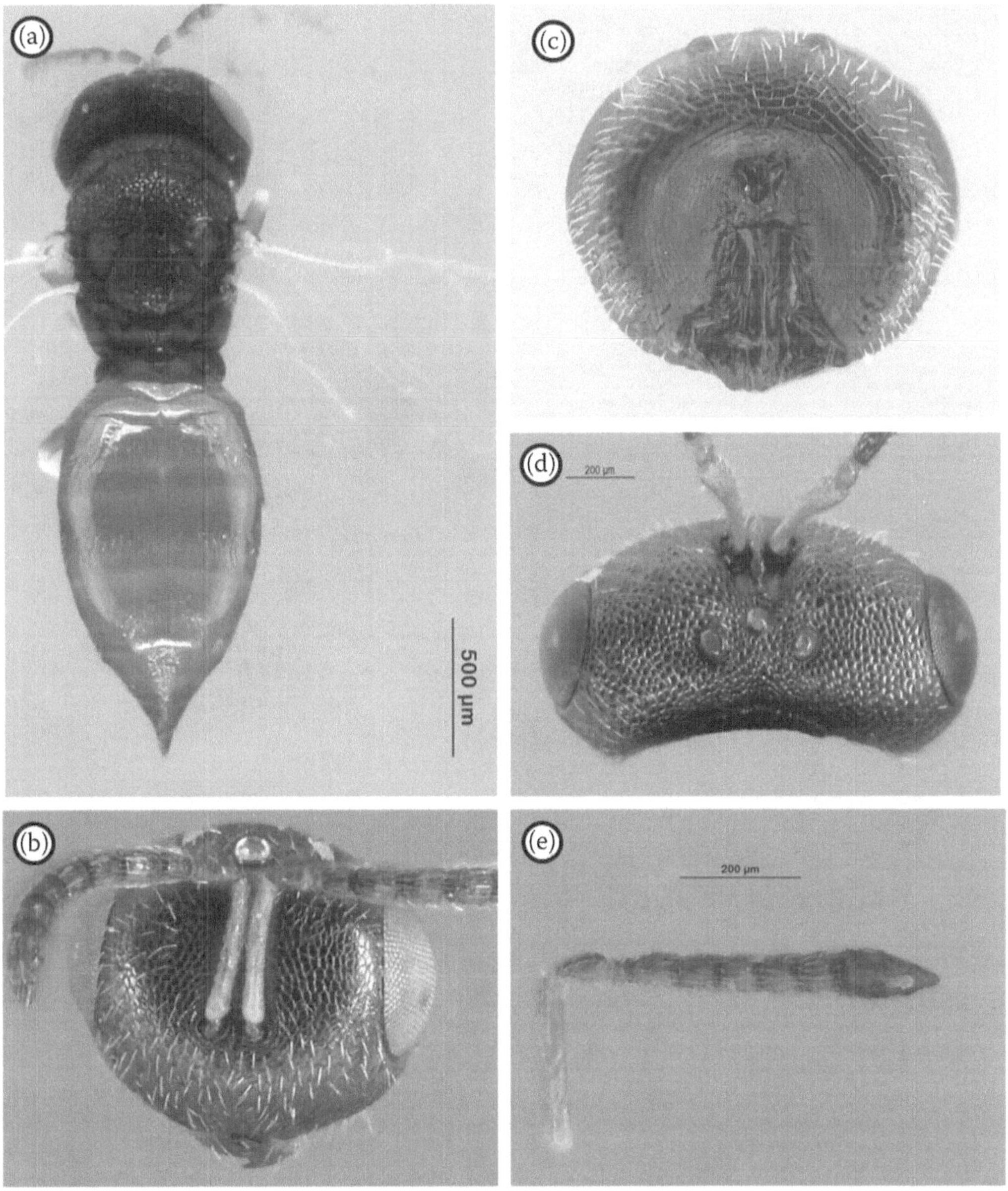

FIGURE 5.1 (a–e): Different body parts of a Pteromailidae. a-Dorsal view; b-front view of head; c-back view of head; d-dorsal view of head; e-antenna of a Pteromalidae wasp.

subfamilies of Pteromalidae. Ashmead (1904) has placed the species included in Pteromalidae into 3 families: Miscogasteridae (with 4 subfamilies), Cleonymidae (2 subfamilies) and Pteromalidae (6 subfamilies). Later based on A.B. Gahan views, Peck (1951) united these 3 families in Pteromalidae with subfamilies Spegigasterinae and Pteromalinae. Graham (1969) classified the European Pteromalidae, which has had a major impact on the classification of the family Pteromalidae and 15 subfamilies were recognised in Europe and 12 subfamilies were represented in the Australian region. Apart from Graham's monograph (1969), keys to European genera were provided by Boucek (1988). Recently, altogether 28 subfamilies were described by Boucek (1988) in his monograph "Australian Chalcidoidea". Presently, the family Pteromalidae is considered as

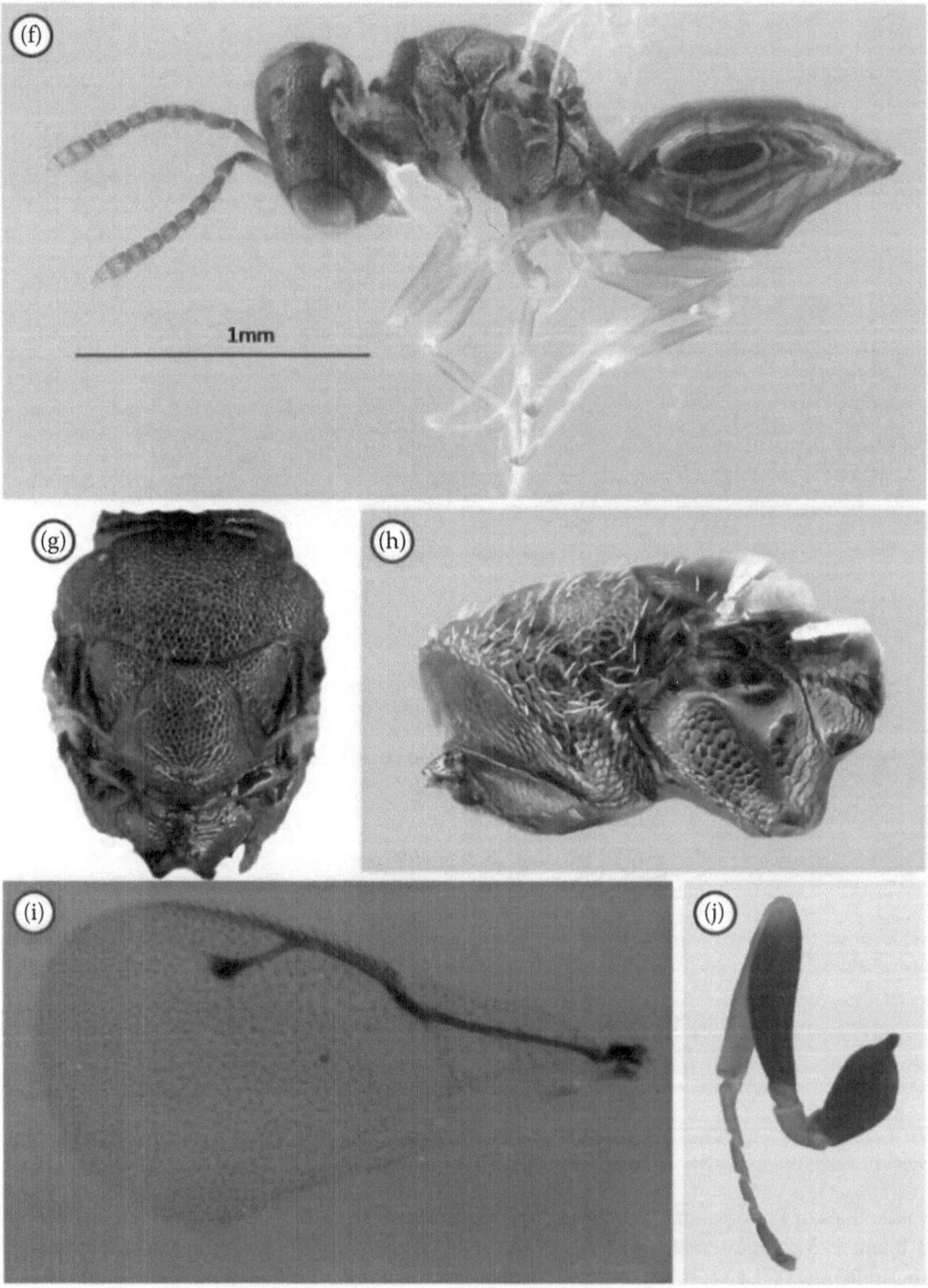

FIGURE 5.2 (f–j): Different body parts of a Pteromalidae wasp; f-lateral view of a wasp; g-dorsal view of thorax; h-lateral view of thorax; i-fore wing; j-hindleg.

TABLE 5.1
Subfamilies of Pteromalidae Recorded Worldwide

Sl.No.	Subfamilies	Sl.No.	Subfamilies
1	Asaphinae	18	Herbertinae
2	Austrosystasinae	19	Keiraninae
3	Austroterobiinae	20	Leptofoeninae
4	Ceinae	21	Louriciinae
5	Cerocephalinae	22	Macromesinae
6	Chromeurytominae	23	Miscogastrinae
7	Cleonyminae	24	Neodiparinae
8	Coelocybinae	25	Neofoeninae
9	Colotrechninae	26	Ormocerinae
10	Cratominae	27	Panstenoninae
11	Diparinae	28	Parasaphodinae
12	Ditropinotellinae	29	Pireninae
13	Elatoidinae	30	Pteromalinae
14	Epichrysomallinae	31	Spalangiinae
15	Erotolepsiinae	32	Storyinae
16	Eunotinae	33	Sycoecinae
17	Eutrichosomatinae	34	Sycoryctinae

one of the largest families in the order Hymenoptera with 34 subfamilies worldwide (Table 5.1). Pteromalidae is classified as follows:

Kingdom-Animalia
Phylum- Arthropoda
Class- Insecta
Order- Hymenoptera
Sub-order- Apocrita (Parasitica)
Super family- Chalcidoidea
Family- Pteromalidae

5.2 DIVERSITY OF PTEROMALIDAE

Insects are the most species-rich taxonomic group on the Earth and about 950,000 insect species have been described and many millions of species are yet to be explored (Groombridge, 1992). The order Hymenoptera has diversified into various morphological forms and ways of life and is one of the largest orders of insects and the family Pteromalidae is one of the largest parasitic families of super-family Chalcidoidea (Gaston, 1993; Austin and Dowton, 2000). The taxonomic assessment of the family Pteromalidae was initiated by Carolus Linnaeus with his species names description of *Ichneumon puparum* and *Sphex colon*. These were later erected as genus *Pteromelus* with the type of species *Ichneumon puparum* Linn. rechristened as *Pteromalus puparum* and *Dinotiscus colon,* respectively (Swederus, 1795). The genus *Pteromalus* is the oldest one and this is also the most diversified genera; most of the species are recorded from Germany among over 480 species recorded worldwide. The Pteromalidae diversity has been described by many workers in different regions of the world and described about 4222 species from more than 200 countries under 640 genera worldwide(Noyes, 2021). Pteromalinae is the largest subfamily of this family. The regional diversity of Pteromalid fauna is also well categorised and presented in Table 5.2.

TABLE 5.2
Regional Diversity of Pteromalid Species

Sl No.	Regions	No. of Species
1	Afrotropical	572
2	Australasian	664
3	Europian	6443
4	Indo-Australian	348
5	Malagasy	108
6	Nearctic	775
7	Neotropical	1028
8	Oriental	535
9	Palaearctic	2885

The diversity of Afrotropical pteromalid species is well studied and described by J. Risbec, L. Masi, Z. Bouček, K.J. Hedqvist and other taxonomists during the 19th century and they described about 81 species. Later with the contributions of many other workers linked with recent taxonomists, Mitroiu recorded 241 species under 103 genera (Mitroiu, 2011; Noyes, 2021). Maximum species in this region are recorded from South Africa (100 species), followed by Zimbabwe (43 species), Yamen (40 species), Senegal (36 species), Uganda (23 species), etc.

Australian region is also rich in pteromalid diversity and the major study has been carried out by F. Walker, A.A. Girault, M.W.R. de V. Graham, Z. Boucek, J. LaSalle. Among them, the work of Z. Boucek is memorable in which about 235 genera and about 71 new genera were erected from the region. The highest number of species are recorded from Australia (596 species) followed by New Zealand (56 species).

The European region is one of the largest geographical regions in terms of floral and faunal diversity. Maximum records of pteromalid species from the region have been made possible by European taxonomists, like F. Walker, M.W.R. de V. Graham, Grissell, Z. Baucek, J.Y. Rasplus, Delucchi, Rosen Hedqvist and J. LaSalle. One of the most outstanding works in the family was done by Graham (1969) who dealt with more than 800 species and described 4 new genera and 87 new species under 15 subfamilies in his monograph on Pteromalidae of North Western Europe. In Europe, Germany has a maximum of 736 species records, followed by Sweden (695 species), the United Kingdom (638 species), Czech Republic (447 species), Hungary (405 species), Spain (339 species), Romania (310 species), Russia (210 species), Italy (208 species), Bulgaria (202 species), Ireland (196 species), Switzerland (158 species), Austria (149 species), and several other countries have a good contribution. Similarly, in the United State of America has 555 species records, followed by Canada (210 species) and Greenland (10 species)in the Nearctic region. Similarly, in the Neotropical region, Brazil has the maximum of 156 species records, followed by Argentina (107 species), Chile (102 species), Montserrat (68 species), the major species recorded countries till now.

The Palaearctic region of Asian countries, majorly studied by K. Grissell, E.E. Kamijo, H. Xiao, D.W. Huang and door Yang, has a major contribution. A total of 46 species with 30 genera within the subfamily Pteromalinae have been recorded in Korea since 1955. Recently 25 unrecorded species of 13 genera have been described from South Korea (Guk-Hyang *et al.*, 2018). In this region major diversity of Pteromalid species is recorded from China (452 species), followed by Belgium (239 species), Czechoslovakia (347 species), Kazakhistan (301 species), Moldova (180 species), Turkey (180 species), Japan (162 species), Serbia (117 species), Iran (111 species) and USSR (104 species).

The Oriental region is one of the biologically rich regions and the major contribution in the Pteromalid diversity study was made by Ferrière (Ferrière, 1930), G. G. Saraswat, S.I. farooqi,

B.R. Subha Rao, M.S. Mani, Z. Boucek, T.C. Narendran and P.M. Sureshan. The remarkable contribution in the taxonomic study of Pteromalidae in the Oriental region has been made by Boucek (1978), who reported 82 genera and 86 species from India and adjacent countries. Subsequently, a review of 72 genera of Pteromalidae of the oriental region was published by Farooqi and Subba Rao (1986). Recent contribution to Pteromalidae taxonomy has been made by T.C. Narendran and P. M. Sureshan from India. They have prepared a checklist of Pteromalidae from Indian Subcontinent. They have listed 84 genera with 185 identified species and 12 genera with unidentified species. Thus, they have added more than 100 species from this region, and currently, India has 339 species records followed by Sri Lanka (56 species) and Taiwan (42 species) from the region, apart from miner contributor countries.

5.3 BIOLOGY

The morphological diversity among the species reflects the diversity in the biological attributes of the species. Additionally, the ecological distribution of the species also throws light on the diversified biology of pteromalid fauna. The biology of the majority of the species is known, but many species are still unknown. The closely related genera are biologically similar to each other. Some of the species completing their life cycle on plants have been grouped into phytophagous category. There are some species, which cause plant galls, seed borers or as inquilines in galls caused by other insects. Some of the species of subfamily Ormoceranae are known to be gall formers (some *Systasis* sp.). Some of the pteromalids are inquilines rather than being parasites. They feed on galls produced by Diptera (Cecydomyiidae) and Hymenoptera (Cynipidae) (Boucek and Rasplus, 1991). Some of the species are also recorded as pollinators and seed feeders in *Ficus* species, *viz. Odonotofroggatia corneri. O. galili, Walkerella microcarpae, Elisabethiella baijnathi, Elisabethiella stuckenbergi, Sycophila punctum, Otitesella uluzi, Platyscapa quadraticeps* (Noort *et al.*, 2013). The genus *Stenetra* Masi is found to be the seed feeder of *Sphora sp.*; *Alopecuroides* also *Cyrtoptyx* cf. *latipes* Rondani reared from the seed of *Pistacia atlantica* (Lotfalizadeh and Gharali, 2008; Mitroiu *et al.*, 2011).

Some pteromalid species are egg parasitoids and they lay eggs on or inside the egg of their host insects. The parasitoid larvae develop inside the eggs and mature adults emerge out from the eggs. Though there are poor records but some of them, like *Caenocerepis bothynoderi* (Gromakov) and *C. arenocola* (Thomson) are egg parasitoids of weevil *Peridesmia* species; *Caenocrepis bothynoderi,* a egg parasitoid of *Pachycerus cordiger, Acroclisoides sinicus* (Huang & Liao) and *Halyomorpha halys, Peridesmia discus* (walker) are egg parasitoids of *Hypera postica* (Baucek & Rasplus, 1991;Mitroiu 2012; Peverieri, *et al.*, 2019; Elijah, 2019).

Some pteromalid species are recorded to be predaceous on the eggs of certain hosts. The adult females lay eggs in the gravid coccids/ scale insects and their larvae feed on the eggs internally in the body of coccids/scale insects, *viz. Scutelista cyanea Ceroplastes floridensis, Cephalita nirupama, Cephaleta australiensis, Cephaleta brunniventris, C. tripathi, Cavitas concave, Eunotus aequalivena, Ophelosia crawfordi, Moranila californica, Eunotus strenus, Eunotus cretaceous, Eunotus acqualivena, Eunotus applanatus, Eunotus acutus, Eunotus cretaceous, Eunotus obscuras, Scutellista caerulea, Cephaleta elongata, Scutellista hayati, Ophelosia indica, Ophelosia maculate, Ophelosia crawfordi* and *Mesopeltita truncatipennis*. (Minshawy *et al.*, 1978; Narendran and Mini, 1998-1999, ;Xiao and Huang, 2001a; Sureshan and Narendran, 2005; Sureshan *et al.*, 2011).

Pteromalid species are also recorded to complete their life cycle on adult insects, though there are very few examples in this category, *viz. Tomicobia pityophthori, T. seitneri* and *Tomicobia tibialis* adult parasitoids of bark beetles, *Pityogenes chalcographus, Otiorthynchus* and *Ips* sp. respectively; *Pachyneuron aphidis* (Bouché) adult parasitoid of *Aphisgossypii* and *Brevicoryne brassicae* (Hasani and Madjdzadeh, 2012; Wegensteiner *et al.*, 2015).

Pteromalidae parasitoids have good host-searching ability and many of the species are excellent larval parasitoids. Adult females search host insects and lay their eggs on the larval stages or near

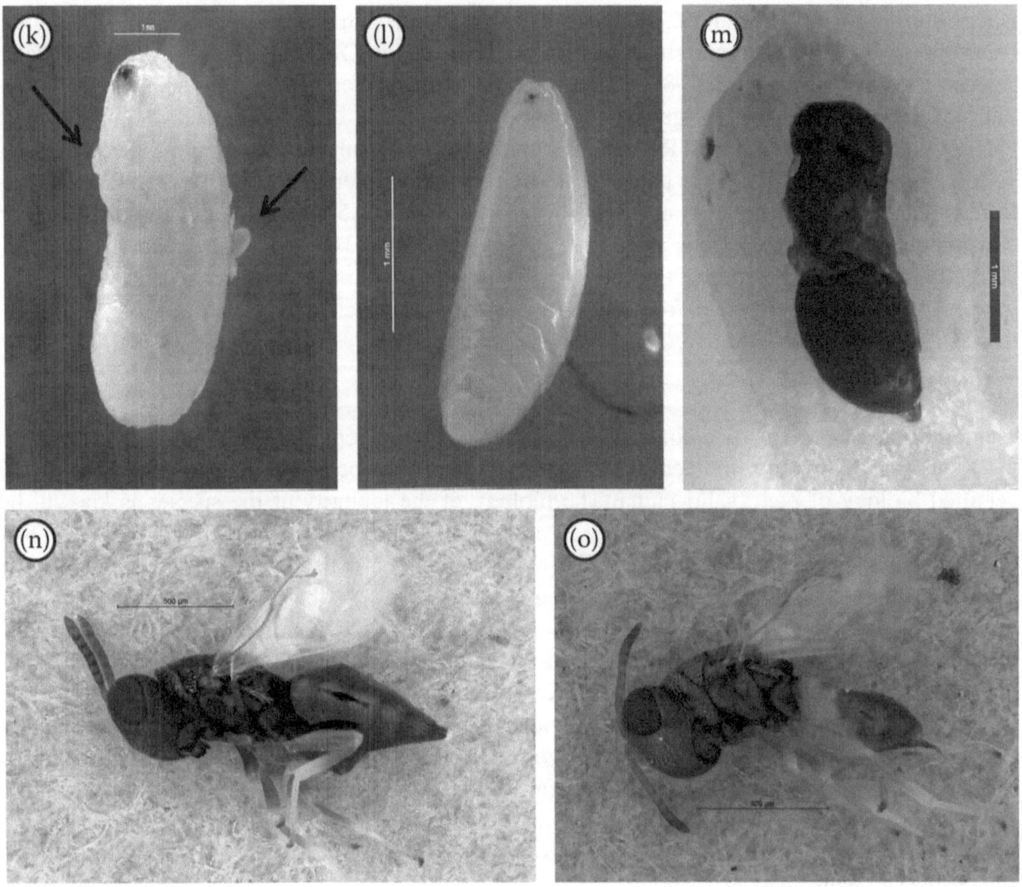

FIGURE 5.3 (k–o): Different biological stages of *Dinarmus basalis* Rond. k-egg and larvae attached with the *Callosobruchus maculates* larvae; l-Larvae; m-pupa; b-Female adult; o-male adult.

the larvae and after emergence parasitoid larvae are attached to the host larvae and complete their life cycle (Figure 5.3). Pteromalid species *Anisopteromalus calandrae, Catolaccus hunter, Catolaccus grandis, Catolaccus hunter, Dinarmus basalis. Lariophagus distinguendus* and *Theocolax elegans* are the best larval parasitoids of bostrichid and curculionid insect hosts that feed inside cereal grain kernels (Hagstrum and Subramanyam, 2006).

Pteromalus puparum Linn. is the most diverse cosmopolitan endoparasitoid of prepupal and pupal stages of Lepidoptera. The female assesses the size of host pupa and accordingly oviposits the number of eggs. The size of emerging adults decreases with an increase in the number of parasitoid larvae and host-pupae (Takagi, 1987). *Pteromalus puparum* is one of the polyphagus gregarious pupal parasitoid of the many butterfly pupae *viz. Catopsilia pyranthe, Graphium agamemnon, Danaus chrysippus, Pachilopta aristolochiae, Papilio polytes, Pterourus multi-caudatus* and *Euploea core* (Santhosh and Basavarajappa, 2017; Edith *et al.*, 2020). *Sposophagus lepidopterae* (Risbec) is a pupal endoparasitoid of lepidopteran pupa of *Spodoptera littoralis* (Delvare and Rasplus, 1994).

There are many studies on the biology of pteromalid parasitoids under laboratory conditions and their developmental period has also been worked out (Table 5.3). The biology of a very effective parasitoid *Anisopteromalus calandrae* has been investigated on the stored grain pest hosts, viz. *Rhizopertha domeinica, Sitophilus zeamais, Lasioderma serricorne* (Ahmad, 1996; Visarathanonth *et al.*, 2010; Oanh *et al.*, 2017).

TABLE 5.3
Developmental Period (days) of Some Important Pteromalid Parasitoids

Species	Egg incubation period	Larval period	Pupal period	Life cycle	Reference
Anisopteromalus calandrae	1.3	6.9	10.6	18.9	Ahmad, 1996
Anisopteromalus calandrae	1.0	4.1	6.3	11.4	Visarathanonth *et al.*, 2010
Catolaccus grandis	1.4	6.6	5.3	13.3	Juan *et al.*, 1992
Catolaccus hunteri	1.17	4.47	5.52	13.18	Esteban *et al.*, 2000
Dinarmus basalis Rond	–	–	–	13.27	Islam, 1994
Pteromalus cerealellae	2	5–9	4–5	12	Onagbola, and Fadamiro, 2008

Catolaccus hunteri is a good larval ectoparasitoid of *Callosobruchus maculatus* F. and its detailed biology has been studied. A single female lays about 24 eggs in the life span (Esteban *et al.*, 2000), *Catolaccus grandis* (Burks) a natural parasitoid of boll weevil, *Anthonomus grandis* Boheman, larvae has a good host-searching ability in the field (Morales- Morales-Ramos *et al.*, 1998). A single female parasitic wasp oviposits about 25 eggs and after oviposition parasitic larvae move towards the host larvae and remain attached to the body and complete their life cycle (Juan *et al.*, 1992).

Dinarmus basalis Rond, a larval parasitoid, is highly diverse and occurs in tropical and subtropical environments. It can be collected from any leguminous seed infested with bruchid beetle. This is a solitary, idiobiont ectoparasitoid of immature and even recently formed adults of several bruchid species that develop on legumes (Verma, 1991). This wasp can be reared in the laboratory and female take more time (308.14 h) to develop than the male wasp (284 h). The female progeny ratio always dominated (Islam and Kabir, 1991, Islam, 1994). *Pteromalus cerealellae* (Ashmead) is another effective larval ectoparasitoid of weevils and beetles under storage conditions (Onagbola and Fadamiro, 2008).

Lariophagus distinguendus, is a solitary ectoparasitoid of larval, pre-pupal, and pupal host stages of beetle, *Sitophilus granarius.* The female wasps effectively search the host larvae in the grain using chemical cues of host faeces (Hase, 1924; Ryabov, 1926; Kashef, 1955; Steidle and Scholler, 1997; Steidle, 2000; Steidle & Scholler, 2002). After encountering an infested seed or cocoon, female wasps show a distinct sequence of behaviors and determine the presence of a suitable host stage (Hase, 1924); thereafter female injects the venom and paralyzes the host for egg-laying (Ryabov, 1926, Isla and Kabir, 1991). The parasitic female wasp lays a single egg in/on each larva of the host and the parasitic larva feeds on the host larva and then molts, pupates, and the adult emerges. Female takes more developmental time and mating takes place just after adult emergence (Gonen and Kugler, 1970).

Larval ectoparasitoids, *Catolaccus hunteri* Crawford, of pepper weevil, *Anthonomus eugenii* and boll weevil, *A. grandis,* the biology of these parasitoids have been studied on *Callosobruchus maculatus* (Esteban *et al.*, 2000).

5.4 EFFECTIVE BIOCONTROL AGENTS

Biological control is a safe and sustainable approach in IPM that takes advantage of natural enemies, such as predators, parasitic insects or pathogens to suppress the pest populations. This is a possible alternative to chemical insecticides for pest control without affecting the growth and development of plants. Hymenopteran parasitoid wasps are a large evolutionary group of insects and are well-known biocontrol agents for arthropod pests in agricultural and forest ecosystems (Wang *et al.*, 2019). Pteromalidae parasitoids have been extensively studied to control the insect pests of agriculture and veterinary and to manage invasive plant species worldwide. The management of weevils, *Sitophilus granarius*, *S. oryzae* (Coleoptera: Curculionidae), and many other storage beetles with the pteromalid wasp, *Lariophagus distinguendus,* has been suggested. The use of this wasp as a biocontrol agent was already reported by Hase in 1919 (Niedermayer *et al.*, 2016).

The genus *Anisopteromalus* Ruschka has many parasitic species for noxious beetles and weevils, *viz. A. calandrae*is an effective biocontrol agent of *Lasioderma serricorne*, *Rhyzopertha dominica, Sitophilus zeamais, Sitophilus granaries, S. oryzae, Trogoderma granarium* and *Callosobruchus chinensis* distributed worldwide and parasitize under storage conditions (Kapil and Chaudhary, 1973; Sweetman, 1964; Okamoto, 1971; Williams and Floyd, 1971).

Catolaccus grandis (Burks) is an effective biocontrol agent of cotton ball weevil, *Anthonomus grandis* (Morales-Ramos *et al.*, 1992). Another pteromalid wasp,*Monoska dorsiplana* is effective in the biocontrol of bruchid beetle *Pseudopachymeria spinipes* (Danielle *et al.*, 2007). *Catolaccus hunteri* Crawford an effective biocontrol agent of *Anthonomus vestitus* Boheman, *A. grandis* Boheman, *A. eugenii* Cano, and *A. macromalus,* which are important pests of cotton (*Gossypium hirsutum*), pepper (*Capsicum annuum*), and acerola, *Malpighia glabra*, respectively (Cross and Mitchell, 1969, Cross and Chesnut, 1971; Hunsberger and Pena, 1997; Esteban, 2000).

Dinarmus basalis (Rondani) is an effective biocontrol agent of *Callosobruchus chinensis, C. analis, C. maculatus* F. (Coleoptera: Bruchidae), maxican bean weevil, *Zabrotes subfasciatus* Bohman and maize weevil, *Sitophilus zeamais, Acanthoscelides macrophthalmus, Bruchidius lineatopygus*, and *Caryedon pallidus* (Islam *et al.*, 1985; Verma, 1990; Monge and Huignard, 1991; Bolévane, 2004; Effowe *et al.*, 2010).

Pteromalus puparum Linn. has a very good potential for insect pest management in fields (Lee and Heimpel, 2005). *P. puparum* was used as a biocontrol agent for the management of *Pieris rapae,* and more than 70 % effective parasitization was recorded (Chamberlin and Kok, 1986; Lasota & Kok, 1986a). The females of *Pteromalus cerealellae* were another effective parasitoid of storage insect pest *Sitotroga cerealella* (Olivier), *Callosobruchus maculatus* (Fabricius), *Lasioderma serricorne* (Fabricius), *Prostephanus truncates* (Horn) and *Sitophilus* spp. under stored conditions (Mbata, 2004; Onagbola and Fadamiro, 2008; George and Warsi, 2019).

Scutellista cyanea (Motsch) wasp lays eggs on the body of a soft scale and was recorded as the most effective biocontrol agent of *Ceroplastis rusci* worldwide. A parasitization of 83.2% was recorded in India (Kumar, 2013), 75% parasitisation in Jordan (Awamaleh *et al.*, 2009) and 65% parasitisation in Italy (Inserra, 1971). *Similarly, S.caerulea* (Fonscolombe) recorded 40.9% effectiveness against a soft body scale, *Saissetia oleae* in Egypt (Badary and Shaaban, 2011).

These filth flies *Musca domestic* and *Stomoxys calcitrans* and over 90% population can be controlled with the very effective pupal parasitoid *Spalangia* spp. (Hogsette, 1981; Machtinger, 2011; Pitzer *et al.*, 2011b; Machtinger *et al.*, 2013). *Spalangia cameroni* Perkins dominant parasitoid on equine farms and found effective up to 86% of species and cause 32.5 % parasitization in filth flies population (Machtinger, 2011; Pitzer *et al.*, 2011b; Tormos *et al.*, 2012). Similarly, *Spalangia endius* was recorded as effective (33.9% and 27.3%), *S. cameroni* (27.9% and 40.6%), and *S. nigroaenea* (21.0% and 24.8%), against house fly and stable fly, respectively, in Florida (Romero *et al.*, 2010). *Spalangia cameroni, Spalangia nigroaena, Trichomalopsis sarcophagae* and *Urolepis rufipes* have also been recorded to be successful on pupae of house fly, *Musca domestica* and stable flies (Meyer, 1983; Greene, 1990; Seymour and Campbell, 1993; Andress and Campbell, 1994).

Parasitoids, *Muscidifurax raptor*, *M. raptorellus* and *M. zaraptor*are effective parasitoids of house flies and stable flies and the fecundity of these parasitoids ranges from 69 to 211 (Morgan *et al.*, 1989; Geden *et al.*, 1992c; Lysyk, 2001a) and *Muscidifurax raptorellus* was recorded 34% effective against house flies and stable flies (Floate *et al.*, 2000). Similarly, *Nasonia vitripenni, Pachycrepoideus vindemmiae* and *Splangia nigroaenea* were appropriate parasitoids of synanthropic flies (Khoobdel *et al.*, 2019).

Pteromalid wasp, *Narbanus scabriculus* (Nees) is an effective parasitoid of wheat pest *Cephus pygmeus*, and *Pachycrepoides vindemmiae* (Rondani) parasite of *Dacus oleae.* Similarly,

Sceptrothelys deione (Walker) is a parasitoid of *Coleophora* spp. in Europe and *S. intermedia* Graham of curculionid *Hypera* spp. *Tritneptis klugii* (Ratz.) is an effective parasitoid of sawflies, *Trichomalopsis inops* (Walker) of *Hypera postica* on Clover. *Trichomalopsis peregaria* (Graham) is an effective parasitoid of *Euproctis chrysorrhoea* and it has been introduced to its management in America from Europe (Boucek and Rasplus, 1991).

Trichomalus perfectus (Walker) is a very prevalent ectoparasitoid of weevil and its large-scale application of biological management of oilseed rape pest, *Ceutorhynchus obstrictus* (Coleoptera: Curculionidae) has been done in Sweden, Denmark, England and Germany (Gibson *et al.*, 2006; Mason *et al.*, 2011; Haye *et al.*, 2018; Eurostat 2018). Additionally, it has also been found very effective against *C. turbatus* Schultze, *C. cardariae* Korotyaev, *C. omissus* Fall and *C. querceti* (Gyllenhal) (Haye *et al.*, 2015).

5.4.1 Mass Rearing

Mass rearing involves the multiplication of the parasitoid population in many folds for the management of insect pests under laboratory or field conditions. This technique is underutilised in the Pteromalidae for many promising parasitoids. Though the mass rearing and release technique has been successfully concluded for weeds and insect pests with other groups of natural enemies, the mass rearing and release of pteromalid parasitoids have been successfully performed in many cases worldwide.

Pteromalid wasp, *Cyrogaster vulgaris* Walker, was mass-reared and released for the control of Diptera leaf minor of *Ilex aquifolium* and *Halticoptera daci*was released against fruit fly (Diptera: Tephritidae) in the year 1914 (Baucek & Rasplus, 1991).

Anisopteromalus calandrae is one of the most distributed and effective parasitoids, used as an effective biocontrol agent for stored grain pests, *Rhizopertha dominica* and *Sitophilus oryzae* in many countries (Perez-Mendoza *et al.*, 1999), and successful suppression (95.3%) was achieved with weevil ratios of 1:24 (Press *et al.*, 1984). Visarathanonth *et al.* (2010) used *A. calandrae* against the stored grain pest, *Sitophilus zeamais* in Rice. *A.apiovorus* (Cheong *et al.*, 2016) and *A. calandrae* (Oanh 2017) were mass-reared and were good candidates for the biocontrol of stored grain pest, *Lasioderma serricorne.*

The selective ectoparasitoid, *Catolaccus grandis* (Burks) is the major parasitoid of cotton ball weevil, *Anthonomus grandis* (Elzen and King 1999), and reared successfully on *A. grandis*larvae (King *et al.*, 1995). The cotton ball weevil field population was suppressed by augmentative releases of *C. grandis* (Morales-Ramos *et al.*, 1992; Roberson and Harsh, 1993; Summy *et al.*, 1995). About 90% mortality of boll weevil was recorded at weekly releases of 500 females per acre for 6 consecutive weeks (Morales-Ramos *et al.*, 1994; King *et al.*, 1995; Summy *et al.*, 1995; Coleman *et al.*, 1996).Mass rearing of *Dinarmus basalis* can be done by mass rearing of its host insect bruchid species, *viz. Callosobruchus chinensis, C. maculatus, Bruchidius lineatopygus* (Coleoptera: Bruchidae), *Acanthoscelides macrophthalmus* (Schaeffer), or *Caryedon pallidus* (Sexena, 1996; Bolévane, 2004). Mass rearing of these host beetles can be easily done on Laguminacae seeds, *viz. Vigna mungo, Vigna radiata, Cajanus cajan, Cicer arietinum,* etc. Islam (1998) described the mass-rearing technique first time for *Dinarmus basalis* and mass cultured on *Callosobruchus maculatus* population in a sealed room and a release of 1000 parasitoids produced > 35,000 parasitoid wasps per week. The female of *D. basalis* has produced 35.5 progeny with the introduction of 5 pairs and they were found to be capable to suppress about 85% pulse beetle population. Later *D. basalis* parasitoid was successfully mass-reared on another alternate host, seed beetle *A. macrophthalmus* (Bolevane *et al.*, 2006; Effowe *et al.*, 2010), which easily fed with low-cost *Leucaena leucocephala* (Lamark) (Fabaceae) seeds (Raghu *et al.*, 2005; Tuda *et al.*, 2009).These mass-reared parasitoids were used for the biocontrol of bruchid beetle. *D. basalis* wasp was introduced to Japan from India in 1980 (Fujii and Wai, 1990), which suppressed about 100% population of *C.chinensis* under storage at the introduction level of 50 pairs (Islam and Kabir, 1995). *Dinarmus acutus*was mass cultured and released for the management of *Bruchus* species in USA. *Dibrachoides dynasres* (Forster) was mass multiplied and released for the management of *Hypera postica.*

Mass rearing of parasitoid, *Monoska dorsiplana* was discovered on the seeds of *Acacia caven* attacked by the bruchid *Pseudopachymeria spinipes* (Danielle *et al.*, 2007). The mass rearing of *M. dorsiplana* was carried out in a population cage where the density of egg-laying females per infested seed increased from 1:1 on the first day to 5:1 on the last (fifth) day. Under these experimental conditions egg-clutch size per host increased, and at the same time, the mortality of eggs laid also increased.

Erdoesina alboannulata (Ratzeburg) parasitoid was multiplied and mass released against *Panolis flammea* (Denis & Sch.). *Mesopolobus modestus* (Sil.) was successfully mass multiplied and released for the management of *Dacus oleae* in Italy. Another species, *M. subfumatus* (Ratze.) was used in Canada for the management of spruce sawfly, *Gilpinia hercyniae.*

Two more parasitoides, *Muscidifurax raptor* Girault and Saunders and *M. raptorellus* Kogan & Legner were were mass reared in the laboratory and released for the management of veterinary pest filth flies and parasitization was recorded about 38% and 80%, respectively, in filth flies (Floate, 2002; Phillip *et al.*, 2012).

The gregarious nature of *Pteromalus puparum* makes the parasitoid more favourable for its mass multiplication and a single female lays eggs in groups of up to 300 in the same pupa (Muggeridge, 1943). The gregarious egg-laying nature is quite helpful for its storage. It can be successfully stored for four months at 10°C under puparia without any emergence loss (McDonald and Kok, 1990).

5.5 CONCLUSIONS

Application of chemical pesticides for the management of insect pests is not feasible in many cases, for instance, in the forestry ecosystem, natural plantations, and against the insect pest feeding inside the stem or bark. Additionally, chemical management can't be preferred or recommended in organic farming, veterinary and medical insect pest management, and also it is environmentally unsafe. Biocontrol is a safe and sustainable method of insect pest management in agriculture, forestry, and veterinary and is also medically important. There are many successful stories of classical biological control in the world using parasitoids and predators. The family Pteromalidae has lots of contribution *viz.* introduction of *Cheiropachus quadrum* Gromakv introduced in North America against scolytid beetle before 1900, in the year 1919 Hase has also suggested, *Lariophagus distinguendus* against *Sitophilus granarius* and *S. oryzae*. Additionally, the contribution of *Anisopteromalus calandrae, Catolaccus grandis, Dinarmus basalis, Mucidifurax raptor* and *Pteromalus puparum* also cannot be overlooked. As the species of this family are largely distributed in 200 countries world over and the species are parasitic to a wide range of insect pests. They are parasitic to the egg, larvae prepupal, pupal and even adult stages of insects, even to the insect stages concealed in the plant parts or under the bark. Therefore, the use of pteromalid species in biocontrol is not a new concept, but it is not preferred due to many reasons. To thrive in the future biocontrol using pteromalid parasitoids must be practical, cost-effective, biologically sound and sustainable. To achieve this goal, the biological attributes of more numbers of parasitoids should be studied in detail in different environmental .conditions. The studies on standardisation of mass rearing technique of more numbers of host species, preferably on artificial diets should be strengthened, development of mass multiplication techniques should be developed for more numbers of effective parasitoids. The development of environmental and insecticidal-resistant strains of successful biocontrol agents is required.

5.6 POINTS TO REMEMBER

1. Pteromalidae is one of the largest family of Hymenoptera and their species are diversified parasitoids of insect pest stages, like egg, larvae, pupa and adult.

2. The diversity of this family in more than 200 countries of the world and more than 4222 species are recorded tillto date.
3. Mass multiplication of *Anisopteromalus Calandrae for* the management of storage insect pests can be easily made. A sustainable rearing technique should be evolved.
4. *Dinarmus basalis* has a wide range of host insects and this parasitoid is very effective for the management of bruchid beetles.
5. Mass multiplication of *Mucidifurax raptor* may be encouraged and promoted for the management of medical and veterinary insect pests.

REFERENCES

Ahmed, K.S. (1996). Studies on the ectoparasitoid, *Anisopteromalus calandrae* How. (Hymenoptera:Pteromalidae) as a biocontrol agent against the lesser grain borer, *Rhyzopertha dominica* (Fab.) in Saudi Arabia. *Journal of Stored Products Research*, 32: 137–140.

Andress E.R. & Campbell, J.B. (1994) Innundative release of Pteromalid parasitoids (Hymenoptera: Pteromalifdae) for the control of stable flies, *Syomoxys calcitrans* (L.) (Diptera: Muscidae) at confined cattle installation in West Central Nebraka. L. *Econo. Entomol.*, 87(3): 714–722.

Ashmead, W.H. (1904). Classification of the chalcid flies of the superfamily Chalcidoideadea, with description of a new species in the Carnegie Museum, collected in south America by Herbrt H. *Smith. Mem. Carneg. Mus.*, 1 (4): i–xi, 225–551.

Austin, A.D. & Dowton, M. (2000). The Hymenoptera: an introduction. In *Hymenoptera: Evolution, Biodiversity and Biological Control*, eds. A.D. Austin and M. Dowton, pp. 3–7. Collingwood, Victoria, Australia: CSIRO Publishing.

Awamleh, R.A., Al-Antary, T.M. & Bilal, H.M. (2009). Survey of Natural Enemies of Fig Wax Scale Ceroplastes rusci L. (Homoptera: Coccidae) and Seasonal Abundance of the Parasitoid *Scutellista cyanea* Fonscolombe (Hymenoptera: Pteromalidae) in Jordan. *Jordan Journal of Agricultural Sciences*, 5: 434–445.

Badary, H. & Abd-Rabou, S. (2011). Role of pteromalid parasitoid *Scutellista caerulea* (Fonscolombe) (Hymenoptera:Pteromalidae) for biological control of the soft scale insects (Hemiptera:Coccidae) in Egypt. *Egypt. Acad. J. biolog. Sci.*, 4 (1): 49–58.

Bolévane, O.F.S. (2004). Recherché d'un hôte de substitution au développement de *Dinarmus basalis* Rond (Hymenoptera: Pteromalidae) dans l'entomofaune séminivore des légumineuses spontanées et subspontanées en zone guinéenne au sud du Togo. Thèse de Doctorat. Université de Lomé, Lomé (Togo).

Bolévane-Ouantinam, F. S., Amevoin, K., Nuto, Y., Monge, J.P. and Glitho, I.A. (2006). Comparaison de quelques caractéristiques biologiques de Dinarmus basalis Rond (Hymenoptera: Pteromalidae) élevé soit sur hôte habituel *Callosobruchus maculatus* F. (Coleoptera: Bruchidae) soit sur *Acanthoscelides macrophthalmus* Schaef. ou *Bruchidius lineatopygus* Pic. identifiés comme hôtes de substitution.*Tropicultura*, 24(2): 101–106.

Boucek, Z. & Rasplus, J.Y. (1991). Illustrated key to West-Palaearctic genera of Pteromalidae (Hymenoptera: Chalcidoidea). INRA Editions, série Techniques et Pratiques, *Institute National de la Recherche Agronomique,* série Techniques et Pratiques, Paris, France. 140 pp.

Boucek, Z. (1978). A generic key to Perilampinae (Hymenoptera, Chalcidoidea) with a revision of *Krombeiniusn*.gen. and *Euperilampus* Walker. *Entomologica Scandinavica*, 9: 299–307.

Boucek, Z. (1988). *Australasian Chalcidoidea* (Hymenoptera). A biosystematic revision of genera of fourteen families, with a reclassification of species. CAB International, Wallingford, Oxon, U.K., Cambrian News Ltd; Aberystwyth, Wales. pp. 357, 410, 458.

Chamberlin, J.R. & Kok, L.T. (1986). Cabbage lepidopteraus pest and their parasite in southwestern Virginia. *Journal of Econ. Entomol.*, 12: 187–197.

Cheong, S.W. & Yoon, C.K. (2016). Method for controlling *Lasioderma serricorne* using *Anisopteromalus apiovorus*. U.S. Patent Application No. 14/491, 975.

Coleman, R.J., Morales-Ramos J.A., King, E.G.&Wood, L.A. (1996). Suppression of the boll weevil in organic cotton by release of Catolaccus grandis as part of the Southern Rolling Plain boll weevil eradication program, p. 1094. *In* Proceedings Beltwide Cotton Conferences 1996, Vol. 2, National Cotton Council of America, Memphis, TN.

Cross, W.H. & Chesnut, T.L. (1971). Arthropod parasites of the boll weevil, *Anthonomus grandis:* 1. An annotated list. *Ann. Entomol. Soc. Am.*, 64: 516–527.

Cross, W.H. & Mitchell, H.C. (1969). Distribution and importance of *Heterolaccus grandis* as parasite of the boll weevil. *Ann. Entomol. Soc. Am.*, 62: 235–236.

Dalman, J.W. (1820). Forsok till up stallning of insect familjen Pteromalini, I synnerhet med afseende pa de I sverige funne arter. *K. VetenskAkad. Handl.*, 41: 123–174, 177–182, 340–385.

Danielle, R.R., Poitrineau, K. & Basso, C. (2007). The potential of mass rearing of *Monoska dorsiplana* (Pteromalidae) a native gregarious ectoparasitoid of *Pseudopachymeria spinipes* (Bruchidae) in South America. *Biological Control*, 41(3): 348–353.

Delvare, G. & Rasplus, J.Y. (1994). *Spodophagus*, a new genus of Pteromalidae (Hymenoptera), for and important parasite of *Spodoptera littoralis* (Lepidoptera: Noctuidae) in Madagascar. *Bulletin of Entomological Research*, 84: 191–197.

Edith, J.G., Álvaro, C.V., Jesús S.P.R., Sotero, A.M., Jorge, V.C. & Juana C.B.M. (2020). First record of the parasitoid *Pteromalus puparum* L. (Hymenoptera: Pteromalidae) associated with pupae of *Pterourus multicaudatus* (Kirby) (Lepidoptera: Papilionidae) in Mexico. *Revista de la Sociedad Entomológica Argentina*, 79(2): 43–47. 10.25085/rsea.790206

Effowe, T.Q., Amevoin, K., Nuto, Y., Mondedji, D. & Glitho, I.A. (2010). Reproductive Capacities and Development of a Seed Bruchid Beetle, *Acanthoscelides macrophthalmus*, a Potential Host for the Mass Rearing of the Parasitoid, *Dinarmus basalis*. *Journal of Insect Science*, 10:129: 1–14.

Elijah, J.T. (2019). Surveys of stink bug egg parasitism in Asia, Europe and North America, morphological taxonomy, and molecular analysis reveal the Holarctic distribution of *Acroclisoides sinicus* (Huang & Liao) (Hymenoptera, Pteromalidae). *J. Hym. Res.*, 74: 123–151. 10.3897/jhr.74.46701

Elzen & King, E.G. (1999). Periodic Release and Manipulation of Natural Enemies. In *Handbook of biological control*. Academic Press, London, 252–264.

Esteban, R.G.L., Jorge, L.L., Victoria, GO´ Mez, Nina M. Ba Rcenas & Gary W. Elzen (2000). Biology of *Catolaccus hunteri* (Hymenoptera: Pteromalidae), A Parasitoid of Pepper Weevil and Boll Weevil (Coleoptera: Curculionidae). *Ann. Entomol. Soc. Am.*, 93(4): 862–868.

Eurostat, (2018). Main annual crop statistics 2015. http://ec.europa.eu/eurostat/statisticsexplained/index.php/Main_annual_crop_statistics#Oilseeds (accessed August, 2020).

Farooqi S.I. & Subba Rao, B.R. (1986). Family Pteromalidae. *In* Shubha Rao, B.R. and Hayat, M. (Edts). The Chalcidoidea (Insecta: Hymenoptera) of India and adjacent countries. *Oriental Insects*, 20: 279–320.

Ferrière, C. (1930). *In* Beier, M. – ZoologischeForschungreisenachdemJonichenInseln und Peloponnes. XIII. Teil Hymenoptera Parasitica. *Sber. Akad. Wiss. Wien (Abteilung 1)*, 139 (5,6): 393–406.

Floate, K.D. (2002). Production of Filth Fly Parasitoids (Hymenoptera: Pteromalidae) on Fresh andon Freeze-killed and Stored House Fly Pupae. *Biocontrol Science and Technology*, 12: 5, 595–603.

Floate, K.D., Coghlin, P. & Gibson G.A.P. (2000). Dispersal of the Filth Fly Parasitoid *Muscidifurax raptorellus* (Hymenoptera: Pteromalidae) Following Mass Releases in Cattle Confinements. *Biological Control*, 18: 172–178.

Forster, A. (1856). Hymenopterologgische studien. 2 Heft. Chalcididae und Proctotrupii., 152 pp Aachen.

Fujii, K., & Wai, K.M. (1990). Sex-ratio determination in three wasp species ectoparasitic on bean weevil larvae. In *Bruchids and Regumes: Economics, Ecology and Coevolution*, eds. K. Fujii, A.M.R. Gatehouse, C.D. Johnson, R. Mitchel & T. Yoshida, pp. 331–340. Dordrecht: Kluwer Academic Publishers, 407 pp.

Gaston, K.J. 1993. Spatial patterns in the description and richness of the Hymenoptera. In *Hymenoptera and Biodiversity*, eds. J. LaSalle and I.D. Gauld, pp. 277–293. Wallingford, UK: CAB International.

Geden, C.J., Rutz, D.A., Scott, J.G. & Long, S.J. (1992c). Susceptibility of house flies (Diptera, Muscidae) and five pupal parasitoids (Hymenoptera: pteromalidae) to abamectin and seven commercial insecticides. *J. Econ. Entomol*, 85: 435–440.

George, N.M. & Warsi, S. (2019). *Habrobracon hebetor* and *Pteromalus cerealellae* as tools in post-harvest integrated pest management. *Insects*, 10(4): 85. 10.3390/insects10040085.

Gibson, G.A.P., Gillespie, D.R. & Dosdall, L. (2006). The species of Chalcidoidea (Hymenoptera) introduced to North America for biological control of the cabbage seedpod weevil, and the first recovery of *Stenomalina gracilis* (Chalcidoidea: Pteromalidae). *The Canadian Entomologist*, 138: 285–291.

Gonen, M. & Kugler, J. (1970). Notes on the biology of *Lariophagus distinguendus* (Förster) (Hym. Pteromalidae) as a parasite of *Sitophilus oryzae* (L.) (Col. Curculionidae). *Isr. J. Entomol.*, 5: 133–140.

Graham, M.W.R. de V. (1969). The Pteromalidae of north-western Europe (Hymenoptera: Chalcidoidea). *Bulletin of the British Museum (Natural History) (Entomology)* 16: 1–908.

Greene, G.L. (1990). Biological control of filth flies in confined cattle feedlots using pteromalid parasites. In *Biological control of arthropods affecting livestock and poultry*, eds. Rutz, D.A. & Patterson, R.S., pp. 29–42. Boulder, Colorado: Westview.

Groombridge B. (1992). *Global biodiversity: status of the Earth's living resources*. London: Chapman & Hall.

Guk-Hyang, Ko, Duk-Young, Park & Jong-Wook Lee (2018). Taxonomic review of subfamily Pteromalinae (Hymenoptera,Chalcidoidea) with 25 newly recorded species in South Korea. *Journal of Asia-Pacific Biodiversity*, 11: 87–122. 10.1016/j.japb.2017.08.006

Hagstrum, D.W. & Subramanyam, B. (2006). Biological Control. *Fundamentals of Stored-Product Entomology*. 151–156. 10.1016/B978-1-891127-50-2.50017-X

Haliday (1844). Contribution towards the classification of the Chalcididae. *Trans. Ent.soci. Lond*, 3: 295–301.

Hasani, A. & Madjdzadeh, S.M. (2012). Contribution to the knowledge of the Pteromalidae (Hymenoptera: Chalcidoidea) from Khorasan Razavi province, Northeastern Iran. *Iranian Journal of Animal Biosystematics*, 8 (1): 57–69.

Hase, A.Z. (1924). Kenntnis wirtschaftlich wichtiger Tierformen. I. Über den Stech-und Legeakt, sowie den Wirtswechsel von *Lariophagus distinguendus. Chalcididae. Pteromlaidae. Die Naturwiss*, 20: 377–384. (In German)

Haye, T., Fisher, S., Zhang, J. and Gariepy, T. (2015). Can native egg parasitoids adopt the invasive brown marmorated stink bug, *Halyomorpha halys* (Heteroptera: Pentatomidae), in Europe. *J Pest Sci*, 88(4): 693–705.

Haye, T., Olfert, O., Weiss, R., Peter, G.M., Gibson, G., Gariepy, T.D. & Gillespie, D.R. (2018). Bioclimatic analyses of Trichomalus perfectus and Mesopolobus morys (Hymenoptera: Pteromalidae) distributions, two potential biological control agents of the cabbage seedpod weevil in North America. *Biological Control*, 124 (2018): 30–39.

Hayea, T., Masonb, P.G., Gillespiec, D.R., Miallb, J.H., Gibsonb, G.A.P., Diaconud, A.A., Braunerb, M. & Kuhlmanna, U. (2015). Determining the host specificity of the biological control agent *Trichomalus perfectus* (Hymenoptera: Pteromalidae): the importance of ecological host range. *Biocontrol Science and Technology*, 25(1): 21–47. 10.1080/09583157.2014.945900

Hogsette, J.A. (1981). Fly control by composting manure at south Florida equine facility, In *Status of biological control of filth flies*, eds. R.S. Patterson, P.G. Koehler, Morgan & R.L. Harria, pp. 105–113, USDA, ARS, A106.2.:F64. New Orlenas, LA.

Hunsberger, A.G. & Pena, J.E. (1997). Catolaccus hunter (Hymenoptera: Pteromalidae), a parasite of *Anthonomus macromalus* (Coleoptera: Curculionidae) in South Florida. *Florida Entomologist*, 80: 301–304.

Inserra, S. (1971). *Ceroplastes rusci* L. in the citrus groves of the province of Catania. *Bollettino del Laboratorio di Entomologia graria 'Flippo Silvestri' Portici*, 28: 77–97.

Islam, W. & Kabir, M.H.S. (1995). Biological control potential of *Dinarmus basalis* (Rond.) (Hymenoptera: Pteromalidae), a larval-pupal ectoparasitoid of the pulse beetle. *Callosobruchus chinensis (L.),* Crop protection, 14(6): 439–443.

Islam, W. (1994). Effect of host age on rate of development of *Dinarmus basalis* (Rond.) Hyme., Pteromalidae). *Journal of Applied Entomology*, 118: 392–398.

Islam, W. (1998). Rearing and release of the pulse weevil parasitoid *Dinarmus basalis* (Rond.) (Hymenoptera: Pteromalidae). *Tropical Agricultural Research and Extension*, 1(2): 131–135.

Islam, W. & Kabir, S.M.H. (1991). Effect of temperature and relative humidity on the development and progeny production of *Anisopteromalus calandrae* (Hymenoptera: Pteromalidae). *Bangladesh J. Entomol.*, 18(1): 45–49.

Islam, W., Ahmed, K.N. & Malek, M.A. (1985). Hymenopterous parasites on insect pest of two stored pulses. *Bangladesh J. Zoo*, 13: 81–85.

Juan, A., Morales-Ramos & James, R.C. (1992). Laboratory Determination of Age-Dependent Fecundity, Development, and Rate of Increase of *Catolaccus grandis* (Burks) (Hymenoptera: Pteromalidae). *Annals of the Entomological Society of America*, 85(4): 469–476.

Kapil, R.P. & Chaudhary, J.P. (1973). Record of nine new hymenopterous parasites for Khapra beetle, *Trogoderma granarium* Everts. (Coleoptera, Dermestidae). *Indian Journal of Entomology*, 35: 353–355.

Kashef, A. (1955). Étude biologique de Stegobium paniceum L. et de son parasite: Lariophagus distinguendus Först. *Ann.Société Entomol. France*, 124: 1–88. 16.

Khoobdel, M., Hossein, S., Dehghan, D., Kamran, A. & Ehsan R. (2019). Natural host preferences of parasitoid wasps (Hymenoptera: Pteromalidae) on synanthropic flies. *Eur J. Transl. Myol.*, 29 (2): 118–123.

King, E.G., Coleman, R.J., Wood, L., Wendel, L., Greenberg, S., & Scott, A.W. (1995). Suppression of boll weevil in commercial cotton by augmentative releases of a wasp parasite, *Catolaccus grandis* (pp. 26–30). *In* Proceedings of the Beltwide Cotton Product Research Conference National Cotton Council, Memphis, TN.

Kumar, A. (2013). Fig wax scale, *Ceroplastes rusci* an emerging pest of *Dalbergia sissoo* and its parasitisation in India *Int. J. Curr. Sci.*, 8: E 106–114.

Lasota, J.A. and Kok, L.T. (1986). Parasitism and utilization of imported cabbage worm pupas by *Pteromalus puparum* (Hymenoptera: Pteromalidae). *Environmental Entomology*, 15: 994–998.

Lee, J.C. & Heimpel, G.E. (2005). Impact of Xowering buckwheat on Lepidopteran cabbage pests and their parasitoids at two spatial scales. *Biological Control*, 34: 290–301.

Lotfalizadeh, H. & Gharali, B. (2008). Pteromalidae (Hymenoptera: Chalcidoidea) of Iran: New records and a preliminary checklist. *Entomofauna*, 29 (6): 93–120.

Lysyk, T.J. (2001a). Relationships between temperature and life history parameters of Muscidifurax raptor (Hymenoptera: Pteromalidae). *Environ. Entomol.*, 29: 596–605.

Machtinger, E.T. (2011). Studies on the field ecology, breeding biology, and parasitism of house flies, *Musca domestica*, and stable flies, *Stomoxys calcitrans*, (Diptera: Muscidae) to improve integrated pest management for north Florida small equestrian farms. Masters Thesis, University of Florida, Gainesville, FL.

Machtinger, E.T., Leppla, N.C. & Sanders, C. (2013). Pest management perceptions and practices for equine farms in north and central Florida (IFAS publication number ENY-2028). University of Florida, Gainesville, FL. (http://edis.ifas.ufl.edu/pdffiles/IN/IN98300.pdf).

Mason, P.G., Miall, J.H., Bouchard, P., Gillespie, D.R., Broadbent, A.B. & Gibson, G.A.P. (2011). The parasitoid communities associated with an invasive canola pest, *Ceutorhynchus obstrictus* (Coleoptera: Curculionidae), in Ontario and Quebec, Canada. *The Canadian Entomologist*, 143(5): 524–537. 10.4039/n11-041

Mbata, G.N., Shu, S., Phillips, T.W. & Ramaswamy, S.B. (2004). Semiochemical cues used by *Pteromalus cerealellae* (Hymenoptera: Pteromalidae) to locate its host, *Callosobruchus maculatus* (Coleoptera: Bruchidae). *Ann. Entomol. Soc. Am.*, 97: 353–360.

McDonald, R.C. & Kok, L.T. (1990). Post refrigeration viability of *Pteromalus puparum* (Hymenoptera: Pteromalidae) prepupae within Host Chrysalids. *Journal of Entomological Science*, 25(3): 409–413.

Meyer, J.A. & J.J. Petersen. (1983). Characterization and seasonal distribution of breeding sites of stable flies and house flies on eastern Nebraska feedlots and dairies. *J. Econ. Entomol.*, 76: 103–108.

Minshawy, A.M.E., Saad, A.H. & Hammad, S.M. (1978). Efficacy of natural enemy *Scutelista cyanea* Motsch (Hym., Pteromalidae) on *Saissctia coffee* Walk S.oleae (Bern.) and *Cerplastes floridensis* Comst. (Hom.,Coccidae) *Zeitsschrift fur angewandte Entomologie*, 85: 31–37.

Mitroiu, M.D., Abolhassanzadeh, F. & Madjdzadeh, S.M. (2011). New records of Pteromalidae (Hymenoptera: Chalcidoidea) from Iran, with description of a new species. *North Western Journal of Zoology*, 7(2): 243–249.

Mitroiu, M.D. (2012). The genus *Caenocrepis* Thomson (Hymenoptera: Pteromalidae) in the Afrotropical region, with a key to world species. *Zootaxa*, 3557: 49–55.

Monge, J.P. & Huignard, J. (1991). Population fluctuation of two bruchid species *Callosobruchus maculatus* F. and *Bruchidius atrolineatus* Pit (Coleoptera: Bruchidae) and their parasitoids *Dinarmus basalis* (Rond.) and *Eupelmus vuilleti* (Crw) in a storage situation in Niger. *J. Afr. Zoo.*, 105: 187–196.

Morales-Ramos J.A., Rojas, M.G., Coleman, R.J. & King, E.G. (1998). Potential Use of In Vitro-Reared *Catolaccus grandis* (Hymenoptera: Pteromalidae) for Biological Control of the Boll Weevil (Coleoptera: Curculionidae). *J. Econ. Entomol.*, 91(1): 101–109.

Morales-Ramos, J.A., Rojas, M.G., Roberson, C.J., Jones, R., King, E.G. & Brazzel, J.R. (1994). Suppression of the boll weevil first generation by augmentative releases of *Catolaccus grandis* in Aliceville, AL (pp. 958–965). In Proceedings of the Beltwide Cotton Production Research Conference. National Cotton Council, Memphis, TN.

Morales-Ramos, J.A., Summy, K.R., Raulston, J., Cate, J.R. & King, E.G. (1992). Feasibility of mass-rearing Catolaccus grandis (pp. 723- 726). In Proceedings of the Beltwide Cotton Production Research Conference. National Cotton Council, Memphis, TN.

Morgan, P.B., Hoyer, H. & Patterson, R.S. (1989). Life history of *Spalangia cameroni* (Hymenoptera: Pteromalidae), a microhymenopteran pupal parasite of muscoid flies.

Muggeridge, J. (1943). white butterfly (Pieris rapae L.) II. Parasites of the butterfly. *New Zealand Journal of Science and Technology*, 25: 01–18.

Narendran, T.C. & Mini, T.V. (1998-99). A review of the genera of oriental Eunotinae (Hymenoptera: Pteromalidae). *Journal of Zool. Soc. Kerala*, 8&9: 15–28.

Niedermayer, S., Marie, P. & Johannes L.M. Steidle. (2016). *Lariophagus distinguendus* (Hymenoptera: Pteromalidae) (Förster)—Past, Present, and Future: The History of a Biological Control Method Using *L. distinguendus* against Different Storage Pests. *Insects*, 7: 39. 10.3390/insects7030039

Noort, S. van., Wang, R. & Compton, S.G. (2013). Fig wasp (Hymenoptera: Agaonidae: Pteromalidae) associated with Asian fig tree (Fivus, Moraceae) in Southern Africa: Asian Flowers and African Colonists. *African Invertibrate*, 54(2): 382–400.

Noyes, J.S. (2021). Universal Chalcidoidea Database. World Wide Web electronic publication. Available online: http://www.nhm.ac.uk/our-science/data/chalcidoids/database/#sthash.IJ8aiVKT.dpuf (accessed on April 2021).

Oanh, N.T., Tran, N.L. & Truong, X.L. (2017). Egg-lying behavior of *Anisopteromalus calandrae* (howard), an ectoparasitoid of *Lasioderma serricorne* (fabricius). *Tap Chi Sinh Hoc*, 39(4): 416–420.

Okamoto, K. (1971). The synchronization of the life cycles between *Callosobruchus chinensis* (L.) and the parasite *Anisopteromalus calandrae* Howard. *Japanese Journal of Ecology*, 21: 233–237.

Onagbola, E.O. & Fadamiro, H.Y. (2008). Morphology and development of *Pteromalus cerealellae* (Ashmead) (Hymenoptera: Pteromalidae) on *Callosobruchus maculatus* (F.) (Coleoptera: Chrysomelidae). *BioControl.*, 53: 737–750. 10.1007/s10526-007-9117-x

Peck, O. (1951). Superfamily Chalcidoidea, In *Hymenoptera of America north of Maxico*, eds. Muesebeck, Krombein & Townes, pp. 410–594. Synoptic Catalog. US Deptt of Agric. Monogr., 21420.

Perez-Mendoza, J., Baker, J.E., Arthur, F.H. & Flinn P.W. (1999). Biological Control Effects of Protect-It on Efficacy of Anisopteromalus calandrae (Hymenoptera: Pteromalidae) Parasitizing Rice Weevils (Coleoptera: Curculionidae) in Wheat. *Environmental Entomology*, 28(3): 529–534.

Peverieri, S.G., Mitroiu, M.D., Bon, M.C., Balusu, R., Benvenuto, L., Bernardinelli, L., Fadamiro, H., Falagiarda, M., Fusu, L., Grove, E. (2019). Surveys of Stink Bug Egg Parasitism in Asia, Europe and North America, Morphological Taxonomy, and Molecular Analysis Reveal the Holarctic Distribution of *Acroclisoides sinicus (*Huang & Liao) (Hymenoptera, Pteromalidae). *J. Hymenopt. Res.* 74: 123–151.

Phillip, E. K., Strong, C., Waldron, J. K. And Rutz, D. A. (2012). Individual and Combined Releases of Muscidifurax raptorand M. raptorellus (Hymenoptera: Pteromalidae) as a Biological Control Tactic Targeting House Flies in Dairy Calf Facilities. *J. Med. Entomol.*, 49(5):1059–1066.

Pitzer, J. B., Kaufman, P. E., Hogsette, J., Geden, C. J. and Tenbroeck, S. H. (2011). Seasonal Abundance of Stable Flies and Filth Fly Pupal Parasitoids (Hymenoptera: Pteromalidae) at Florida Equine Facilities. *Journal of Economic Entomology,* 104(3): 1108–1115.

Press, J.W., Cline, I.D. & Flaherty, B.R. (1984). Suppression of residual populations of the rice weevil, *Sitophilus oryzae*, by the parasitic wasp. *Anisopteromalus calandrae. J. Georgia Entomol. Sot.*, 19: 110–113.

Raghu, S., Wiltshire, C. & Dhilleepan, K. (2005). Intensity of pre-dispersal seed predation in the invasion legume *Leucaena leucocephala* is limited by the duration of pod retention. *Austral Ecology*, 30: 310–318.

Roberson, J.L. & Harsh, D.K. (1993). Machanized production process to encapsulate ball weevil larvae (*Anthonomus grandia*) for mass production of *Catolaccus grandis* (Burks) pp 922–923. In proc. Beltwide cotton conference 1993. 2-National Cotton Council of America Maemphis, TN.

Romero, A., Hogsette, J.A. & Coronado, A. (2010). Distribution and Abundance of Natural Parasitoid (Hymenoptera: Pteromalidae) Populations of House Flies and Stable Flies (Diptera: Muscidae) at the University of Florida Dairy Research Unit. *Neotropical Entomology*, 9(3): 424–429.

Ryabov, M.A. (1926). The possibilities of applying the parasitic method of control in the case of granary pests. *Bull. N. Caucasian Plant Prot. Sta.*, 1: 19–50. 15.

Santhosh, S. & Basavarajappa, S. (2017). Record of natural enemies of few butterfly species amidst agriculture ecosystems of Chamarajanagar district, Karnataka, India. *RJLBPCS*, 2(5) 10.26479/2017.0205.02

Sexena, R. (1996). Utilization of *Dinarmus baslis* Rondani (Hymenoptera:Pteromalidae) in biocontrol of *Callosobruchus analis* F. (Coleoptera: Bruchidae). *Insect Environment*, 2(3): 77–79.

Seymour, R.C. & Campbell, J.B. (1993). Predators and parasitoides of house flies and stable flies (Diptera: Muscidae) in cattle confinements in west central Nebraska. *Environ. Entomo.*, 22: 212–219.

Steidle, J.L.M., (2000). Host recognition cues of the granary wevil parasitoid Lariophagus distinguendus (Hymenoptera:Pteromalidae). *Entomologia Experimentalis et Applicata*, 95: 185–192.

Steidle, J.L.M., Scholler, M. (1997). Olfactory host location and learning in the granary weevil parasitoid Lariophagus distinguendus (Hymenoptera: Pteromalidae). *Journal of Insect Behaviour*, 10: 331–342.

Steidler, J.L.M. & Scholler, M. (2002). Fecundity and ability of the parasitoid *Lariophagus distinguendus* (Hymenoptera: Pteromalidae) to find larvae of the granary weevil *Sitophilus granarius* (Coleoptera: Curculionidae) in bulk grain. *Journal of Stored Products Research*, 38: 43–53.

Summy, K.R., Morales-Ramos, J.A. & King, E.G. (1995). Suppression of boll weevil infestations on South Texas cotton by augmentative releases of the exotic parasite *Catolaccus grandis* (Hymenoptera: Pteromalidae). *Biological Control*, 5: 523–529.

Sureshan, P. M. (2003). Pteromalinae (Pteromalidae:Chalcidoidea:Hymenoptera). *Zoological Survey of India Occasional paper*. 100:1–2, 1–170.

Sureshan, P.M. & Narendran, T.C. (2005). A new species of *Theocolax* Westwood (Chalcidoidea: Pteromalidae) parasitizing wood boring beetles from India. *Records of the Zoological Survey of India*, 104(1–2): 141, 143–145.

Sureshan, P.M., Dhanya, B., Bijoy, C. & Ramesh Kumar, A. (2011). A review of the Oriental Eunotinae (Hymenoptera: Chalcidoidea: Pteromalidae) with description of a new *Cephaleta* and first record of *Mesopeltita truncatipennis* (Waterston) from the Oriental region. *Hexapoda*, 18(2): 98.

Swederus, N.S. (1795). Beskrifning paetnytt genus Pteromalus ibland Insecterna, hoerande til Hymenoptera. *K. svenska vetensk Akad Hanndl*, 16: 201–205, 216–222.

Sweetman, H.L. (1964). *The Principles of Biological Control*. Iowa, USA: Brown Comp.

Takagi, M. 1987. The reproductive strategy of the gregarious parasitoid, Pteromalus puparum (Hymenoptera: Pteromalidae). 2. Host size discrimination and regulation of the number and sex ratio of progeny in a single host. *Oecologia*, 70: 321–325.

Thomson, C.G. (1876). Hymenoptera Scandinaviae. Tom. IV. Pteromalus (Svederus)., 259pp. 1pl., Lundae.

Thomson, C.G. (1878). Hymenoptera Scandinaviae. Tom. IV. Pteromalus (Svederus)., 307pp. 1pl., Lundae.

Tormos, J., Asís, J.D., Sabater-Muñoz, B., Baños, L., Gayubo, S., and Beitia, F. (2012). Superparasitism in laboratory rearing of Spalangia cameroni (Hymenoptera: Pteromalidae), a parasitoid of medfly (Diptera: Tephritidae). *Bulletin of Entomological Research*, 102:51–61.

Tuda, M., Wu, L.H., Tateishi, Y., Niyomdham, C., Buranapanichpan, S., Morimoto, K., Wu, W.J., Wang, C.P., Chen, Z.Q., Zhu, H.Y., Zhang, Y.C., Murugan, K., Chou, L.Y., & Johnson, C.D. (2009). A novel host shift and invaded range of a seed predator, *Acanthoscelides macrophthalmus* (Coleoptera: Chrysomelidae: Bruchinae), of an invasive weed, *Leucaena leucocephala*. *Entomological Science*, 12(1): 1–8.

Verma, R. (1990). Host habitat location and host location by *Dinarmus hasalis,* a parasite of bruchids of stored legumes. *Indian J. Exp. Biology*, 28: 179–184.

Verma, R. (1991). Life history and some aspects of biology of *Dinarmus basalis* (Hymenoptera, Pteromalidae), a parasite of stored legume bruchids. *Acta Entomologica Bohemoslovaca*, 88: 359–366.

Visarathanonth, P., Kengkanpanich, R., Uraichuen, J. & Thongpan, J. (2010). Suppression of *Sitophilus zeamais* Motschulsky by the ectoparasitoid, *Anisopteromalus calandrae* (Howard). 10th International Working Conference on Stored Product Protection. Julius-Kühn-Archiv, 425, 755–759. 10.5073/jka.2010.425.196.

Wang, Z., Liu, Y., Shi, M., Huang, J. & Chen, X. (2019). Parasitoid wasps as effective biological control agents. *Journal of Integrative Agriculture*, 18 (4): 705–715.

Walker, F. (1833). Monographia Chalcidum (continued). *Entomology Monthly Magazine*, 1: 115–142: 367–384.

Wegensteiner, R., Beat, W. & Matthias, H. (2015). Natural Enemies of Bark Beetles: Predators, Parasitoids, Pathogens, and Nematodes, Bark Beetles. 10.1016/B978-0-12-417156-5.00007-1

Westwood, J.O. (1839). Synopsys of genera of the British insects, An introduction to the modern classification of insects. *London, Longman*. 13: 49–80.

Williams, R.N. & Floyd, E.H. (1971). Effect of low temperature on hymenopterous parasites *Choetospila elegans* and *Anisopteromalus calandrae* of the maize weevil. *Journal of Economic Entomology*, 64: 1438–1439.

Xiao, H. & Huang, D.W. (2001a). A taxonomic study on genus *Cheiropachus* from China, with description of two new species (Hymenoptera: Pteromalidae). *Acta Zootaxonomica*, 44(1): 102–108.

6 Encyrtid Parasitoids

P.S. Shera and Sudhendu Sharma
Department of Entomology, Punjab Agricultural University, Ludhiana, Punjab, India

Prasun Karmakar
Rain Forest Research Institute, Jorhat, Assam, India

Sudhir Singh
Forest Entomology Discipline, Forest Protection Division, Forest Research Institute, Dehradun, Uttarakhand, India

CONTENTS

LEARNING OBJECTIVES

1. To learn taxonomy, basic characteristics, biology and life cycle of encyrtids.
2. To know the diversity of the encyrtid parasitoids, their utilisation potential and mass multiplication procedures.
3. The case studies of encyrtid wasps as biological control agents of insect pests.

DOI: 10.1201/9781003354239-6

6.1 INTRODUCTION

6.1.1 Systematic Position

Kingdom	:	Animalia
Phylum	:	Arthropoda
Class	:	Insecta
Order	:	Hymenoptera
Superfamily	:	Chalcidoidea
Family	:	Encyrtidae

Encyrtidae is one of the most speciose families of chalcid parasitoid wasps. They are important biocontrol agents of various insect species especially mealybugs and scales insects (Hemiptera) and help in maintaining necessary equilibrium among populations of various insect species in the ecosystem. Presently in the family 456 genera and 3826 species placed in 2 subfamilies, i.e., Encyrtinae and Tetracneminae, have been listed in Species 2000 & ITIS Catalogue of Life (Noyes 2020; Table 6.1). Hosts of the species belonging to the subfamily Tetracneminae are restricted to the family Pseudococcidae (Hemiptera), whilst species of subfamily Encyrtinae have a broad host range including insects from many orders, mites, ticks and spiders.

The taxonomy of chalcid started in the middle of 18th century. Latrielle, 1809, laid the foundation of encyrtid taxonomy by describing several species of genus *Encyrtus*. Walker (1837) created the family name Encyrtidae on the basis of the genus *Encyrtus*, which he credited to Dalman (1820) instead of Latrielle (1809). Westwood (1840) kept encyrtids under "Encyrtides" as one of its subfamilies of his family Chalcididae and attributed 17 genera, some of which belonged to families: Aphelinidae, Trichogrammatidae and Euplemidae of today. Foerster (1856) renamed the family as "Encyrtoidae" while describing 12 new genera. He was the first worker who laid down certain characters for the family. Some outstanding contributions to encyrtid taxonomy have been made by workers during the 19th century (Foerster, 1856, Mayr, 1876; Thomson, 1876; Howard, 1881, 1885; Westwood, 1840; Ashmead, 1899, 1900a, b). During the early 20th century, Ashmead (1904) added some valuable findings and raised the subfamily Encyrtinae into four tribes: Encrytini, Ectromini, Mirini and Arrhenophagini. Girault (1915) gave a brief description of about 200 genera from Australia, Nearctic region and Southeast Asia. Later, Mercet (1921) greatly improved the classification of Encyrtidae and described 93 genera and 214 species, his notable work is considered a landmark in the history of Encyrtidae taxonomy. Several species from Encyrtidae were also described from Hawaii (Tiberlake, 1920); and Japan (Ishii, 1928; Tachikawa, 1963). Hoffer (1955) and Erdos and Novicky (1955) contributed to the classification of the family into different subfamilies. Compere and Annecke(1960, 1961) contributed to the taxonomy of the African Encyrtidae. Trjapitziri's (1973a, 1973b) classification of the family up to tribes is universally accepted. Noyes and Hayat (1994) did a comprehensive classification of Tetracneminae and reduced it into 6 tribes.

The keys to the genera belonging to the family Encyrtidae from different regions of the world are given, i.e., North America (Trjapitzin and Gordh, 1978a & b), Palearctic region (Trjapitzin, 1989), Neotropical region (Noyes, 1980, 2000, 2004, 2010), Indo-Pacific region (Noyes and Hayat, 1984) and Ethiopian region (Prinsloo and Annecke, 1979).

The contribution to the knowledge of the family Encyrtidae has been made by several workers from different countries: Argentina (De Santis, 1964), Australia (Dahms and Gordh, 1997), China (Liao et al., 1987; Shi et al., 1994; Yanzohu and Dawei, 2004; Li et al., 2020), Egypt (Abd-Rabou, 2001; Evans and Abd-Rabou, 2013), Hawaiian island (Beardsley, Jr. 1976), Iran (Guerrieri and Ghahari, 2018), Japan (Tachikawa, 1963), Korea (Jung Suh, 2019), Macaronesia (Trjapitzin, 2008), Madagascar (Risbec, 1951); Mexico (Rodríguez et al., 2010), New Zealand (Noyes, 1988), Turkey (Oncuer, 1991; Trjapitzin and Doganlar, 1997; Kaydan and Japoshvili, 2010; Japoshvili, 2012) and USA (Triapitsyn et al., 2015; Triapitsyn et al., 2020), European part of erstwhile USSR (Nikol'skaya, 1952).

TABLE 6.1
List of Species Belonging to the Family Encyrtidae Worldwide

Subfamily	Genera (Number of species)
Encyrtinae	Aamna (1), Acerophagus (18), Achalcerinys (4), Adelencyrtoides (14), Adelencyrtus (28), Adencyrtus (3), Admirencyrtus (1), Aenasiella (6), Aenasomyiella (3), Aesaria (1), Aethognathus (5), Agarwalencyrtus (2), Agekianella (2), Ageniaspis (14), , Agromyzaphagus (1), Allencyrtus (1), , Aloencyrtus (18), Amauroencyrtus (1), Ameromyzobia (2), Amicencyrtus (1), Amicroterys (4), Amira (3), Ammonoencyrtus (2), Anagyrodes (7), Anusemion (1), Andinoencyrtus (1), Anicetus (38), Anisophleps (1), Anthemus (20), Aphidencyrtoides (1), Aphycinus (1), Aphycoides (7), Aphycomastix (1), Aphycomorpha (1), Aphycopsis (1), Aphyculus (7), Aphycus (27), Apsilophrys (4), Archinus (1), Argutencyrtus (1), Arhopoidiella (1), Arketypon (1), Arrhenophagoidea (5), Arrhenophagus (2), Aschitus (24), Aseirba (1), Asterolecanobius (1), Astymachus (5), Atelaphycus (1), Atropates (1), Australanusia (2), Australaphycus (1), Austrochoreia (3), Austroencyrtoidea (1), Austroencyrtus (3), Austromira (1), Avernes (1), Avetianella (9), Baeoanusia (3), Baeocharis (2), Baeoencyrtus (1), Beethovena (1), Bennettisca (10), Blanchardiscus (2), Blastothrix (26), Bolangera (1), Borrowella (2), Bothriocraera (2), Bothriophryne (8), Bothriothorax (30), Boucekiella (1), Brachyencyrtus (2), Brachyplatycerus (1), Brethesiella (7), Caenohomalopoda (1), Caldencyrtus (3), Carabunia (10), Casus (1), Ceballosia (2), Centencyrtus (1), Cerapteroceroides (8), Cerapterocerus (12), Ceraptroceroideus (2), Cerchysiella (26), Cerchysius (13), Cercobelus (7), Charitopsis (1), Cheiloneurella (2), Cheiloneuromyia (3), Cheiloneurus (129), Cheilopsis (1), Choreia (4), Chorotega (1), Chrysomelechthrus (1), Cibdeloencyrtus (1), Cicoencyrtus (1), Cirrhencyrtus (3), Clivia (1), Coagerus (1), Coccidaphycus (1), Coccidencyrtus (30), Coccidoctonus (8), Coccopilatus (4), Coelopencyrtus (28), Comones (2), Comperia (7), Comperiella (9), Conchynilla (1), Copidosoma (156), Copidosomopsis (9), Copidosomyia (2), Cowperia (4), Cranencyrtus (1), Cryptanusia (7), Cyderius (1), Deilio (1), Deloencyrtus (1), Diaphorencyrtus (2), Diasula (3), Dionencyrtus (2), Discodes (39), Diversinervus (12), Doddanusia (1), Ebito (1), Echthrobaccella (1), Echthroplexiella (28), Echthroplexis (1), Ectroma (10), Encyrtoalces (1), Encyrtoidea (2), Encyrtus (64), Eocencyrtus (1), Epiblatticida (7), Epicerchysius (1), Epiencyrtus (3), Epistenoterys (2), Epitetracnemus (6), Epitetralophidea (3), Eremencyrtus (2), Erencyrtus (6), Ethoris (1), Eucoccidophagus (5), Eugahania (9), Euogus (1), Eupoecilopoda (2), Euscapularia (1), Eusemion (2), Exoristobia (9), Forcipestricis (5), Formicencyrtus (1), Fulgoridicida (5), Gahaniella (5), Gentakola (1), Ginsiana (8), Globulencyrtus (1), Gonzalezia (1), Grissellia (1), Gwala (1), Habrolepis (15), Habrolepoidea (1), Habrolepopteryx (1), Hadrencyrtus (1), Hadzhibeylia (1), Haligra (1), Helegonatopus (11), Helygia (1), Hemencyrtus (4), Hemileucoceras (3), Hengata (1), Hesperencyrtus (2), Heterococcidoxenus (2), Hexacladia (7), Hexacnemus (1), Hexencyrtus (3), Homalopoda (1), Homalotyloidea (7), Homalotylus (44), Homosemion (1), Hoplopsis (2), Iceromyia (1), Ilicia (1), Inbiaphycus (1), Indaphycus (2), Islawes (1), Isodromoides (1), Isodromus (20), Ixodiphagus (6), Kataka (1), Koenigsmannia (1), Kurdjumovia (1), Laccacida (1), Lakshaphagus (8), Lamennaisia (4), Leefmansia (1), Leiocyrtus (1), Leurocerus (2), Lirencyrtus (1), Lochitoencyrtus (1), Lohiella (1), Lombitsikala (1), Mahencyrtus (12), Manmohanencyrtus (1), Mariola (1), Mashhoodiella (1) Mayrencyrtus (5), Mayridia (30), Melys (1), Meniscocephalus (7), Merlen (1), Meromyzobia (13), Mesanusia (2), Mesastymachus (1), Mesocalocerinus (1), Mesorhopella (3), Metablastothrix (2), Metanotalia (1), Metaphycus (436), Metapsyllaephagus (11), Microterys (156), Moorella (2), Mozartella (1), Mucrencyrtus (8), Muluencyrtus (1), Nassauia (2), Nathismusia (1), Neabrolepoideus (1), Neapsilophrys (1), Neastymachus (11), Neblatticida (3), Neocladella (2), Neocladia (10), Neococcidencyrtus (18), Neocyrtus (4), Neoplatycerus (3), Neperpolia (1), Nerissa (1), Nezarhopalus (1), Oesol (1), Olypusa (1), Oobius (8), Ooencyrtus (183), Orianos (1), Oriencyrtus (2), Ovaloencyrtus (1), Ovidoencyrtus (1), Paksimmondsius (1), Papaka (1), Papuna (1), Parablastothrix (14), Parablatticida (11) Parachalcerinys (3), Paracladella (4), Paraenasomyia (3), Paramucrona (8), Paraphaenodiscus (16), Paraphycus (1), Parasauleia (1), Parastenoterys (1), Paratetracnemoidea (5), Paratetralophidea (1), Parechthrodryinusi (12), Parectromoides (2), Parencyrtomyia (1), Parencyrtus (1), Pareupelmus (1), Pareusemion (1), Pasulinia (1), Pawenus (1),

(Continued)

TABLE 6.1 (Continued)
List of Species Belonging to the Family Encyrtidae Worldwide

Subfamily	Genera (Number of species)
	Pentacladocerus (2), *Pentelicus* (7), *Perpolia* (1), *Phauloencyrtus* (1), *Philosindia* (2), *Pistulina* (1), *Plagiomerus* (9), *Platencyrtus* (1), *Prionomastix* (12), *Prionomitoides* (1), *Prionomitus* (7), *Prochiloneurus* (27), *Profundiscrobis* (1), *Proleuroceroides* (1), *Proleurocerus* (3), *Protyndarichoides* (5), *Pseudaphycus* (23), *Pseudectroma* (8), *Pseudencyrtoides* (1), *Pseudencyrtus* (10), *Pseudhomalopoda* (1), *Pseudococcobius* (11), *Pseudorhopus* (3), *Psilophryoidea* (1), *Psilophrys* (4), *Psyllaephagus* (215), *Psyllaphycus* (1), *Pulexencyrtus* (1), *Quadrencyrtus* (1), *Raffaellia* (1), *Rhopalencyrtoidea* (4), *Rhytidothorax* (16), *Ruandella* (4), *Ruskiniana* (1), *Saera* (1), *Sancarlosia* (1), *Sanghalia* (1), *Saprencyrtus* (1), *Sarisencyrtus* (1), *Satureia* (1), *Saucrencyrtus* (1), *Sauleia* (2), *Scotteus* (1), *Sectiliclava* (4), *Semen* (1), *Shenahetia* (1), *Simmondsiella* (1), *Solenaphycus* (1), *Solenoencyrtus* (1), *Spaniopterus* (1), *Stemmatosteres* (4), *Subprionomitus* (5), *Syrphophagus* (75), *Szelenyiola* (1), *Tachardiaephagus* (5), *Tachardiobius* (2), *Tachinaephagus* (10), *Tanyencyrtus* (1), *Tassoniai* (3), *Teleterebratus* (4), *Tetarticlava* (1), *Tetracyclos* (1), *Thomsonisca* (7), *Tineophoctonus* (4), *Tobiasiai* (1), *Tonkinencyrtus* (1), *Trechnites* (18), *Tremblaya* (5), *Trichomasthus* (42), *Trjapitzinellus* (6), *Tropidophryne* (4), *Tyndarichus* (8), *Vietmachus* (1), *Viggianiola* (1), *Whittieria* (1), *Xenoencyrtus* (5), *Xenostryxis* (7), *Xerencyrtus* (1), *Xylencyrtus* (2), *Zaomma* (14), *Zaommoencyrtus* (6), *Zarhopaloides* (5), *Zdenekiella* (1), *Zelaphycus* (1), *Zelencyrtus* (1), *Zooencyrtus* (2), *Zozoros* (2)
Tetracneminae	*Acerophagoides* (5), *Adektitopus* (2), *Aenasius* (41), *Aeptencyrtus* (1), *Aglyptus* (1), *Alamella* (5), *Allocerellus* (12), *Amasyxia* (1), *Amernscocephalus* (1), *Anagyrus* (247), Anagyrietta (1), *Ananusia* (2), Anikera (1), *Anomalencyrtus* (2), *Anomalicornia* (1), Anusia (1), *Anusioptera* (2), *Apoleptomastix* (6), *Aquaencyrtus* (2), Arabencyrtus (), *Asencyrtus* (1), Asitus (1), *Bactritopus* (1), *Blepyrus* (18), *Bureshiella* (1), *Callaincyrtus* (1), *Callipteroma* (4), *Ceraptrocerella* (1), *Charitopus* (15), *Chrysoplatycerus* (5), *Cladiscodes* (3), *Clausenia* (12), *Coccidoxenoides* (1), *Cryptoplatycerus* (1), *Cyrtocoryphes* (1), *Dicarnosis* (8), *Dinocarsiella* (2), *Dinocarsis* (3), *Dusmetia* (6), *Ectromatopsis* (1), *Eocencnemus* (1), *Eotopus* (2), *Eremophasma* (2), *Ericydnus* (26), *Euzkadiella* (1), *Extencyrtus* (1), *Epanusia* (4), *Gafsa* (1), *Gavria* (1), *Grandiclavula* (1), *Gyranusoidea* (41), *Hambletonia* (9), *Hipponactis* (1), *Hofferencyrtus* (1), *Holanusomyia* (1), *Holcencyrtus* (10), *Incisencyrtus* (3), *Leptomastidea* (22), *Leptomastix* (30), *Lutherisca* (1), *Lyka* (2), *Manicnemus* (1), *Marxella* (1), *Mashhoodia* (1), *Metaphaenodiscus* (9), *Mira* (6), *Mohelencyrtus* (2), *Mohelniella* (2), *Monodiscodes* (2), *Monstranusia* (2), *Moraviella* (1), *Neocharitopus* (3), *Neodusmetia* (1), *Neorhopus* (1), *Nesebaria* (1), *Notodusmetia* (1), *Odiaglyptus* (1), *Parapyrus* (1), *Parectromoidella* (9), *Pelmatencyrtus* (1), *Peneax* (1), *Phasmocephalon* (1), *Phasmocera* (2), *Phasmopoda* (1), *Praleurocerus* (1), *Pseudleptomastix* (4), *Paraclausenia* (1), *Paracopidosoma* (3), *Paraenasioidea* (1), *Paramasia* (1), *Paranathrix* (2), *Rhopus* (50), *Ruanderoma* (1), *Sakencyrtus* (2), *Savzdargia* (1), *Schilleriella* (2), *Taftia* (2), *Tetracnemoidea* (14), *Tetracnemus* (31), *Tricladia* (1), *Tshudo* (1), Trjapitzion(1), *Vosleria* (1), *Xanthoectroma* (1), *Xenanusia* (2), *Yasumatsuiolai* (1), *Zaplatycerus* (5), *Zarhopalus* (8)

Source: Noyes 2020 (www.catalogue of life); www.nhm.ac.uk.

6.1.2 India

Cameron (1913) was the first study on Indian Ecyrtidae, when a new genus *Lissencyrtus* with species *clavicornis* was described. Gahan (1919) recorded some and described a new species *Aphycus fuscidorsum* followed by Ayyar (1932, 1934) who proposed the species *Tetracnemus indicus* and *Comperiella indica*. Mani (1935) described a new genus and species, *Krishnierieila ceroplastodi*. Later Mani (1938) recorded 19 species under 16 genera from India in his catalogue. After a gap of about 27 years, Agarwal (1965) gave first consolidated work on Indian Encyrtidae by describing 17 genera and 22 species.

Hayat et al. (1975) dealt with 21 genera and 55 species, then Shafee et al. (1975) included 29 genera and 59 species. Agarwal et al. (1984) presented a key to 110 encyrtid genera. Kaul and Agarwal (1986) covered 15 genera and 17 species. Hayat (1979a, b; 1981a, b; 1989; 1999; 2006b, 2012) has made substantial contributions to Indian encyrtid taxonomy. Noyes and Hayat (1994) studied 20 genera belonging to the tribe Anagyrini from the Oriental region, they also provided biology and biocontrol use for every genus and identification keys for the known Oriental species. Singh and Agarwal (1993) studied 52 species under 22 genera from Northeastern India. Hayat (2006a) in his book on Indian Encyrtidae gave the key for the identification of 106 genera of the Encyrtinae and 37 genera of the Tetracneminae. Hayat also gave keys to the Indian species under each genus.

Some other contributions to Indian Encyrtidae are as follows: Mani (1935, 1939, 1941, 1989); Narayanan et al. (1957a, b); Narayanan and Subba Rao (1960); Subba Rao (1966); Subba Rao and Rai (1970); Alam (1961, 70); Mani and Kaul in Mani et al., (1973); Mani and Saraswat in Mani et al., (1974); Saraswat and Mukherjee (1975), Subba Rao and Hayat (1985, 1986); Singh and Agarwal (1991); Rao, 1965; Fatima & Shafee (1994); Singh (1996, 2004, 2007, 2016a & b); Singh and Hayat (2002); Singh and Singh (2011); Manickavasagamand Ramesh kumar (2012); Hayat et al. (2013); Singh et al. (1991, 2014) Rameshkumar et al. (2015); Krishnachaitanya and Manickavagam (2016); Fatima (2018); Kumar et al., (2019) and Kaneria and Singh, (2020).

6.2 CHARACTERISTICS OF ENCYRTIDAE

Encyrtids are minute insects with an average length ranging from 1.1 to 2.5 mm. They are distinguished by the presence of a broad convex mesopleura without any groove which may be present in most of the chalcids. Habitus of some encyrtid species is shown in Figure 6.1 A-M. The family Encyrtidae can be distinguished from other chalcids, especially Eupelmidae and Tanaostigmatidae by the following characteristics (Noyes and Hayat, 1994; Hayat, 2006; Noyes, 2020):

i. The antennae in females are usually 11-segmented (9-segmented in males) – scape, pedicel, 6-segmented funicle and 3-segmented club (one segmented in males). Reduced number of funicle and claval segments are not uncommon, especially in females. The true anellus is absent in both sexes.
ii. The pronotum is transverse and is visible in the dorsal view of the thorax. Mesoscutum transverse and without notauli, or if notauli are present, then they are very shallow and curved, never deep and straight.
iii. The axillae are transverse and triangular and often meet or approach each other in the middle.
iv. Mesopleuron enlarged, often occupying more than half of mesosoma in lateral view (Figure 6.1 J). Mid coxae inserted at or before the middle of mesopleuron (In Eupelmidae and Tanaostigmatidae middle coxae are inserted after the middle).
v. The forewing distinctly bears an oblique line extending across the wing called linea calva.
vi. Tarsi mostly are pentamerous but in some tetramerous condition also prevails.
vii. Sexual dimorphism is pronounced through body colour, size and antennal structure; females are bigger and stouter than males.

6.3 BIOLOGY

Encyrtids parasitize a variety of insect hosts belonging to the orders Lepidoptera, Neuroptera, Homoptera – Coccoidea (Figure 6.1 B–D), Aphidoidea and Psylloidea (Figure 6.1 E–H), and some are hyperparasitoids via other Encyridae or Aphelinidae, Pteromalidae, Braconidae, Dryinidae, etc. They usually develop as internal parasitoids, but larvae of some *Microterys* species develop as egg

PLATE 6.1 A–M: Encyrtid parasitoids and their hosts. A. Invasive alien pseudococcid P*henacoccus solenopsis* Tinsley infesting a wild plant, it is heavily parasitized by *Aenasius arizonensis* (Girault) which is

predators (Debach, 1939). Encyrtids reproduce sexually but thelytokous parthenogenesis is also common (De Santis, 1964). A number of genera are hyperparasitoids, whereas a few are polyembryonic, in which case the hosts are usually lepidopterous insects, but occasionally also aculeate Hymenoptera. In the case of polyembryony, the female lays a single egg, which by way of multiple divisions produces hundreds (or even thousands) of larvae (De Santis, 1964). Species of the genus *Copidosoma*, *Pseudorhopus,* etc. are known polyembryonic parasitoids whereas *Epiencyrtus* are polyembryonic hyperparasitoids (Noyes, 1990; Voinovich and Sugonyaev, 1993). After completing development, encyrtid larvae pupate inside the hosts. When adult encyrtids emerge, they chew a round hole in the host integument and exit. Encyrtid parasitoids lay encyrtiform eggs, bearing an external stalk, which protrudes from the host scale integument. The eggs are attached to the host integument, having an aeroscopic plate, which serves as a respiratory mechanism for the early larval instars (Maple, 1947). Though in a few species, eggs may be laid in the hemolymph and do not have stalk (Ben-Dov and Hodgson, 1997). Also, some species belonging to *Diversinervus* and *Microterys* lay their eggs in the hindgut or in the hosts' subesophageal ganglion (Bartlett and Lagace, 1961; Bartlett and Medved, 1966). The female parasitoids are able to discriminate between parasitized and un-parasitized hosts (Tena et al., 2008). Typically, three to four larval instars, in addition to the prepupal stage have been reported (Saakyan-Baranova, 1966; Wright, 1986; Tena et al., 2011), though *Anarhopus sydneyensis* Timberlake has also been reported as the single species with two larval instars (Compere and Flanders, 1934).

When the parasitoid larvae attain maturity within the host, the body of the latter becomes much distended, often being twice or more of its normal size. In the case of solitary parasitoids, an inflation of the host body occurs that is circular in cross-section, cylindrical and smoothly rounded at both ends. The shell interior is often smooth and polished. These parasitized hosts are generally referred to as "mummies". The gregarious encyrtids, like *Acerophagus notativentris*, produce an inflated condition and during parasitization, the host is partitioned into cells, wherein each brood is isolated individually (Bartlett and Ball, 1964; Kapranas et al., 2009). These surface cells are outlined distinctly externally. The parasitoid adults emerge from a hole that is chewed through the host dorsum (Ben-Dov and Hodgson, 1997). The encyrtid wasps generally emerge without eggs and thereafter continue to mature eggs throughout their entire life (Flanders, 1942; Weseloh, 1969; Lampson et al., 1996). The pre-oviposition period of these wasps differs among various species, some emerge with mature eggs (Smith, 1933), thereby starting oviposition within a short span of emergence, while a few start laying eggs after one or two days (Avidov and Podoler, 1968; Kapranas et al., 2011).

Adult encyrtids usually feed on honeydew (sweet excretions) secreted by their homopteran hosts or do not feed at all (De Santis, 1964; Noyes, 1990). Some species feed directly on the hosts' body fluids which ooze out at the time of oviposition.

The life span of the polyembryonic encyrtids depends on that of the host, as larval development may or may not be completed until the host is in its final instar (Askew, 1971; Cruz, 1981).

---

also an introduced species in India. **B.** Waxy scale, *Ceroplastis* sp., with emergence holes made by *Anicetus* sp. **C.** Scales of *Hemilecanium* sp. with emergence holes by parasitoid *Adelencyrtus* sp. **D.** Scales of *Lecanium* sp. on *Cycas* leaves, parasitized by *Encyrtus* sp. **E.** *Terminalia arjuna* leaves galled by psyllid *Trioza hirsuta* (Crawford) which is parasitized by *Psyllaephagus phylloplectrae* Sushil and Khan. **F.** Nymph of leaf pit gall forming psyllid of *Mallotus philippinensis*(Lam.), with emergence hole by an undet. Species of *Psyllaephagus.* **G.** *Garuga pinnata* leaves galled by psyllid *Phacopteron lentiginosum* Buckt., which is parasitized by two species of *Psyllaephagus- P. phacopteron* Singh and *P. lentiginosum* Singh. **H.** *Mallotus philippinensis* galled by a psyllid parasitized by a undet species of *Psyllaephagus.* **I.** Female of *Cheiloneurus zeyai* Anis & Hayat, in lateral view. **J.** Female of *Psyllaephagus phylloplectrae* Sushil and Khan, in lateral view. **K.** Female of *Encyrtus aurantii* (Geoffroy), in dorsal view. **L.** Female of *Ooencyrtus utethesiae* (Risbec), in lateral view. Genus *Ooencyrtus* parasitizes mostly eggs of lepidopterous insects. **M.** Female of *Adelencyrtus* sp. in dorsal view.

Likewise, the stage of host development also affects the parasitoid life cycle. As in *Hunterellus hookeri* Howard a parasitoid of ixodid ticks, the parasitoid deposits the eggs in the nymphal instar of the host. The development of the larvae is withheld until the host becomes ready; the phenomenon called 'latent parasitism'. Under certain situations, this period may extend up to six months, i.e, between the oviposition and parasitization of the host. This obligatory diapause, shown by the parasitoid larvae, appeared due to a lack of nutrition out of the early stages of the host (Cooley, 1927; Smith and Compere, 1928; Cooley and Kohls, 1933; Clausen, 1962).

Many species listed in Table 6.1 have one generation per year, though some have two or more generations per year. The latter usually have a summer brood cycle, from egg to adult, of approximately 30 days. In many cases, when encyrtid wasps are provided with rich host's carbohydrates sources, it results in a longer life span (Bernal et al., 1999). This mechanism is of much importance in augmentative biocontrol programmes, wherein the mass-produced parasitoids could be stored for long periods before the field releases.

Encyrtids can be collected with a fin-meshed sweeping net or by collecting and rearing their hosts. Various techniques for collection, preservation and slide preparation for taxonomic studies are described by Noyes (1982).

6.4 EFFECTIVE ENCYRTID BIOCONTROL AGENTS

Parasitic Hymenoptera has been used in applied biocontrol and proved successful in reducing the population of pest species below the economic threshold level. Over 800 species have been associated with biocontrol programmes. Two families in particular, Aphelinidae and Encyrtidae, have proven extremely successful in the biocontrol of insect pests, although species of most other chalcidoid families have also been successfully utilised. Encyrtids are second, to the Aphelinidae in the ratio of the actual number of introductions to the number of cases in which full economic control of the pest was obtained [11:1 for the Encyrtidae and 5:1 for the Aphelinidae] Clausen (1978) and Noyes (1985). More than 400 species of Encyrtidae have been used for checking the population of insect pests of various crops (Noyes 1985; Greathead 1986; Neuenschwander et al., 1990; Grissell and Schauff, 1990; Yasnosh, 1998; Japoshvili et al., 2006; 2008). Some Encyrtid genera, like *Anicetus, Aloencyrtus, Blastothrix, Diversinervus, Discodes, Encyrtus, Metaphycus*, and *Microterys*, are well recognised as the most important groups of biocontrol agents of soft scale pests (Prinsloo, 1996, 1997). Many *Metaphycus* species, such as *M. eruptor, M. flavus, M. helvolus, M. lounsburyi, M. stanleyi* and *M. zebratus* were introduced to different parts of the world as biocontrol agents against *Ceroplastes* (Fig. B) or soft scales (Noyes and Hayat, 1994; Noyes, 2004). *Ooencyrtus kuvanae* has proved to be an efficient biocontrol agent for controlling gypsy moth (*Lymatria dispar*) in North America and Europe (Huang and Noyes, 1994). Furthermore, many encyrtid species belonging to the genera *Anagyrus, Leptomastix, Leptomastidea, Gyranusoidea, Coccidoxenoides* and *Acerophagus* are important biocontrol agents used across the world (Moore, 1988).

Encyrtids have been utilised in classical biocontrol programmes of soft scales insects and mealybugs. One of the most successful biocontrol programmes is the control of the cassava mealybug, *Phenacoccus manihoti* Matile-Ferrero (Hemiptera: Pseudococcidae) by the encyrtid *Anagyrus lopezi* De Santis in Africa during the early 1980s. The biocontrol programme provided a benefit: cost ratio of 199:1 based on the world market price of cassava crops (Zeddies et al., 2001). The annual economic benefits from the classical biocontrol programmes of the papaya mealybug, *Paracoccus marginatus* Williams and Granara de Willink by the three encyrtid wasps *Acerophagus papayae* (Noyes and Schauff), *Anagyrus loecki* Noyes & Menezes, and *Pseudleptomastix mexicana* Noyes and Schauff ranged from $121 million to $309 million in India (Myrick et al., 2014). The genera of the family Encyrtidae parasitic on insect pests are listed in Table 6.2. *Habrolepis dalmanni* has provided substantial or useful levels of control of golden oak scale, *Asterolecanium variolosum*, a pest of *Quercus* species in New Zealand, Australia and Chile

TABLE 6.2
Effective Encyrtid Biocontrol Agents and Hosts

Parasitoid name	Host name	Reference(s)
Acerophagus papayae Noyes and Schauff	*Paracoccus marginatus* William & Granara de Willink	Manickavasagam and Rameshkumar (2012)
Adelencyrtus mayurai Subba Rao	*Melanaspis glomerata* (Green)	Manickavasagam and Rameshkumar (2012)
Aenasius indicus Narayanan & Subba Rao	*Icerya formicarum* Newstead, *Nipaecoccus viridis* (Newstead), *Ferrisia virgata* Cockerell, *Coccus viridis* (Green), *Phenacoccus indicus* Avasthi and Shafee, *Planococcus citri* Risso, *Planococcus lilacinus* Cockerell	Joshi et al. (2017)
Aenasius advena Compere	*F. virgata, P. citri*	Manickavasagam and Rameshkumar (2012)
Aenasius arizonensis Girault	*Phenacoccus solenopsis* Tinsley	Fallahzadeh et al. (2014) Shera et al. (2017)
Anagyrus chrysos Noyes and Hayat	*Rastrococcus iceryoides* Noyes and Hayat	Manickavasagam and Rameshkumar (2012)
Anagyrus dactylopii (Howard)	*F. virgata, Maconellicoccus hirsutus* Green, *N. viridis, P. solenopsis, P. citri*	Manickavasagam and Rameshkumar (2012)
Anagyrus kamali Moursi	*F. virgata, M. hirsutus, N. viridis, P. solenopsis, P. citri*	Manickavasagam and Rameshkumar (2012)
Anagyrus mirzai Agarwal & Alam	*P. solenopsis*	Vennila et al. (2010) Nagrare et al. (2011)
Anagyrus loecki Noyes and Menezes	*P. marginatus, Phenacoccus madeirensis* Green	Joshi et al. (2017)
Anagyrus ranchiensis Shamim and Shafee	*N. viridis*	Manickavasagam and Rameshkumar (2012)
Anicetus Howard	*Ceroplastes* Gray, *Ceroplastodes* Cockerell *Chloropulvinaria* Borchsenius, *Coccus* L., *Eucalymnalus* Cockerell, *Lichtensia* Sigoret *Parasaissetia* Takahashi, *Parthenolecanium* Sulc, *Pulvinaria* Targioni Tozzetti, *Saissetia* Deplanche, *Toumeyella* Cockerell, *Vinsonia* Signoret, *Waxiella* De Lotto	Prinsloo (1997)
Aloencyrtus Prinsloo	*Ceroplastes* sp., *Coccus* sp., *Cryptinglisia* Cockerell, *Inglisia* Maskell, *Parasaissetia* sp., *Saissetia* sp., *Waxiella* sp.	Prinsloo (1997)
Apoanagyrus lopezi De Santis	*Phenacoccus manihoti* Matile-Ferrero	Norgaard (1988)
Blepyrus insularis Cameron	*F. virgata, M. hirsutus*	Manickavasagam and Rameshkumar (2012)
Cerapteroceroides Ashmead	*Ceroplastes* sp., *Chloropulvinaria* Borchsenius, *Ericerus* Guérin-Méneville, *Eulecanium* Cockerell, *Metaceronema* Takahashia, *Pulvinaria* sp., *Takahashia* Cockerell	Prinsloo (1997)
Cerapterocerus Westwood	*Ceroplastes* sp., *Coccus* sp., *Didesmococcus* Borchsenius, *Eriopeltis* Signoret, *Parthenolecanium* sp., *Sphaerolecanium* Sulc	Prinsloo (1997)
Cheiloneurus Westwood	*Ceroplastes* sp., *Ceroplastodes* sp., *Coccus* sp., *Didesmococcus* sp., *Ericerus* sp., *Eriopeltis* sp., *Eulecanium* sp., *Filippia* Targioni Tozzetti,	Fuchs et al. (1991), Prinsloo (1997)

(Continued)

TABLE 6.2 (Continued)
Effective Encyrtid Biocontrol Agents and Hosts

Parasitoid name	Host name	Reference(s)
	Luzulaspis Cockerell, *Parasaissetia* sp., *Parthenolecanium* sp., *P. solenopsis*, *Physokermes* Targioni Tozzetti *Protopulvinaria* Cockerell, *Pulvinaria* sp. *Pulvinariella* Borchsenius, *Pulvinarisca* Borchsenius, *Rhodococcus* Borchsenius, *Saissetia* sp., *Sphaerolecanium* sp., *Takahashia* sp., *Waxiella* sp.	
Coccidoctonus Crawford	*Ceroplastes* sp., *Coccus* sp., *Parasaissetia* sp., *Protopulvinaria* sp., *Saissetia* sp.	Prinsloo (1997)
Comperiella bifasciata Howard	*Aonidiella aurantii* Maskell	Orphanides (1996)
Copidosoma desantisi Annecke & Mynhardt	*Phthorimaea operculella* Zeller	Mandoor et al. (2008)
Discodes Förster	*Acantholecanium* Borchsenius, *Acanthopulvinaria* Borchsenius, *Eriopeltis* sp., *Eulecanium* sp., *Rhizopulvinaria* Borchsenius, *Rhodococcus* sp., *Sphaerolecanium* sp.	Prinsloo (1997)
Diversinervus Silvestri	*Ceronema* Maskell, *Ceroplastes* sp., *Chloropulvinaria* sp., *Coccus* sp., *Ctenochiton* Maskell, *Inglisia* sp., *Parasaissetia* sp., *Protopulvinaria* sp., *Pulvinaria* sp., *Pulvinariella* sp., *Saissetia* sp.	Prinsloo (1997)
Encyrtus Latreille	*Acantholecanium* sp., *Acanthopulvinaria* sp. *Ceroplastes* sp., *Chloropulvinaria* sp., *Coccus* sp., *Ericerus* sp., *Eucalymnalus* sp. *Eulecanium* sp., *Hemilecanium* Newstead, *Lichtensia* sp., *Parasaissetia* sp., *Parthenolecanium* sp., *Pulvinaria* sp., *Pulvinariella* sp., *Saissetia* sp., *Takahashia* sp., *Udinia* sp., *Waxiella* sp.	Kapranas and Tena (2015)
Encyrtus aurantii (Geoffroy)	*Ceroplastes actiniformis* Green, *Coccus viridis* (Green), *Eulecanium tiliae* (Linnaeus), *Parasaissetia nigra* (Nietner), *Parthenolecanium fletcheri* (Cockerell), *P. solenopsis*, *Pulvinaria* sp., *Saissetia coffeae* (Walker)	Vennila et al. (2010) Nagrare et al. (2011) Joshi et al. (2017)
Homalotylus albiclavatus (Agarwal)	*P. solenopsis*	Vennila et al. (2010) Nagrare et al. (2011)
Leptomastix dactylopii Howard	*Planococcus citri* Risso	Krishnamoorthy & Singh (1987) Joshi et al. (2017)
Microterys Thomson	*Acantholecanium* sp., *Acanthopulvinaria* sp. *Anapulvinaria* sp., *Ceronema* sp., *Ceroplastes* sp., *Chloropulvinaria* sp., *Coccus* sp., *Didesmococcus* sp., *Ericerus* sp. *Eulecanium* sp., *Filippia* sp., *Lichtensia* sp. *Messinea* sp., *Parasaissetia* sp., *Parthenolecanium* sp., *Physokermes* sp., *Protopulvinaria* sp., *Pulvinaria* sp., *Pulvinariella* sp., *Rhizopulvinaria* sp., *Rhodococcus* sp., *Saissetia* sp., *Sphaerolecanium* sp., *Stotzia* sp., *Toumeyella* sp.	Prinsloo (1997)

TABLE 6.2 (Continued)
Effective Encyrtid Biocontrol Agents and Hosts

Parasitoid name	Host name	Reference(s)
Metaphycus Mercet	*Acanthopulvinaria* sp., *Ceronema* sp., *Ceroplastes* sp., *Chloropulvinaria* sp., *Cissocossus* Cockerell, *Coccus* sp., *Cryptinglisia* sp., *Ctenochiton* sp., *Didesmococcus* sp., *Ericerus* sp., *Eriopeltis* sp., *Eulocanium* sp., *Filippia* sp., *Hemilecanium* Newstead, *Idiosaissetia* Brain, *Inglisia* sp., *Lichtensia* sp., *Luzulaspis* sp., *Messinea* De Lotto, *Parasaissetia* sp., *Parthenolecanium* sp., *Physokermes* sp., *Pulvinaria* sp., *Pulvinariella* sp., *Rhodococcus* sp., *Saissetia* sp., *Sphaerolecanium* sp., *Stotzia* Marchai, *Takahashia* sp., *Waxiella* sp.	Prinsloo (1997) Nagrare et al. (2011) Kapranas and Tena (2015)
Prochiloneurus aegyptiacus (Mercet)	*F. virgata* , *P. citri*, *P. solenopsis*, *N. viridis* *R. iceryoides*	Joshi et al. (2017)
Prochiloneurus pulchellus Silvestria	*Coccidohystrix insolita* (Green), *F. virgata*, *P. citri*, *P. solenopsis*, *N. viridis*, *R. iceryoides*	Vennila et al. (2010) Nagrare et al. (2011) Suroshe et al. (2013) Wang et al. (2014) Joshi et al. (2017)

(Clausen, 1978). Species of *Ixodiphagus* are used to control ixodid ticks transmitting Lime disease and encephalitis (Mwangi et al., 1991).

6.5 MASS REARING OF IMPORTANT ENCYRTID PARASITOIDS

i. *Acerophagus papayae* (Noyes and Schauff), *Anagyrus loecki* Noyes & Menezes, and *Pseudleptomastix mexicana* Noyes and Schauff on *Paracoccus marginat*us Williams and Granara de Willink

The colonies of *A. papayae*, *A. loecki* and *P. mexicana* are maintained in separate plexi glass cages (30 × 30 × 30 cm) using sprouted red potatoes infested with third instar nymphs of *P. marginatus* (Amarasekare, 2007). The mealybug-infested potatoes are exposed separately to each parasitoid species for parasitization. After 24 hours of exposure, potatoes with parasitized mealybugs are moved to new cages for emergence. A cotton soaked in 10% honey diet (1 part of honey: 9 parts of water) is kept in each cage as food for the newly emerged adult parasitoids. Colonies of *A. papayae*, *A. loecki* and *P. mexicana* are maintained in insectary at a temperature of 25 ± 2°C, 65 ± 2% relative humidity and 12:12 (L:D) photoperiod.

ii. *Adelencyrtus mayurai* Subba Rao on *Melanaspis glomerata (Green)*

The mass culture of *A. mayurai* is maintained on sugarcane setts infested with *M. glomerata* nymphs in the insectary (Anonymous, 2020). After 30 to 35 days of infestation by *M. glomerata*, the sugarcane sets are exposed to freshly emerged *A. mayurai* at a host: parasitoid ratio of 15:1. The parasitoid completes its development in about 15 days. The parasitoids are collected daily by an aspirator and kept in separate glass tubes plugged with cotton wool. A cotton soaked in 10% honey diet (1 part of honey: 9 parts of water) is kept in each glass tube as food for the newly emerged adult parasitoids.

iii. *Aenasius arizonensis* Girault on *Phenacoccus solenopsis* Tinsley

The field-collected mummies of *P. solenopsis* (Fig. A) (parasitized mealybugs) are kept individually in glass tubes (3 × 4 cm) covered with cotton plugs until adult emergence. The mated adult parasitoids are released in glass jars containing healthy colonies of mealybugs maintained on sprouted potatoes (Shera et al., 2017; Karmakar and Shera, 2018). After parasitization, mummies are collected and kept in glass tubes plugged with cotton wool. A swab of cotton soaked in 10% honey diet (1 part of honey: 9 parts of water) is kept in each glass tube as food for the newly emerged adult parasitoids.

iv. *Comperiella bifasciata* Howard on *Aonidiella aurantii* Maskell

The 2nd instar nymphs of *A. aurantii* are preferred for parasitization by *C. bifasciata* (Anonymous, 2020). The parasitized scale nymphs are allowed to colonise on ripe pumpkins in cages (45 × 45 × 50 cm). The parasitoid completes its development in about 20 days. The emerging adults are collected and used either for field release or for the continuation of the culture. A cotton soaked in 10% honey diet (1 part of honey: 9 parts of water) is kept as food for the newly emerged adult parasitoids.

v. *Copidosoma koehleri* Blanchard and *Copidosoma desantisi* Annecke & Mynhardt on*Phthorimaea operculella* Zeller

The freshly laid eggs of *P. operculella* are placed between two layers of muslin cloth. The mated female adults of *C. koehleri*/ *C. desantisi* (fed with honey droplets on wax paper) are exposed to the eggs of *P. operculella* for parasitization (Anonymous, 2020). The parasitised eggs are collected after 24 hours. After hatching, the parasitised *P. operculella* larvae are maintained on potato tubers. The parasitoid grubs form mummies within the host carcass of *P. operculella* larvae. Each mummified larva may yield up to 60–100 adults of the parasitoid.

vi. *Leptomastix dactylopii* Howard on *Planococcus citri* Risso

Ripe pumpkins infested with 15–20-day old *P. citri* are exposed to 1–2-day old *L. dactylopii* females in the host-parasitoid ratio of 40:1 (Anonymous, 2020). After 24 hours of exposure, the parasitoids are aspirated and the pumpkins infested with parasitized mealybugs are placed in separate cages. The newly emerged parasitoids are collected daily by an aspirator starting from 15 days after exposure. A cotton soaked in 10% honey diet (1 part of honey: 9 parts of water) is kept in each cage as food for the newly emerged adult parasitoids. The parasitoids are then either released in the field or used for the continuation of the culture.

6.6 CASE STUDIES

i. Biocontrol of Papaya mealybug – Success story

One of the most successful biocontrol programmes in India was the control of the papaya mealybug, *Paracoccus marginatus* (Williams and Granara de Willink) (Hemiptera: Pseudococcidae) by the three exotic encyrtid wasps, *Acerophagus papayae* (Noyes and Schauff), *Anagyrus loecki* (Noyes and Menezes) and *Pseudleptomastix mexicana* (Noyes and Schauff) (Hymenoptera: Encyrtidae) in 2010 (Shylesha et al., 2018). In 2008, the papaya mealybug emerged as a serious threat to papaya, cassava and mulberry cultivation in the states of Karnataka, Maharashtra and Tamil Nadu in Southern India. Several insecticides were used for its management, but the farmers failed to check the rapid population build-up of the mealybug pests. The farmers were compelled to abandon the cultivation of papaya, cassava and mulberry in Southern India.

ICAR-National Bureau of Agricultural Insect Resources (ICAR-NBAIR) in collaboration with the United States Department of Agriculture (USDA) imported the three encyrtid wasps of the papaya mealybug from Animal and Plant Health Inspection Services (APHIS) in Puerto Rico. The parasitoids were mass multiplied in large numbers by ICAR Institutes, State Agricultural Universities (SAUs), Krishi Vigyan Kendras (KVKs) and other stakeholders, and released in the farmers' field @ 100 parasitoids of each species/field in more than 2500 villages in Karnataka, Maharashtra and Tamil Nadu. The biocontrol programme successfully controlled the papaya mealybug within a period of six months and the farmers were encouraged to take up cultivation of papaya, cassava and mulberry. The annual economic savings from the classical biocontrol programmes of the papaya mealybug in Karnataka, Maharashtra and Tamil Nadu was approximately Rs 1623 crores (Shylesha et al., 2018).

ii. Biocontrol of Cassava mealybug – Success story

The classical biocontrol programme for the control of the cassava mealybug, *Phenacoccus manihoti* Matile-Ferrero (Hemiptera: Pseudococcidae) by the encyrtid parasitoid *Apoanagyrus lopezi* De Santis in the African continent during the early 1980s proved to be a spectacular success story. The cassava mealybug was accidentally introduced from South America to the African continent in the early 1970s. In Africa, this mealybug pest spread rapidly to more than 25 countries and became the major insect pest of cassava (Herren, 1981).

In the early 1980s, a large-scale biological control programme was undertaken by the International Institute of Tropical Agriculture (IITA) in Nigeria to control the cassava mealybug. The parasitoid *Apoanagyrus lopezi* (= *Epidinocarsis lopezi* De Santis) was imported from South America and was introduced into Nigeria in 1981 (Herren and Lema, 1982). By July 1988, the parasitoid was successfully established in 19 African countries and had spread over more than 5 million square kilometre area (Neuenschwander and Herren, 1988). By the late 1980s, *A. lopezi* was established in 90% of the cassava-growing region of sub-Saharan Africa, and the extent of losses due to this mealybug pest was minimised to a great extent. The successful biocontrol of the cassava mealybug increased cassava production, averted widespread famine and saved the lives of 20 million people across sub-Saharan Africa (Chakupurakal et al., 1994; Maredia and Raitzer, 2010). The annual economic benefits from the classical biocontrol programmes of the cassava mealybug in the African continent were more than US$120 billion (Maredia and Raitzer, 2010). The biocontrol programme provided a benefit: cost ratio of 199:1 based on the world market price of cassava crops (Zeddies et al., 2001).

6.7 CONCLUSIONS

Encyrtidae is one of the most speciose families of chalcidoid wasps. Presently, it includes 456 genera and 3826 species placed in 2 subfamilies, i.e., Encyrtinae and Tetracneminae. They are distributed worldwide and ubiquitous in all habitats. Encyrtids are extremely important biocontrol agents of insect species from order Hemiptera, particularly mealybugs and scales insects, though eggs or larvae of other host species from Blattaria, Coleoptera, Diptera, Hymenoptera, Lepidoptera, Neuroptera and Orthoptera are also attacked. The important genera in the family Encyrtidae include *Apoanagyrus*, *Anagyrus*, *Aenasius*, *Comperiella*, *Ooencyrtus*, *Hunterellus*, *Leptomastix*, *Leptomastidea*, *Gyranusoidea*, *Coccidoxenoides* and *Acerophagus*. The encrytids have been widely recognised in various biocontrol programs against mealybugs worldwide. The most successful example is the biological management of cassava mealybug, *Phenacoccus manihoti* Matile-Ferrero by parasitoid, *Apoanagyrus* (*Epidinocarsis*) *lopezi* De Santis in African countries. In India, significant success has been achieved through classical biocontrol of an alien invasive species, papaya mealybug, *Paracoccus marginatus* Williams and Granara de Willink using three exotic parasitoids imported with the help of USDA-APHIS namely *Acerophagous papaya* Noyes and Schauff, *Anagyrus loecki* Noyes and Menezes and *Pseudleptomastix mexicana* Noyes and Schauff. Another encyrtid parasitoid, *Aenasius arizonensis* (Girault) (= *Aenasius bambawalei* Hayat) has played a key role in checking the menace of another alien invasive species, solenopsis mealybug, *Phenacoccus solenopsis* Tinsley on cotton crop and hundreds of other plants. The documented case studies of successful biocontrol programmes reveal the importance of encyrtid parasitoids as potential biological control agents of insect pests and their direct role in pest control and annual savings of billions of dollars.

6.8 POINTS TO REMEMBER

1. Encyrtids are important endoparasitoid of insect species, particularly mealybugs and scale insects.
2. Wasps in the family Encyrtidae are characterised by the migration of cerci forward on the metasoma and the presence of an enlarged mesopleuron with anteriorly positioned mesocoxae.
3. The host range of encyrtids includes insect species from the order Blattaria, Coleoptera, Diptera, Hemiptera, Hymenoptera, Lepidoptera, Neuroptera and Orthoptera.
4. Maximum (about 90%) work on the taxonomy of Indian Encyrtidae is credited to the Department of Zoology, Aligarh Muslim University where Professors S.M. Alam, M.M. Agarwal, S.A. Shafee, Mohd. Hayat and co-workers researched and produced many Ph.Ds in this field. Prof. Hayat has worked extensively and is still active and working to bring out the revised book on Indian Encyrtidae. Presently, apart from AMU, some work on taxonomy is carried out at Zoological Survey of India, Kolkotta National Bureau of Agricultural Insect Resources, Banguluru; Forest Research Institute, Dehradun; Dept of Zoology Annamalai University, Annamalainagar, etc.
5. The important genera in the family Encyrtidae include *Apoanagyrus*, *Anagyrus*, *Aenasius*, *Comperiella*, *Ooencyrtus* and *Hunterellus*, *Leptomastix*, *Leptomastidea*, *Gyranusoidea*, *Coccidoxenoides* and *Acerophagus*.
6. The classical biocontrol of cassava mealybug, *Phenacoccus manihoti* by *Apoanagyrus (Epidinocarsis) lopezi* in Africa and papaya mealybug, *Paracoccus marginatus* by *Acerophagous papaya*, *Anagyrus loecki* and *Pseudleptomastix mexicana* in India are the most successful examples of biological control of insect pests by Encyrtids.

REFERENCES

Abd-Rabou, S. (2001). The species of *Anagyrus* Howard (Hymenoptera: Encyrtidae) recorded from Egypt. *Egyptian Journal of Agricultural Research*, 79: 463–470.

Agarwal, M. M. (1965). Taxonomy of encyrtid parasites (Hymenoptera: Chalcidoidea) of Indian Coccoidea. *Acta Hymenopterologica*, Fukuoka, 2: 37–97.

Agarwal, M. M., Kaul, K. K. & Agarwal, S. (1984). Key to encyrtid genera of India (Hymenoptera: Chalcidoidea: Encyrtidae). *Bulletin of Entomology*, New Delhi, 25: 58–75.

Alam, S. M. (1961). Undescribed encyrtid parasites (Hym.: Chalcidoidea) of Indian coccids. *Verhandlungen, XI Internationaler Kongress für Entomologie, Vienna*, 1: 235–240.

Alam, S. M. (1970). Encyrtid parasites reared from *Laccifer lacca* (Kerr.) (Hymenoptera). *Mushi*, 43: 137–142.

Amarasekare, K. S. (2007). Life history of papaya mealybug (*Paracoccus marginatus*), and the effectiveness of three introduced parasitoids (*Acerophagus papayae*, *Anagyrus loecki*, and *Pseudleptomastix mexicana*). PhD dissertation. University of Florida, Florida. pp. 59.

Anonymous. (2020). E-resources on mass production protocols for some agriculturally important insects in India. ICAR-NBAIR, Bengaluru, India. Available at https://databases.nbair.res.in//IOBC-mp/parasitoids.html

Ashmead, W. H. (1899). Classification of the old family Chalcididae. *Proceedings of the Entomological Society of Washington*, 4: 242–249.

Ashmead, W. H. (1900a). Notes on some New Zealand and Australian parasitic Hymenoptera, with descriptions of new genera and species. *Proceedings of the Linnean Society of New South Wales*, 25: 327–360.

Ashmead, W. H. (1900b). On the genera of chalcid-flies belonging to the subfamily Encyrtinae. *Proceedings of the United States National Museum*, 22: 323–412.

Ashmead, W. H. (1904). Classification of the chalcid flies of the superfamily Chalcidoidea, with descriptions of new species in the Carnegie Museum collected in South America by Herbert H. Smith. *Memoirs of the Carnegie Museum*, 1: i–xi, 225–551.

Askew, R. R. (1971). *Parasitic Insects*. American Elsevier Publishing Co. Inc. 316 pp.

Ayyar, T. V. R. (1932). A new chalcid with branched antennae from south India (*Tetracnemus indicus* sp. nov.). *Records of the Indian Museum*, 34: 287–288.

Ayyar, T. V. R. (1934). First record of the chalcid genus *Comperiella* Howard from India with description of a new species. *Records of the Indian Museum*, 36: 219–221.

Avidov, Z. & Podoler H. (1968). Studies on the life history of Metaphycus flavus (How.) (Encyrtidae). *Israel Journal of Entomology*, 3: 1–15.

Bartlett, B.R. & Ball, J.C. (1964). The developmental biologies of two encyrtid parasites of Coccus hesperidium and their intrinsic competition. *Annals of Entomological Society of America*, 57: 496–503.

Bartlett, B.R. & Lagace, C.F. (1961). A new biological race of Microterys flavus introduced into California for the control of lecaniine coccids, with an analysis of its behavior in host selection. *Annals of Entomological Society of America*, 54: 222–227.

Bartlett, B.R. & Medved, R.A. (1966). The biology and effectiveness of Diversinervus elegans (Encyrtidae: Hymenoptera), an imported parasite of lecaniine scale insects in California. *Annals of Entomological Society of America*, 59: 974–976.

Beardsley Jr, J.W. (1976). A synopsis of the Encyrtidae of the Hawaiian Islands with keys to general and species (Hymenoptera: Chalcidoidae). Proceedings of Hawaiian Entomological Society Vol XXII, No. 2, Hawaiian.

Ben-Dov Y. & Hogson C. J. (1997). *Soft scale insects: their biology, natural enemies and control*. Elsevier Science B.V., Amsterdam, The Netherlands.

Bernal, J. S., Luck, R.F. & Morse, J.G. (1999). Host influences on sex ratio, longevity, and egg load in two *Metaphycus* species parasitic on soft scales: implications for insectary rearing. *Entomologia Experimentalis Et Applicata*, 92: 191–204.

Cameron, P. (1913). On some new and other species of Hymenoptera in the collections of the Zoological Branch of the Forest Research Institute Dehra Dun Part I. On the Parasitic Hymenoptera reared at Dehradun, northern India, from the lac *(Tachardia)* and sal insects. *Indian Forest Records*, 4: 91–110.

Chakupurakal, J., Markham, R.H., Neuenschwander, P., Sakala, M., Malambo, C., Mulwanda, D., Banda, E., Chalabesa, A., Bird, T. & Haug, T. (1994). Biological control of the cassava mealybug, *Phenacoccus manihoti* (Homoptera: Pseudococcidae), in Zambia. *Biological Control*, 4: 254–262.

Clausen, C. P. (1962). *Entomophagous insects*. Hefner Publishing Company, New York.

Clausen, C. P. (1978). Scarabaeidae. In *Introduced Parasites and Predators of Arthropod Pests and Weeds: a World Review* ed. C. P. Clausen, pp. 277–292. Washington, DC: United States Department of Agriculture Handbook No. 480.

Compere, H. & Annecke, D. P. (1960). A reappraisal of *Aphycus* Mayr, *Metaphycus* Mercet and related genera (Encyrtidae). *Journal of the Entomological Society of Southern Africa*, 23: 375–389.

Compere, H. & Annecke, D. P. (1961). Descriptions of parasitic Hymenoptera and comments (Hymenoptera: Aphelinidae. Encyrtidae, Eulophidae). *Journal of the Ecological Society of South Africa*, 24: 17–71.

Compere, H. & Flanders, S.E. (1934). *Anarhopus sydneyensis* Timb., an encyrtid parasite of *Pseudococcus longispinus* (Targ.) recently introduced into California from Australia. *Journal of Economic Entomology*, 27: 966–997.

Cooley, R. A. (1927). Tick Parasites. *Medical Sentinel*, 35: 805–815.

Cooley, R. A. & Kohls, G.M. (1933). A Summary on Tick Parasites. *Proceedings of 5th Pacific Science Congress*, 5: 3375–3381.

Cruz, Y. P. (1981). A sterile defender morph in a polyembryonic hymenopterous parasite. *Nature*, 294: 446–447.

Dahms, E.C. & Gordh, G. (1997). A review of the genera of Australian Encyrtidae (Hymenoptera: Chalcidoidea) described from Australia by A.A. Girault with a checklist of included species. *Memoirs on Entomology, International*, 9: 192–193.

Dalman, J. W. (1820). FoMk till Uppstdlning af Inwt-familjen Pteromdini, i s~merhet med afseen de pd de i Sverige funne Arter. *Kungliga SwnsRa Vetenskapsakadoniens Hundlingar*, 41(1): 123–174, 177–182.

De Bach, P. (1939). *Microterys titani* Gir., an egg predator of *Lecanium corni* Bouch. *Journal of Economic Entomology*, 32: 728–729.

De Santis, L. (1964). Encirtidos de la Republica Argentina (Hymenoptera: Chalcidoidea). Anales de la Comisi6n de Investigaci6n Cientifica Provincia de Buenos Aires, *Gobernación*, 4: 9–422.

Erdos, J. & Novicky, S. (1955). Genera *Encyrtidam* regionis palaearcticae. *Beitrage zur Entmnologie*, 5: 165–202.

Evans, G.A. & Abd-Rabou, S. (2013). An Annotated List of the Encyrtids of Egypt (Hymenoptera: Chalcidoidea: Encyrtidae). *Acta Phytopathologica et Entomologica Hungarica*, 48(1): 107–128.

Fallahzadeh, M., Abdimaleki, R. & Saghaei, N. (2014). New records of Tetracneminae (Hymenoptera, Chalcidoidea, Encyrtidae) from Iran. Turkish Journal of Zoology, 38: 515–518.

Fatima, F. (2018). A Revision of Indian Genera of the Tribe Habrolepidini Hymenoptera Chalcidoidea Encyrtidae Encyrtidae Encyrtinae. Ph.D. Thesis, Aligarh Muslim University, Aligarh.

Fatima, A. & Shafee, S. A. (1994). Studies on the taxonomy of the Indian encyrtids (Hymenoptera: Encyrtinae). *Aligarh Muslim University Publications (Zoological Series) on Indian Insect Types*, 15: 141.

Flanders, S.E. (1942). Oosorption and ovulation in relation to oviposition in the parasitic Hymenoptera. *Annals of Entomological Society of America*, 35: 251–266.

Foerster, A. (1856). Hymenopterologische Studien. 2. *Chalcidiae und Prcctotmpii*. 152 pp. Aachen.

Fuchs, T. W., Stewart, J. W., Minzenmayer, R. & Rose, M. (1991). First record of Phenacoccus solenopsis Tinsley in cultivated cotton in the United States. Southwestern Entomologist, 16: 215–221.

Gahan, A. B. (1919). Report on a small collection of Indian parasitic Hymenoptera. *Proceedings of the United States National Museum*, 56: 513–524.

Girault, A. A. (1915). Australian Hymenoptera Chalcidoidea-MI. The family Encyrtidae with descriptions of new genera and species. *Memoirs of the Queensland Museum*, 4: 1–184.

Greathead, D. J. (1986). Parasitoids in classical biological control, In: *Insect parasitoids* ed. Waage J. K. & D. Greathead, pp. 290–318. London: Academic Press, 548 pp.

Grissell, E. E. & Schauff, M. E. (1990). A handbook of the families of Nearctic Chalcidoidea (Hymenoptera). *Entomological Society of Washington Handbook*, 1: 85 pp.

Guerrieri, E. & Ghahari, H. (2018) New records, descriptions and notes on Encyrtidae (Hymenoptera: Chalcidoidea) from Iran. *Zootaxa*, 4444 (3): 316–326.

Hayat, M. (1979a). Taxonomic notes on Indian Encyrtidae (Hym.: Chalcidoidea). I. *Journal of Natural History*, 13: 315–326.

Hayat, M. (1979b). Indian species of *Anagyrus* (Hym.: Encyrtidae) - I. *Oriental Insects*, 13: 167–188.

Hayat, M. (1981a). Taxonomic notes on Indian Encyrtidae (Hymenoptera: Chalcidoidea). III. *Colemania*, 1: 13–34.

Hayat, M. (1981b). Taxonomic notes on Indian Encyrtidae (Hym.: Chalcidoidea). II. *Journal of Natural History*, 15: 17–29.

Hayat, M. (1989). Taxonomic notes on Indian Encyrtidae (Hymenoptera: Chalcidoidea) - IV. *Oriental Insects*, 23: 275–285.

Hayat, M. (1999). Taxonomic notes on Indian Encyrtidae (Hymenoptera: Chalcidoidea) - V. *Oriental Insects*, 33: 349–407.

Hayat, M. (2006a). Indian Encyrtidae (Hymenoptera: Chalcidoidea): Department of Zoology, Aligarh Muslim University, India.

Hayat, M. (2006b). Descriptions of eight new species of encyrtidae from India, with some records (Hymenoptera: Chalcidoidea). *Oriental Insects*, 40: 303–315.

Hayat, M. (2012). Taxonomic notes on Indian Encyrtidae (Hymenoptera: Chalcidoidea) - VI. *Oriental Insects*, 46(2): 163–181.

Hayat, M., Alam, M. & Agarwal, M. M. (1975). Taxonomic survey of encyrtid parasites (Hymenoptera: Encyrtidae) in India. *Aligarh Muslim University Publication, Zoological Series on Indian Insect Types*, 9: iii + 112.

Hayat, M., Zeya, S.B. & Veenakumari, K. (2013). On some brachypterous Encyrtidae (Hymenoptera: Chalcidoidea) from India, with description of four new species. *Zootaxa*, 3716 (2): 259–276.

Herren, H.R. (1981). Biological control of the cassava mealybug. In: Tropical Root Crops, Research Strategies for the 1980s (Ed. E.R. Terry *et al.*). Proceedings of the First Triennial Root Crops Symposium, 8–12 September 1980, Ibadan, Nigeria. International Institute of Tropical Agriculture. pp. 79–80.

Herren, H.R. & Lema, K.M. (1982). CMB-first successful releases. *Biocontrol News Information*, 3: 185.

Hoffer, A. (1955). The phylogeny and taxonomy of the family Encyrtidae (Hym., Chalcidoidea). *Acta Musei Nationalis Pragae*, 11(B): 1–22.

Howard, L. O (1881). Report of the parasites of Coccidae in the collections of the U. S. Department of Agriculture Part IU. Report of United States Department of &culture. Washington. Entomology, 1880: 350–372.

Huang, D. W. & Noyes, J. S. (1994). A revision of the Indo-Pacific species of *Ooencyrtus* (Hymenoptera: Encyrtidae), parasitoids of the immature stages of economically 363 important insect species (mainly Hemiptera and Lepidoptera). *Bulletin of the Natural History Museum, London (Entomology)*, 63: 1–136.

Ishii, T. (1928). The Encyrtinae of Japan. *I. Bulletin of the Imperial Agricultural Experiment Station of Japan*, 3: 79–160.

Japoshvili, G. (2012). New data on encyrtid (Hymenoptera: Chalcidoidea: Encyrtidae) parasitoids of Coccoids (Hemiptera: Coccoidea) from Turkey, with description of five new species. *Entomol Fennica*, 23: 72–82.

Japoshvili, G., Ay, R., Karaca, I., Gabroshvili, N., Barjadze, Sh. & Chaladze, G. (2008). Studies on the parasitoid complex attacking the globose scale *Sphaerolecanium prunastri* (Fonscolombe) (Hemiptera: Coccoidea) on *Prunus* species in Turkey. *Journal of the Kanssa Entomological Society*, 81: 339–344.

Japoshvili, G. & Noyes, J. (2006). The Western Palaearctic species of *Psilophrys* Mayr (Hymenoptera, Chalcidoidea: Encyrtidae), parasitoids of kermesids (Hemiptera, Coccoidea: Kermesidae) attacking oaks (*Quercus* spp.). *Journal of Natural History*, 40: 1783–1800.

Joshi, S., Rameshkumar A. & Mohanraj, P. (2017). New host-parasitoid associations for some coccids (Hemiptera: Coccoidea) from India. Journal of Entomological Research, 41: 177–182.

Jung Suh, S. (2019). Korean encyrtids (Hymenoptera: Encyrtidae) associated with scale insects (Hemiptera: Coccomorpha). *Insecta Mundi*, 0719: 1–9.

Kaneria, M. & Singh, S. (2020). Description of a new species of *Ericydnus* (Haliday, 1832) (Hymenoptera: Encyrtidae: Tetracneminae) from India. *Journal of Entomological Research Society*, 22(1): 75–82.

Kapranas, A., Lo Giudice, D., Morse, G.P. & Luck, R.F. (2011). Biology and behaviour of Metaphycus angustifrons Compere (Hymenoptera: Encyrtidae), a newly established parasitoid of soft scale insects (Hemiptera: Coccidae) in California. *Biological Control*, 56: 139–144.

Karmakar, P. & Shera, P.S. (2018). Seasonal and biological interactions between the parasitoid, *Aenasius arizonensis* (Girault) and its host, *Phenacoccus solenopsis* Tinsley on cotton. *Phytoparasitica*, 46: 661–670.

Kapranas, A. & Tena, A. (2015). Encyrtid parasitoids of soft scale insects: biology, behaviour and their use in biological control. Annual Review of Entomology, 60: 195–211.

Kapranas, A., Wanjberg, E. & Luck, R.F. (2009). Sequences of sex allocation and mortality in clutches of *Metaphycus* parasitoids of soft scale insects and the prevalence of all-female broods. *Ecological Entomology*, 34: 652– 662.

Kaul, K. & Agarwal, M. M. (1986). Taxonomic studies on encyrtid parasitoids (Hymenoptera: Chalcidoidea) of India. *Aligarh Muslim University Publications (Zoological Series) on Indian Insect Types*, 13: viii + 89.

Kaydan, M. B. & Japoshvili, G. (2010). The aphelinid and encyrtid parasitoids (Hmenoptera: Chalcidoidea) of Pseudococcidae (Hemiptera: Coccoidea) in the Van Lake basin of Turkey. *Turkish Journal of Entomology*, 34 (4): 465–476.

Krishnachaitanya, T. & Manickavagam, S. (2016). Description of two new species of Tetracneminae Howard (Hymenoptera: Encyrtidae) from India with key to the species. *Journal of Insect Biodiversity*, 4(8): 1–11.

Krishnamoorthy, A. A. & Singh, S. P. (1987). Biological control of citrus mealybug, Planococcus citri with an introduced parasite, Leptomastix dactylopii in India. Entomophaga, 3: 143–148.

Kumar, A., Manickavasagam, S., & Krishnachaitanya, T. (2019). New records of fourteen genera and twelve species of encyrtid (Chalcidoidea: Encyrtidae) from Bihar, India. *Journal of Entomology and Zoology Studies*, 7(3): 677–682.

Lampson, L. J., Morse, J.G. & Luck, R.F. (1996). Host selection, sex allocation, and host feeding by *Metaphycus helvolus* (Hymenoptera: Encyrtidae) on Saissetia oleae (Homoptera: Coccidae) and its effect on parasitoid size, sex, and quality. *Environmental Entomology*, 25: 283–294.

Latreille, P. A. (1809). Genera Crustaceorurn et Insectorurn 4: 399pp. A. Kcenig, Parisius et Argentorati.

Li, Z., Yao, T. Xu, Z., Meng, L. & Li, B. (2020). A new species of Cheiloneurus Westwood *(Hymenoptera, Encyrtidae)* as a hyperparasitoid of the invasive cotton mealybug, Phenacoccus solenopsis Tinsley in China. *Zookeys*, 974: 23–29.

Liao, D.X., Li, X.L., Pang, T.L. & Chen, T.L. (1987). Hymenoptera: Chalcidoidea. *Economical Insect Fauna China*, 34: 241.

Mandoor, N.S., Mahmoud, M.F., Osman, M.A.N. & Qiu, B. (2008). Efficiency, intrinsic competition and interspecific host discrimination of Copidosoma desantisi and Trichogramma evanescens, two parasitoids of Phthorimaea operculella. Biocontrol Science and Technology, 18: 903–912.

Mani, M. S. (1935). A new encyrtid chalcid genus *Krishnieriella* gen. novo from India. *Records of the Indian Museum*, 37: 421–423.

Mani, M. S. (1938). *Catalogue of Indian Insects. Part 23. Chalcidoidea*. Government of India, Delhi. iii+ 174 pp.

Mani, M. S. (1939). Descriptions of new and records of some known chalcidoid and other hymenopterous parasites from India. *Indian Journal of Entomology*, 1: 69–99.

Mani, M. S. (1941). Studies on Indian parasitic Hymenoptera I. *Indian Journal of Entomology*, 3: 25–36.

Mani, M. S. (1989). *The Fauna of India and the adjacent countries. Chalcidoidea (Hymenoptera). Part I. Zoological Survey of India*. Government of India. xxvi+ 1067 pp.

Mani, M. S., Dubey, O. P., Kaul, B. K. & Saraswat, G. G. (1973). On some Chalcidoidea from India. *Memoirs of the School of Entomology, St. John's College, Agra*, 2: 1–128.

Mani, M. S., Dubey, O. P., Kaul, B. K. & Saraswat, G. G. (1974). Descriptions of some new and new records of some known Chalcidoidea (Hymenoptera) from India. *Memoirs of the School of Entomology, St. John's College, Agra*, 3: 1–108.

Manickavasagam, S. & Rameshkumar, A. (2012). New distributional records of Encyrtidae (Hymenoptera: Chalcidoidea) from Puducherry, *India. Check List*, 8: 1337–1343.

Maple, J. D. (1947). The eggs and first instar larvae of Encyrtidae and their morphological adaptations for respiration. *University of California Publications in Entomology*, 8: 25–122.

Maredia, M.K. & Raitzer, D.A. (2010). Estimating overall returns to international agricultural research in Africa through benefi-cost analysis: a "best evidence" approach. *Agricultural Economics*, 41: 81–100.

Mayr, G. (1876). Ihe europa~chen Encyrtiden. *Verhandlungen deT hlogisch-Botanischen Gesellscha ft in Wim*, 25: 675–778.

Mercet, R. G. (1921). *Fauna Iberica. Himenoptem Farn. Encyrtidos*. Mws Nacional de Cienw Naturales, Madrid, 727 pp.

Moore, D. (1988). Agents used for biological control of mealybugs (Pseudococcidae). *Biocontrol News and Information*, 9(4): 209–225.

Mwangi, E. N., Dipeolu, O. O., Newson, R. M., Kaaya, G. P., & Hassan, S. M. (1991). Predators, parasitoids and pathogens of ticks: A review. *Biocontrol Science and Technology*, 1(3): 147–156. 10.1080/0958315 9109355195

Myrick, S., Norton, G.W., Selvaraj, K.N., Natarajan, K. & Muniappan, R. (2014). Economic impact of classical biological control of papaya mealybug in India. *Crop Protection*, 56: 82–86.

Nagrare, V., Kranthi, S., Kumar, R., Dhara Jothi, B., Amutha, M., Deshmukh, A., Bisane, K. & Kranthi, K. (Eds.). (2011). Compendium of Cotton Mealybugs. CICR, Shankar Nagar, Nagpur, India. pp 29.

Narayanan, E. S. & Subba Rao, B. R. (1960). New species of encyrtid and braconid parasites. *Indian Journal of Entomology*, 22: 75–79.

Narayanan, E. S., Subba Rao, B. R. & Mathur, R. B. (1957a). Some known and new records of parasites of sugarcane scale insects from India. *Indian Journal of Entomology*, 19: 144–146.

Narayanan, E. S., Subba Rao, B. R. & Sangwan, H. S. (1957b). New species of parasites of Rhodes grass scale from Indian Union. *Indian Journal of Entomology*, 19: 65–66.

Neuenschwander P., Hammond, W. N. O., Ajuonu, O., Gado, A., Echendu, N., Bokononganta, A. H., Allomasso, R. & Okon, I. (1990). Biological control of the cassava mealybug, *Phenacoccus manihoti* (Hom.: Pseudococcidae) by *Epidinocarsis lopezi* (Hym.: Encyrtidae) in West Africa, as influenced by climate and soil. *Agriculture, Ecosystems and Environment*, 32: 39–55.

Neuenschwander, P. & Herren, H. R. (1988). Biological control of the cassava mealybug, *Phenacoccus manihoti,* by the exotic parasitoid *Epidinocarsis lopezi* in Africa. *Philosophical Transactions of the Royal Society B*, 318: 319–333.

Nikol'skaya, M. (1952). Chalcids of the fauna of the USSR (Chalcidoidea). *Opredeliteli po Faune SSSR*, 44: 348 Zoologicheskim Institutom Akademii Nauk SSSR, Moscow and Leningrad.

Norgaard, R. B. (1988). The biological control of cassava mealybug in Africa. American Journal of Agricultural Economics, 70: 366–371.

Noyes, J. S. (1980). A review of the genera of Neotropical Encyrtidae (Hymenoptera: Chalcidoidea). *Bulletin of the British Museum (Natural History) Entomology*, 41: 107–53.

Noyes, J. S. (1982). Collecting and preserving chalcid wasps (Hymenoptera: Chalcidoidea). *Journal of Natural History*, 16: 315–334.

Noyes J. S. (1985). Chalcidoids and biological control. *Chalcid Forum*, 5: 5–10.

Noyes, J. S. (1988). Encyrtidae (Insecta: Hymenoptera). *Fauna of New Zealand*, 13: 131–132.

Noyes, J. S. (1990). Encyrtidae. In: *Armored Scale Insects their Biology, Natural Enemies and Control. Volume 4B* eds. D. Rosen, pp. 133–166. Rehovot, Israel: The Hebrew University of Jerusalem, Faculty of Agriculture. 688 p.

Noyes, J. S. 2000. Encyrtidae of Costa Rica (Hymenoptera: Chalcidoidea), 1. The subfamily Tetracneminae, parasitoids of mealy bugs (Homoptera: Pseudococcidae). *Memoirs of American Entomological Institute*, 62: 1–355.

Noyes, J. S. (2004). Encyrtidae of Costa Rica (Hymenoptera: Chalcidoidea), 2. Metaphycus and related genera, parasitoids of scale insects (Coccoidea) and whiteflies (Aleyrodidae). *Memoirs of the American Entomological Institute*, 73: 1–459.

Noyes, J. S. (2010). Encyrtidae of Costa Rica (Hymenoptera: Chalcidoidea), 3. Subfamily Encyrtinae: Encyrtini, Echthroplexiellini, Discodini, Oobiini and Ixodiphagini, parasitoids associated with bugs (Hemiptera), insect eggs (Hemiptera, Lepidoptera, Coleoptera, Neuroptera) and ticks (Acari). *Memoirs of the American Entomological Institute*, 84: 848.

Noyes J. (2020). UCD: Universal Chalcidoidea Database (version Sep 2007). In: Roskov Y., Ower G., Orrell T., Nicolson D., Bailly N., Kirk P. M., Bourgoin T., DeWalt R. E., Decock W., van Nieukerken E. J., Penev L. (eds.). Species 2000 & ITIS Catalogue of Life, 2020-12-01. Digital resource at www.catalogueoflife.org. Species 2000: Naturalis, Leiden, the Netherlands.

Noyes, J. S. & Hayat, M. (1984). A review of the genera of Indo-Pacific Encyrtidae (Hymenoptera: Chalcidoidea). *Bulletin of the British Museum (Natural History) Entomology*, 48: 131–395.

Noyes, J. S. & Hayat, M. (1994). *Oriental mealybug parasitoids of the Anagyrini (Hymenoptera: Encyrtidae)*, CAB International, Oxon, UK. pp. 554.

Orphanides, G. M. (1996). Establishment of Comperiella bifasciata (Hymenoptera: Encyrtidae) on Aonidiella aurantii (Homoptera: Diaspididae) in Cyprus. Entomophaga, 41: 53–57.

Öncüer, C. (1991). *Acatalogue of the parasites and predators of Turkey. Publication of Ege University*, Faculty of Agriculture No: 505, Izmir. 354 pp. [In Turkish.]

Prinsloo, G. L. (1996). The genus *Comperiella* Howard (Hymenoptera: Encyrtidae) in southern Africa: parasitoids of armoured scale insects (Homoptera: Diaspididae). *African Entomology*, 4: 153–160.

Prinsloo, G. L. (1997). The Role of Encyrtids as Natural Enemies of Soft Scale Insects. In: *Soft Scale Insects-Their Biology, Natural Enemies and Control.* eds. Ben-Dov, Y. & Hodgson, C.J., pp. 69–109. Amsterdam: Elsevier Science.

Prinsloo, G. L. & D. P. Annecke. (1979). A key to the genera of Encyrtidae from the Ethiopian region, with descriptions of three new genera (Hymenoptera: Chalcidoidea). *Journal of Entomological Society of South Africa*, 42: 349–382.

Rameshkumar, A., Poorani, J. & Naveen, V. (2015). Additions to the Encyrtidae and Mymaridae (Chalcidoidea) of India with new distribution and host records for some species. *Biodiversity Data Journal*, 3: e5216. 10.3897/BDJ.3.e5216

Rao, S. B. R. (1965). A key to species of *Anicetus* Howard, 1896 (Hymenoptera: Encyrtidae) and descriptions of new species from India. *Proceedings of the Royal Entomological Society of London. Series B, Taxonomy*, 10.1111/j.1365-3113.1965.tb01658.x

Risbec, J. (1951). Les chalcidoides de 1' Afrique occidentale francaise. IL Les Microgasterinae de 1' Afrique occidentale francaise. *Memoires de I' Institute Francais d' Afrique Noire*, 13: 7–409.

Rodríguez, J. M., Velez, B.R., Caballero, S.Z., Martínez, F.A.N., Soriano, E.G. & García, E.R. (2010). Diversity of Encyrtidae (Hymenoptera: Chalcidoidea) collected with Malaise traps in the tropical dry forest of San Javier, Sonora, Mexico. *Revista Mexicana de Biodiversidad*, 81: 813–822.

Saakyan-Baranova, A. A. (1966). The life cycle of *Metaphycus luteolus* Timb. (Hymenoptera, Encyrtidae), parasite of *Coccus hesperidum* L. (Homoptera, Coccoidea), and the attempt of its introduction into the USSR. *Entomological Review*, 45: 414–423.

Saraswat, G. G. & Mukerjee, M. K. (1975). Records of some known and descriptions of new species of chalcids (Hymenoptera) from India. *Memoirs of the School of Entomology, St. John's College, Agra*, 4: 35–62.

Shafee, S. A., Alam, S. M. & Agarwal, M. M. (1975). Taxonomic survey of encyrtid parasites (Hymenoptera: Encyrtidae) in India. *Aligarh Muslim University Publications (Zoological Series) on Indian Insect Types*, 10: iii+ 1–125.

Shera, P. S., Karmakar, P., Sharma, S. & Sangha, K.S. (2017). Suitability of different mealybug species as hosts to solitary endoparasitoid, *Aenasius arizonensis* (Girault) (=*Aenasius bambawalei* Hayat). *International Journal of Pest Management*, 63: 280–288.

Shi, Z. Y., Xu, G. & Si, S.G. (1994). A key to the known genera, species of Encyrtidae from Henan. *Fauna and Taxonomy of Insects in Henan*, 1: 52–54.

Shylesha, A. N. Joshi S., Rabindra, R.J., Kumar P.S., Sreedevi, K. & Ballal, C.R. (2018) Biocontrol Bears Fruit: Saving Papaya from the Mealybug. *ICAR–National Bureau of Agricultural Insect Resources*. Hebbal, Bengaluru, India, p. 2.

Singh, S. (1996). Two new species of *Psyllaephagus* Ashmead (Hymenoptera: Chalcidoidea: Encyrtidae) attacking *Mycopsylla* sp. (Homoptera: Psyllidae) infesting *Ficus religiosa* in Mizoram, India. *Oriental Insects*, 30: 155–166.

Singh, S. (2004). A new species of *Caenohomalopoda* Tachikawa (Hymenoptera: Encyrtidae) from Mizoram, India. *Zoo's Print Journal*, 19(9): 1613–1615.

Singh, S. (2007). Two new species of Genus *Cerchysiella* Girault (Hymenoptera: Chalcidoidea: Encyrtidae) From Northern India. *Ann Entomol*, 25(1-2): 17–21.

Singh, S. (2016a). Description of a new species of *Psyllaephagus* Ashmead (Hymenoptera: Encyrtidae) parasitizing pit gall forming Psyllid *Trioza Fletcheri* (Hemiptera: Psylloidea: Triozidae) on *Terminalia Arjuna* from India. *Annales Zoologici (Warszawa)*, 66(3): 393–402.

Singh, S. (2016b). A new species of *Paraphaenodiscus* Girault (Hymenoptera: Encyrtidae) from India parasitizing *Coccus* sp. (Hemiptera: Coccidae). *Journal of Insect Biodiversity*, 4(4): 1–13.

Singh, S. & Agarwal, M. M. (1991). Descriptions of new species of *Cerapteroceroides* and *Cerapterocerus* (Hymenoptera: Encyrtidae) from north-eastern India. *Oriental Insects* 25: 211–220.

Singh, S. & Agarwal, M. M. (1993). Taxonomic studies on Indian encyrtid parasites (Hymenoptera: Encyrtidae) from north-eastern region. *Aligarh Muslim University Zoological Publications, Indian lnsect Types*, 14: 180.

Singh, S., Agarwal, M. M. & Basha, M. C. (1991). Descriptions of three new species of *Neastymachus* (Hymenoptera: Encyrtidae) from north-eastern India. *Oriental Insects*, 25: 221–230.

Singh, S. & Hayat, M. (2002). New species and new record of *Microterys* (Hymenoptera: Chalcidoidea: Encyrtidae) from north-eastern India. *Oriental Insects*, 36: 405–421.

Singh, S., Rema Devi, O. K. & Srinivasa, Y. B. (2014). Description of a new genus and three species of Encyrtidae (Hymenoptera: Chalcidoidea) from the Western Ghats of Karnataka, India. *Zootaxa*, 3814(3): 369–384.

Singh, S. & Singh, K. P. (2011). Description of two new species of *Psyllaephagus* Ashmead (Hymenoptera: Encyrtidae) parasitizing *Phacopteron lentiginosum* Buckton (Hemiptera: Psyllidae), a leaf gall-former of *Garuga pinnata* Roxburgh (Burseraceae). *Zootaxa*, 2885: 33–43.

Smith, H. S. (1933). The influence of civilization on the insect fauna by purposeful introduction. *Annals of Entomological Society of America*, 26: 518–528.

Smith, H. S. & Compere, H. (1928). A preliminary report on the insect parasites of the black scale *Saissetia oleae* (Bernard). *University of California Publications in Entomology*, 4: 231–334.

Subba Rao, B. R. (1966). *Comperiella aspidiotiphaga* new species from India (Encyrtidae: Hymenoptera). *Mushi*, 39: 137.

Subba Rao, B. R. & Hayat, M. (1985). [eds]. The Chalcidoidea (Insecta: Hymenoptera) of India and the adjacent countries. Part I. *Oriertal lnsects*, 19: 163–310+310a-o.

Subba Rao, B. R. & Hayat, M. (1986). [eds]. The Chalcidoidea (Insecta: Hymenoptera) of India and the adjacent countries. Part II. A catalogue of Chalcidoidea of India and the adjacent countries. *Oriental Insects*, 20: 1–430.

Subba Rao, B. R. & Rai, P. S. (1970). Studies on the Indian species of the genus *Anagyrus* Howard, 1896 (Hymenoptera: Encyrtidae). *Beitrage zur Entomologie*, 20: 85–96.

Suroshe, S. S., Gautam, R. D. & Fand, B. B. (2013). Natural enemy complex associated with the mealybug, Phenacoccus solenopsis Tinsley (Hemiptera: Pseudococcidae) infesting different host plants in India. Journal of Biological Control, 27: 204–210.

Tachikawa, T. (1963). Revisional studies of the Encyrtidae of Japan (Hymenoptera: Chalcidoidea). *Memoirs of Ehime University*, 6(9) 1–264.

Tena, A., Kapranas, A., Garcia-Mari, F. & Luck, R.F. (2008). Host discrimination, superparasitism and infanticide by a gregarious endoparasitoid. *Animal Behavior*, 76: 789–799.

Tena, A., Kapranas, A., Walker, G.P., Garcia-Mari, F. & Luck, R.F. (2011). Larval morphology of *Metaphycus flavus* and the role of host attachment in larval cannibalism. *Bulletin of Entomological Research*, 101: 365–372.

Thomson, C. G. (1876). *Skandinaviens Hymenoptera*. 4: 192 pp. Lund.

Timberlake, P. H. (1920). Descriptions of new genera and species of Hawaiian Encyrtidae (Hymenoptera). II. *Proceedings of the Hawaiian Entomological Society*, 4: 409–437.

Triapitsyn, S. V., Andreason, SA, Power, N, Ganjisaffar, F, Fusu, L, Dominguez, C & Perring, TM (2020). Two new species of *Ooencyrtus* (Hymenoptera, Encyrtidae), egg parasitoids of the bagrada bug Bagrada hilaris (Hemiptera, Pentatomidae), with taxonomic notes on Ooencyrtus telenomicida. *Journal of Hymenoptera Research*, 76: 57–98.

Triapitsyn, S. V., Petrice, T.R., Gate, M.W. & Bauer, L.S. (2015). Two new species of Oobius Trjapitzin (Hymenoptera, Encyrtidae) egg parasitoids of *Agrilus* spp. (Coleoptera, Buprestidae) from the USA, including a key and taxonomic note on other congeneric Nearctic taxa. *ZooKeys*, 498: 29–50.

Trjapitzin, V. A. (1973a). The classification of the family Encyrtidae (Hymenoptera, Chalcidoidea). Part 1. Survey of the systems of classification. The subfamily Tetracneminae Howard, 1892. *Entomlogical Obozrenie*, 52(1): 163–175.

Trjapitzin, V. A. (1973b). Classification of the parasitic Hymenoptera of the family Encyrtidae (Chalcidoidea). Part II. *Subfamily Encyrtinae Walker. 1837. Entmlogicheshoe Obozrenie*, 52(2): 416–429.

Trjapitzin, V. A. (1989). *Parasitic Hymenoptera of the Fam. Encyrtidae of Palaearctics. (in Russian) Opredeliteli fv Faune SSSR, Izhvaemie & logicheskim lnstitutonr Akahii Nauk SSR*. 158: 1–489. Leningrad, Nauka.

Trjapitzin, V. A. (2008). A review of encyrtid wasps (Hymenoptera, Chalcidoidea, Encyrtidae) of Macaronesia. *Entomological Review*, 88: 218–232.

Trjapitzin, V. A. & Doganlar, M. (1997). A review of encyrtids (Hymenoptera, *Encyrtidae*) of Turkey. *Entomologicheskoe Obozrenie*, 76(1): 213–222.

Trjapitzin, V. A. & Gordh, G. (1978a). *Review of the genera of Nearctic Encyrtidae (Hymenoptera, Chalcidoidea) I (in Russian) Entomologicheskce Obozrenie*, 57(2): 364–385.

Trjapitzin, V. A. & Gordh, G. (1978b). *Review of genera of Nearctic Encyrtidae (Hymenoptera, Chalcidoidea) II (in Russian) Entomologicheskce Obozrenie*, 57(3): 636–653.

Vennila, S., Ramamurthy, V.V., Deshmukh, A., Pinjarkar, D.B., Agarwal, M., Pagar, P.C., Prasad, Y.G., Prabhakar, M., Kranthi, K.R. & Bambawale, O.M. (Eds.). (2010). A Treatise on Mealybugs of Central Indian Cotton Production System. Technical Bulletin 24. National Centre for Integrated Pest Management, New Delhi, India. pp 1–39.

Voinovich, N. D. & Sugonyaev, E.S. (1993). Peculiarities of parasitism of chalcids (Hymenoptera, Chalcidoidea) on coccids (Homoptera, Coccoidea): V. Biology and preimaginal phases of *Pseudorrhopus testaceus* Ratzeburg, a parasite of *Physokermes hemicryptus* Dalm. *Entomologicheskce Obozrenie*, 72: 549–557.

Walker, F. (1837). Monographia Chalciditum. *Entomological Magazine*, 4: 439–461.

Wang, Z.H., Huang, J. & Xu, Z.H. (2014). Chinese species of the genus Prochiloneurus Silvestri with description of a new species (Hymenoptera: Encyrtidae). Zoological Systematics, 39: 424–432.

Weseloh, R. M. (1969). Biology of *Cheiloneurus noxius*, with emphasis on host relationships and oviposition behavior. *Annals of Entomological Society of America*, 62: 299–305.

Westwood, J. O. (1840). Synopsis of the genera of British insects. In: *A Introduction to the Modem Classification of Insects*, 2(Appendix): 158. London.

Wright, E. J. (1986). Immature stages of *Encyrtus saliens* (Hymenoptera: Encyrtidae), an imported parasite of ice plant scales (Homoptera: Coccidae) in California. *Annals of Entomological Society of America*, 79: 273–279.

Yanzohu, Z. & Dawei, H. (2004). *A Review and an Illustrated Key to Genera of Encyrtidae (Hymenoptera: Chalcidoidea)* from China. Science Press, China.

Yasnosh, V. A. (1998). Japanese wax scale and natural enemies in Tbilisi. *Bulletin of the Georgian Academy of Science*, 157: 132–134.

Zeddies, J., Schaab, R.P., Neuenschwander, P. & Herren, H.R. (2001). Economics of biological control of cassava mealybug in Africa. *Agricultural Economics*, 24: 209–219.

7 Chalcidoid Parasitoids

P. M. Sureshan and P. Girish Kumar
Western Ghat Regional Centre, Zoological Survey of India, Kozhikode, Kerala, India

CONTENTS

LEARNING OBJECTIVES

i. To learn about chalcid parasitoids and their importance in the biocontrol of pests.
ii. To understand the diversity and distribution of chalcidoid wasps.
iii. To learn about the common and effective parasitoids and their hosts in the Indian context, including some successful biocontrol Programmes in India utilising chalcid wasps.
iv. To understand the basic characteristics of effective biocontrol agents.
v. To learn about the rearing techniques, mass production and field application in biocontrol programmes.
vi. To understand the threat and conservation status of chalcid wasps.

DOI: 10.1201/9781003354239-7

7.1 INTRODUCTION

The superfamily Chalcidoidea (popularly called the 'chalcidoid wasps' or briefly 'chalcids') is one of the largest and species-rich groups of insects belonging to the order Hymenoptera. Owing to their diverse body colour and structure, they are also popularly called as 'microscopic jewels'. They are taxonomically the most difficult, ecologically most complex and economically perhaps the most important group of insects. Most members of the group are parasitic, attacking major insect orders, many of which are agricultural and forest pests of economic importance. By their parasitic mode of life, they destroy populations of potential pest species in the field, regulating their numbers, thus playing an important role in the biocontrol of pests. The population numbers and activities of chalcid parasitoids will be high in the crop fields, which are maintained in an organic farming practice without the use of synthetic pesticides. The chalcids are parasitoids because their immature stages are only parasitic, which kill and consume the host during development (Huber, 2017). Chalcidoidea also contains some phytophagous species, which directly or indirectly influence the health of crop plants by forming galls or by feeding on soft tissues. Chalcids are characterised by their minute size ranging from about 0.13 mm to almost 25 mm and are readily distinguished from other such groups of Hymenoptera by their reduced forewing venation without closed cells and pronotum not touching tegula (the sclerite on which forewings are attached), due to the presence of a sclerite called pre-pectus. The other diagnostic characteristics of chalcids are the geniculate antennae with 13 or fewer segments, biarticulate trochanters, sclerotised metasomal segments and ovipositor issuing from the ventral surface far anteriorly to the anus.

7.2 DIVERSITY AND DISTRIBUTION OF CHALCIDS

Chalcid wasps are one of the most diverse groups of Hymenoptera numerically, structurally, and biologically. They range in size from the smallest insect known, *Dicopomorpha echmepterygis* (Mymaridae), at about 130 microns (0.13 mm), to over 30 mm, including bizarre as well as beautiful winged and wingless forms. About more than 22,700 valid species have been described in about more than 2100 genera worldwide, but these numbers represent only a fraction of the true chalcid diversity and estimates of about 60,000 to 1,00,000 species worldwide do not seem unreasonable. Even though chalcids have a wider distribution, they are poorly known in the colder parts of the country. The reason for the less common occurrence of chalcids at higher altitudes may be attributed to the less availability of host insects, which in turn may be due to the decreased floral diversity. Low temperature may also act as a limiting factor for the less abundance of these minute parasitoids in such areas. The number, definition and limits of the superfamily Chalcidoidea have always been in a state of flux and according to the most recent estimates, the superfamily is divided into 25 families to accommodate more than 2000 genera and more than 22,700 species worldwide (Aguiar et al., 2013; Noyes, 2019). Members of 19 families of Chalcidoidea are known from India, which includes approximately more than 3,000 species in about 528 genera (Noyes, 2019). This forms about 27% and about 13% respectively of the world genera and species. Chalcids also show fascinating diversity in biology exhibiting both entomophagy and phytophagy. Adult chalcids are free-living, feeding on nectar, honeydew and in some cases little water. The females (parasitic forms) insert their eggs into the body of the host with the help of their needle-like ovipositor, which develops inside the body of the host and the ensuing larvae consume the host tissue, pupate and emerge out as adults. Chalcids are distributed in all zoogeographical regions of the world and the majority of them are parasitoids attacking a large range of insect orders in their various stages of development, such as eggs, larvae, pupae or even adults, and occasionally attack spider eggs, and also associated with some acari and other arachnids. The phytophagous species either cause galls on plants or are seed eaters. Some species feed on gall tissue but do not cause galls and are called inquilines and they coexist with the gall inducers. In host selection, parasitic Chalcidoidea are specialists, generalists or opportunists. Mymaridae and Trichogrammatidae are specialists being strictly internal parasitoids of insect eggs. Other examples of specialists are found in the families, such as Eucharitidae, parasitizing ant larvae and pupae; and Leucospidae parasitizing bees and wasps. Chalcid parasitoids are solitary or gregarious, monophagous, oligophagous or polyphagous or they are primary (attacking only a phytophagous species), secondary or super-parasitic (attacking a host already attacked by another primary parasite). Hyperparastitism may reach even tertiary or quaternary levels in some species and are either ectoparasites or endoparasites. Chalcid parasitoids are either idiobionts (the attacked host is often immediately prevented from further growth and feeding) or koinobionts (the larvae of the parasite co-exist with the feeding host for some time). Sexual dimorphism is prominent in chalcids and the sex ratio varies very much, and in some species, males are fewer in numbers.

7.3 COMMON AND EFFECTIVE PARASITOIDS AND THEIR HOSTS

In the field of insect pest control, increasing attention is being paid to eco-friendly and environmentally safe methods. Chalcids represent one of the most effective biocontrol agents owing to their wide range of host associations and diverse and often fascinating ways of biology, ecology and behaviour (Huber, 2017). Many species have been utilised successfully in biocontrol programmes worldwide and as far as India is concerned, the group has received relatively little attention in the field of biocontrol of pests. Most experiments and studies in India were devoted to the egg parasite, *Trigchogramma,* whose host range included more than 150 species of insects. Many prospective parasitoids of scale insects are found in the family Aphelinidae. In the family Encyrtidae, there are several effective parasitoids of coccids. The other chalcid families with biocontrol potential are Mymaridae, Pteromalidae, Chalcididae and Eupelmidae. Fig wasps (Agaonidae), being the sole pollinators of fig plants act as keystone mutualists in nature. The relationship between fig wasps and fig plants is an example of mutualism and coevolution and the fig plants can only be pollinated by these wasps and their existence in nature depends only upon them and vice versa. By controlling the populations of herbivorous insects, chalcids not only maintain high species diversity but also prevent herbivorous insects from decimating their plant hosts. Phytophagous chalcids (families Eurytomidae, Torymidae, Agaonidae, Pteromalidae, Tanaostigmatidae and Eulophidae) may assume the status of minor pests since they are mainly gall inducers and seed eaters. Due to the formation of galls, the plant stem becomes lignified and brittle, snapping easily in the wind. Seed and tissue eaters and fruit breeders cause damage to growing parts of crops, seeds and fruits.

7.3.1 Some Successful Biocontrol Parasitoids in India

7.3.1.1 Family Aphelinidae

1. *Encarsia guadeloupae* Viggiani (Chalcidoidea: Aphelinidae) is an accidentally introduced parasitoid of spiralling whitefly, *Aleurodicus dispersus* Russell, which is well established in India, acting as a substantial biological control agent. Whiteflies, which are often polyphagous, feed on phloem sap and reduce the growth of their host plants. Whitefly numbers usually increase very rapidly and the damage due to them to the crops and ornamental plants could be severe (Geetha, 2000; Banjo, 2010).
2. *Aphelinus mali* (Haldeman) (Chalcidoidea: Aphelinidae) is an endoparasitoid of the woolly apple aphid, *Eriosoma lanigerum* (Hausmann) (Hemiptera: Pemphigidae). This parasitoid is established in all the apple-growing areas of India and is found more effective in the valleys than in mountain slopes. During the period of activity of other predators, the population of this exotic parasitoid is diminished. In general, *A. mali* has emerged as a key population regulatory agent of *E. lanigerum* in the areas of its successful establishment (Singh, 2004).

7.3.1.2 Family Encyrtidae

3. *Leptomastix dactylopii* Howard (Chalcidoidea: Encyrtidae) is an exotic parasitoid of *Planococcus* spp., which is well established in southern India. This parasitoid is introduced from West Indies in 1983. It is capable of parasitising *Planococcus citri* and *P. lilacinus*. It was found to be very effective in controlling *P. citri* on acid lime, lemon, mandarins, sweet oranges, guava, coffee, etc. It was recovered in the same locations even after 20 years of release (Singh, 2004).

7.3.1.3 Family Eulophidae

4. *Trichospilus pupivorus* (Chalcidoidea: Eulophidae) has been proven to be a very effective biocontrol agent against the coconut black–headed caterpillar, *Opisina arenosella* (previously known as *Nepantis serinopa*) in different parts of south India. *N. serinopa* in its larval stages causes serious damage to the plant, while the adult moth having a short life span is harmless. The first record of this pest in India was from Coimbatore in 1907 on Palmyra plants, while it was first reported on coconut palms from Bapatala, Andhra Pradesh in 1909 (Rao et al., 1948). Often the pest was found to occur in alarming numbers along the coastal belts where the plants are abundant. With increasing cultivation of the crop in unconventional areas, the pest has also been found to occur in new areas, including interior places (Nasser and Abdurahiman, 2001).
5. The egg parasitoid, *Tetrastichus pyrillae* Crawford (Chalcidoidea: Eulophidae) is an effective biocontrol agent against *Pyrilla perpusilla* (Walker) (Hemiptera: Lophopidae), commonly known as sugarcane leaf hopper. *Pyrilla perpusilla* is one of the most serious pests widely distributed on various agricultural crops, such as wheat, barley, oats, maize, sorghum and a number of grass species near sugarcane fields during outbreak situations. It is one of the most destructive pests in agroecosystems of India and appears periodically. The damage caused by this major pest has been reported up to 28% in the potential sugarcane yield and 2–34% reduction in the sucrose content of sugarcane (Chand et al., 2018). *P. perpusilla* feeding cause poor growth, which also causes difficulty in the milling of affected canes (Kumarsinghe and Wratten, 1996). Both nymphs and adults cause heavy damage to the plant and excrete thick transparent liquid known as honeydew, which acts as a good medium for the growth of black mould. The mould reduces the photosynthetic activity of the leaves and reduces about 25% of the sugar yield. The egg parasitoid, *Tetrastichus pyrillae* (Hymenoptera: Eulophidae) parasitizes 20 to 50 per cent *Pyrilla* eggs (Lakshmanan et al., 1942). There was a severe outbreak of sugarcane *Pyrilla perpusilla* during 1972–73 in the states of Punjab, Haryana, U.P. and Bihar, which was successfully controlled by the utilisation of potential biocontrol agents, like egg parasitoid *Tetrastichus pyrillae*. This saved the Government Exchequer to the tune of Rs.11.00 crores. Similarly, during 1987, there was again an outbreak of this pest in some sugarcane-growing states of the country, when again utilisation of potential bioagent saved Govt. exchequer to the tune of Rs.16.00 crores. In 1994, the severe incidence of this pest in Karnataka was successfully controlled by the potential biocontrol agents (Directorate of Plant Protection, Quarantine & Storage, Ministry of Agriculture & Farmers Welfare, Government of India website).
6. *Tetrastichus howardi* (Olliff) (Chalcidoidea: Eulophidae) is one of the common pupal parasitoids of several lepidopteran pests (Baitha et al., 2004; Prasad et al., 2007) including *Diatraea saccharalis* in sugarcane (Vargas et al., 2011) and maize (Cruz et al., 2011). In addition, *T. howardi* was recorded from larvae and pupae of *D. saccharalis* (Vargas et al., 2011), and it can locate and parasitize this stem borer inside the galleries formed within sugarcane stems (Kfir et al., 1993).

7. *Oomyzus sokolowskii* (Kurdjumov) (Chalcidoidea: Eulophidae) is an effective pupal parasitoid of the diamondback moth, *Plutella xylostella* (Linnaeus) (Lepidoptera: Plutellidae), which is a major pest of Brassicaceae (Fitton and Walker, 1992). This gregarious parasitoid is adapted to high-temperature conditions and has been introduced in tropical and subtropical zones to control the population of diamondback moth (Talekar and Hu, 1996), where it has been established as an effective parasitoid (Ooi, 1988; Liu et al., 1997).
8. *Trichospilus pupivorus* Ferriere (Chalcidoidea: Eulophidae) is a gregarious pupal parasite of several lepidopterous pests (Tavares et al., 2012, 2013; Silva et al., 2016), which has been proven to be a very effective biocontrol agent against the coconut black headed caterpillar, *Opisina arenosella* in different parts of south India (Remadevi et al., 1980); also hyperparasitic through tachinids. Even though it is a pupal parasitoid, it has been used to control pest in South India along with other mechanical control methods (Anantanarayanan, 1934). The characteristics of the parasitoid have been described in detail by Ferriere (1930) and Narendran (2011). Extensive studies on various aspects of *T. pupivorus* have been done by previous workers, such as the biological aspects (Jayaratnam, 1941; Rao et al., 1948; Dharmaraju, 1952; Nirula, 1956), factors affecting the realisation of the biotic potential of the parasitoid (Dharmaraju and Pradan, 1976), its hosts' range (Tavares et al., 2012; Tavares et al., 2013; Silva et al., 2016), and its parasitic behaviour (Nor Ahya et al., 2018).

7.3.1.4 Family Trichogrammatidae

9. The species *Trigchogramma chilonis* Ishii (Chalcidoidea: Trichogrammatidae) is an egg parasitoid of several lepidopterous hosts, commonly produced by Indian insectaries for augmentative releases in crops, like sugarcane, cotton, etc. It has been effectively utilised in the biocontrol of sugarcane borers in Karnataka, Tamil Nadu and Gujarat. The success of Integrated Pest Management strategies involving the mass releases of egg parasitoid, *T. chilonis* against sugarcane tissue borers has been well documented (Srikanth et al., 2016).
10. *Trichogramma japonicum* Ashmead (Chalcidoidea: Trichogrammatidae) is an effective egg parasitoid of several lepidopterous hosts. This species is recommended against graminaceous stem and tissue borers on rice, sugarcane, etc. Inundative releases of egg parasitoid, *T. japonicum* against sugarcane tissue borers has been well documented (Srikanth et al., 2016).
11. *Trichogramma pretiosum* Riley (Chalcidoidea: Trichogrammatidae) is an introduced egg parasitoid of several lepidopterous hosts on sugarcane, cotton, tomato, etc. *Tuta absoluta* (Meyrick), one of the most destructive pests of tomato, has recently invaded India and *T. pretiosum* was found very effective against the pest (Manohar et al., 2020). *T. pretiosum* was found very effective in the control of the population of *Helicoverpa armigera* infesting tomato (Navik & Richa, 2018).
12. *Trichogrammatoidea bactrae* Nagaraja (Chalcidoidea: Trichogrammatidae) is an indigenous egg parasitoid of several lepidopterous hosts, like *Bactra venosana*, *Chilo* spp., *Helicoverpa armigera*, *Pelopidas mathias*, etc. *T. bactrae* has been also used effectively for the control of Rice leaf folder, *Cnaphalocrocis medinalis* (Guenee) (Lepidoptera: Pyralidae) (Perera et al., 2015).

7.4 CHARACTERISTICS OF EFFECTIVE BIOCONTROL AGENTS

An effective biological control agent should have a narrow host range. Generalised parasitoids may be good natural enemies but they do not kill enough pests when other types of prey are also available in an ecosystem. The climatic adaptability of parasitoids is also a very important factor. Natural enemies must be able to survive the extremes of temperature and humidity that they will encounter in the new habitat. Synchrony of the life cycle of parasitoids with that of the host is a factor for an effective biocontrol agent. The parasitoid should be present when the pest first emerges or appears in an ecosystem. The high reproductive potential of parasitoids is extremely significant. Good biocontrol agents produce large numbers of offspring. Ideally, a parasitoid completes more than one generation during each generation of the pest. In order to survive, effective natural enemies must be able to locate their host or prey even when it is scarce. In general, better ability in search results in lower pest population densities. Natural enemies that consume prey rapidly or lay eggs quickly have more time to locate and attack other members of the pest population. Small populations of efficient natural enemies may be more effective biocontrol agents than larger populations of less efficient species. If a natural enemy is too efficient, it may eliminate its own food supply and then starve to death. The most effective biocontrol agents reduce a pest population below its economic threshold and then maintain it at this lower equilibrium level.

7.5 REARING TECHNIQUES, MASS PRODUCTION AND FIELD APPLICATION

Mass rearing of biocontrol agents includes the production of thousands/millions of insects to control some insect pests. The importance of chalcid parasitoids in controlling pest populations has been well established. However, usually adequate numbers of individuals are not always present or available to provide adequate control in ecosystems. Biocontrol by augmentation of parasitoids that have been mass-produced in insectaries provides a means of ensuring the availability of individuals that are adequate to maintain the populations of various pests at the desired level. Production of chalcid parasitoids for augmentation has been of considerable interest to workers in biocontrol since the beginning of the 20th century. Significant advances have been made in the recent past in mass rearing and handling procedures for insect parasitoids (Singh, 2004; Elbehery et al., 2020).

The mass production and release of chalcid parasitoids are affected by several factors, such as supplemental food, environmental conditions and the host. The main obstacle to the widespread application of the augmentation method of pest control was the difficulty in mass-rearing of parasitoids at economical costs. The lack of effective techniques for mass-producing parasitoids on an artificial diet as a major limiting factor in the utilisation of this approach. Also, economic constraints and a lack of techniques that simulate natural conditions result in selection pressures that lead to the genetic deterioration of mass-reared arthropods (Mackauer, 1972). Such deterioration can mean the loss of behavioural traits basic to the effectiveness of released natural enemies (Boiler, 1972), thereby jeopardizing the success of a biocontrol program.

One goal in pest control is to develop managing strategies that kill insect pests without harming the environment or other organisms. An ideal method would be to enhance larval parasitoids to control some insect pest species (Elbehery et al., 2020). This can be accomplished by the mass release of biocontrol agents for controlling many agricultural insect pests.

7.6 THREAT AND CONSERVATION STATUS

Chalcidoidea is one of the most vulnerable insect groups to environmental changes. No species of Chalcidoidea has so far been recorded as extinct or any taxon has been categorised under the IUCN Red Data book. However, several species of chalcids were introduced by man in different parts of the world. Species introductions mainly occurred in search of natural enemies for biocontrol programmes. Species introductions may also occur along with the dispersal of pest-affected plant material, such as fruits, vegetables and seeds, which are parasitized by chalcids. Several species of *Trichogramma* (Trichogrammatidae) were introduced in different parts of India in biocontrol programmes against sugarcane and cotton pests and many species of Encyrtidae has also reached different parts of the country as coccid and mealybug parasitoid. Introduced species are also found in the families, such as Aphelinidae, Eulophidae and Pteromalidae due to their greater utilisation in biocontrol programmes. The major threats faced by chalcids are habitat destruction, excessive use of pesticides and environmental pollution. Severe habitat destruction emanating from faulty agricultural practices, urbanisation and climate change lead to the loss of floral diversity, which in turn affects chalcid populations. Uncontrolled uses of nonselective pesticides will eliminate parasitoid populations from agroecosystems. Secondary enemies or hyperparasitoids can also disrupt the activity of chalcid parasitoids in nature.

The importance of chalcids in nature makes urgent conservation measures imperative in order to protect their populations. As most of its reduction in the ecosystems is due to the application of pesticides, large-scale use of pesticides should be controlled. Large-scale destruction of natural habitats (deforestation, monocultures) leading to the loss of floral diversity should be controlled. Increasing the adjacent vegetational diversity around the cultivated fields will enhance the beneficial activities of chalcid parasitoids. The integrated pest management (IPM) strategies adopted recently in pest control may be of some help in protecting useful insects in nature.

7.7 CONCLUSIONS

In recent decades, increased awareness of the impact of pesticides on the environment and human health has resulted in efforts to reduce reliance on chemical control. The use of biocontrol agents, like chalcid parasitoids, is the most suitable and eco-friendly management strategy, which minimises the disturbance of the ecosystem.

7.8 POINTS TO BE REMEMBER

i. Chalcidoidea (Insecta: Hymenoptera) are an important group of parasitoids both in terms of number and action.
ii. Chalcids are quite successful in the biocontrol of insect pests.
iii. For successful biocontrol programmes the chalcids must possess certain basic characteristics, like host specificity, good searching ability, shorter generation time than the host, etc.
iv. For effective and timely application in the field, their rearing techniques, mass stock production and field application of such biocontrol agents are essential.
v. The threat and conservation status of chalcid wasps should also be taken into the consideration.

ACKNOWLEDGEMENTS

The authors are grateful to Dr. Kailash Chandra, Director, Zoological Survey of India, Kolkata, for providing facilities and encouragement.

REFERENCES

Aguiar, A.P., Deans, A.R., Engel, M.S., Forshage, M., Huber, J.T., Jennings, J.T., Johnson, N.F., Lelej, A.S., Longino, J.T., Lohrmann, V., Miko, I., Ohl, M., Rasmussen, C., Taeger, A. & Yu, D.S.K. (2013). Order Hymenoptera. *Zootaxa*, 3703(1): 051–062.

Anantanarayanan, K.P. (1934). On the bionomics of a Eulophid (*Trichospilus pupivora* Ferr.) a natural enemy of the coconut caterpillar (*Nephantis serinopa* Meyr.) in South India. *Bulletin of Entomological Research*, 25(1): 55–61.

Baitha, A., Jalali, S.K., Rabindra, R.J., Venkatesan, T. & Rao, N.S. (2004). Parasitizing efficiency of the pupal parasitoid, *Tetrastichus howardi* (Olliff) (Hymenoptera: Eulophidae) on *Chilo partellus* (Swinhoe) at different exposure periods. *Journal of Biological Control*, 18: 65–68.

Banjo, A.D. (2010). A review of *Aleurodicus dispersus* Russel (spiralling whitefly) [Hemiptera: Aleyrodidae] in Nigeria. *Journal of Entomology and Nematology*, 2: 1–6.

Boiler, E. (1972). Behavioral aspects of mass-rearing of insects. *Entomophaga*, 17: 9–25.

Chand, P., Kumar, A. & Kumar, N. (2018). Potential of natural enemy, *Epiricania melanoleuca* and *Tetrastichus pyrillae* against leaf hopper under sugar cane agro ecosystem in Bihar. *Journal of Entomology and Zoology Studies*, 6(4): 1220–1225.

Cruz, I., Redoan, A.C., Silva, R.B., Figueiredo, M.L.C. & Penteado-Dias, A.M. (2011). New record of *Tetrastichus howardi* (Olliff) as a parasitoid of *Diatraea saccharalis* (Fabricius) on maize. *Scientia Agricola*, 68: 252–254.

Dharmaraju, E. (1952). The biological control of the black headed caterpillar *Nephantis serinopa* M., in the East Godavari district of Madras State. *Indian Coconut Journal*, 4: 171–176.

Dharmaraju, E. & Pradhan, S. (1976). Factors affecting the realisation of the biotic potential in *Perisierola nephantidis* Muesebeck and *Trichospilus pupivora* Ferriere, parasites of the coconut leaf caterpillar, *Nephantis serinopa* Meyrick. *Indian Journal Entomology*, 38: 370–376.

Elbehery, H., Saleh, M. & El-Wakeil, N. (2020). Propagation and Application of Larval Parasitoids. *In*: *Cottage Industry of Biocontrol Agents and Their Applications. Practical Aspects to Deal Biologically with Pests and Stresses Facing Strategic Crops*. eds. El-Wakeil, N., Saleh, M. & Abu-hashim, M., pp. 49–71. Switzerland: Springer Nature.

Ferriere, C. (1930). Notes on Asiatic Chalcidoidea. *Bulletin of Entomology Research*, 21(3): 353–360.

Fitton, M. & Walker, A. (1992). Hymenopterous parasitoids associated with diamondback moth: the taxo nomic dilemma. In: *Diamond-back Moth and other Crucifer Pests. Proceedings of the Second International Workshop*, ed. Talekar, N.S., pp. 225–232. Tainan, Taiwan: Asian Vegetable Research Development Centre.

Geetha, B. (2000). Biology and management of spiralling whitefly, *Aleurodicus dispersus* (Russell) (Homoptera: Aleyrodidae). PhD thesis, Tamil Nadu Agricultural University

Huber, J.T. (2017). Biodiversity of Hymenoptera. In: *Insect Biodiversity: Science and society*, ed. Foottit, R.G. and Adler, P.H., Vol. I. second Edition. John Wiley & Sons Ltd.

Jayaratnam, T.J. (1941). A study of the control of the coconut caterpillar, *Nephantis serinopa* Meyr. in Ceylon with special reference to its eulophid parasite, *Trichospilus pupivora* Ferr. *Tropical Agriculturist*, 96: 3–21.

Kfir, R., Gouws, S. & Moore, S.D. (1993). Biology of *Tetrastichus howardi* (Olliff) (Hymenoptera: Eulophidae): a facultative hyperparasitoid of stem borers. *Biocontrol Science and Technology*, 3: 149–159.

Kumarsinghe, N.C. & Wratten, S.D. (1996). The Sugarcane lophopid plant hopper, *Pyrilla perpusilla* (Homoptera: Lophopidae), a review of its biology, pest status and control. *Bulletin of Entomological Research*, 86: 485–498.

Lakshmanan, K., Mutiyil, J.A. & Singh, R. (1942). The effects of healthy and parasitoid eggs of *Pyrilla* spp. in wire gauze cages on the population of pest and its natural enemies. *Indian Journal of Entomology*, 4: 221–223.

Liu, S.S., Wang, X.G., Guo, S.J., He, J.H. & Song, H.M. (1997). A survey of insect parasitoids of *Plutella xylostella* and the seasonal abundance of the major parasitoids in Hangzhou, China. In: *The Management of Diamondback Moth and other Crucifer Pests. Proceedings of the 3rd International Workshop*, eds. Sivapragasan, A., Loke, W.H., Hussan, A.K. & Lim, G.S., pp. 61–66. Mardi, Kuala Lumpur, Malaysia.

Mackauer, M. (1972). Genetic aspects of insect control. *Entomophaga*, 17: 27–48.

Manohar, T.N., Sharma, P.L., Verma, S.C., Sharma, K.C. & Chandel, R.S. (2020). Functional response of indigenous *Trichogramma* spp. to invasive tomato leafminer, *Tuta absoluta* (Meyrick) under laboratory conditions. *International Journal of Tropical Insect Science*, 40: 101–107.

Narendran, T.C. (2011). Fauna of India and adjacent countries. *Eulophinae (Hymenoptera: Eulophidae)*. Kolkata, India: Zoological Survey of India.

Nasser, M. & Abdurahiman, U.C. (2001). Biological Control of the Coconut Caterpillar Opisina arenosella (Lepidoptera: Xylorictidae): Achievements and Prospects. In: *Biocontrol Potential and its Exploitation in Sustainable Agriculture*, eds. Upahyay, R.K. *et al*, pp. 285–305. New York: Kluwer Academic/ Plenum Publishers.

Navik, O.S. & Richa Varshney (2018). Utilization of trichogrammatid egg parasitoid in pest management. *Acta Scientific Agriculture*, 2(3): 49–53.

Nirula, K.K. (1956). Investigations on the pests of coconut palm. Part III. *Nephantis Serinopa Meyrick, Control. Indian Coconut Journal*, 9: 174–201.

Nor Ahya, M., Tajul Ariffin, A.Y. & Rozeita, L. (2018). Searching and parasitism behaviour of *Trichospilus pupivorus* (Hymenoptera: Eulophidae), a potential parasitoid for coconut black-headed caterpillar, *Opisina arenosella* (Lepidoptera: Oecophoridae). *Serangga*, 24(1): 80–90.

Noyes, J.S. (2019). Universal Chalcidoidea Database. World Wide Web electronic publication. http://www.nhm.ac.uk/chalcidoids (last updated March, 2020).

Ooi, P.A.C. (1988). Laboratory studies of *Tetrastichus* soko-lowskii. *Entomophaga*, 33: 145–152.

Perera, M.C.D., Hemachandra, K.S. & Sirisena, U.G.A.I. (2015). *Trichogrammatoidea bactrae* (Hymenoptera: Trichogrammatidae): A Potential Biocontrol Agent of Rice Leaf Folder. *Tropical Agricultural Research*, 26(3): 537–546.

Prasad, K.S., Aruna, A.S., Kumar, V. & Kariappa, B.K. (2007). Feasibility of mass production of *Tetrastichus howardi* (Olliff), a parasitoid of leaf roller (*Diaphania pulverulentalis*), on *Musca domestica* (L.). *Indian Journal of Sericulture*, 46: 89–91.

Rao, Y.R., Cherian, M.C. & Ananthanarayanan, K.P. (1948). Infestations of Nephantis serinopa Meyrick in South India and their control by the biological method. *Indian Journal of Entomology*, 10: 205–247.

Remadevi, O.K., Abdurahiman, U.C., Mohamed, U.V.K. & Beena, C.G. (1980). Biology and behaviour of *Trichospilus pupivora* (Hymenoptera: Eulophidae) with a note on its field parasitism of *Nephantis serinopa* (Lepidoptera: Xylorictidae). *Ceylon Coconut Quartelity*, 31: 19–126.

Silva, I.M., Pereira, K.D.S., Spranghers, T., Zanuncio, J.C. & Serrao, J.E. (2016). Antennal sensilla and sexual dimorphism of the parasitoid *Trichospilus pupivorus* (Hymenoptera: Eulophidae). *Microscopy and microanalysis*, 22: 913–921.

Singh, S.P. (2004). *Some Success Stories in Classical Biological Control in India*. Asian and Pacific Coconut Community, *APCC*, 73 pp. APAARI/FAO-RAP, Bangkok, Thailand.

Srikanth, J., Earwaramoorthy, S. & Jalali, S.K. (2016). A 100 years of biological control of sugarcane pests in India. *CAB Rev*, 11: 1–32.

Talekar, N.S. & Hu, W.J. (1996). Characteristics of parasitism of *Plutella xylostella* (Lep., Plutellidae) by *Oomyzus sokolowskii* (Hym., Eulophidae). *Entomophaga*, 41: 45–52.

Tavares, W.S., Hansson, C., Serrão, J.E. & Zanuncio, J.C. (2012). First report of *Trichospilus pupivorus* (Hymenoptera: Eulophidae) parasitizing pupae of *Anticarsia gemmatalis* (Lepidoptera: Noctuidae). *Entomology Generalis*, 33(4): 281– 282.

Tavares, W.S., Hansson, C., Serrão, J.E. & Zanuncio, J.C. (2013). *Trichospilus pupivorus* (Hymenoptera: Eulophidae): First report of parasitism on *Thagona tibialis* (Lepidoptera: Lymantriidae) in Brazil. *Studies on Neotropical Fauna and Environment*, 48(2): 104–105. doi: 10.1080/01650521.2013.837348

Vargas, E.L., Pereira, F.F., Tavares, M.T. & Pastori, P.L. (2011). Record of *Tetrastichus howardi* (Hymenoptera: Eulophidae) parasitizing *Diatraea* sp. (Lepidoptera: Crambidae) in sugarcane crop in Brazil. *Entomotropica*, 26: 143–146.

LIST OF INDIAN CHALCID PARASITES WITH ITS KNOWN HOSTS

(Adopted from Noyes, 2019)

Sl. No.	Chalcid parasites	Host
		Family: Aphelinidae
1	*Aphelinus abdominalis* (Dalman)	Parasites of many species of aphids (Hemiptera: Aphididae).
2	*Aphelinus albipodus* Hayat & Fatima	Parasites of many species of aphids (Hemiptera: Aphididae) and also on Aleyrodidae (Hemiptera).
3	*Aphelinus ancer* Hayat	Parasites of *Greenidea artocarpi* (Hemiptera: Aphididae).
4	*Aphelinus asychis* Walker	Parasites of many species of aphids (Hemiptera: Aphididae) and also on Agromyzidae (Diptera).
5	*Aphelinus desantisi* Hayat	Parasites of some species of aphids (Hemiptera: Aphididae).
6	*Aphelinus gossypii* Timberlake	Parasites of many species of aphids (Hemiptera: Aphididae) and also on (Hemiptera. Aleyrodidae).
7	*Aphelinus mali* (Haldeman)	Parasites of many species of aphids (Hemiptera: Aphididae), scale insects (Hemiptera: Diaspididae) and mealy bugs Hemiptera: Pseudococcidae) and also on (Lepidoptera: Coleophoridae and Pyralidae).
8	*Aphelinus semiflavus* Howard	Parasites of many species of aphids (Hemiptera: Aphididae) and also on *Heliothrips haemorrhoidalis* (Thysanoptera: Thripidae).
9	*Aphytis alami* Agarwal	Parasites of *Planococcus citrii* and *Pseudococcus citrii* (Hemiptera: Pseudococcidae).
10	*Aphytis bangalorensis* Rosen and Debach	Parasites of *Pseudaulacapsis barberi* (Hemiptera: Diaspididae).
11	*Aphytis chrysomphali* Mercet	Parasites of *Pseudaulacapsis barberi, Chloropulvinaria psidii* (Hemiptera: Coccidae) and many species of scale insects (Hemiptera: Diaspididae). Parasites of *Protargionia larreae* (Hymenoptera: Diprionidae).
12	*Aphytis diaspidis* (Howard)	Parasites of many species of scale insects (Hemiptera: Diaspididae) and parasitoids of *Marietta carnesi* (Hymenoptera: Aphelinidae).
13	*Aphytis fisheri* Debach	Parasites of some species of scale insects (Hemiptera: Diaspididae).
14	*Aphytis gordoni* Debach and Rosen	Parasites of some species of scale insects (Hemiptera: Diaspididae).
15	*Aphytis hispanicus* (Mercet)	Parasites of *Coccus hesperidum* (Hemiptera: Coccidae) and many species of scale insects (Hemiptera: Diaspididae).
16	*Aphytis holoxanthus* Debach	Parasites of many species of scale insects (Hemiptera: Diaspididae) and parasitoids of *Marietta leopardina* (Hymenoptera: Aphelinidae).
17	*Aphytis lepidosaphes* Compere	Parasites of some species of scale insects (Hemiptera: Diaspididae) and parasitoids of *Marietta javensis* (Hymenoptera: Aphelinidae).
18	*Aphytis lingnanensis* Compere	Parasites of many species of scale insects (Hemiptera: Diaspididae).
19	*Aphytis maculicornis* (Masi)	Parasites of many species of scale insects (Hemiptera: Diaspididae).
20	*Aphytis melinus* Debach	Parasites of many species of scale insects (Hemiptera: Diaspididae).
21	*Aphytis mytilaspidis* (Le Baron)	Parasites of *Astero lecanium pustulans* and *A. variolosum* (Hemiptera: Asterolecaniidae). Parasites of many species of scale insects (Hemiptera: Diaspididae).
22	*Aphytis paramaculicornis* Debach and Rosen	Parasites of many species of scale insects (Hemiptera: Diaspididae).
23	*Aphytis philippinensis* Debach and Rosen	Parasites of some species of scale insects (Hemiptera: Diaspididae).

Sl. No.	Chalcid parasites	Host
24	*Aphytis pinnaspidis* Debach and Rosen	Parasites of some species of scale insects (Hemiptera: Diaspididae).
25	*Aphytis proclia* (Walker)	Parasites of *Dialeurodes citri* (Hemiptera. Aleyrodidae) and *Coccus hesperidum* (Hemiptera: Coccidae). Parasites of many species of scale insects (Hemiptera: Diaspididae).
26	*Aphytis sankarani* Rosen and Debach	Parasites of some species of scale insects (Hemiptera: Diaspididae).
27	*Centrodora azizi* Hayat	Parasites of some species of tree hoppers (Hemiptera: Membracidae).
28	*Centrodora idioceri* Ferriere	Parasites of some species of leaf hoppers (Hemiptera: Cicadellidae) and some species of tree hoppers (Hemiptera: Membracidae).
29	*Coccobius aspidiellae* Hayat	Parasites of *Aspidiella hartii* (Hemiptera: Diaspididae).
30	*Coccobius comperei* (Hayat)	Parasites of some species of scale insects (Hemiptera: Diaspididae).
31	*Coccobius confuses* (Compere and Annecke)	Parasites of *Aonidiella* species (Hemiptera: Diaspididae).
32	*Coccobius flaviceps* (Girault and Dodd)	Parasites of *Pinnaspis strachani* and *Unaspis citri* (Hemiptera: Diaspididae).
33	*Coccobius fulvus* (Compere and Annecke)	Parasites of some species of scale insects (Hemiptera: Diaspididae) Parasites of *Planococcus citrii* (Hemiptera: Pseudococcidae).
34	*Coccobius phoebus* Hayat	Parasites *Duplachionaspis* species of (Hemiptera: Diaspididae).
35	*Coccobius reticulatus* (Compere and Annecke)	Parasites of *Cerococcus hibisci* (Hemiptera: Cerococcidae), *Coccus hesperidum* (Hemiptera: Coccidae) and *Aonidiella orientalis* (Hemiptera: Diaspididae).
36	*Coccobius subflavus* (Annecke and Insley)	Parasites of scale insects *Aulacaspis tegalensis* and *Melanaspis glomerata* (Hemiptera: Diaspididae).
37	*Coccophagoides orientalis* (Agarwal)	Parasites of *Aonidiella orientalis* (Hemiptera: Diaspididae).
38	*Coccophagoides utilis* Doutt	Parasites of *Parlatoria oleae* (Hemiptera: Diaspididae).
39	*Coccophagus anthracinus* Compere	Parasites of some scale insects (Hemiptera: Coccidae).
40	*Coccophagus atratus* Compere	Parasites of many scale insects (Hemiptera: Coccidae) and parasitoids of *Coccophagus atratus* (Hymenoptera: Aphelinidae).
41	*Coccophagus bivittatus* Compere	Parasites of many species of scale insects (Hemiptera: Coccidae), *Anomalococcus crematogastri* (Hemiptera: Asterolecaniidae).
42	*Coccophagus bogoriensis* (Koningsberger)	Parasites of many species of scale insects (Hemiptera: Coccidae) and *Chrysomphalus pinnulifer* (Hemiptera: Diaspididae).
43	*Coccophagus ceroplastae* (Howard)	Parasites of *Aclerda sacchari* (Hemiptera: Aclerdidae) and many species of scale insects (Hemiptera: Coccidae).
44	*Coccophagus chloropulvinariae* Hayat	Parasites of *chloropulvinaria polygonata* (Hemiptera: Coccidae).
45	*Coccophagus cowperi* Girault	Parasites of many species of scale insects (Hemiptera: Coccidae), *Nipaecoccus viridis* (Hemiptera: Pseudococcidae).
46	*Coccophagus dius* Hayat	Associated with mango tree.
47	*Coccophagus indochraceus Hayat*	Parasites of *Ceroplastes pseudoceriferus* (Hemiptera: Coccidae).

(Continued)

Sl. No.	Chalcid parasites	Host
48	*Coccophagus longiclavatus* Shafee	Parasites of *Coccus* species (Hemiptera: Coccidae) and some lac insects (Hemiptera: Kerriidae).
49	*Coccophagus longipedicellus* Shafee	Parasites of Coccus species and *Parasaissetia nigra* (Hemiptera: Coccidae).
50	*Coccophagus lutescens* Compere	Parasites of some scale insects (Hemiptera: Coccidae).
51	*Coccophagus lycimnia* (Walker)	Parasites of many scale insects (Hemiptera: Coccidae, Asterolecaniidae, Diaspididae), etc.
52	*Coccophagus ochraceus* Howard	Parasites of many scale insects (Hemiptera: Coccidae).
53	*Coccophagus parlobatae* Hayat	Parasites of *Paratachardina lobate* (Hemiptera: Kerriidae).
54	*Coccophagus prinslooi* Hayat	Parasites of *Chloropulvinaria psidii* (Hemiptera: Coccidae).
55	*Coccophagus pseudococci* Compere	Parasites of *Pulvinaria maxima* (Hemiptera: Coccidae), *Icerya aegyptiaca* (Hemiptera: Margarodidae) and many other scale insects (Hemiptera: Pseudococcidae).
56	*Coccophagus sexvittatus* Hayat	Parasites of *Rastrococcus iceryoides* (Hemiptera: Pseudococcidae).
57	*Coccophagus shafeei* Hayat	Parasites of *Coccus* species (Hemiptera: coccidae).
58	*Coccophagus silvestrii* Compere	Parasites of many scale insects (Hemiptera: Coccidae).
59	*Coccophagus tschirchii* Mahdihassan	Parasites of some scale insects (Hemiptera: Coccidae, Kerriidae).
60	*Encarsia acaudaleyrodis* Hayat	Parasites of some white flies (Hemiptera: Aleyrodidae).
61	*Encarsia aseta* Hayat and Polaszek	Parasites of some white flies (Hemiptera: Aleyrodidae).
62	*Encarsia aurantii* (Howard)	Parasites of *Aleuroplatus coronata* (Hemiptera: Aleyrodidae), many scale insects (Hemiptera: Coccidae, Diaspididae).
63	*Encarsia azimi* Hayat	Parasites of some white flies (Hemiptera: Aleyrodidae) and *Thrips tabaci* (Thysanoptera: Thripidae).
64	*Encarsia bimaculata* Heraty and Polaszek	Parasites of some white flies (Hemiptera: Aleyrodidae).
65	*Encarsia brimblecombei* (Girault)	Parasites of many scale insects (Hemiptera: Diaspididae).
66	*Encarsia cibcensis* Lopez-Avila	Parasites of some white flies (Hemiptera: Aleyrodidae).
67	*Encarsia citrina* (Craw)	Parasites of many scale insects (Hemiptera: Coccidae, Diaspididae, Eriococcidae), some white flies (Hemiptera: Aleyrodidae), etc.
68	*Encarsia clypealis* (Silvestri)	Parasites of some white flies (Hemiptera: Aleyrodidae).
69	*Encarsia dialeurodis* Hayat	Parasites of *Dialeurodes* species and *Kanakrajiella cardamom* (Hemiptera: Aleyrodidae).
70	*Encarsia dispersa* Polaszek	Parasites of some white flies (Hemiptera: Aleyrodidae).
71	*Encarsia divergens* (Silvestri)	Parasites of some white flies (Hemiptera: Aleyrodidae).
72	*Encarsia elegans* Masi	Parasites of many white flies (Hemiptera: Aleyrodidae).

Sl. No.	Chalcid parasites	Host
73	*Encarsia flavoscutellum* Zehntner	Parasites of some species of aphids (Hemiptera: Aphididae).
74	*Encarsia guadeloupae* Viggiani	Parasites of some white flies (Hemiptera: Aleyrodidae).
75	*Encarsia inaron* (Walker)	Parasites of many species of white flies (Hemiptera: Aleyrodidae), some scale insects (Hemiptera: Coccidae, Diaspididae) and some moths, *Spodoptera littoralis* (Lepidoptera: Noctuidae).
76	*Encarsia interstrica* Huang and Polaszek	Parasites of the white fly, *Dialeuropora decempuncta* (Hemiptera: Aleyrodidae).
77	*Encarsia issaci* Mani	Parasites of some white flies (Hemiptera: Aleyrodidae).
78	*Encarsia lahorensis* (Howard)	Parasites of some white flies (Hemiptera: Aleyrodidae).
79	*Encarsia levo* Hayat	Parasites of the white fly, *Dialeuropora decempuncta* (Hemiptera: Aleyrodidae).
80	*Encarsia lipaleyrodis* Krishnan and Vasantharaj David	Parasites of *Lipaleyrodes euphorbiae* (Hemiptera: Aleyrodidae).
81	*Encarsia lutea* (Masi)	Parasites of many species of white flies (Hemiptera: Aleyrodidae), scale insect, *Parthenolecanium corni* (Hemiptera: Coccidae) and some moths (Lepidoptera: Noctuidae).
82	*Encarsia macroptera* Viggiani	Parasites of the white fly *Aleurolobus barodensis* (Hemiptera: Aleyrodidae).
83	*Encarsia merceti* (Silvestri)	Parasites of some white flies (Hemiptera: Aleyrodidae).
84	*Encarsia muliyali* Mani	Parasites of the white fly, *Aleurolobus barodensis* (Hemiptera: Aleyrodidae).
85	*Encarsia obtusiclava* Hayat	Parasites of the scale insect, *Aulacaspis tubercularis* (Hemiptera: Diaspididae).
86	*Encarsia opulent* (Silvestri)	Parasites of some white flies (Hemiptera: Aleyrodidae).
87	*Encarsia orangae* Raina,Khurad and Rathod	Parasites of the white fly, *Aleurocanthus nagpurensis* (Hemiptera: Aleyrodidae).
88	*Encarsia perniciosi* (Tower)	Parasites of many scale insects (Hemiptera: Diaspididae).
89	*Encarsia Perplexa* Huang and Polaszek	Parasites of some white flies (Hemiptera: Aleyrodidae).
90	*Encarsia septentrionalis* Hayat	Parasites of some white flies (Hemiptera: Aleyrodidae).
91	*Encarsia smithi* (Silvestri)	Parasites of some white flies (Hemiptera: Aleyrodidae) and scale insect, *Aonidiella aurantii* (Hemiptera: Diaspididae).
92	*Encarsia Sophia* (Girault and Dodd)	Parasites of many white flies (Hemiptera: Aleyrodidae), and aphid, *Aphis sacchari* (Hemiptera: Aphididae).
93	*Encarsia strenua* (Silvestri)	Parasites of many white flies (Hemiptera: Aleyrodidae).
94	*Encarsia tinctoriae* Krishnan and Vasantharaj David	Parasites of the white flies *Aleurocanthus woglumi* and *Lipaleyrodes euphorbiae* (Hemiptera: Aleyrodidae).
95	*Encarsia tristis* (Zehntner)	Parasites of some white flies (Hemiptera: Aleyrodidae) and scale insect, *Parlatoria ziziphi* (Hemiptera: Diaspididae).

(*Continued*)

Sl. No.	Chalcid parasites	Host
96	*Encarsia udaipuriensis* (Shafee)	Parasites of some white flies (Hemiptera: Aleyrodidae), scale insect *Kerria lacca* (Hemiptera: Kerriidae).
97	*Eretmocerus breviclavus* Subba Rao	Parasites of the white flies *Aleurocanthus woglumi* and *Aleurocanthus zizyphi* (Hemiptera: Aleyrodidae).
98	*Eretmocerus corni* Haldeman	Parasites of many white flies (Hemiptera: Aleyrodidae) some scale insects (Hemiptera: Kerriidae).
99	*Eretmocerus delhiensis* Mani	Parasites of some white flies (Hemiptera: Aleyrodidae).
100	*Eretmocerus haldemani* Howard	Parasites of many white flies (Hemiptera: Aleyrodidae).
101	*Eretmocerus hayati* Zolnerowich and Rose	Parasites of some white flies (Hemiptera: Aleyrodidae).
102	*Eretmocerus indicus* Hayat	Parasites of the white flies *Aleurocanthus woglumi* and *Aleurocanthus* species (Hemiptera: Aleyrodidae).
103	*Eretmocerus longiscapus* Hayat	Parasites of some white flies (Hemiptera: Aleyrodidae).
104	*Eretmocerus mundus* Mercet	Parasites of many white flies (Hemiptera: Aleyrodidae).
105	*Eretmocerus paulistus* Hempel	Parasites of some white flies (Hemiptera: Aleyrodidae).
106	*Eretmocerus serius* Silvestri	Parasites of some white flies (Hemiptera: Aleyrodidae).
107	*Eretmocerus trialeurodis* Hayat	Parasites of some white flies (Hemiptera: Aleyrodidae).
108	*Marietta carnesi* (Howard)	Parasites of some scale insects in the family Coccidae and many scale insects in the family Diaspididae (Hemiptera).
109	*Marietta leopardina* Motschulsky	Parasites of some flies (Diptera: Cecidomyiidae, Chamaemyiidae), many scale insects (Hemiptera: Coccidae, Diaspididae, etc), and leaf hopper of *Amblerdiella* species (Hemiptera: Cicadellidae).
110	*Marietta picta* (Andre)	Parasites of the scale insect *Aclerda subterranean* (Hemiptera: Aclerdidae), aphid *Sipha maydis* (Hemiptera: Aphididae) and many scale insects in the families, Coccidae, Diaspididae), etc.
111	*Pteroptix albifemur* (Girault)	Parasites of the scale insects, *Aspidiotus camelliae* and *Howardia biclavis* (Hemiptera: Diaspididae).
112	*Pteroptix chinensis* (Howard)	Parasites of the scale insects, *Eulecanium kuwanai* and *Parthenolecanium corni* (Hemiptera: Coccidae) and many other scale insects in Diaspididae (Hemiptera).
		Family: Chalcididae
113	*Antrocephalus hakonensis* (Ashmead)	Parasites of the moth, *Contheyla rotunda* (Lepidoptera: Limacodidae) and other moths in the families Oecophoridae and Pyralidae (Lepidoptera).
114	*Antrocephalus indicus* Husain and Agarwal	Parasites of the moth, *Marasmia trapezalis* (Lepidoptera: Pyralidae).
115	*Antrocephalus maculipennis* Cameron	Parasites of *Opisina arenosella* (Lepidoptera: Oecophoridae).
116	*Antrocephalus mitys* (Walker)	Parasites of some moths (Lepidoptera: Pyralidae).

Sl. No.	Chalcid parasites	Host
117	*Brachymeria albotibialis* (Ashmead)	Parasites of some butterflies in the families, Hesperiidae and Pieridae and some moths in the families Noctuidae and Pyralidae (Lepidoptera).
118	*Brachymeria atteviae* Joseph,Narendran and Joy	Parasites of some moths in the families Noctuidae, Oecophoridae and Pyralidae (Lepidoptera) and the ermine moth, *Atteva fabriciella* (Lepidoptera: Yponomeutidae).
119	*Brachymeria bengalensis* (Cameron)	Parasites of the moths, *Earias vittella* (Lepidoptera: Noctuidae) and *Pieris brassicae* (Lepidoptera: Pieridae).
120	*Brachymeria carinata* Joseph, Narendran and Joy	Parasites of some moths (Lepidoptera: Psychidae).
121	*Brachymeria compestris* Husain and Agarwal	Parasites of the moths *Diacrisia obliqua* and *Spilosoma oblique* (Lepidoptera: Arctiidae).
122	*Brachymeria criculae* (Kohl)	Parasites of some moths in the families Lasiocampidae and Saturniidae (Lepidoptera).
123	*Brachymeria deesensis* (Cameron)	Parasitoids of the fly, *Eozenillia equatorialis* (Diptera: Tachinidae).
124	*Brachymeria euploeae* (Westwood)	Parasites of *Charops obtusus* (Hymenoptera: Ichneumonidae).
125	*Brachymeria excarinata* Gahan	Parasites of the leaf beetle, *Calopepla leayana* (Coleoptera: Chrysomelidae) and many moths and butterflies of the families-Arctiidae, Gelechiidae, Hesperiidae, Noctuidae, Oecophoridae, Pyralidae, Tortricidae and Yponomeutidae.
126	*Brachymeria femorata* (Panzer)	Parasites of the snout beetle, *Cleonus punctiventris* (Coleoptera: Curculionidae) and many moths and butterflies of the families-Arctiidae, Lasiocampidae, Noctuidae, Nymphalidae, Pieridae, and Yponomeutidae (Lepidoptera).
127	*Brachymeria habui* Ozdikmen	Parasites of *Udaspes folus* (Lepidoptera: Hesperiidae) and *Dendrolimus punctatus* (Lepidoptera: Lasiocampidae).
128	*Brachymeria hime* Habu	Parasites of some moths and butterflies (Lepidoptera: Pyralidae and Tortricidae).
129	*Brachymeria jambolana* Gahan	Parasites of *Sarcophaga misera* (Diptera: Sarcophagidae) and some butterflies and moths (Lepidoptera: Danaidae, Lymantriidae, Noctuidae and Papilionidae).
130	*Brachymeria lasus* (Walker)	Parasites of many moths and butterflies of the families Arctiidae, Lasiocampidae, Noctuidae, Nymphalidae, Pieridae, Pyralidae, Yponomeutidae, etc. (Lepidoptera).
131	*Brachymeria lugubris* (Walker)	Parasites of *Exorista sorbillans* (Diptera: Tachinidae), silk moth, *Bombyx mori* (Lepidoptera: Bombycidae) and parasitoids many flies (Diptera: Tachinidae).
132	*Brachymeria lymantriae* Joseph, Narendran and Joy	Associated with sugar cane plants.
133	*Brachymeria marmonti* (Girault)	Parasites of *Charops* species (Hymenoptera: Ichneumonidae) and some moths in the families Gelechiidae, Noctuidae and Psychidae.
134	*Brachymeria masoodii* Jamal Ahmad	Parasites of Parnara guttata (Lepidoptera: Hesperidae).
135	*Brachymeria megaspila* (Cameron)	Parasites of *Opisina arenosella* (Lepidoptera: Oecophoridae) and some butterflies (Lepidoptera: Pieridae).

(*Continued*)

Sl. No.	Chalcid parasites	Host
136	*Brachymeria menoni* Joseph, Narendran and Joy	Parasites of the leaf beetle, *Cassida circumdata* and *Metriona circumdata* (Coleoptera: Chrysomelidae).
137	*Brachymeria minuta* (Linnaeus)	Parasites of some flies (Diptera: Calliphoridae and Sarcophagidae), some saw flies (Hymenoptera: Cimbicidae and Diprionidae) and moths and butterflies (Lepidoptera: Arctiidae, Lasiocampidae, Pieridae, Tortricidae, Yponomeutidae).
138	*Brachymeria nephantidis* Gahan	Parasites of the fly, *Stomatomyia bezziana* (Diptera: Tachinidae) *Angitia* species (Hymenoptera: Ichneumonidae) and some moths in the families Noctuidae, Oecophoridae, Pyralidae and the butterfly, *Deudorix Isocrates* (Lepidoptera: Lycaenidae).
139	*Brachymeria nosatoi* Habu	Parasites of some moths in the families Oecophoridae, Pyralidae, *Pectinophora gossypiella* (Gelechiidae) *Petrova cristata* (Tortricidae) in the order Lepidoptera.
140	*Brachymeria olethria* (Waterston)	Parasites of some moths in the families Pyralidae, Gelechiidae and the mompha moth *Pyroderces simplex* (Lepidoptera).
141	*Brachymeria phya* (Walker)	Parasites of some moths in the families Gelechiidae, Tortricidae, Yponomeutidae (Lepidoptera).
142	*Brachymeria porthetriali* Joseph, Narendran and Joy	Parasites of the moths *Lymantria ampla* and *Porthetria dispar* (Lepidoptera: Lymantriidae).
143	*Brachymeria rufotibialis* Husain and Agarwal	Parasites of the moths *Diacrisia obliqua* and *Spilosoma oblique* (Lepidoptera: Arctiidae), *Antigastra catalaunalis* in the family Pyralidae and the butterfly *Parnara mathias* (Lepidoptera: Hesperiidae).
144	*Brachymeria tibialis* (Walker)	Parasites of the gall midge, *Resseliella piceae* (Diptera: Cecidomyiidae), some flies (Diptera: Tachinidae), *Brachymeria intermedia* (Hymenoptera: Chalcididae) and many moths and butterflies in the families Arctiidae, Hesperiidae, Lymantriidae, Gelechiidae, Tortricidae, Zygaenidae, etc.
145	*Brachymeria wiebesina* Joseph, Narendran and Joy	Associated with sweet potato.
146	*Cratocentrus tomentosus* (Nikolskaya)	Parasites of the jewel beetle, *Chrysobothris parvipunctata* (Coleoptera: Buprestidae).
147	*Epitranus albipennis* Walker	Parasites of the moth, *Euzophera perticella* (Lepidoptera: Pyralidae).
148	*Epitranus erythrogaster* Cameron	Parasites of some moths (Lepidoptera: Pyralidae).
149	*Epitranus indicus* Husain and Agarwal	Parasites of the moth, *Leucinodes orbonalis* (Lepidoptera: Pyralidae).
150	*Epitranus ramnathi* (Mani and Dubey)	Parasites of some moths (Lepidoptera: Pyralidae) and *Crypsithris longicornis* (Lepidoptera: Tineidae).
151	*Hockeria atra* Masi	Parasites of the moth *Spilosoma* species (Lepidoptera: Arctiidae) and *Corcyra cephalonica* (Lepidoptera: Pyralidae).
152	*Hockeria nikolskayae* Husain and Agarwal	Parasites of the moths *Diacrisia obliqua* and *Spilosoma oblique* (Lepidoptera: Arctiidae).
153	*Proconura minusa* Narendran	Parasites of the moth, *Sitotroga cerealella* (Lepidoptera: Gelechiidae).
154	*Psilochalcis soudanensis* Steffan	Parasites of some moths (Lepidoptera: Pyralidae) and *Busseola fusca* (Lepidoptera: Noctuidae).
155	*Uga menoni* Kerrich	Parasites some beetles (Coleoptera: Coccinellidae).

Sl. No.	Chalcid parasites	Host
		Family: Eucharitidae
156	*Chalcura deprivata* (Walker)	Parasites of the ant *Odontomachus haematodes* (Hymenoptera: Formicidae).
157	*Neolosbanus laeviceps* (Gahan)	Associated with *Artocarpus* plants.
		Family: Eulophidae
158	*Aceratoneuromyia indica* (Silvestri)	Parasites of many fruit flies (Diptera: Tephritidae) and parasitoids of *Sarcophaga auricaudata* (Diptera: Sarcophagidae).
159	*Anselmella kerrichi* (Narayanan, Subba Rao and Patel)	Collected from crop field.
160	*Aprostocetus ajmerensis* (Khan and Shafee)	Parasites of the mealy bug *Coccidohystrix insolita* (Hemiptera: Pseudococcidae).
161	*Aprostocetus atomelli* (Khan)	Parasites of *Tropicomyia atomella* (Diptera: Agromyzidae).
162	*Aprostocetus bangaloricus* Narendran	Parasites of *Laccifer lacca* (Hemiptera: Kerriidae).
163	*Aprostocetus coimbatorensis* (Rohwer)	Parasites of Cecidomyiidae unspecified and *Stenodiplosis sorghicola* (Diptera: Cecidomyiidae).
164	*Aprostocetus diplosidis* Crawford	Parasites of some gall midges (Diptera: Cecidomyiidae).
165	*Aprostocetus fasciatus* (Ashmead)	Parasites of some gall midges (Diptera: Cecidomyiidae).
166	*Aprostocetus gala* (Walker)	Parasites of some weevils (Coleoptera: Curculionidae) and some gall midges (Diptera: Cecidomyiidae).
167	*Aprostocetus jaipurensis* Ozdikmen	Parasites of the scale insect *Pulvinaria maxima* (Hemiptera: Coccidae) and the mealy bug *Coccidohystrix insolita* (Hemiptera: Pseudococcidae).
168	*Aprostocetus lasallei* Khan, Agnihotri and Sushil	Parasites of Coccidae unspecified species (Hemiptera: Coccidae).
169	*Aprostocetus nainitalensis* (Khan and Shafee)	Parasites of *Kerria lacca* (Hemiptera: Kerriidae).
170	*Aprostocetus neglectus* (Domenichin)	Parasites of many beetles (Coleoptera: Coccinellidae) and the aphid *Myzus cerasi* (Hemiptera: Aphididae).
171	*Aprostocetus nigricornis* (Khan and Shafee)	Parasites of *Nipaecoccus vastator* (Hemiptera: Pseudococcidae).
172	*Aprostocetus percaudatus* (Silvestri)	Parasites of *Oecanthus* species and *Oecanthus pellucens* (Orthoptera: Gryllidae).
173	*Aprostocetus psyllidis* (Khan and Shafee)	Collected from crop fields.
174	*Aprostocetus purpureus* (Cameron)	Parasites of *Ceroplastes rubens* and *Pulvinaria psidii* in the family Coccidae, many other scale insects in the family Diaspididae, some in the family Kerriidae and some mealy bugs in the family Pseudococcidae (Hemiptera).

(*Continued*)

Sl. No.	Chalcid parasites	Host
175	*Aprostocetus sankarani* Boucek	Parasites of Cynipidae unspecified species (Hymenoptera: Cynipidae).
176	*Aprostocetus santalinus* Narendran	Parasites of *Ceroplastes actiniformis* (Hemiptera: Coccidae).
177	*Aprostocetus yoshimotoi* Khan, Agnihotri and Sushil	Associated with mango tree.
178	*Aroplectrus contheylae* Narendran	Parasites of *Contheyla rotunda* (Lepidoptera: Limacodidae).
179	*Aroplectrus dimerus* Lin	Parasites of *Parasa bicolor* and *Penthocrates* species (Lepidoptera: Limacodidae).
180	*Baryscapus chrysopae* (Crawford)	Parasites of *Melissopus latiferreanus* (Lepidoptera: Tortricidae) and some green lacewings (Neuroptera: Chrysopidae).
181	*Baryscapus galactopus* (Ratzeburg)	Parasites of *Clubiona* species (Araneae: Clubionidae), *Phytomyza orobanchia* (Diptera: Agromyzidae) and some moths and butterflies in the families Tortricidae, Notodontidae, Pieridae, Yponomeutidae and Lymantriidae of the order Lepidoptera.
182	*Ceranisus menes* (Walker)	Parasites of *Braueriella phillyreae* (Diptera: Cecidomyiidae), *Callirhytis glandium* (Hymenoptera: Cynipidae) and thrips (Thysanoptera: Thripidae).
183	*Ceratoneura indi* Girault	Parasites of *Asphondylia sphaera* and Cecidomyiidae unspecified species (Diptera: Cecidomyiidae).
184	*Chrysocharis horticola* Mani	Parasites of *Phytomyza horticola* (Diptera: Agromyzidae).
185	*Chrysocharis nitida* Hansson	Associated with *Sorghum* plants.
186	*Chrysocharis pubicornis* (Zetterstedt)	Parasites of many leaf miner flies in the family Agromyzidae, *Cystiphora sonchi* and *Cystiphora taraxaci* in the family Cecidomyiidae etc. in the order Hemiptera and moths in the families Gracillariidae, Lyonetiidae, Nepticulidae, Tischeriidae and Tortricidae of the order Lepidoptera.
187	*Chrysonotomyia appannai* (ChandyKurian)	Parasites of some *Epilachna* beetles (Coleoptera: Coccinellidae).
188	*Chrysonotomyia hyalinipennis* (Khan and Shafee)	Associated with papaya plant.
189	*Chrysonotomyia obesula* Boucek	Associated with mango tree.
190	*Chrysonotomyia postmarginaloides* (Saraswat)	Parasites of *Procontarinia matteiana* (Diptera: Cecidomyiidae).
191	*Chrysonotomyia rexia* Narendran	Parasites of *Liriomyza* species and *Liriomyza trifolii* (Diptera: Agromyzidae).
192	*Chrysonotomyia younusi* Hayat and Perveen	Associated with mango tree.
193	*Cirrospilus brevicorpus* Shafee and Rizvi	Parasites of some leaf miner flies (Diptera: Agromyzidae).
194	*Cirrospilus ingenuus* Gahan	Parasites of Agromyzidae unspecified species (Diptera: Agromyzidae), some moths in the family Gracillaridae and *Leucoptera coffeella* in Lyonetiidae of the order Lepidoptera.

Sl. No.	Chalcid parasites	Host
195	*Citrostichus phyllocnistoides* (Narayanan)	Parasites of *Trioza obsolete* (Hemiptera: Triozidae), *Phyllocnistis citrella* in the family Gracillaridae and *Acalyptris minimella* and *Stigmella* species in the family Nepticulidae (Lepidoptera).
196	*Closterocerus agromyzae* Narayanan,Subba Rao and Ramachandra	Parasites of some leaf miner flies (Diptera: Agromyzidae).
197	*Closterocerus insignis* Waterston	Parasites of some leaf miner flies (Diptera: Agromyzidae), some grass flies (Diptera: Chloropidae) and the moth *Lithocolletis* species (Lepidoptera: Gracillaridae).
198	*Closterocerus longiscapus* (Boucek)	Parasites of *Procontarinia matteiana* (Diptera: Cecidomyiidae).
199	*Closterocerus phytomyzae* Mani	Parasites of some leaf miner flies (Diptera: Agromyzidae).
200	*Closterocerus pulcherrimus* (Kerrich)	Parasites of *Procontarinia matteiana* (Diptera: Cecidomyiidae).
201	*Diglyphus horticola* Khan	Parasites of *Chromatomyia horticola* (Diptera: Agromyzidae).
202	*Diglyphus indicus* Arifa and Khan	Parasites of *Pseudonapomyza asiatica* (Diptera: Agromyzidae).
203	*Diglyphus isaea* (Walker)	Parasites of *Lycopersicon esculentum* (Family: Solanaceae), many leaf miner flies (Diptera: Agromyzidae) and some moths in the families Gracillaridae, Lyonetiidae and Nepticulidae of the order Lepidoptera.
204	*Elachertus piperis* Narendran	Parasites of *Liothrips karnyi* (Thysanoptera: Phlaeothripidae).
205	*Elasmus albopictus* Crawford	Parasites of some moths in the family Pyralidae and *Eucosma critica* in Tortricidae (Lepidoptera).
206	*Elasmus anticles* Walker	Parasites of *Contarinia* species (Diptera: Cecidomyiidae), *Inglisia bivalvata* (Hemiptera: Coccidae) and *Holcocera pulverea* in yhe family Blastobasidae, *Aproaerama modicella* in Gelechiidae and some moths in the family Gracillariidae, etc. of the order Lepidoptera.
207	*Elasmus brevicornis Gahan*	Parasites of many moths in the family Gelechiidae, *Erionota thrax* in Hesperiidae, *Rivula* species in Noctuidae and many moths in the family Pyralidae (Lepidoptera).
208	*Elasmus homonae* Ferriere	Parasites of some moths in the family tortricidae (Lepidoptera).
209	*Elasmus hyblaeae* Ferriere	Parasites of *Conopomorpha cramerella* in the family Gracillariidae, *Antispila anna* in Heliozelidae, *Hyblaeae puera* in Hyblaeidae, *Diaphania indica* and *Noorda moringae* in Pyralidae and *Prays acmonias* in Tponomeutidae (Lepidoptera).
210	*Elasmus indicus* Rohwer	Parasites of *Anomalococcus indicus* (Hemiptera: Asterolecaniidae), *Eublemma* species in the family Noctuidae and some moths in the family Pyralidae (Lepidoptera).
211	*Elasmus johnstoni* Ferriere	Parasites of *Pectinophora gossypiella* and *Platyedra gossypiella* in the family Gelechiidae, *Hyblaeae puera* in Hyblaeidae and some moths in the families Noctuidae and Pyralidae of the order Lepidoptera.
212	*Elasmus nephantidis* Rohwer	Parasites of *Nephantis serinopa* and *Opisina arenosella* in the family Oecophoridae and *Conopia* species in Sesiidae (Lepidoptera).

(*Continued*)

Sl. No.	Chalcid parasites	Host
213	*Elasmus philippinensis* Ashmead	Parasites of *Pectinophora gossypiella* in the family Gelechiidae, *Erionota thrax* and *Erionota torus* in the family Hesperiidae and some moths in the family Pyralidae of the order Lepidoptera.
214	*Elasmus platyedrae Ferriere*	Parasites of *Ceroplastes* species (Hemiptera: Coccidae), *Zeuzera coffea* in the family Cossidae, some moths in the family Gelechiidae, *Euproctis chrysorrhoea* in Lymantriidae, etc. of the order Lepidoptera.
215	*Elasmus zehntneri* Ferriere	Parasites of *Pectinophora gossypiella* in the family Gelechiidae, *Phyllocnistis citrella* in Gracillariidae and many moths in the family Pyralidae.
216	*Entedonomphale postmarginalis* (Shafee, Rizvi and Khan)	Parasites of Aleyrodidae unspecified species (Hemiptera).
217	*Euderus agromyzae* Gangrade	Parasites of *Agromyza obtusa* and *Melanagromyza obtusa* (Diptera: Agromyzidae).
218	*Euderus lividus* (Ashmead)	Parasites of *Oberea ulmicola* and *Saperda inornata* in the family Cerambycidae and the weevil *Ceutorhynchus rapae* in Curculionidae (Coleoptera), *Agromyza obtusa* and *Melanagromyza obtusa* (Diptera: Agromyzidae), *Eulecanium corni* and *Lecanium corni* (Hemipterta: Coccidae) and moths in the families Gelechiidae, Gracillariidae, Lyonetiidae and Pyralidae (Lepidoptera).
219	*Euplectrus coimbatorensis* Ferriere	Parasites of *Chilo partellus* (Lepidoptera: Pyralidae).
220	*Euplectrus Indicus* Ferriere	Parasites of Geometridae unspecified species and Naranga species and *Othreis maternal* in the family Noctuidae (Lepidoptera).
221	*Euplectrus longiscapus* Khan and Agnihotri	Parasites of *Lamprosema indicata* (Lepidoptera: Pyralidae).
222	*Euplectrus longiscapus* Khan, Agnihotri and Sushil	Parasites of *Lamprosema indicata* (Lepidoptera: Pyralidae).
223	*Euplectrus nuperus* Narendran	Parasites of caterpillar of Lepidoptera.
224	*Euplectrus viggianii* Khan and Agnihotri	Parasites of *Marasmia trapezalis* (Lepidoptera: Pyralidae).
225	*Euplectrus viggianii* Khan, Agnihotri and Sushil	Parasites of *Marasmia trapezalis* (Lepidoptera: Pyralidae).
226	*Farooqiella kashmiriensis* Jamal Ahmad	Parasites of *Plutella xylostella* (Lepidoptera: Yponomeutidae).
227	*Goetheana shakespearei* Girault	Parasites of many thrips (Thysanoptera: Thripidae).
228	*Hemiptarsenus indicus* Khan	Parasites of *Chromatomyia horticola* (Diptera: Agromyzidae).
229	*Hemiptarsenus unguicellus* (Zetterstedt)	Parasites of *Ceutorhynchus assimilis* in the family Curculionidae and *Blastophagus piniperda* and *Myelophilus piniperda* in Scolytidae (Coleoptera), many leaf miner flies in Agromyzidae, *Hydrellia griseola* in Ephydridae and some moths in the families Noctuidae, Pyralidae, etc (Lepidoptera).
230	*Hemiptarsenus varicornis* (Girault)	Parasites of many leaf miner flies in the family Agromyzidae (Diptera).

Sl. No.	Chalcid parasites	Host
231	*Melittobia acasta* (Walker)	Parasites of *Coroebus undatus* in the family Buprestidae, *Donacia* species in Chrysomelidae, *Calliphora* species of flies in the family Calliphoridae, some gall midges in the family Cecidomyiidae, Musca species in Muscidae, the fruit fly *Urophora solstitialis* (Tachinidae) and *Diplostichus* and *Lydella* species of flies in the family Tachinidae of the order Diptera and also parasitize many social bees in the family Apidae, solitary wasps in the family Sphecidae, many Vespidae species of wasps, some ants in Formicidae family, some saw flies in Diprionidae and *Chrysis* species in Chrysididae (Hymenoptera) and some moths and butterflies in the families Lasiocampidae, Limacodidae, Lymantriidae, Noctuidae, Pieridae and Tortricidae (Lepidoptera).
232	*Mestocharella javensis* Gahan	Parasites of some moths in the family Gracillariidae of the order Lepidoptera.
233	*Necremnus leucarthros* (Nees)	Parasites of Choragus piceus in the family Anthribidae, some Apion species in the family Apionidae, many leaf beetles in the family Chrysomelidae, many weevils in Curculionidae and *Quedius brevis* in the family Staphylinidae (Coleoptera), also parasitizes some flies species in the families Agromyzidae, Cecidomyiidae and Chloropidae (Diptera) and *Coleophora laricella* in the family Coleophoridae, *Eriocrania* species of moths in the family Eriocraniidae, some moths in the family Tortricidae, and *Yponomeuta padellus* in Yponomeutidae of the order Lepidoptera.
234	*Neoaceratoneura mangiferae* Khan, Agnihotri and Sushil	Parasites of *Acrocercops* species of moths (Lepidoptera: Gracillariidae).
235	*Neochrysocharis horticola* Khan, Agnihotri and Sushil	Parasites of *Chromatomyia horticola* (Diptera: Agromyzidae).
236	*Neochrysocharis indicus* Khan, Agnihotri and Sushil	Parasites of *Liriomyza* species of leaf miner flies (Diptera: Agromyzidae).
237	*Neotrichoporoides delhiensis* (Shafee, Fatma and Kishore)	Parasites of *Atherigona soccata* (Diptera: Anthomyiidae).
238	*Neotrichoporoides mediterraneus* Graham	Associated with rice.
239	*Neotrichoporoides nyemitawus* (Rohwer)	Parasites of some Muscoidea flies (Diptera: Anthomyiidae).
240	*Nesolynx flavipes* Ashmead	Parasites of *Heliothis armigera* (Noctuidae) and *Metisa plana* (Psychidae) of the order Lepidoptera and also parasitoids of flies in the family Tachinidae (Diptera) and wasps in Braconidae family (Hymenoptera).
241	*Nesolynx orientalis* Khan, Agnihotri and Sushil	Associated with *Hibiscus* plant.
242	*Nesolynx thymus* (Girault)	Parasites of *Musca domestica* (Diptera: Muscidae), *Bombyx mori* in the family Bombycidae and also on some other species of moths in the families Gracillariidae, Heliodinidae, Oecophoridae, Pyralidae, Saturniidae and Zygaenidae (Lepidoptera).
243	*Oomyzus liriomyzae* Narendran	Parasites of *Liriomyza* species of leaf miner flies (Diptera: Agromyzidae).
244	*Oomyzus mashhoodi* (Khan and Shafee)	Parasites of Aphididae unspecified species in the order Hemiptera.

(*Continued*)

Sl. No.	Chalcid parasites	Host
245	*Oomyzus scaposus* (Thomson)	Parasites of many species in the family Coccinellidae (Coleoptera) and some species in the family Chrysopidae (Neuroptera).
246	*Oomyzus sokolowskii* (Kurdjumov)	Parasites of the butterfly, *Pieris brassicae* in the family Pieridae and *Plutella* species of moths in the family Tponomeutidae of the order Lepidoptera.
247	*Oomyzus viraktamathi* Narendran	Parasites of *Aleurodicus dispersus* (Hemiptera: Aleyrodidae).
248	*Parachrysocharis javensis* Girault	Parasites of *Cenestra affinis* and *Cinestra affinis* in the family Flatidae, *Pyrilla* species in Lophopidae and the mealy bug *Saccharicoccus sacchari* in the family Pseudococcidae of the order Hemiptera and also on the moth *Chilo infuscatellus* (Lepidoptera: Pyralidae).
249	*Parasecodella pantnagarensis* Khan, Agnihotri and Sushil	Parasites of Agromyzidae unspecified species in the order Diptera.
250	*Pediobius bethylicidus Kerrich*	Parasites of *Sesamia inferens* (Lepidoptera: Noctuidae) and parasitoids of *Goniozus*, and Perisierola species of aculeate wasps in the family Bethylidae of the order Hymenoptera.
251	*Pediobius bhimtalensis* Khan, Agnihotri and Sushil	Associcated with cabbage plant.
252	*Pediobius bruchicida* (Rondani)	Parasites of *Bruchus* species in the family Bruchidae, *Plagiodera versicolora* in Chrysomelidae and *Rhamphus* species of weevils in Curculionidae (Coleoptera), *Dasineura* species of flies (Diptera: Cecidomyiidae), cynipid species (Hymenoptera: Cynipidae) many moths in the family Lymantriidae, some in the family Noctuidae, Yponomeutidae, Tortricidae, also *Pieris rapae* in Pieridae, etc. of the order Lepidoptera.
253	*Pediobius cydiae* Khan	Parasites of *Cydia critica* (Lepidoptera: Tortricidae).
254	*Pediobius elasmi* (Ashmead)	Parasites of *Aspidomorpha* and *Calopepla* species of leaf beetles in the family Chrysomelidae of the order Coleoptera and also moths in the families Psychidae (*Mahasena corbetti* and *Pteroma* species), Pyralidae (*Chilo indicus*) and Zygaenidae (*Artona catoxantha*) of the order Lepidoptera.
255	*Pediobius foveolatus* (Crawford)	Parasites of many ladybird beetles (Coleoptera: Coccinellidae), *Diacrisia punctulata* in the family Arctiidae, some butterflies in Hesperiidae family, some moths in Noctuidae, etc. of the order Lepidoptera.
256	*Pediobius imbrues* (Walker)	Parasites of some leaf beetles in the family Chrysomelidae of the order Coleoptera, *Epicephala* species of moths (Gracillariidae), *Opisina arenosella* (Oecophoridae) and *Artona catoxantha* (Zygaenidae) of the order Lepidoptera.
257	*Pediobius incertulus* Khan, Agnihotri and Sushil	Parasites of *Tryporyza incertulas* (Lepidoptera: Pyralidae).
258	*Pediobius indicus* Khan	Parasites of *Chromatomyia horticola* (Diptera: Agromyzidae).
259	*Pediobius khani* Ozdikmen	Parasites of *Trichoplusia ni* (Lepidoptera: Noctuidae).
260	*Pediobius metallicus* (Nees)	Parasites of many leaf miner flies (Agromyzidae), *Pegomya* species of muscoidea flies (Anthomyiidae), some gall midges (Cecidomyiidae) and *Oscinella frit* (Chloropidae) and also parasites of one sawfly, *Syringophilus ihommei* (Hymenoptera: Tenthredinidae), many moths in the family Gracillariidae, *Elachista* species in the family Elachistidae, some in Nepticulidae, etc. of the order Lepidoptera.

Sl. No.	Chalcid parasites	Host
261	*Pediobius parvulus* (Ferriere)	Parasites of many leaf beetles (Coleoptera: Chrysomelidae) and *Artona catoxantha* (Zygaenidae) of the order Lepidoptera.
262	*Pediobius shafeei* Khan, Agnihotri and Sushil	Associated with mango tree.
263	*Pediobius viggianii* Khan	Parasites of the moth *Nymphula depunctalis* (Lepidoptera: Pyralidae).
264	*Phymastichus coffea* LaSalle	Parasites of some bark beetles (Coleoptera: Scolytidae).
265	*Pomphale striptipennis* Husain, Rauf and Kudeshia	Parasites of *Aleurolobus* species of white flies (Hemiptera: Aleyrodidae).
266	*Quadrastichus ovulorum* (Ferriere)	Parasites of some *Epilachna* and *Henosepilachna* species of ladybird beetles (Coleoptera: Coccinellidae).
267	*Sigmophora brevicornis* (Panzer)	Parasites of the beetles *Mesocoelopus niger* (Anobiidae) and *Apion trifolii* (Apionidae) many gall midges (Cecidomyiidae), *Urophora jaceana* (Tephritidae), some hymenopteran species like Cynipids, *Bruchophagus gibbus*, *Eurytoma onobrychidis* (Eurytomidae) and *Euura atra* (Tenthredinidae) and moths of Cochylidia (Tortricidae) and *Yponomeuta* species (Yponomeutidae) of the order Lepidoptera.
268	*Stenomesius anati* Khan and Jaikishan Singh	Associated with plants of Rosaceae family.
269	*Stenomesius japonicus* (Ashmead)	Parasites of some moths in the family Gelechiidae, Noctuidae, Pyralidae, many other moths in the family Gracillariidae, *Lyonetia clerkella* in Lyonetiidae (Lepidoptera).
270	*Stenomesius modicellus* Khan	Parasites of the moth *Aproaerema modicella* (Lepidoptera: Gelechiidae).
271	*Sympiesis dolichogaster* Ashmead	Parasites of many moths in the family Gracillariidae, some moths in Tortricidae, *Aproaerema* and *Stomopteryx* species of moths in Gelechiidae family, *Hydriomena renunciata* in Geometridae, etc. of thr order Lepidoptera.
272	*Sympiesis jeolikotensis* Khan, Agnihotri and Sushil	Parasites of *Acrocercops* species of moths (Lepidoptera: Gracillariidae).
273	*Sympiesis striatipes* (Ashmead)	Parasites of *Agromyza* species (Diptera: Agromyzidae) and few moths in the family Gracillaridae of the order Lepidoptera.
274	*Tachinobia repanda* Bouček	Parasites of *Pseudogaurax* and *Sarcophaga* species of flies in the families Chloropidae and Sarcophagidae respectively and Megalopyge moth in Megalopygidae family, *Papilio* species in Papilionidae, *Mahasena corbetti* in Psychidae and *Artona catoxantha* in Zygaenidae (Lepidoptera).
275	*Tamarixia radiate* (Waterston)	Parasites of some jumping plant lice in Psyllidae family and *Trioza* species in Triozidae family of the order Hemiptera.
276	*Tetrastichus apanteles* Khan, Agnihotri and Sushil	Parasites of the *Apanteles* species of wasps (Hymenoptera: Braconidae).
277	*Tetrastichus bilgiricus* Narendran	Parasites of *Euthalia aconthea* (Lepidoptera: Nymphalidae).
278	*Tetrastichus dasi* Narendran	Parasites of Cecidomyiidae unspecified species in the order Diptera.
279	*Tetrastichus hindicus* Ozdikmen	Parasites of *Tryporyza incertulas* (Lepidoptera: Pyralidae).

(Continued)

Sl. No.	Chalcid parasites	Host
280	*Tetrastichus howardi* (Olliff)	Parasites of *Tenebrio molitor* in the family (Coleoptera) Tenebrionidae, *Musca domestica* (Diptera: Muscidae), *Nipaecoccus vastator* (Hemiptera: Pseudococcidae), *Apis mellifera* (Hymenoptera: Apidae), many moths in the family Noctuidae and Pyralidae, few ones in Oecophoridae, etc. of the order Lepidoptera and also parasitoids of some flies in the families Tachinidae (Diptera) and Ichneumonidae (Hymenoptera), etc.
281	*Tetrastichus inferens* Yoshimoto	Parasites of *Brachmia modicella* (Gelechiidae), *Sesamia inferens* (Noctuidae) and *Ostrinia furnacalis* (Pyralidae) of the order Lepidoptera.
282	*Tetrastichus Krishnieri* (Mani)	Parasites of *Hypolixus* and *Lixus* species of weevils (Coleoptera: Curculionidae).
283	*Tetrastichus mangifera* Khan and Sushil	Associated with mango tree.
284	*Tetrastichus miser* (Nees)	Parasites of *Cassida* species of leaf beetles (Chrysomelidae), many weevils (Curculionidae) of the order Coleoptera, cockroaches (Dictyoptera: Blattidae) and also on the flies of species *Kaltenbachiola* and *Plemeliella* in the order Diptera.
285	*Tetrastichus neothoracicus* Ozdikmen	Parasites of *Apanteles* species of wasp (Hymenoptera: Braconidae).
286	*Tetrastichus oblique* Khan and Sushil	Parasites of *Spilosoma obliqua* (Lepidoptera: Arctiidae).
287	*Tetrastichus orissaensis* Husain and Khan	Parasites of *Epilachna* and *Subcoccinella* species of beetles (Coleoptera: Coccinellidae) and *Epipyrops* species (Lepidoptera: Epipyropidae).
288	*Tetrastichus pantnagarensis* Khan	Parasites of *Diacrisia obliqua* (Arctiidae) and *Diaphania indica* (Pyralidae) of the order Lepidoptera.
289	*Tetrastichus partellus* Khan and Sushil	Parasites of *Chilo partellus* (Lepidoptera: Pyralidae).
290	*Tetrastichus schoenobii* Ferriere	Parasites of *Sesamia inferens* and *Spodoptera mauritia* (Noctuidae) and many moths (Pyralidae) of the order Lepidoptera.
291	*Thripoctenus javae* (Girault)	Parasites of Coccoidea unspecified species (Hemiptera) and few thrips (Thysanoptera: Thripidae).
292	*Trichospilus diatraeae* Cherian and Margabandhu	Parasites of *Brachmia modicella* (Gelechiidae) some in the family Geometridae, and many other moths in the family Noctuidae, Pyralidae etc. and *Sexava* species (Orthoptera: Tettigoniidae).
293	*Trichospilus pupivorus* Ferriere	Parasites of *Hyblaea puera* (Hyblaeidae), *Thosea cervina* (Limacodidae) *Thagona* species (Lymantriidae), many moths (Noctuidae and Pyralidae) etc. of the order Lepidoptera.
		Family: Eupelmidae
294	*Anastatus bifasciatus* (Geoffroy)	Parasites of *Cinara schimitscheki* in Aphididae family, *Gonocerus* and *Leptoglossus* species in Coreidae, some stink bugs in Pentatomidae, *Psylla* species in Psyllidae, *Eurygaster* species in Scutelleridae of the order Hemiptera and also parasites of some lepidopteran moths in the families Lasiocampidae, Lymantriidae, Notodontidae and butterflies like *Charaxes jasius* (Nymphalidae), *Iphiclides podalirius* (Papilionidae), etc. and *Oxya velox* (Acrididae) and *Phaneroptera* species (Tettigoniidae) of the order Orthoptera.
295	*Anastatus japonicus* Ashmead	Parasites of *Riptortus clavatus* in Alydidae and some stink bugs in Pentatomidae of the order Hemiptera and also parasites of some lepidopteran moths in the families Lasiocampidae, Lymantriidae, Notodontidae and butterflies like *Iphiclides podalirius* (Papilionidae) etc.

Sl. No.	Chalcid parasites	Host
296	*Anastatus ramakrishnai* (Mani)	Parasites of Homoeocerus species in Coreidae family, some stink bugs in Pentatomidae of the order Hemiptera and the moth Halis dentata (Lepidoptera).
297	*Eupelmus australiensis* (Girault)	Parasites of some gall midges (Diptera: Cecidomyiidae) and the moth *Celama sorghiella* (Lepidoptera: Noctuidae).
298	*Eupelmus kashmiricus* Narendran	Parasites of *Scolytus nitidus* beetles (Coleoptera: Scolytidae).
299	*Eupelmus orientalis* (Crawford)	Parasites of many beetles in Bruchidae family, some weevils in Curculionidae (Coleoptera) and the moth *Dactylethra candida* (Lepidoptera: Gelechiidae).
300	*Eupelmus pedatorius* (Ferriere)	Parasites of *Piezotrachelus varium* in Apionidae family, some weevils in Curculionidae (Coleoptera) and *Melanagromyza* and *Diopsis* species of flies in Agromyzidae and Diopsidae family (Diptera).
301	*Eupelmus tenuicornis* Kieffer	Parasites of some gall midges (Diptera: Cecidomyiidae).
302	*Eupelmus terminaliae* Hafiz	Associated with almond plant.
303	*Eupelmus urozonus* Dalman	Parasites of some weevils in Apionidae family, *Atellabus nitens* in Attelabidae, *Bruchidius* and *Bruchus* species of beetles in Bruchidae, many weevils in Curculionidae, many beetles in Scolytidae etc. of the order Coleoptera and also parasites of *Melanagromyza* species of flies in Agromyzidae, many gall midges in Cecidomyiidae, some fruit flies in Tephritidae (Diptera), *Toxoptera* species of aphids (Aphididae), some scale insects in Coccidae family (Hemiptera), many gall wasps in Cynipidae, some wasps in Torymidae (Hymenoptera), some moths in the families, Gelechiidae, Gracillariidae, Tortricidae, etc. (Lepidoptera).
304	*Eupelmus valsus* Narendran	Parasites of *Scolytus* species of beetles (Coleoptera: Scolytidae).
305	*Eupelmus vindex* Erdos	Parasites of *Scolytus* species of beetles (Coleoptera: Scolytidae).
306	*Mesocomys orientalis* Ferriere	Parasites of *Trabala*, *Calliteara*, *Euproctis* and *Cricula* species of moths in the families Lasiocampidae, Lymantriidae and Saturniidae of the order Lepidoptera.
307	*Neanastatus cinctiventris* Girault	Parasites of some gall midges (Diptera: Cecidomyiidae) and parasitoids of *Platygaster oryzae* (Hymenoptera: Platygastridae).
		Family: Eurytomidae
308	*Eurytoma albotibialis* Ashmead	Parasites of *Laccifer lacca* (Hemiptera: Kerriidae), *Angitia* species (Hymenoptera: Ichneumonidae) and moths like *Holcocera pulverea* (Blastobasidae), *Eublemma amabilis* (Noctuidae), *Nephantis serinopa* and *Opisina arenosella* (Oecophoridae), *Corcyra cephalonica* (Pyralidae) and *Artona* species (Zygaenidae) of the order Lepidoptera.
309	*Eurytoma asphondyliae* Narendran	Parasites of *Asphondylia* species (Diptera: Cecidomyiidae).
310	*Eurytoma browni* Crawford	Parasites of *Heliothis armigera* (Lepidoptera: Noctuidae) and also parasitoids of *Campoletis* perdistinctus (Hymenoptera: Ichneumonidae).
311	*Eurytoma chaitra* Narendran	Associated with capsicum plant.
312	*Eurytoma curculionum* Mayr	Parasites of *Apion* species of weevils in Apionidae family, *Agrilus rubucola* in Buprestidae, many weevils in Curculionidae (Coleoptera) and *Lasioptera rubi* (Diptera: Cecidomyiidae).

(*Continued*)

Sl. No.	Chalcid parasites	Host
313	*Eurytoma dentata* Mayr	Parasites of many gall midges (Diptera: Cecidomyiidae).
314	*Eurytoma Indiana* Ozdikmen	Parasites of lepidopteran species like *Parnara* mat*h*ias (Hesperiidae), *Trichoplusia ni* (Noctuidae), *Orsotriaena medus* (Nymphalidae) and *Pieris brassicae* in Pieridae. Also parasitoids of some Braconid wasps (Hymenoptera: Braconidae).
315	*Eurytoma melanagromyzae* Narendran	Parasites of some leaf miner flies (Diptera: Agromyzidae).
316	*Eurytoma morio* Boheman	Parasites of some Cerambycid beetles (*Pogonochaerus hispidus*, *Tetros praeusta*), some weevils in Curculionidae and many Scolytid beetles in the order Coleoptera and also *Exoteleia dodecella* moth (Gelechiidae) and *Aporia crataegi* (Pieridae) of the order Lepidoptera.
317	*Eurytoma samsonowi* Vassilicv	Associated with rose plants of family Rosaceae
318	*Eurytoma setitibia* Gahan	Parasites of *Orseolia oryzae* (Diptera: Cecidomyiidae) and *Trioza fletcheri* (Hemiptera: Triozidae).
319	*Eurytomocharis keralensis* Mukerjee	Parasites of *Melanagromyza hibisci* (Diptera: Agromyzidae).
320	*Philolema braconidis* (Ferriere)	Parasites of Gelechiidae unspecified species, some moths in Noctuidae, *Opisina arenosella* in Oecophoridae, and few moths in Pyralidae families of the order Lepidoptera and parasitoids of many Braconid wasps (Hymenoptera: Braconidae).
321	*Philolema maleena* Narendran	Associated with rice plants.
322	*Sycophila punensis* Sureshan	Associated with java plum tree.
323	*Systole albipennis* Walker	Phytophagous, seed eaters of various plants of Apiaceae family.
		Family: Ormyridae
324	*Ormyrulus gibbus* Bouček	Parasites of *Procontarinia matteiana* (Diptera: Cecidomyiidae).
325	*Ormyrus orientalis* Walker	Parasites of *Melanagromyza obtusa* and *Phytomyza orobanchia* in the family Agromyzidae, *Lasioptera eryngii* in Cecidomyiidae, *Lonchaea* species in Lonchaeidae, some fruit flies in Tephritidae of the order Diptera and also many cynipid wasps (Hymenoptera: Cynipidae).
		Family: Pteromalidae
326	*Agiommatus geethae* Sureshan and Narendran	Parasites of Lepidoptera (Noctuidae).
327	*Anisopteromalus calandrae* (Howard)	Parasites of some beetles in Anobiidae and Bostrychidae families, *Araecerus fasciculatus* in Anthribidae, *Piezotrachelus varium* in Apionidae, many bruchid beetles, some weevils in Curculionidae and Dryophthoridae, *Trogoderma granarium* in Dermestidae, *Carpophilus obsoletus* in Nitidulidae, *Oryzaephilus surinamensis* in Silvanidae and *Tribolium castaneum* in Tenebrionidae (Coleoptera), *Anisopteromalus calandrae* (Hymenoptera: Pteromalidae) and the moth *Sitotroga cerealella* in Gelechiidae and some moths in Pyralidae family (Lepidoptera).
328	*Asaphes suspensus* (Nees)	Parasites of *Liriomyza congesta* and *Rabdophaga saliciootrobiloides* in Agromyzidae and Cecidomyiidae families respectively (Diptera) and many aphids (Hemiptera: Aphididae). Parasitoids of some parasitic wasps in Aphelinidae, many braconid wasps, *Aphidencyrtus aphidivorus* in Encyrtidae, *Alloxysta* species in Figitidae and *Dendrocerus carpenteri* in Megaspilidae family of the order Hymenoptera.

Sl. No.	Chalcid parasites	Host
329	*Asaphes vulgaris* Walker	Parasites of *Anthonomus pomorum* and *Tychius medicaginis* (Coleoptera: Curculionidae), *Liriomyza lutea* (Agromyzidae), *Aphidoletes cucumeris* and *Dasineura ignorata* (Cecidomyiidae), *Syrphus ribesii* (Syrphidae) of the order Diptera, many aphids in Aphididae, *Eulecanium* (Coccidae) and *Aulacaspis* (Diaspididae) species of scale insects, *Phenacoccus* and *Pseudococcus* species of mealybugs (Pseudococcidae), *Psylla mali* and *Psylla pyricola* (Psyllidae) and also the cynipid wasp, *Phaenoglyphis villosa* (Hymenoptera: Cynipidae).
330	*Austroterobia maldica* Narendran and Das	Parasites of *Icerya* species of scale insects (Hemiptera: Margarodidae).
331	*Cephaleta australiensis* (Howard)	Parasites of *Cerococcus indicus* in Cerococcidae family, many scale insects in Coccidae, *Eriococcus* species in Eriococcidae and some mealy bugs in Pseudococcidae of the order Hemiptera and also parasitoids of *Coccophagus indicus* (Hymenoptera: Aphelinidae).
332	*Cephaleta brunniventris* Motschulsky	Parasites of *Asterolecanium* species in Asterolecaniidae, some ornate pit scales in Cerococcidae, many scale insects in Coccidae, *Ferrisia virgata* and *Pseudococcus* species of mealy bugs in Pseudococcidae of the order Hemiptera and also *Acrocercops caerulea* (Lepidoptera: Gracillariidae).
333	*Cephaleta dinoderi* Gahan	Parasites of the beetle, *Dinoderus minutus*, and *Sitophilus oryzae*.
334	*Cheiropachus quadrum* (Fabricius)	Parasites of *Apate* species and *Sinoxylon sexdentatum* in Bostrychidae family, *Tetrops praeusta*, some weevils in Curculionidae and many beetles in Scolytidae of the order Coleoptera, *Zeuzera pyrina* in Cossidae family and *Lymantria dispar* in Lymantriidae in Lepidoptera.
335	*Coruna clavata* Walker	Parasites of numerous aphids (Hemiptera: Aphididae) and parasitoids of *Aphelinus* flavus in Aphelinidae and many Braconid wasps (Hymenoptera).
336	*Cyrtogaster clavicornis* Walker	Parasites of *Agromyza* species in Agromyzidae family, *Pegomya hyoscyami* and *Phorbia* species in Anthomyiidae, some shore flies in Ephydridae and *Lonchoptera* species in Lonchopteridae of Diptera.
337	*Cyrtoptyx latipes* (Rondani)	Parasites of some weevils (Coleoptera: Curculionidae), some fruit flies (Diptera: Tephritidae), *Coleophora stefanii* (Coleophoridae) and *Cydia sinana* (Tortricidae) of the order Lepidoptera.
338	*Dibrachys microgastri* (Bouche)	Parasites of *Epeira* species and *Epeira foliata* in Araneidae family and *Philodromus aureoles* in Thomisidae of the order Araneae, beetles in the families Anobiidae, Bruchidae, Cerambycidae, Coccinellidae, Cucujidae and weevils in Curculionidae, some weevils in Dryophthoridae, etc. and *Forficula auricularia* (Dermaptera: Forficulidae), *Phytomyza affinis* in Agromyzidae, some flies in Calliphoridae, Muscidae, *Pseudococcus citri* (Hemiptera: Pseudococcidae), some bees in Apidae, *Cephus* species in Cephidae, many sawflies in Diprionidae and Tenthredinidae, *Trypoxylon figulus* in Sphecidae, some wasps in Vespidae, some moths in Arctiidae, Coleophoridae and Gracillariidae, *Bucculatrix* species in Bucculatricidae, many moths in Gelechiidae and Geometridae, Lasiocampidae, Tortricidae etc. and also some butterflies in Nymphalidae, Pieridae, etc. some *Chrysopa* species (Neuroptera: Chrysopidae), etc. Parasitoids of many wasps in Braconidae, Ichneumonidae (Hymenoptera) etc.
339	*Dinarmus acutus* (Thomson)	Parasites of many beetles in Bruchidae family (Coleoptera), the scale insect, *Ceroplastes rusci* (Hemiptera: Coccidae) and *Dinarmus rufimanus* (Hymenoptera: Pteromalidae).

(Continued)

Sl. No.	Chalcid parasites	Host
340	*Dinarmus altifrons* (Walker)	Parasites of some beetles in Bruchidae family (Coleoptera).
341	*Dinarmus basalis* (Rondani)	Parasites of *Piezotrachelus varium* in Apionidae, *Cylas puncticollis* in Brentidae, many beetles in Bruchidae and *Trogoderma granarium* in Dermestidae of the order Coleoptera.
342	*Dinarmus colemani* (Crawford)	Parasites of some beetles in Bruchidae family of the order Coleoptera.
343	*Dinarmus vagabundus* (Timberlake)	Parasites of some beetles in Bruchidae family of the order Coleoptera.
344	*Euneura lachni* (Ashmead)	Parasites of *Neocnemodon coxalis* (Diptera: Syrphidae), *Aleurolobus* species in Aleyrodidae, many aphids in Aphididae (Hemiptera) and *Dendrolimus* species of moths in Lasiocampidae (Lepidoptera) and also parasitoids of many Braconid wasps (Hymenoptera: Braconidae).
345	*Gastrancistrus cherry* Boucek	Parasites of *Erosomyia mangiferae* and *Procontarinia matteiana* (Diptera: Cecidomyiidae).
346	*Gastrancistrus muneswari* Yadav	Parasites of *Oligotrophusz mangiferae* and *Rabdophaga mangiferae* (Diptera: Cecidomyiidae).
347	*Halticoptera propinqua* (Waterston)	Parasites of some leaf –miner flies (Diptera: Agromyzidae).
348	*Herbertia indica* Burks	Parasites of some leaf –miner flies (Diptera: Agromyzidae).
349	*Heydenia indica* Narendran	Parasites of *Dinoderus* species of beetles (Bostrychidae) and *Scolytus* species of beetles (Scolytidae) of the order Coleoptera.
350	*Lariophagus distinguendus* (Forster)	Parasites of some beetles in Anobiidae family, *Rhyzopertha* species of beetles in Bostrychidae, some species in Bruchidae, *Pachytychius hordei* in Curculionidae, some weevils in Dryophthoridae, some spider beetles in *Ptinidae* and *Tribolium* species in Tenebrionidae of the order Coleoptera and also parasites of *Mayetiola phalaris* (Diptera: Cecidomyiidae) and the parasitoid wasp, *Lariophagus distinguendus* (Hymenoptera: Pteromalidae).
351	*Macromesus harithus* Narendran	Parasites of *Scolytus* species of beetles (Coleoptera: Scolytidae).
352	*Metastenus concinnus* Walker	Parasites of some small beetles (Coleoptera: Coccinellidae).
353	*Moranila californica* (Howard)	Parasites of *Asterolecanium* species in Asterolecaniidae, many species of scale insects in Coccidae, *Eriococcus* species in Eriococcidae, *Antonina* and *Rosanococcus* species of mealy bugs in Pseudococcidae of the order Hemiptera and also *Sympherobius californicus* (Neuroptera: Sympherobiidae).
354	*Ophelosia crawfordi* Riley	Parasites of *Pulvinaria* species of scale insects, some scale insects in Margarodidae, and some mealy bugs in Pseudococcidae of the order Hemiptera.
355	*Oricoruna arcotensis* (Mani and Kurian)	Parasites of some *Icerya* species of Scale insects in Margarodidae family of the order Hemiptera.
356	*Oricoruna orientalis* (Crawford)	Parasites of *Rodolia fumida* (Coleoptera: Coccinellidae).
357	*Oxysychus coimbatorensis* (Ferriere)	Parasites of some weevils (Coleoptera: Curculionidae).

Sl. No.	Chalcid parasites	Host
358	*Pachycrepoideus vindemmiae* (Rondani)	Parasites of some muscoidea flies (Anthomyiidae), many blow flies (Calliphoridae), *Orseolia oryzae* (Cecidomyiidae), some fruit flies in the genus *Drosophila*, many flies (Muscidae), *Piophila* species of cheese flies (Piophilidae), some flesh flies (Sarcophagidae), many fruit flies (Tephritidae), etc. of the order Diptera, *Anasa tristis* (Hemiptera: Coreidae), *Bombyx mori* (Bombycidae) and some moths (Pyralidae) of the order Lepidoptera.
359	*Pachyneuron aphidis* (Bouche)	Parasites of *Hippodamia* species of beetles (Coleoptera: Coccinellidae), some flies in the families Agromyzidae, Cecidomyiidae and Syrphidae of the order Diptera, many aphids (Hemiptera: Aphididae), some scale insects in Coccidae, Psyllidae (Hemiptera), *Phaenoglyphis villosa* (Hymenoptera: Cynipidae), some moths in the families Gelechiidae and Tortricidae of the order Lepidoptera.
360	*Pachyneuron groenlandicum* (Holmgren)	Parasites of *Oscinella frit* in Chloropidae, *Psila rosae* in Psilidae, many Syrphid flies (Diptera), some aphids in Aphididae, *Physokermes jezoensis* in Coccidae (Hemiptera), some moths (*Dendrolimus* and *Autographa* species) in Lasiocampidae and Noctuidae) and the butterfly of *Pieris* species in Pieridae of the order Lepidoptera and also parasitoids of many braconid wasps (Hymenoptera: Braconidae).
361	*Pachyneuron leucopiscida* Mani	Parasites of some flies (Diptera: Chamaemyiidae) and *Acletoxenus indicus* (Diptera: Drosophilidae), many aphids (Aphididae) and the mealy bug *Dactylopius* species (Pseudococcidae) of the order Hemiptera.
362	*Pachyneuron muscarum* (Linnaeus)	Parasites of *Chilocorus* species of beetles in Coccinellidae and *Ceutorhynchus* species of weevils in Curculionidae of the order Coleoptera,many leaf-miner flies in Agromyzidae, *Dasineura brassicae* in Cecidomyiidae, some grass flies in Chloropidae (Diptera), some aphids in Aphididae, many scale insects in Coccidae, few in Eriococcidae, some mealky bugs in Pseudococcidae, *Acantholyda nemoralis* (Hymenoptera: Pamphiliidae) and some moths in the families Lasiocampidae, Tortricidae and Yponomeutidae).
363	*Pachyneuron solitarium* (Hartig)	Parasites of *Coccinella* species of beetles (Coleoptera: Coccinellidae). *Eutolmus* species of flies (Diptera: Asilidae), many aphids in Aphididae, some scale insects in Coccidae, some mealy bugs in Pseudococcidae, *Psylla pyri* in Psyllidae, some moths in Lasiocampidae and *Dasychira* and *Lymantria* species of moths in Lymantriidae (Lepidoptera).
364	*Propicroscytus mirificus* (Girault)	Parasites of some gall midges in Cecidomyiidae of the order Diptera and *Scirpophaga incertulas* in Pyralidae family of the order Lepidoptera.
365	*Propicroscytus oryzae* (Subba Rao)	Parasites of *Orseolia oryzae* and *Pachydiplosis oryzae* (Diptera: Cecidomyiidae) and also *Schoenobius incertulas* (Lepidoptera: Pyralidae).
366	*Pteromalus puparum* (Linnaeus)	Parasites of *Bruchidius unicolor* (Bruchidae), *Ceutorhynchus obstrictus* (Curculionidae), *Pityocteines curvidens* (Scolytidae), *Oscinella frit* (Diptera: Chloropidae), *Lepidosaphes malicola* (Hemiptera: Diaspididae), some wasps (Braconidae, Cynipidae, Sphecidae, Vespidae), many butterflies in the families Nymphalidae, Papilionidae, Pieridae, etc. and some moths in Tortricidae, Noctuidae, Lymantriidae, Lasiocampidae (Lepidoptera), etc.
367	*Pteromalus semotus* (Walker)	Parasites of *Ernobius abietis* in Anobiidae, *Apion pomonae* in Apionidae, *Oulema gallaeciana* in Chrysomelidae, some weevils in Curculionidae of the order Coleopteran, *Neodiprion sertifer* (Hymenoptera: Diprionidae) and also many moths in Coleophoridae, Tortricidae, some moths in Gracillariidae and Yponomeutidae, etc. of the order Lepidoptera.

(*Continued*)

Sl. No.	Chalcid parasites	Host
368	*Rhaphitelus maculatus* Walker	Parasites of some weevils in Curculionidae family, many beetles in Scolytidae (Coleoptera) and *Zeuzera pyrina* (Lepidoptera: Cossidae).
369	*Scutellista caerulea* (Fonscolombe)	Parasites of *Cerococcus hibisci* in Cerococcidae, many scale insects in Coccidae, *Quadraspidiotus* species in Diaspididae and *Phenacoccus* species of mealy bugs in Pseudococcidae family of the order Hemiptera.
370	*Spalangia endius* Walker	Parasites of some flies in Anthomyiidae, Calliphoridae and Sarcophagidae, many muscidae flies, fruit flies, and also *Bombyx mori* (Bombycidae) and *Diatraea* species of moths (Pyralidae) of the order Lepidoptera.
371	*Spalangia gemina* Boucek	Parasites of *Mimegralla* species (Micropezidae), *Musca domestica* (Muscidae), some flies in Tephritidae and also moths like *Adisura* (Noctuidae) and *Cryptophlebia* (Tortricidae) species (Lepidoptera).
372	*Sphegigaster brunneicornis* (Ferriere)	Parasites of some leaf-miner flies (Diptera: Agromyzidae).
373	*Sphegigaster stepicola* Boucek	Parasites of many leaf-miner flies (Diptera: Agromyzidae).
374	*Systasis dasyneurae* Mani	Parasites of *Dasineura* and *Erosomyia* species of gall midges (Diptera: Cecidomyiidae).
375	*Theocolax elegans* (Westwood)	Parasites of *Lasioderma* and *Stegibium* species of beetles (Anobiidae), some beetles in Bostrychidae, Bruchidae, *Caulophilus* species of weevils in Curculionidae and some other weevils in Dryophthoridae of the order Coleoptera and the moth *Sitotroga cerealella* (Lepidoptera: Gelechiidae).
376	*Thinodytes cyzicus* (Walker)	Parasites of some leaf-miner flies (Diptera: Agromyzidae).
377	*Trichomalopsis apanteloctena* (Crawford)	Parasites of *Dicladispa armigera* and *Oulema oryzae* (Coleoptera: Chrysomelidae), some fly species in Diptera (Agromyzidae, Cecidomyiidae and Ephydridae) and also moths in Hesperiidae, Noctuidae, Pyralidae, etc of the order Lepidoptera.
378	*Trichomalopsis deplanata* Komijo and Grissell	Parasites of *Oulema species* and *Oulema oryzae* (Coleoptera: Chrysomelidae), *Agromyza yanonensis* in Agromyzidae, Tachinidae unspecified species in Tachinidae in the order Diptera, *Parnara* species and *Parnara guttata* in Hesperiidae and *Ostrinia furnacalis* in Pyralidae of the order Lepidoptera. Also parasitoids of *Thecocarcelia thrix* (Diptera: Tachinidae) and *Apanteles glomeratus* (Hymenoptera: Braconidae).
379	*Trichomalopsis neelagastra* Sureshan and Narendran	Collected from rice fields, parasites of variuous paddy pests belonging to the order Lepidoptera and Coleoptera.
380	*Trichomalopsis ovigastra* Sureshan and Narendran	Collected from rice fields, parasites of variuous paddy pests belonging to the order Lepidoptera and Coleoptera.
381	*Trichomalopsis thekkadiensis* Sureshan and Narendran	Collected from rice fields, parasites of variuous paddy pests belonging to the order Lepidoptera and Coleoptera.
		Family: Signiphoridae
382	*Chartocerus bengalensis* Hayat	Parasites of *Aspidiotus* species of scale insects (Hemiptera: Diaspididae).
383	*Chartocerus hyalipennis* Hayat	Parasites of some mealy bugs (Hemiptera: Pseudococcidae) and also parasitoids of some parasitic wasps (Hymenoptera: Encyrtidae).
384	*Chartocerus Kerrichi* (Agarwal)	Parasites of *Cerococcus* species (Cerococcidae), *Eriococcus greeni* (Eriococcidae), and some mealy bugs (Pseudococcidae) of the order Hemiptera.

Sl. No.	Chalcid parasites	Host
385	*Chartocerus kurdjumovi* (Nikolskaya)	Parasites of *Leucopis* species (Diptera: Chamaemyiidae), some aphids in Aphididae, *Ericoccus* species in Eriococcidae, some mealy bugs in Pseudococcidae, and *Psylla pyri* in Psyllidae of the order Hemiptera.
386	*Chartocerus walkeri* Hayat	Parasites of Aphididae unspecified species and *aphis nerii* in Aphididae, *Rastrococcus* species of mealy bugs in Pseudococcidae and *Diaphorina cardiae* in Psyllidae of the order Hemiptera. Also parasitoids of *Trioxys indicus* in Braconidae and *Diaphorencyrtus diaphorinae* in Encyrtidae of the order Hymenoptera.
387	*Thysanus ater* Walker	Parasites of some scale insects in Coccidae and many scale insects in Diaspididae of the order Hemiptera. Also parasitoids of *Aphytis mytilaspidis* in Aphelinidae and *Coccidencyrtus dissimilis* and *Euaphycus duplus* in Encyrtidae of the order Hymenoptera.
		Family: Tanaostigmatidae
388	*Tanaostigmodes cajaninae* LaSalle	Parasitoids of *Neanastatus grallarius* in Eupelmidae and *Pseudotorymus* species in Torymidae of the order Hymenoptera.
		Family: Torymidae
389	*Chrysochalcissa oviceps* Boucek	Parasites of *Paradasynus rostratus* (Hemiptera: Coreidae).
399	*Mangostigmus amraeus* (Chandy Kurian)	Parasites of some gall midges (Diptera: Cecidomyiidae).
391	*Mangostigmus bengalicus* Narendran and Sudheer	Parasites of Cecidomyiidae unspecified species (Diptera: Cecidomyiidae).
392	*Mangostigmus malabaricus* Narendran and Vijayalakshmi	Associated with mango trees, probably parasites of gall causing Diptera.
393	*Megastigmus brevivalvus* (Girault)	Parasites of *Bruchophagus fellis* and *Eurytoma fellis* (Hymenoptera: Eurytomidae).
394	*Monodontomerus aereus* Walker	Parasites of *Megachile muraria* in Apidae, *Diplolepis rosae* in Cynipidae, *Diprion pini* in Diprionidae, *Trichiocampus ulmi* in Tenthredinidae (Hymenoptera) and also *Hyphantria cunea* in Arctiidae, some moths in Lasiocampidae, Yponomeutidae and butterflies in Pieridae, many moths in Lymantriidae and Tortricidae, *Nymphalis polychloros* in Nymphalidae etc. of the order Lepidoptera. Parasitoids of *Sarcophaga* species in Sarcophagidae and many flies in Tachinidae (Diptera), many wasps in Braconidae, Ichneumonidae, some in Chalcididae, etc. (Hymenoptera).
395	*Monodontomerus minor* (Ratzeburg)	Parasites of *Megachile* and *Osmia* species in Apidae, some sawflies in Cimbicidae and Diprionidae, *Hyphantria cunea* in Arctiidae, *Cystidia couggaria* in Geometridae, many moths in Lasiocampidae, Lymantriidae, Tortricidae, and some butterflies in Pieridae, Nymphalidae unspecified species in Nymphalidae etc. of the order Lepidoptera. Parasitoids of some flies in Tachinidae (Diptera), many wasps in Braconidae, Ichneumonidae (Hymenoptera), *Hyalophora cecropia* (Lepidoptera: Saturniidae), etc.
396	*Podagrionella indarbelae* Narendran and Sureshan	Parasites of *Indarbella tetraonis* (Lepidoptera: Metarbelidae).
397	*Pseudotorymus harithavarnus* Narendran	Assoicated with pigeon pea.
398	*Torymus calcaratus* Nees	Parasites of some Cynipid wasps and many Sphecidae wasps in Hymenoptera.

(*Continued*)

Sl. No.	Chalcid parasites	Host
399	*Torymus varians* (Walker)	Parasites of *Diplolepis* species of wasps (Hymenoptera: Cynipidae).
		Family: Trichogrammatidae
400	*Brachygrammatella indica* Viggiani and Hayat	Parasites of some tree hoppers (Hemiptera: Membracidae).
401	*Chaetostricha magniclavata* Yousuf and Shafee	Parasites of *Bruchus* species of beetles (Coleoptera: Bruchidae).
402	*Mirufens afrangiata* Viggiani and Hayat	Parasites of some tree hoppers (Hemiptera: Membracidae).
403	*Mirufens longifunuculata* Viggiani and Hayat	Parasites of *Erosomyia indica* (Diptera: Cecidomyiidae).
404	*Mirufens mangiferae* Viggiani and Hayat	Parasites of *Oxyrhachis* species of tree hoppers (Hemiptera: Membracidae).
405	*Oligosita naias* Girault	Parasites of *Nephotettix* species of leaf hoppers in Cicadellidae, and some plant hoppers in Delphacidae of the order Hemiptera.
406	*Oligosita sanguinea* (Girault)	Parasites of Asphondylia species of gall midges (Diptera: Cecidomyiidae) and the scale insect, *Unaspis citri* (Hemiptera: Diaspididae).
407	*Paracentrobia ajmerensis* Yousuf and Shafee	Associated with *Sorghum* plants.
408	*Paracentrobia andoi* (Ishii)	Parasites of some leaf hoppers in Cicadellidae and some plant hoppers in Delphacidae of the order Hemiptera.
409	*Poropoea bella* Hayat and Poorani	Associated with java plum tree.
410	*Pseudoligosita aesopi* (Girault)	Parasites of some plant hoppers in Delphacidae of the order Hemiptera.
411	*Pseudoligosita brevicilia* (Girault)	Associated with rice plant.
412	*Pseudoligosita tachikawai* (Yashiro)	Parasites of *Nilaparvata lugens* and *Sogatella furcifera* in Delphacidae of the order Hemiptera.
413	*Trichogramma achaeae* Nagaraja and Nagarkatti	Parasites of *Atherigona soccata* (Diptera: Anthomyiidae), *Pectinophora* and *Sitotroga* species of moths in Gelechiidae, *Boarmia variegata* in Geometridae, many moths in Noctuidae, some moths in Pyralidae, *Clostera* species in Notodontidae, and butterflies like *Ergolis merione* in Nymphalidae, *Catopsilia* and *Eurema* species in Pieridae, etc. of the order Lepidoptera.
414	*Trichogramma agriae* Nagaraja	Parasites of *Corcyra cephalonica* in Pyralidae and *Agrius convolvuli* in Sphingidae of the order Lepidoptera.
415	*Trichogramma australicum* Girault	Parasites of *Atherigona soccata* (Diptera: Anthomyiidae), *Diacrisia oblique* in Arctiidae, some moths in Gelechiidae, Tortricidae, many in Noctuidae, Pyralidae, etc. of the order Lepidoptera.
416	*Trichogramma brevifringiata* Yousuf and Shafee	Parasites of the moth *Chilo infusecatellus* (Lepidoptera: Pyralidae).
417	*Trichogramma californicum* Nagaraja and Nagarkatti	Parasites of some butterflies in Lycaenidae,*Hemerocampa* and *Orgyia* species of moths in Lymantriidae and *Cydia* and *Epinotia* species in Tortricidae (Lepidoptera).

Sl. No.	Chalcid parasites	Host
418	*Trichogramma chilonis* Ishii	Parasites of *Hippotion celerio* in Sphindidae (Coleoptera), *Atherigona soccata* (Diptera: Anthomyiidae), some moths in Arctiidae, Gelechiidae, Lymantriidae, many in Noctuidae, Pyralidae, Tortricidae, etc. and also butterflies in Lycaenidae, Papilionidae, Pieridae, etc. and *Chrysopa* species of green lacewings (Neuroptera: Chrysopidae).
419	*Trichogramma chilotraeae* Nagaraja and Nagarkatti	Parasites of *Phragmataecia gummata* in Cossidae, *Pelopidas mathias* in Hesperiidae, *Deudorix Isocrates* in Lycaenidae, some moths in Noctuidae, many in Pyralidae, *Agrius convolvuli* in Sphingidae and *Bactra* and *Tetramoera* species in Tortricidae of the order Lepidoptera.
420	*Trichogramma convolvuli* Nagaraja	Parasites of *Corcyra cephalonica* in Pyralidae and *Agrius convolvuli* in Sphingidae of the order Lepidoptera.
421	*Trichogramma cuttackense* Nagaraja	Parasites of *Psalis* species (Lepidoptera: Lymantriidae).
422	*Trichogramma danausicida* Nagaraja	Parasites of *Danaus chrysippus* in Danaidae and *Corcyra cephalonica* in Pyralidae of the order Lepidoptera.
423	*Trichogramma dendrolimi* Matsumura	Parasites of *Deporaus betulae* (Coleoptera: Rhynchitidiae), *Acantholyda posticalis* (Hymenoptera: Pamphiliidae), *Hyphantria cunea* in Arctiidae, many moths in Lasiocampidae, Noctuidae, Pyralidae and Tortricidae, some moths in Limacodidae, Lymantriidae, Notodontidae, *Inacbis io* and *Inachis io* in Nymphalidae, *Opisina arenosella* in Oecophoridae, *Papilio* species in Papilionidae, *Pieris* species in Pieridae, etc. in Lepidoptera.
424	*Trichogramma embryophagum* (Hartig)	Parasites of some saw flies in Pamphiliidae, *Nematus melanaspis* in Tenthredinidae (Hymenoptera), *Pectinophora* and *Sitotroga* species in Gelechiidae, some moths in Geometridae, Lasiocampidae, Lymantridae, many moths in Noctuidae, Pyralidae, Tortricidae, etc. of the order Lepidoptera.
425	*Trichogramma evanescens* Westwood	Parasites of *Bruchus* obtectus in Bruchidae, some beetles in Chrysomelidae, *Rhynchaenus* species of weevils Curculionidae, *Triboleum castaneum* in Tenebrionidae (Coleoptera), some muscoidea flies in Anthomyiidae, flower flies in Syrphidae, horse flies in Tabanidae (Diptera), *Cimex* species in Cimicidae, some sawflies in Pamphilildae, Tenthredinidae, some moths in Arctiidae, *Rondotia menciana*, *Danaus chrysippus* in Danaidae, some moths in Gelechiidae, some moths in Geometridae, Lasiocampidae, Notodontidae, some butterflies in Lycaenidae, many moths in Lymantriidae, Noctuidae, Tortricidae, etc. and *Sialis lutaria* (Megaloptera: Sialidae) and some green lacewings (Neuroptera: Chrysopidae).
426	*Trichogramma exiguum* Pinto and Platner	Parasites of some moths in Gelechiidae, *Deudorix* and *Strymon* species of butterflies in Lycaenidae, *Euproctis lunata* in Lymantriidae, many moths in Noctuidae and Pyralidae, some in Sphingidae,Tortricidae, etc. of the order Lepidoptera.
427	*Trichogramma fasciatum* (Perkins)	Parasites of *Gnorimoschema* and *Sitotroga* species in Gelechiidae, *Oxydia* in Geometridae, *Lymantria* and *Nygmia* species in Lymantriidae, many moths in Noctuidae and Pyralidae, *Erinnyis* species in Sphingidae and *Acleris* and *Carpocapsa* species in Tortricidae of the order Lepidoptera.
428	*Trichogramma flandersi* Nagaraja and Nagarkatti	Parasites of some moths in Pyralidae, and *Agrius convolvuli* in Sphingidae of the order Lepidoptera.
429	*Trichogramma hebbalensis* Nagaraja	Parasites of *Corcyra cephalonica* in Pyralidae of the order Lepidoptera.

(*Continued*)

Sl. No.	Chalcid parasites	Host
430	*Trichogramma hesperidis* Nagaraja	Parasites of *Pelopidas mathias* in Hesperiidae and *Corcyra cephalonica* in Pyralidae of the order Lepidoptera.
431	*Trichogramma japonicum* Ashmead	Parasites of *Dicladispa armigera* (Coleoptera: Chrysomelidae), some shore flies in Ephydridae, marsh flies in Sciomyzidae (Diptera), some moths in the families Noctuidae, Gelechiidae, and many in Pyralidae, some butterflies in Hesperiidae, *Eurema* species in Pieridae, etc. of the order Lepidoptera.
432	*Trichogramma kashmiricum* Nagaraja, Ahmad, and Gupta	Parasites of Sciomyzidae unspecified species in Diptera.
433	*Trichogramma manii* Nagaraja and Gupta	Parasites of *Deudorix isocrates* (Lepidoptera: Lycaenidae).
434	*Trichogramma minutum* Riley	Parasites of *Xylotrechus* species in Cerambycidae, some beetles in Chrysomelidae, *Xyleborus* species in Scolytidae, *Hippotion* species in Sphindidae (Coleoptera) and some flies in the families Agromyzidae, Anthomyiidae, Tabanidae (Diptera), hoppers and scale insects in the families Aleyrodidae, Cicadellidae, Diaspididae, Kerriidae, Miridae, sawflies in Cimbicidae, Pamphiliidae and Tenthredinidae (Hymenoptera) and also some moths in Arctiidae, Coleophoridae, Gelechiidae, Geometridae, many in Noctuidae, Lymantriidae, Tortricidae, and butterflies in Pieridae, Nymphalidae, etc. of the order Lepidoptera. Parasitoids of *Bracon gelechiae* (Hymenoptera: Braconidae).
435	*Trichogramma pallidiventris* Nagaraja	Parasites of *Corcyra cephalonica, Scirpophaga incertulas* and *Tryporyza incertulas* (Lepidoptera: Pyralidae).
436	*Trichogramma perkinsi* Girault	Parasites of some moths in Noctuidae, Arctiidae, Gelechiidae, Lymantriidae, Pyralidae, Tortricidae, and the butterflies *Deudorix epijarbas* (Lycaenidae), *Vanessa tammeamea* (Nymphalidae) of the order Lepidoptera.
437	*Trichogramma plasseyense* Nagaraja	Parasites of *Helicoverpa armigera* in Noctuidae and some other moths in Pyralidae of the order Lepidoptera.
438	*Trichogramma poliae* Nagaraja	Parasites of Lepidoptera unspecified species, *Clostera cupreata* in Notodontidae, some other moths in Pyralidae, and *Plutella xylostella* in Yponomeutidae of the order Lepidoptera.
439	*Trichogramma pretiosum* Riley	Parasites of *Anthonomus grandis* (Coleoptera: Curclionidae) some bugs in Anthocoridae and Lygaeidae families, moths in the families, Arctiidae, Gelechiidae, Geometridae, Noctuidae, etc. and butterflies in Nymphalidae, Lycaenidae, Papilionidae, Pieridae, etc. and many green lacewings (Neuroptera: Chrysopidae).
440	*Trichogramma raoi* Nagaraja	Parasites of *Corcyra cephalonica* in Pyralidae, *Eurema* species in Pieridae, *Achaea janata* in Noctuidae, etc. of the order Lepidoptera.
441	*Trichogramma sankarani* Nagaraja	Parasites of *Corcyra cephalonica* in Pyralidae and *Agrius convolvuli* in Sphingidae of the order Lepidoptera.
442	*Trichogramma semblidis* (Aurivillius)	Parasites of *Rhynchites* species in Rhynchitidae, some beetles in Scolytidae, many flies in Tabanidae, some sawflies in Pamphiliidae, moths in the families Pyralidae, Noctuidae, etc. of the order Lepidoptera.
443	*Trichogramma semifumatum* (Perkins)	Parasites of *Oediopalpa guerini* (Coleoptera: Chrysomelidae), *Estigmene acraea* in Arctiidae, some moths in the families Pyralidae, Gelechiidae, and many in Noctuidae, some butterflies in Nymphalidae, Pieridae, etc. of the order Lepidoptera.
444	*Trichogrammatoidea armigera* Manjunath	Parasites of some moths in Noctuidae, Pyralidae, *Opisina arenosella* in Oecophoridae and Lampides boeticus in Lycaenidae of the order Lepidoptera.

Sl. No.	Chalcid parasites	Host
445	*Trichogrammatoidea bactrae* Nagaraja	Parasites of *Atherigona soccata* (Diptera: Anthomyiidae), *Pectinophora* and *Sitotroga* species of moths in Gelechiidae, many moths in Noctuidae, Pyralidae, etc. of the order Lepidoptera.
446	*Trichogrammatoidea brasiliensis* (Ashmead)	Parasites of *Amsacta moorei* in Arctiidae, some moths in Gelechiidae, Pyralidae and many in Noctuidae etc. of the order Lepidoptera.
447	*Trichogrammatoidea eldanae* Viggiani	Parasites of *Helicoverpa armigera* and *Spodoptera mauritia* in Noctuidae and some other moths in Pyralidae of the order Lepidoptera.
448	*Trichogrammatoidea fulva* Nagaraja	Parasites of *Cassida* species of beetles (Coleoptera: Chrysomelidae), *Cryptophelbia* species of moths in Tortricidae and *Corcyra cephalonica* in Pyralidae of the order Lepidoptera.
449	*Trichogrammatoidea fumata* Nagaraja	Parasites of *Corcyra cephalonica* in Pyralidae, some moths in Tortricidae, etc. of the order Lepidoptera.
450	*Trichogrammatoidea lutea* Girault	Parasites of *Sitotroga cerealella* in Gelechiidae, many species of moths in Noctuidae, Pyralidae, and Tortricidae of the order Lepidoptera.
451	*Trichogrammatoidea nana* (Zehntner)	Parasites of *Brontispa* species of beetles (Coleoptera: Chrysomelidae), *Sepedon sauteri* in Sciomyzidae (Diptera), *Kerria lacca* in Kerriidae (Hemiptera), many species of moths in Pyralidae, and Tortricidae, some in Noctuidae, etc. of the order Lepidoptera.
452	*Trichogrammatoidea robusta* Nagaraja	Parasites of *Heliothis* species of moths in Noctuidae and some other moths in Pyralidae of the order Lepidoptera.
453	*Tumidiclava tenkasiensis* Yousuf and Shafee	Associated with rice plant.
454	*Uscana mukerjii* (Mani)	Parasites of many beetles (Coleoptera: Bruchidae).

Some chalcid wasps.

Brachymeria bengalensis (Cameron), female.

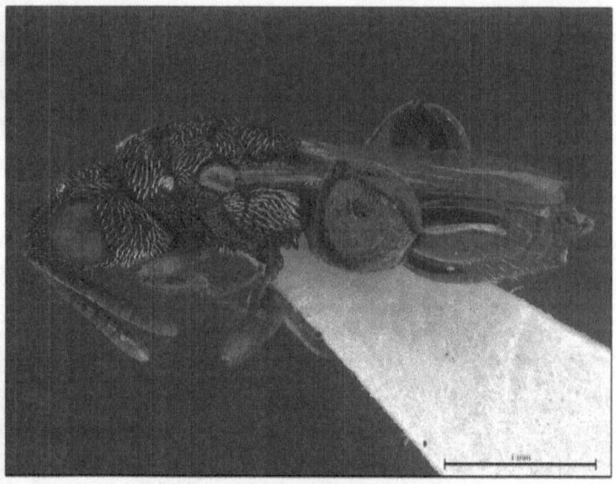

Dirhinus anthracia Walker, female.

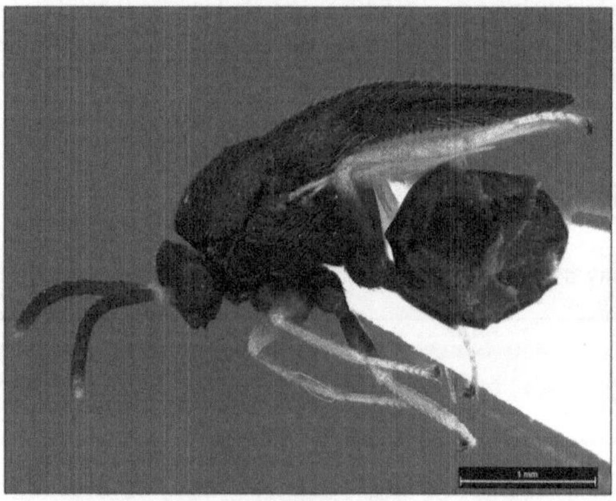

Cherianella narayani Narendran, female.

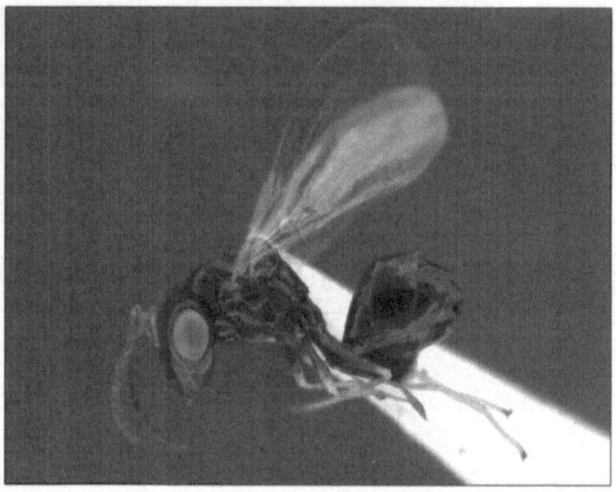

Neolosbanus palgravei (Girault), female.

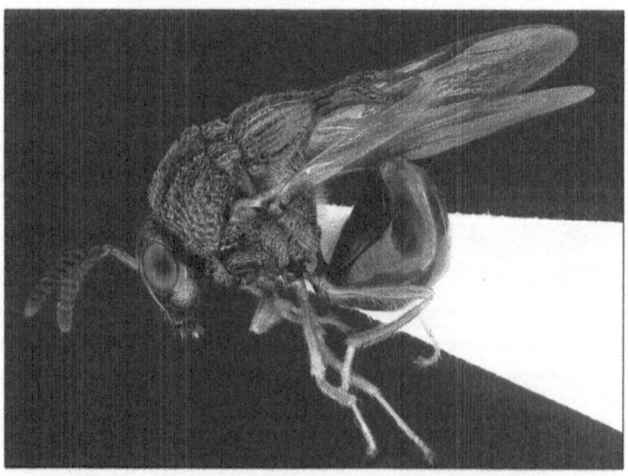

Schizaspidia brevifuniculata Narendran, female.

Diglyphomorphomyia ebifurcata Narendran, female.

Some chalcid wasps.

Elachertus zellus Narendran, female.

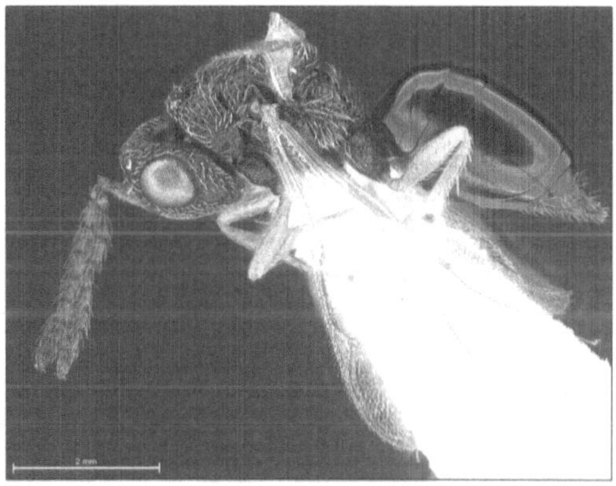

Eurytoma porensis Mukerjee, female.

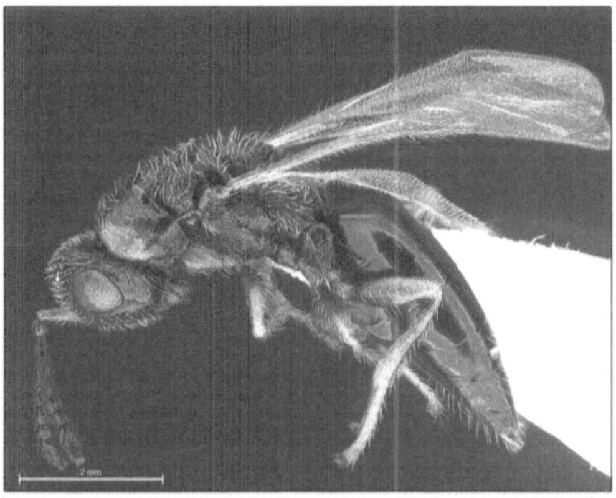

Tetramesa vadana Narendran, female.

Leucospis histrio Maindron, female.

Leucospis pediculata Guérin-Méneville, female.

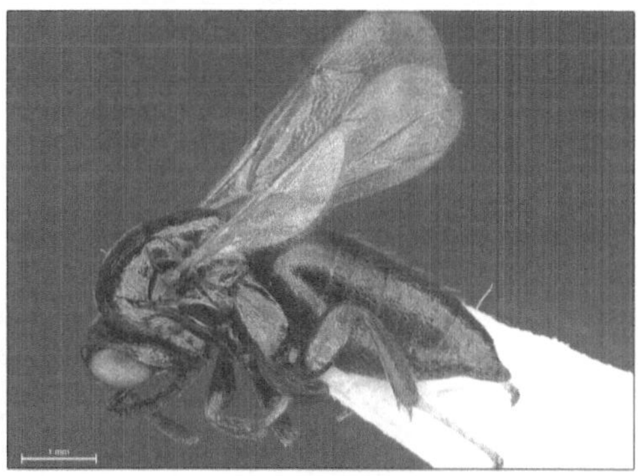

Ormyrus langlandi Girault, female.

Some chalcid wasps.

Anisopteromalus calandrae (Howard), female.

Cerocephala dinoderi Gahan, female.

Netomocera nigra Sureshan & Narendran, female.

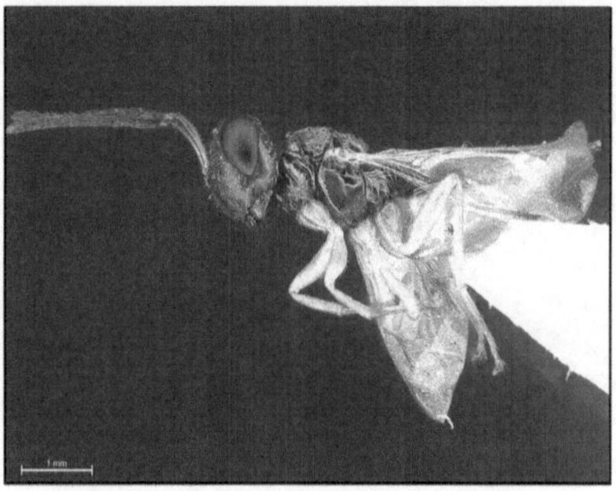

Propicroscytus oryzae Subba Rao, female.

Trichomalopsis apanteloctena (Crawford), female.

Torymoides amabilis Walker, female.

8 Eulophid Parasitoids

Meena Agnihotri and J. M. Samraj
Department of Entomology, G B Pant University of Agriculture and Technology, Pantnagar, Uttarakhand

Arvind Kumar
Forest Entomology Division, Forest Research Institute, Dehradun, Uttarakhand

CONTENTS

LEARNING OBJECTIVES

1. To learn the biology and diversity of eulophid parasitoids and their host range.
2. To understand the status of euophoid research in India and in the World.
3. To recognize the potential of eulophids in the biocontrol of insect pests.
4. To learn mass production techniques of commercially used eulophid parasitoids.
5. To know the successful biocontrol agents used in biocontrol programmes.
6. To emphasise the research gap in exploring eulophids for agricultural pest management.

DOI: 10.1201/9781003354239-8

8.1 INTRODUCTION

With the sharp rise in the need for food crops furthered by the growing human population, agriculture in India is being intensified. One of the major effects of intensive cropping is the use of chemicals in pest management. In India, around 272 chemicals are registered as pesticides and are readily available in the market (Mooventhan et al., 2020). Indiscriminate use of pesticides causes pollution of natural resources, residue in plant parts, the resurgence of pests, resistance to insecticides and health complications for humans and wildlife. These ill effects drive us to explore eco-friendly modes of pest management. The use of hymenopteran parasitoids as biocontrol agents in pest management is one of the promising and sustainable tactics applied in controlling crop pests. Hymenoptera is one of the largest orders of insects. Approximately 130,000 species have been described in Hymenoptera so far and it is estimated that anywhere from 300,000 to 2.5 million species may be present (Everson, 2014). If the numbers are extrapolated, Hymenoptera would form 10% of the entire life on the planet (Sharkey, 2007). Parasitic Hymenoptera probably accounts for between 6.5% and 20% of insect species, and a range of between 170,000 and 6 million species of parasitic Hymenoptera has been proposed based on various estimates of the number of insect species. Parasitic Hymenoptera in a given ecosystem indirectly represents the diversity of the hosts they attack. Hence, they act as good bioindicators of environmental health (Sharkey, 2007). Parasitoid hymenopterans are unique bioindicators because they are representative of the diversity of the hosts that they attack, and these constitute a great part of the diversity of all arthropods. The order Hymenoptera is especially rich in families exhibiting the entomophagous habit and most of the parasitic species utilised in biological control belong to this order. Among the parasitic Hymenoptera, wasps in the superfamily Chalcidoidea are ecologically and economically the most important insects for the biocontrol of other insect populations and they represent 33% of the entire parasitic Hymenoptera (LaSalle, 1993). Noyes (2003) estimated that there could be 500,000 chalcidoid species present worldwide. Eulophidae is a large family of the superfamily Chalcidoidea comprising about 300 genera and more than 3000 species and their number is constantly increasing. Knowledge of eulophid fauna is rather scanty considering the vastness, diverse climatic conditions and vegetation of the subcontinent. Being very small in size and microscopic, these insects usually undergo undetected and this remains the main cause that very little work has been done on them.

8.2 BASIC CHARACTERS FOR IDENTIFICATION

Eulophids are small to medium-sized chalcids, which can be easily recognised by the four segmented tarsi, antennae generally up to four funicular segments (ranging from two to four funicle segments), straight tibial spur of the foretibia, and non-pedunculate wings. Mesoscutum with or without well-marked notauli; axilla often produced forward of the scuto-scutellar suture. Metasoma is constricted basally, not broadly attached to mesosoma (Plate 8.1; Fig. 1–12). Eulophids are often soft-bodied and may shrivel and collapse after death; colour varying from metallic to black to yellow.

8.3 DIVERSITY AND BIOLOGY

Eulophidae comprises nearly 6000 species worldwide in 324 genera grouped under 4 subfamilies, namely Entedoninae, Euderinae, Eulophinae and Tetrastichinae (Kosheleva, 2019). Eulophidae exhibits a great range of lifeways and are of immense importance in biocontrol as having a great suppressive effect on pest populations. Most eulophids are larval parasitoids but some species are egg and pupal parasitoids. They are host specific and habitually attack a variety of hosts, either as ectoparasitoids, endoparasitoids or facultative hyperparasitoids. Eulophids attack a wide range of

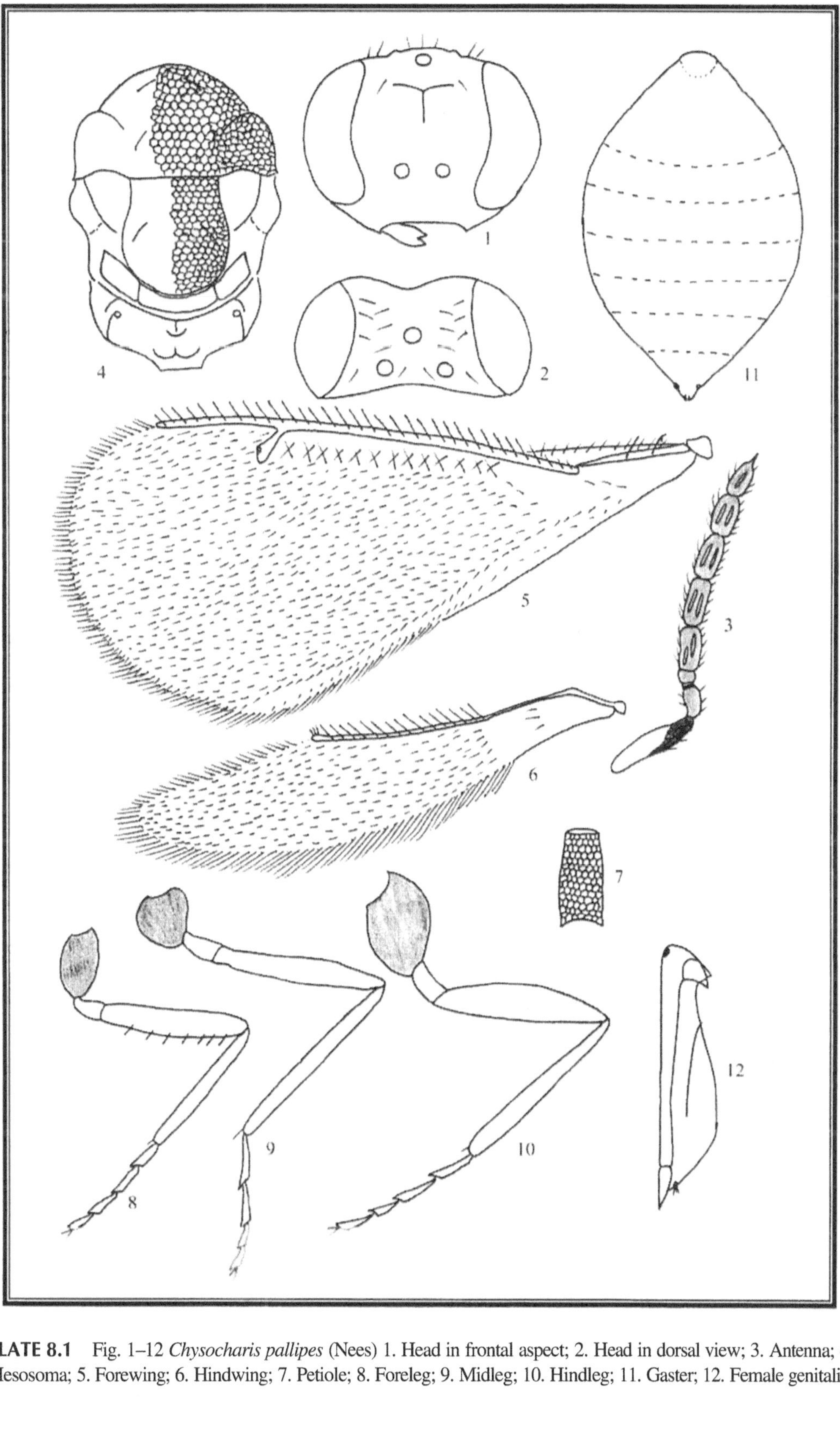

PLATE 8.1 Fig. 1–12 *Chysocharis pallipes* (Nees) 1. Head in frontal aspect; 2. Head in dorsal view; 3. Antenna; 4. Mesosoma; 5. Forewing; 6. Hindwing; 7. Petiole; 8. Foreleg; 9. Midleg; 10. Hindleg; 11. Gaster; 12. Female genitalia.

hosts, *viz.* Lepidoptera, Diptera, Coleoptera, Hemiptera (Gadallah et al., 2015). Host spectrum may expand on further studies. Although most species of the Eulophidae are parasitoids, some are phytophagous and predatory species. Several species are parasitoids of aquatic insects. Some Eulophids are parasitoids of spiders. Two species of eulophid are known to be phoretic and one species acts as a pollinator of orchids (Gauthier et al., 2000). Eulophids have provided a series of extremely valuable experimental model systems for investigating a wide variety of questions in ecology and evolution (Godfray and Godfray, 1994). A total of 272 species have been reported in the subfamily Tetrastichinae from India (Narendran, 2007). Similarly, 251 species have been reported from the subfamily Eulophinae. Subfamilies of Eulophidae have interesting biology. Eulophines are solitary or gregarious, idiobiont ectoparasitoids of the larvae, or sometimes the pupae of leaf-miners or similarly concealed hosts, such as leaf-folders, case-bearers, gall-makers and stem-borers. Entedonines are principally solitary or gregarious primary or secondary endoparasites of concealed eggs, larvae or pupae. Many of the genera have well-defined host ranges. For example, most *Entedon* species attack beetles; *Chrysocharis* species parasitize lepidopterous, dipterous and hymenopterous leaf miners (Figure 8.1). Tetrastichines are usually primary endoparasitoids of the eggs, larvae or pupae of Diptera, Hymenoptera or Lepidoptera, through a specific association with small gallicolous hosts. Some even attack eriophyid mites or gall-forming nematodes. Several species develop as ectoparasitoids or as facultative or obligate hyperparasitoids (Noyes, 2003). Others feed as predators on the eggs of various insects. Some species are primary ectoparasitoids of concealed larvae of Lepidoptera or Coleoptera. The blue gum chalcid Leptocybe invasa is an invasive pest species on *Euclayptus* (Zheng et al., 2014).

Eulophids are very promising agents for the biocontrol of insect pests of agricultural importance. Even though they are usually of minute size, eulophids have a great suppressive effect on pest populations. Eulophids (Citrostichus phyllocnistoides, Closterocercus citrella and Cirrospilus ingenuus, are parasitoids of Citrus leaf miner, Phyllocnistis citrella) have been successfully utilised for the biocontrol of many pests of agricultural importance the world over Figure 8.1, Figure 8.2 (Figure 8.3).

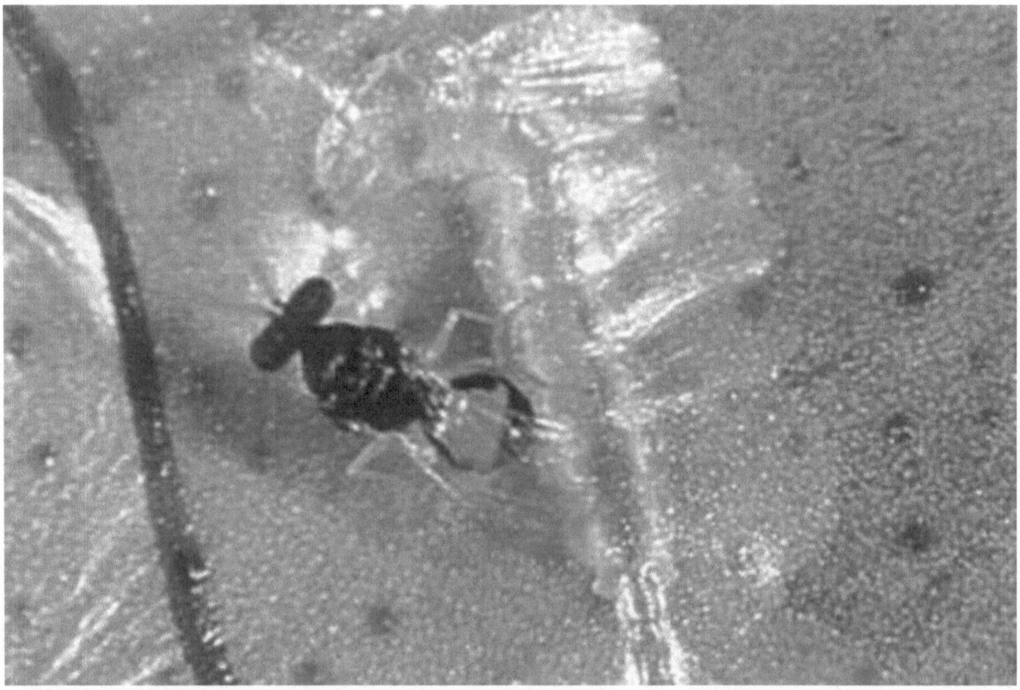

FIGURE 8.1 *Citrostichus phyllocnistoides* parasitizing Citrus leafminer *Phyllocnisitis citrella.*

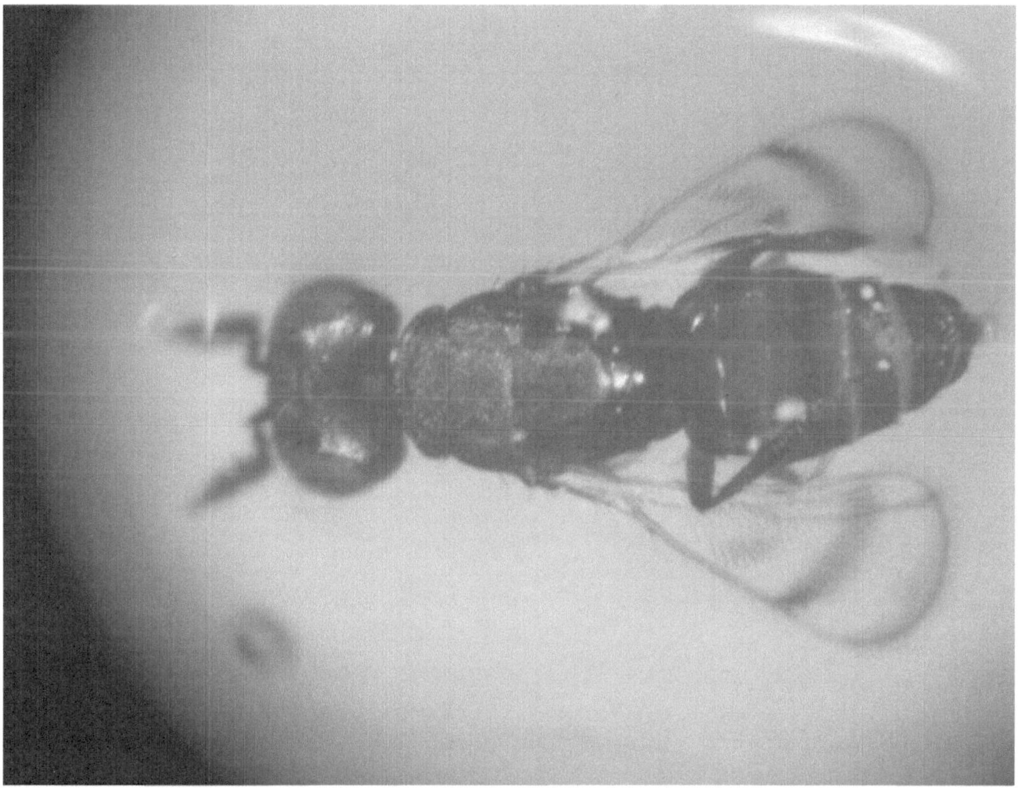

FIGURE 8.2 *Closterocercus citrella*, a parasitoid of Citrus leaf miner, *Phyllocnistis citrella*.

8.4 HISTORY OF EULOPHIDAE RESEARCH

8.4.1 Taxonomic Position of Eulophidae

The family Eulophidaewas established based on the genus Eulophus in the year 1791 (Olivier, 1791). Several notable works in the second half of 19th century played a key role in the euophid taxonomy. Eulophid parasitoids were initially placed under the subfamily Eulophidae of the family Chalcididae (Westwood, 1840). Genera of eulophids were then proposed as families, *viz*. Elachistoidea, Eulophidae, Entedonoidea and Tetrastichoidea (Forster, 1856), which were later renamed as Elachistidae, Entedonidae, Eulophidae and Tetrastichidae (Walker, 1872) and further demoted to the rank of tribes as Elachistina, Eulophina, Entedonina and Tetrastichina (Thomson, 1878).

Elachistinae, Eulophinae, Entedoninae and Tetrastichinaewere are recognised as subfamilies of Chalcididae (Ashmead, 1894). Later, the family status of Eulophidaewas revived and included Elachertinae (= Elachistoidae Forster), Eulophinae, Entedoninae, Tetrastichinae and Aphelininae as its subfamilies (Ashmead, 1904). Further, the subfamilies were divided into tribes *viz*., 1. Elachertinae (Euplectrini, Ophelinini and Elachertini), 2.Eulophinae (Eulophini and Hemiptarsenini), 3. Entedoninae (Tetracampini, Entedonini, Omphalini and Pediobini), 4. Tetrastichinae (Tetrastichini and Ceratoneurini) and 5. Aphelininae (Aphelinini and Pteroptricini). Later, the subfamily Aphelininae was raised to Aphelinidae (Vireck, 1916). Erdos (1944) recognized four subfamilies, *viz*. Elacheritinae, Eulophinae, Entedontinae and Tetrastichinae. Whereas, three subfamilies, *viz*. Elachertinae, Entedoninae and Eulophinae were recognized by Risbec (1951). The subfamily Entedoninae was divided into two tribes: Tetracampini and Omphalina, and the subfamilies Elacheritinae and Eulophinae remain undivided. In the following year, 1952, Risbec again

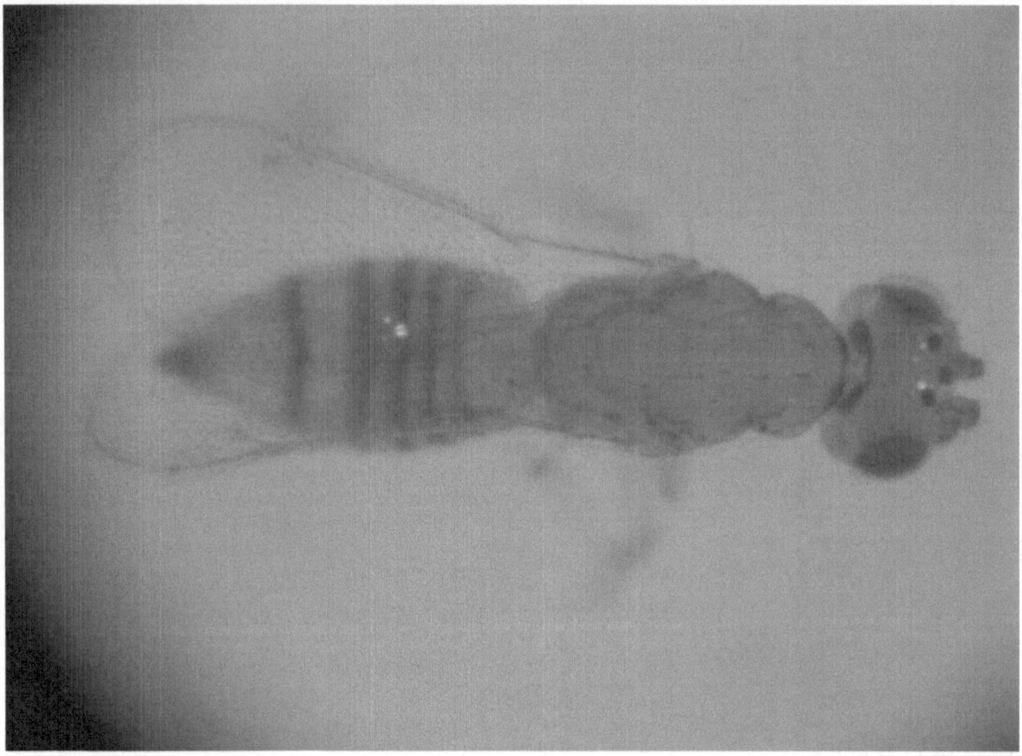

FIGURE 8.3 *Cirrospilus ingenuus* (Gahan).

demoted the subfamilies and tribes (excluding aphelinids) as recognised by Ashmead (1904) under the family Eulophidae to the ranks of tribes and subtribes, respectively. Tribe Euderini was raised to the rank of subfamily Euderinae (Graham, 1959).

The taxonomic importance of the morphological structures, especially the abdominal characters were significantly highlighted in dividing the subfamilies and genera (Domenichini, 1953; Erdos, 1944). Excellent illustrations on the male genitalia of eulophid parasitoids were provided by Miller (1962, 1964) but, he did not make any attempt to show their taxonomic significance. The widely accepted classification of Eulophidae is based on the revision of Eulophidae by Gauthier et al., (2000), in which the family is divided into four subfamilies, Entedoninae, Euderinae, Eulophinae and Tetrastichinae, all of which are cosmopolitan in distribution. Five subfamilies, *viz*. Entiinae, Eulophinae, Entedoninae, Opheliminae and Tetrastichinae were recognized by Noyes (2003).

8.4.2 Studies on Euophid Fauna of the World

Eulophid fauna of Asiatic region (Ferriere, 1930, 1931b), African region (1931a, 1936, 1938), Oriental and Australian region (Ferriere, 1940a, 1940b), and Europe (Ferriere, 1941), were studied in the early decades of the 20th century. An outstanding contribution to the chalcid fauna of Russia was given by Nikol'skaya (1952). In the later years, the British and other European fauna of Eulophidae were also explored (Boucek, 1958; 1959a; 1959b; Graham, 1959).

Eulophid fauna of Japan was documented by Kamijo (1977a; 1977b; 1977c; 1979). Similarly, the Korean fauna of Eulophidae was studied by Ikeda (1996a, 1996b). Note that 471 species of eulophid parasitoids under 79 genera were recorded from Neartic region by Peck (1963). Key to the genera of Australian Eulophidae was provided under three subfamilies (Boucek, 1988).

Hansson (1990, 1994, 1996) made excellent studies on the eulophid fauna of the Palearctic and Nearctic regions. LaSalle and Schauff (1990) gave new subfamily placement for some North American eulophids and they (1992) also made excellent preliminary studies on Neotropical eulophid fauna based on Ashmead, Cameron, Howard and Walker species. Taxonomic work on the family Eulophidae is poorly known from Sri Lanka and in this direction Wijesekara and Schauff have made a good attempt in 1994 and 1997. In the year 1994, Wijesekara and Schauff revised the genera and species of the tribe Euplectrini (Eulophidae) from Sri Lanka and recognized the following genera, *viz. Euplectrus*, *Euplectromorpha*, *Metaplectrus* and *Platyplectrus*.

Several individual genera, such as *Euderus* in America and North of Mexico, *Chrysocharis* in North America and Canada (Yoshimoto, 1971), Neartic species of *Elachertus* (Schauff, 1985a; 1985b), *Neochrysocharis* and *Chrysonotomyia* (Hansson, 1990) of Paleartic region, *Sigmophora* and *Quadrastichodella* of the world (Ikeda, 1999) were studied and documented. Schauff (1985a, 1985b, 1987, 1988, 1991) made excellent and exhaustive studies on the world fauna of Eulophidae. Later, Schauff and LaSalle (1993) and Schauff et al. (1998) added some more remarkable contributions to the global fauna. Contemporarily, LaSalle (1989, 1990a, 1990b, 1994) has made a valuable contribution to the knowledge of the eulophid fauna of the world.

8.4.3 Eulophid Research in India

The first consolidated work on Indian chalcids was done by Mani in 1938, who published the catalogue on Chalcidoidea of India, Ceylon and Burma, in which he covered 67 species under 29 genera of the family Eulophidae. Besides Mani, the fauna of the Indian subcontinent was studied by various authors in the following decades. Cameron (1913) studied the hymenopterans in the George Town museum. Chalcidoidwasps from Coimbatore were documented by Rohwer (1921). Cherian and Margabundhu (1942) described a new species of *Trichospilus*, *T. diatraea* host reared from the pupae of *Diatreavenosata*, a crambid moth. Studies on host relationships were also done by Subba Rao and Sharma (1966) on gelechid moths and Dubey (1974) on leaf miners. Description of new species and records of known chalcidoids from India was done by Bhatnagar (1952), Kurian (1953, 1954), Narayanan (1960) and Kaul and Saraswat (1974). The genus *Tetrastichus* from India was studied by Saraswat (1975, 1978). Several remarkable studies on Indian eulophidae were done by Khan and Shafee (1980a, 1980b, 1981a, 1981b, 1981c, 1982) and Shafee et al. (1983a). New species were described from Andaman Islands by Shafee et al. (1983b) and several Tetrastrichine species were described by Shafee and Rizvi (1984, 1985). An informative and working key to the genera of the family Eulophidae of India and adjacent countries was given by Hayat (1985).

Hussain and Khan (1986) have elaborately contributed to our knowledge of Eulophidae of India, Pakistan, Nepal, Burma, Bangaladesh and Sri Lanka, through a catalogue covering 46 genera. They have also cited a list of synonymy and erected a number of new combinations under each genus. Narendran and Krishna Kumar (1995) studied *Ceratoneuraindi*, a eulophid pest of Brinjal and Sweet Pepper. Various contributions, such as by Singh and Khan (1994a, 1994b, 1995, 1996, 1997, 1998) on the genus *Omphale*, *Mestocharella* and *Closterocerus*, Surekha and LaSalle (1996) on *Sympiesis* a parasitoid of Teak defoliator *Hyblaeapuera*, Sushil and Khan (1997) on a new genus *Shardiella*, Khan (1982, 1985a, 1985b, 1992, 1993, 1995, 1996) on some Tetrastrichines and Eulophines fostered the eulophid research in the country. Studies by Khan et al. (1986), Khan and Singh (1994), Khan and Sushil (1992, 1998), Narendran et al. (2001, 2002a), Sureshan and Narendran (2001), Narendran and Sinu (2003), Khan et al. (2005) also made remarkable contributions to the eulophid fauna of India by way of describing new species, a taxonomic revision of genus and subfamilies.

Studies on Indian eulophidae are gaining momentum over the last two decades. Several recent and noteworthy studies are briefly discussed here. Twenty-five species of Eulophidae were described from India (Hayat and Shahi, 2004). Narendran et al. (2005) described a new species in the subfamily Tetrastichinae *Tamaraxia sheebae* from Kerala on the host reared from leaf galls of

Terminalia arjuna. Narendran and Sinu (2003) described a new genus *Anumanniola* by monotypy with a description of a new species *A. lasalle*. Narendran et al. (2005) described a new genus *Kiggaella* Narendran from rice fields of Karnataka with a description of a new species *K. oryzae*. Narendran and Hayat (2007) described a new species of eulophid parasitoid *Euderomphale stomale* from Hoskote in Karnataka. A key to the Indian species of the genus Euderomphale has been given. Narendran (2007) published a monograph on the Tetrastichinae subfamily. The voluminous work compiled 34 genera and 272 species in detail. Notes on biology, distribution, host plants and behavior of the tetrastichines were discussed. A detailed pictorial key for the Indian genera and the species has been given in the study. It included the description of several new species. Each species carries a detailed description along with an illustration of taxonomically important morphological characters. Narendran et al. (2007b) described two new species of the genus *Deutereulophus, viz. D. iunctatus* Narendran and *D. malabarensis* Narendran, from India. Narendran et al. (2007) revised the genus *Pediobomyia*. The genus *Bomyiabius* was synonymised with *Pediobomyia* along with a description of two new species. Khan and Hayat (2011) discussed the collection of *Elasmus* from various states of the country. A total of 368 species were collected, of which, 293 were identified in the study. Hayat et al. (2004) in their second contribution to the taxonomic status of the Indian Eulophidae dealt with the types of 26 species belonging to the Eulophinae (8 species), Entedoninae (15 species) and Euderinae (3 species) along with several new combinations.

Narendran et al. (2013) reviewed the genus *Chrysocharis* Foerster of the Indian subcontinent. A total of 13 species were described in detail including the description of several new species. Pant et al. (2019) identified *Elasmus brevicornis* Gahan as a parasitoid of Banana Skipper, *Erionota torus* for the first time. Kazmi and Kumar (2013) studied the genus *Elasmus* from West Bengal. Fifteen species of *Elasmus* were recorded from West Bengal of which, six were previously unreported. Budhachandra et al. (2013) described a new species *Pediobius clavatruncus* from Uttarakhand along with the first report of *Ceranisus menes* (Walker) and *Sympiesis hyblaeae* Surekha from the state. Gupta et al. (2015a) described two new species of Eulophid parasitoids, *Tetrastichuthetisae* Gupta (Hymenoptera: Eulophidae) reared from the pupa of *Curetisthetis* (Drury) (Lepidoptera: Lycaenidae) and *Sympiesis thyrsisae* Gupta (Hymenoptera: Eulophidae) reared from the caterpillar of *Gangarathyrsis* (Fabricius) (Lepidoptera: Hesperiidae). Shree and Singh (2015) studied the diversity of eulophid fauna from Doon valley, Uttarakhand. The study recorded a total of 45 species from three different ecosystems, *viz*. forest, agriculture and horticulture.

Parshuram and Agnihotri (2017) described a new species *Chrysonotomyia ricini*, host reared from the pupae of *Liriomyza trifolii* (Agromyzidae: Diptera), a leaf-mining fly attacking the leaf of Castor *Ricinus communis*. Yousuf et al. (2017) gave an overview of the infestation of *Eucalyptus* gall wasp from Northern India. The study also included an attempt for the biocontrol of *Eucalyptus* gall wasp wherein, continuous application of biocontrol agents for 4 years resulted in a reduction of infestation below 10%. Jamali et al. (2017) described a new species in the genus *Anumanniola* named *A. narendrani* from India. A key to all the species and additional characters for the diagnosis of *Anumanniola* were given. Jamali et al. (2018) described a new species of the subfamily Entedoninae, namely *Pediobius hebbalensis* from Karnataka, a state in India. The parasitoid was reared from the egg sacs of the spider. Kazmi and Sheela (2018) presented an annotated checklist of *Elasmus* from the Indian region with a total of 58 species of *Elasmus* Westwood (1834) given together with the Indian records in different states and Union territories.

Singh (2018) described a new species of *Aprostocetus doonensis* Singh from northern India. The new species is a parasitoid of mango leaf gall midge, *Procontarinia mangiferae* (Felt) (Diptera: Cecidomyiidae). Some observations on biology and parasitization rate are also given in the report. Rao et al. (2019) tested the field efficacy of eulophid parasitoid *Pediobius imbrio* on the coconut slug caterpillar, *Macroplectra nararia* (Lepidoptera: Limacodidae). The study revealed the suitability of this parasitoid for laboratory rearing. Mass rearing and field release of the parasitoid yielded up to 50.31% of control of the pest within 30 days of release. Jamali (2020) reported the genus *Ionympha* for the first time from India and the Oriental region with the description of new species *I. lenis* from Uttar Pradesh. Khan (2020) studied 19 species of the genus *Elasmus* Westwood collected from different parts of India. Out of all the recorded species, ten species were recorded first time from the seven different states of India, such as Andaman Islands, Himachal Pradesh, Karnataka, Kerala, Odisha, Uttarakhand and Uttar Pradesh, amounting to 13 new state records. All the recorded species were presented in the study with host specificity and geographical distribution. Ekka et al. (2020) studied the distribution of *Elasmus* sp. in Chattisgarh state of India. Kumar et al. (2020) conducted a study to assess the diversity of eulophid fauna and their host-parasitoid complex on Arjuna tree (*Terminalia arjuna*). One species of the genus *Tamarixia* emerged from the irregular spherical leaf galls produced by a Psyllid (Triozidae). The parasitoid species is identified and redescribed as *Tamarixia sheebae* Narendran.

8.4.4 Exploration of Eulophids as Biocontrol Agents

A study entitled "The North American Parasitic wasps of the Genus *Tetrastichus*-A Contribution to Biological Control of Insect Pests" (Burks, 1943), drew the attention of many towards the potency of eulophid parasitoids as biocontrol agents. The study made very significant efforts towards the understanding of its species and host relationship. Twenty-three new species under the genus *Tetrastichus* were described for the first time. LaSalle and Schauff (1994) provided an exhaustive review of eulophid parasitoids of whiteflies, with comments on their systematic placement and gave a generic key. They placed all the eulophid whitefly parasitoids in the tribe Euderomphalini in the subfamily Entedoninae and described three new genera and the rest of the seven genera were divided into two distinct genetic groups, viz. the *Euderomphale* group including *Baeoentedon* Girault, *Euderomphale* Girault, *Neopomphale* gen. n.; and *Pomphale*Husainand the *Entedonone cremnus* groups including *Aleuroctonus* gen.nov., *Dasyompale* gen. nov. and *Entedononecremnus*Girault. LaSalle and Pena (1997) described *Galeopsomyia fausta* as fortuitous parasitoids of the gracillariid *Phyllocnistiscitrella,* which is widely distributed in the Neotropics, being known from Mexico and Puerto Rio to Argentina.

Triapitsyn and Headrick (1995) made valuable contributions on thrips attacking Eulophidae (subfamily Entedoninae) from the Nearctic region and described one new species, *Ceranisus loomansi.* Kamarudin et al. (1996) reared 18 species of hymenopterous parasitoids from oil palm bagworms, *Metisa plana* Walker and *Mahasena corbetti* Tams, out of which two eulophid parasitoids, *Sympiesis*sp. and *Tetrastichus*sp., were associated with these pests, while the rest of the eulophid parasitoids, *viz. Nesolynx flavipes*Ashmead, *Pediobius anomalius*(Gahan), *Pediobius imbrues* (Walker) and *Teleopterus* sp. were reared only from *M. plana* and *P. elasmi* Ashmead was associated with *Mahasena corbetti.* They also recorded the host range and their distribution in the Indo-Australasian region.

8.5 HOST PARASITOID INDEX OF EULOPHID PARASITOIDS IN INDIA

This is provided in Table 8.1.

8.6 MASS REARING OF IMPORTANT EULOPHID PARASITOIDS

Mass multiplication of a parasitoid is the primary step in establishing it as a promising biocontrol agent. Eulophids are being utilised as biocontrol agents of various agriculturally important pests as discussed under previous headings of this chapter.

8.6.1 Mass Multiplication of *Tamarixia Radiata*

Tamarixia radiata is an important species-specific parasitoid of the citrus psyllid, *Diaphorina citrii. Tamarixia radiata* is exclusively reared on the nymphs of *Diaphorina citri.* It involves three

TABLE 8.1
Host-Parasitoid Index of Eulophid Parasitoids in India

Parasitoids	Host	Distribution	Reference
Subfamily: Eulophinae			
Aroplectrus contheylae Narendran	*Contheyla rotunda* Hampson	India: Kerala, kayamkulan	Narendran et al., 2002a
Arunus indicus Singh and Khan	*Trioza falcheri* (leaf galls) Hemiptera: Psyllidae	India: Uttarakhand; Matkota	Singh and khan, 1997
Cirrospilus quadrisriata (Subba Rao and Ramamani)	Unidentified Lep.	India:Rajasthan: Udaipur	Subba Rao and Ramamani, 1965; Batra and Sandhu, 1981; Boucek, 1988; Hussain and Khan, 1986
Dicladocerus viggianii Khan	*Agromyza* sp. Diptera:Agmyzidae	India: Uttarakhand; Bhimtal	Khan, 1995
Dicladocerus antennalis Khan	*Phytobia humeralis* Diptera:Agromyzidae	India; Uttarakhand: Dehradun	Khan, 1995
Dicladocerus indicus Khan	*Tropicomyia vigneae* (Seguy) Diptera: Agromyzidae	India: Uttar Pradesh; Ramnagar	Khan, 1995
Dicladocerus liriomyza Khan	*Liriomyza compositella*(Malloch) Diptera: Agromyzidae	India: Uttarakhand: Kathgodam	Khan, 1995
Diglyphus horticola Khan	*Chromatomyia horticola* Diptera: Agromyzidae	India: Uttarakhand: Pantnagar	Khan, 1995
Diglyphus funicularis Khan	*Melanoagromyza obtusa*(Malloch) Diptera: Agromyzidae	India: Uttarakhand: Nainital: Jeolikot	Khan, 1995
Diglyphus mandibularis Khan	*Melanoagromyza obtuse* (Malloch) Diptera: Agromyzidae	India: Uttarakhand: Nainital: Jeolikot	Khan, 1995
Diglyphus indicus Arifa and Khan	*Pseudonapomyza asiatica* Diptera: Agromyzidae	India: Uttarakhand: Nainital: Kashipur	Arifa and Khan, 1991
Diglyphus frontolatus Arifa and Khan	*Liriomyza taraia* Garg Diptera: Agromyzidae	India: Uttar Pradesh: Lakhimpur Kheri, (Palia)	Arifa and Khan, 1991
Diglyphus kashmiricus Jamal et al.	*Chromotomyia horticola* (Goureu)	India: Kashmir	Ahmad et al., 2013
Dimmockia sp.	Unidentified	India : Karnataka	Kerrich, 1969; Hussain and Khan, 1986.

Euplectrus bussyi Crawford	Unidentified Lepidoptera	India: Delhi; Tamil Nadu	Kerrich, 1969; Hussain and Khan, 1986 Crawford, 1909; Ayyar, 1925–1927; Mani, 1938; Pruthi and Mani, 1940 Hussain and Khan, 1986
Euplectrus ceylonicus Howard (*E. insulanus* Crawford)	*Euproctis flava* Bremer *E. fraternal* Moore Lepidoptera: Lymantridae	India : Tamil Nadu	Howard and Ashmead, 1896 Crowford, 1911; Ferriere, 1941; Hussain and Khan, 1986.
Euplectrus coimbatorensis Ferriere	Unidentified caterpillar	India : Tamil Nadu	Hussain and Khan, 1986
Euplectrus euplexiae Rohwer	*Prospalta conducta* (Walker) Lepidoptera: Noctuidae *Selepa docilis* Butler Lepidoptera: Noctuidae *Amyna axis* (Guenee) Lepidoptera: Noctuidae *Chrysodeixis eriosoma* Doubleday Lepidoptera: Noctuidae	India: Tamil Nadu; Rajasthan.	Rohwer, 1921; Hussain and Khan, 1986; Santhosh et al., 2020.
Euplectrus gopimohani Mani	*Spodoptera exigua* Hubner Lepidoptera: Noctuidae Unidentified caterpillar	India: Uttar Pradesh	Mani, 1941; Khan and Shaffe, 1980a; Hussain and Khan, 1986.
Euplectrus himalayaensis Mani	Unidentified	India: Darjeeling	Mani, 1938; Hussain and Khan, 1986
Euplectrus indicus Ferriere	*Apanteles obliqua niger* Wilk Hymenoptera: Braconidae Unidentified Geometrid larvae	India: Uttarakhand: Dehradun	Hussain and khan, 1986
Euplectrus latifrons Shafee et al.	Unidentified	India: Andaman Nicobar Islands	Hussain and Khan, 1986
Euplectrus leucostomus Rohwer	*Achaea janata* Linnaeus Lepidoptera: Noctuidae *Trabala vishnou* Lefebure Lepidoptera: Lasiocampidae	India : Tamil Nadu	Rohwer, 1921; Cherian and Basheer, 1947; Hussain and Khan, 1986
Euplectrus materus Bhatnagar	*Eudocima fullonica* *E. materna*	India : Uttar Pradesh: Kanpur	Bhatnagar, 1952; Hussain and Khan, 1986
Euplectrus mathuri Bhatnagar	Agaristid larva	India: Uttarakhand: Dehradun	Bhatnagar, 1952; Hussain and Khan, 1986
Euplectrus nyctemerae Crawford	*Nyctemera latcincta* Lepidoptera: Arctiidae *Pelopidas mathias* Lepidoptera: Hesperiidae	India: Karnataka	Crawford, 1914; Mani, 1938; Hussain and Khan, 1986

(Continued)

TABLE 8.1 (Continued)
Host-Parasitoid Index of Eulophid Parasitoids in India

Parasitoids	Host	Distribution	Reference
Euplectrus parvulus Ferriere	*Ascotis sealenaria imparata* Walker Lepidoptera: Geometridae *Pyrilla* sp. *Tephrina disputaria* Guenee Lepidoptera: Geometridae *Elasmus sp.* Hymenoptera: Elasmidae *Tetrasticus sp.* Hymenoptera: Eulophidae	India: Uttar Pradesh : Allahabad; Uttarakhand: Dehradun	Chatterjee, 1941–1944; Cherian and Pillai, 1941; Mathur, 1942; Hussain and Khan, 1986
Euplectrus plecopterae Mani	*Plecopteran reflexa* Guenee Lepidoptera: Noctuidae	India: uttar Pradesh Allahabad, Uttrakhand Dehradoon	Mani, 1941; Hussain and khan, 1986
Euplectrus petiolatus Ferriere	*Unidentified larvae*	India :Uttrakhand :Dehradun	Hussainand Khan, 1986
Euplectrus spodoptera Bhatnagar	*Mythimna sp.* Lepidoptera: Noctudiae *Spodoptera Mauritia Boisduval* Lepidoptera: Noctudiae *Utethesiapluchella* Lepidoptera: Arctiidae	India :Tamilnadu ;Pakam , Travancore	Bhatnagar, 1952; Hussain and Khan, 1986
Elachertus piperi Narendran	*Liothrips karnyi* (Bagnall) Thysanoptera: Thripidae	India: Kerala	Narendran and Bindu, 2009
Elasmus brevicornis Gahan	*Biloba subsecivela*; *Cnaphalocrocis medinalis*; *Diaphania indica*; *Hapalia machaeralis* on *Tectona grandis*; *Lamprosema indicata*; *Lygropia quarternalis* defoliating *Helictares isora*; *Marasma suspicalis*; *Nausinoe geometralis*. (Lepidoptera); Braconidae: *Apanteles machaeralis*	India	Kazmi and Sheela, 2017
Euplectr usutethesia Mani and Kurian	*Utethesia pluchella* Lepidoptera: Arctiidae	India: Karnataka: Bangalore	Mani and Kurian, 1953; Hussain and Khan, 1986

Euplectromorpha natadae Kurian	*Darna nararia Moore* Lepidoptera: Limacodidae	India : Tati Paka	Kurian, 1954; Hussain and Khan, 1986
Euplectromorpha viridiceps Ferriere	*Dasychira sp.* Lepidoptera: Lymantriidae	India : Uttarakhand : Dehradun	Ferriere, 1940; Hussain and Khan, 1986
Eulophomorpha flavicornis Dodd	Egg case of unidentified mantid	India: Coimbatore	Boucek, 1988; Kamijo, 1996
Ginsiella indica Arifa and Khan	*Phytobia humeralis* Diptera: Agromyzidae	India : Uttarakhand: Udham Singh Nagar	Arifa and Khan, 1991
Guptaiellaindica Sushil and Khan	*Chromatomyia horticola* Diptera: Agromyzidae	India : Uttarakhand: Pantnagar	Sushil and Khan, 1997
Hemiptarsenus indicus Khan	*Chromatomyia horticola* Diptera: Agromyzidae	India : Uttarakhand: Pantnagar	Khan, 1985b; Hussain and Khan, 1986
Hemiptarsenus semialbiclava Girault		India : Uttarakhand: Pantnagar	Kerrich, 1969; Hussain and Khan, 1986
Mohaniella indica Khan	Unidentified Lepidopterous Leaf miner	India : Uttarakhand Mussoorie; Chakrauta	Khan, 1995
Necremnus leucarthros Nees	*Hyper postica* (Gyllenhal) Coleoptera: Curculionidae	India : Delhi	Nees, 1834; Subba Roa et al., 1967; Hussain and Khan, 1986
Shardiella savitri Sushil and Khan	*Unidentified leaf miner* Diptera: Agromyzidae	India : Uttarakhand: Kaladungi	Sushil and Khan, 1997
Stenomesius japonicas (Ashmead) (*Sympiesomorpha japonica* Ashmead =*Stenelachistus impressus* Masi =*Euryscotolinx ciombatorensis* Rohwer =*Stenomesiodeus ashmeadi* Subba Rao and Sharma)	*Acrocercopes sp.* Lepidoptera: Gracillaridae	India : Andhra Pradesh; Bihar; Delhi; Karnataka; Tamil Nadu; Uttarakhand: Pantnagar	Ashmead, 1904; Masi, 1917; Rohwer, 1921; Subba Rao and Sharma, 1966; Boucek, 1976; Kerrich, 1974a; Hussain and Khan, 1986; Khan, 1992
Stenomesius modicellus Khan	*Aproaerema modicella* (Deventer) Lepidoptera: Gelechiidae	India : Tamil Nadu: Padappai	Khan, 1992
Stenomesius anati Khan and Singh	*Unidentified larvae* Lepidoptera	India : Uttarakhand; Nainital: Jeolikot	Khan and Singh, 1994
Setalacher fasciatus Boucek	Bagworm cocoon on mango	India: Bihar: Pusa	Boucek, 1988

(Continued)

TABLE 8.1 (Continued)
Host-Parasitoid Index of Eulophid Parasitoids in India

Parasitoids	Host	Distribution	Reference
Sympiesis dolicogaster Ashmead *Asymbiesiella india* Girault	*Caloptilia soyella* (Van Deventer) Lepidoptera: Gracillariidae *Aproaerema modicella* Lepidoptera: Gelechidae	India : Bihar; Maharashtra; Uttar Pradesh	Ashmead, 1888; Ayyar, 1925- 1927; Dutt, 1921; Mani, 1938; Pruthi and Mani, 1940; Subba Rao et al., 1967; Kamijo, 1976; Hussain and Khan, 1986; Surekha et al., 1996.
Sympiesis hyblaeae Surekha	*Hyblaea puera* (Cramer) Lepidoptera: Hyblaeidae	India : Kerala, Calicut	Surekha et al., 1996
Sympiesis striatipes Ashmead	*Acerocercops sp.* Lepidoptera: Gracillariidae	India	Ashmead, 1904; Kamijo, 1976
Sympiesis thyrissae Gupta	*Gangara thyrsis* Lepidoptera: Hesperiidae	India	Gupta et al., 2015a
Trichospilus diatraeae Cherian and Margabandhu	*Chilo sacchariphagus indicus* (Kapur); *C. sacchariphagus* (Walker) Lepidoptera: Pyralidae *Hypsipyla robusta* Moore Lepidoptera: Pyralidae *Pyrausta machaeralis* (Walker) *Lamprosema annubilobata* (Swinhoe) Lepidoptera: Noctuidae *Sylepta derogata* (Fabricius) Lepidoptera: *Pyralidae* *Sesamia inferens* (Walker) Lepidoptera: Noctuidae *Thosea cervina* (Walker) Lepidoptera: Limacodidae *Tryporyza incertulas* Lepidoptera: pyralidae *Unidentified tachinids* Diptera : *Tachinidae*	India : Karnataka; Uttar Pradesh; Tamil Nadu; West Bengal	Cherian and Maragabandhu, 1942; Boucek, 1976; Hussain and Khan, 1986

Trichospilus pupivora Ferriere	*Lamprosema annubilobata* (Swinhoe) Lepidoptera : Pyralidae *Opisina arenosella* Lepidoptera : Oecophoridae		Ferriere, 1930; Boucek, 1976; Dharmaraju and Pradhan, 1977; Hussain and Khan, 1986
Subfamily: Entedoninae			
Anumanniola lasallei Narendran	*Unknown*	India: Karnataka, Sringeri	Narendran and Sinu, 2003
Baeoentedon sp. Girault	*Ajuga bracteosa*	India: Kerala	Dubey, 1974; Hussain and Khan, 1986
Chrysocharis ajugus Dubey	Unknown	India: Karnataka	LaSalle and Schauff, 1994
Chrysocharis horticola Mani	*Chromatomyia horticola* Diptera: *Agromyzidae*	India: Uttar Pradesh, Aligarh	Mani, 1971; Hussain and Khan, 1986
Chrysocharis johnsoni Subba Rao	*Epilachana viginitioctopuntata* Coleoptera: *Coccinellidae*	India: Kerala; Trivandrum	Subba Rao, 1957; Hussain and khan, 1986
Chrysocharis indicus Khan	*Phytomyza* sp. Diptera: *Agromyzidae*	India:Uttarakhand, Ranikhet	Khan, 1985b; Hussain and Khan, 1986
Chrysocharis funicularis Khan	*Phytomyza sp.* Diptera:Agromyzidae	India:Uttarakhand, Ranikhet	Khan, 1985b; Hussain and Khan, 1986
Chrysocharis sp.	*Acrocercops syngramma* Meyrick Lepidoptera: Gracillaridae	India: Karnataka	Vanitha, 2015
Chrysonotomyia plucherrima (Kerrich)	*Procontarinia matteiana*	India: Uttar Pradesh, Lucknow,Varansi ; Karnataka, Banglore	Kerrich, 1970b–1974b; Khan and Shafee, 1981c; Hussain and Khan, 1986; Boucek, 1986; Narendran et al., 2001
Chrysonotomyia postmarinaloides (Saraswat)	*Procontarinia matteiana*	India:Karnataka, Banglore; Chikballapur, Tamil Nadu, Chennai.	Saraswat, 1975; Boucek, 1986; Narendran et al., 2001; Hussain and Khan, 1986
Chrysonotomyia obesula Boucek	*Procontarinia matteiana*	India:Uttar Pradesh, Gorakhpur	Boucek, 1986; Narendran et al., 2001
Chrysonotomyia rexia Narendran	*Liriomyza* sp. *Dip: Agromyzidae*	India:Maharashtra, Rahuri	Narendran et al., 2001
Chrysonotomyia ricini Parshuram and Agnihotri	*Liriomyza trifolii* Dip: Agromyzidae	India: Uttarakhand	Parshram and Agnihotri, 2017.
Closterocerus agromyzae Narayanan et.al	*Agromyza* sp.; *Chromatomyia horticola* Diptera: Agromyzidae	India:Delhi	Narayanan et.al. 1960; Gokulpure, 1972; Hussain and khan, 1986
Closterocerus insignis Waterston	*Oscinis theae*	India:Tamil nadu	Mani, 1971; Hussain and Khan, 1986

(Continued)

TABLE 8.1 (Continued)
Host-Parasitoid Index of Eulophid Parasitoids in India

Parasitoids	Host	Distribution	Reference
Closterocerus pytomyzae Mani	*Tropicomyia coffeae auctt.* Dip: Agromyzidae *Chromatomyia horticola* *Chromatomyia horticola*	India: Bihar, Kerela;Tamil Nadu; Uttar Pradesh	Mani, 1971; Dubey, 1974; Hussain and Khan, 1986
Closterocerus scapiatus Singh and Khan	Unidentified leaf miner	India: Uttarakhand, Pantnagar	Singh and Khan, 1996
Closterocerus sp.	*Acrocercops syngramma* Meyrick Lep: Gracillaridae	India: Karnataka	Vanitha, 2015
Entedon pempheridis Ferriere	*Hypolixus truncates* (fabricius) Coleoptera: Curculionidae *Apion sp.* Coleoptera: Apionidae *Apion corchori Marshall* Coleoptera: Apiionidae *Lobotrachelus* sp. Coleoptera: Curculionidae *Pempherulus affinis* (Faust) Coleoptera: Curculionidae	India: Tamil Nadu	Ferriere, 1940a; Hussain and Khan, 1986
Euderomphale secundus Mani	*Aleurolobus barodensis* (Maskell) Hemiptera: Aleyrodoidea	India: Delhi	Mani, 1939; Hussain and Khan, 1986
Euderus alcidodes Singh	*Alcidodes ludificator* Curculionidae: Coleoptera	India: Assam	Singh, 2005
Euderus pantnagarensis Mahendran, Agnihotri and Pillai	*Alcidoded bubo* Curculionidae: Coleoptera	India: Uttarakhand	Mahendran et al., 2014
Goetheana thripsiovorous Narayanan et al.	*Thrips* Thysanoptera*: Thripidae*	India: Maharashtra	Narayanan et al., 1960; Hussain and Khan, 1986
Ladna malabarica Surekha and Narendran	Unknown	India: Kerala	Surekha and Narendran, 1992
Ladna elongata Sureshan and Narendran	Unidentified leaf gall	India: Kerala; Eranakulam, Malayatoor	Sureshan and Narendran, 2001
Mangocharis longiscapus Boucek	*Procontarinia matteiana*	India; Karnataka, Chikaballapur; Bangalore; Tamil Nadu, Chennai	Boucek, 1986

Neochrysocharis hyalinipennnis Khan and Shafee	Unidentified Scale	India: Uttar Pradesh	Khan and Shafee, 1980b; Hussain and Khan, 1986
Neochrysocharis metallicus Khan and Shafee	Unidentified Scale	India: Karnataka, Bangalore	Khan and Shafee, 1980b; Hussain and Khan, 1986
Neochrysocharis nr. distatae (Howard)	*Letana sp.* Orthoptera: Tettigonidae	India: Karnataka	Gupta et al., 2015
Neochrysocharis pubipennis Khan and Shafee	*Pulvinaria maxima* Green	India: Uttar Pradesh	Khan and Shafee, 1980b; Hussain and Khan, 1986
Neopediobopsis setosa Narendran	*Prosopis sp.*	India: New Delhi, Lodhi Gardens	Narendran and Sheela, 1993
Omphale narariae Kurian	*Darna navaria* Moore Lepidoptera: Limacodidae	India: Appanapalli	Kurian, 1954
Omphale epilachinae Singh and Khan	*Epilachna vigintiopunctata* Fab. Coleoptera: Coccinelladae	India: Uttarakhand, Pantnagar	Singh and Khan, 1995 Walker, 1839; Kerrich, 1973;
Pediobius acanthi Walker	*Marasmia trapezalis* (Guenee) Lepidoptera: Pyralidae *Phytomyza spp.* Diptera: Agromyzidae	India: Karnataka	Hussain and Khan, 1986
Pediobius agaristae (Cameron) (=*Cluthaira agaristae* Cameron)	*Ophioderes fullonica*	India: Maharashtra	Cameron, 1912; Kerrich, 1973; Hussain and Khan, 1986
Pediobius bethylicidus Kerrich	*Secondary parasite of Parasierola* sp. on *Sesamia inferens* (?) *Goniozus* sp.	India: Uttar Pradesh, West Bengal	Kerrich, 1973; Hussain and Khan, 1986
Pediobius bruchicida (Rondani) (= *Spartiiophilus bruchicida* Rondani; = *Pleurotropis waterstonii* Masi) *Pediobius deplagastrus* Surekha and Narendran	*Pectinophora gossypiella* (Saunders) Lepidoptera: Gelechiidae	India:	Rondani, 1872; Kerrich, 1973; Boucek, 1974; Surekha and Narendran, 1993.
Pediobius elasmi Ashmead	*Chilo saccariphagus indicus* (Kapur) Lep: Pyralidae *Sarcophaga* sp. (probably Scirpophaga) Lepidoptera: Pyralidae	India: Karnataka; Madhya Pradesh	Ashmead, 1904; Kerrich, 1973; Hussain and Khan, 1986
Pediobius foveolatus (Crawford) (= *Pleurotropis foveolatus* Crawford; *Pleurotropis epilachnae* Rohwer)	*Menochilus sexmaculatus* (Fabricius); *Coccinella septempunctata* Linnaeus *Epilachna sp.; Henosepilachna vigintioctopunctata; Epilachna vigintioctopunctata* (Fabricius)	India: Bihar, Karnataka, Orissa, Tamil Nadu, West Bengal	Crawford, 1912; Rohwer, 1921; Kerrich, 1973; Pruthi and Mani, 1940; Reddy and Mandal, 2014.

(Continued)

TABLE 8.1 (Continued)
Host-Parasitoid Index of Eulophid Parasitoids in India

Parasitoids	Host	Distribution	Reference
Pediobius fraternus (Motschulsky) (=*Eulophus fraternus* Motschulsky = *Eulophus mantiechthrus* Motschulsky)	*Mantid oothecae*	India : Karnataka	Kerrich, 1973; Hussain and Khan, 1986
Pediobius gunturensis Shafee and Rizvi	–	India : Andhra Pradesh	Shafee and Rizvi, 1985; Hussain and Khan, 1986.
Pediobius imbreus Walker (=*Entedon (Pediobius) imbreus* Walker; *Pleurotropis detrimentosus* Gahan)	*Opisina arenosella Walker* Lepidoptera: Oecophoridae *Perisierola sp.* *Perisierola nephantidis Musebeck* Hymenoptera: Bethylidae *Pteroma pendula* Lepidoptera: Psychidae *Epicephala chalybacme*	India: Gujarat; Karnataka; Maharastra; Rajasthan; Tamil Nadu	Ayyar and Margabanhu, 1934a; Kerrich, 1973; Hussain and Khan, 1986; Rao et al., 2017.
Pediobius indicus Khan	*Chromatomyia horticola*	India: Uttarakhand, Pantnagar	Khan, 1985a; Hussain and Khan, 1986
Pediobius infuscatipennis Shafee et al.	–	India: Andaman and Nicobar Islands	Hussain and Khan, 1986
Pediobius longicorpus Khan and Shafee	Unidentified Dipterous pupae	India: Uttar Pradesh, Aligarh	Khan and Shafee, 1982; Hussain and Khan, 1986
Pediobius maduraiensis Shafee and Rizvi	–	India: Tamil Nadu, Madurai	Shafee and Rizvi, 1985; Hussain and Khan, 1986
Pediobius pondicherryensis Shafee and Rizvi	–	India : Pondichery	Shafee and Rizvi, 1985; Hussain and Khan, 1986
Pediobius stenochorus Kerrich	*Phyllonorecter* sp. (Lepidoptera: Gracillaridae) *Apanteles sp.*	India: Uttar Pradesh	Kerrich, 1973; Hussain and Khan, 1986
Pediobus thakerei (Subba Rao)	*Chromatomyia horticola* (Goureau) Diptera: Agromyzidae	India: Delhi	Subba Rao, 1957; Hussain and Khan, 1986
Pediobius viggiani Khan	*Nymphula depuntaalis* Guenee Lepidoptera: Pyraustidae	India : Uttar Pradesh, Moradabad	Khan, 1996

Pediobius ni Khan	*Trichopulsia ni* Lepidoptera: Plusinae		Khan, 1996
Pediobius cydiae Khan	*Cydia critica* (Meyrick) Lepidoptera : Eucosmidae	India: Uttarakhand, Rishikesh	khan, 1996
Pediobius salicifolii	*Japanogromyza salicifolii* (Collin) Diptera: *Agromyzidae*	India: Uttar Pradesh	Khan, 1996
Pomphale striptipennis Husain et al.	*Aleurolobus* sp.	India: Uttar Pradesh	Hussain and Khan, 1986; LaSalle and Schauff, 1994
Pomphale setosipennis Hayat and Zeya	Unknown	India	LaSalle and Schauff, 1994
Thripobius setosipennis Boucek	*Heliothrips* sp. Thysanoptera: Thripidae *Panchaetothrips indicus* Bagnall	India : Karnataka	Boucek, 1976
Subfamily: Tetrastichinae			
Acerotoneuromyia indica (Silvestri) (=*Syntomoshyrum indium* Silvestri)	*Dacus caudatus* Fabricius Diptera: Tephritidae	India: Karnataka, Bangalore	Silvestri, 1910; Ayyar, 1925; Ferriere, 1960; Mani, 1938; Hussain and Khan, 1986
Aprostocetus hagenowii (Silvestri) (=*Entedon hagenowii* Razerburg) (Subgenus Tetratrichodes)	*Periplanata americana* *P. australasiae* *P. brunnea*	South India	Ratzeburg, 1852; Boucek, 1979; Hussain and Khan, 1986
Aprostocetus asphondylae Mani and Kurian	*Asphondylia pongamiae* Felt Diptera: Cecidomyiidae *Hypolixus truncatulus* (Fabricius) Coleoptera: Curculionidae	India: Tamil Nadu	Mani, 1941; Krishna Ayyar, 1942; Hussain and Khan, 1986.
Aprostocetus krishnieri Mani	*Hypolyxus truncatulus* (Fabricius) Coleoptera: Curculionidae	India: Tamil Nadu	Mani, 1941; Hussain and Khan, 1986
Aprostocetus epilachne Kurian	*Epilachna* sp.	India: Orissa	Kurian, 1954; Hussain and Khan, 1986
Aprostocetus coorgensis Kurian	*Opisinia arenosella* Walker Lepidoptera: Oecophoridae Pyralid larva	India: Karnataka	Kurian, 1954; Hussain and Khan, 1986
Aprostocetus doonensis Singh	*Procontarinia mangiferae* Diptera: Cecidomyidae	India: Dehradun	Singh, 2013
Aprostocetus malloticolae Singh	*Trioza malloticola* Hemiptera: Triozidae	India: Uttarakhand	Singh, 2017

(*Continued*)

TABLE 8.1 (Continued)
Host-Parasitoid Index of Eulophid Parasitoids in India

Parasitoids	Host	Distribution	Reference
Aprostocetus ficicolae Singh	*Pauropsylla ficicola* Hemiptera: Triozidae	India: Uttarakhand	Singh, 2017; Kumar et al., 2018
Aprostocetus sp.	*Acrocercops syngramma* Meyrick Lep: Gracillaridae	India: Karnataka	Vanitha, 2015
Aprostocetus obtusae Narendran	*Melanagromyza obtusa* Diptera: Agromyzidae	India: Kerala	Narendran et al., 2005
Neotrichoporoides nyemitawus (Rohwer)	*Antherigona conigera* Emden *Antherigona soccata* Rondani Diptera: Muscidae	India: Tamil Nadu, Kerala	Narendran et al., 2006
Aprostocetus sankarani Boucek	*Procontariana matteiana* Hemiptera: Cecidomyiidae	India: Karnataka; Uttar Pradesh: Gorakhpur	Boucek, 1986
Ceratoneura indi Girault	Phytophagous eulophid	India: Karnataka; Tamil Nadu; Madhya Pradhes; Maharashtra	Giralut, 1913; Rohwer, 1921; Narendran and Krishnakumar, 1995
Cirtostichus phyllocnistoides (Narayanan) = *Cirrospilus phyllocnistoides* Narayanan = *Tetrastichus phylllocnistoides* Batra & Sandhu	*Phyllocnistis citrella* Stainton Lepidoptera: Phylocnistidae	India: Delhi	Narayanan, 1960; Subba Rao and Ramamani, 1965; Batra and Sandhu, 1981; Hussain and Khan, 1986; Boucek, 1988; Ujiye and Adachi, 1995.
Melitobia acasta (Walker) (=*Cirrospilus acasta* Walker)	*Shionphprus alkae* (Ellinger & Sachtleben) Hymenoptera: Ichenumonidae	India	Walker, 1839; Hussain and Khan, 1986
Mestocharella javensis Gahan	*Caloptilia theivora* (Walsingham) Lepidoptera: Gracillaridae	India: Uttarakhand, Pantnagar	Singh and Khan, 1995
Nesolynx thymus (Girault)	Dipterous larvae; Braconid cocoons	India	Boucek, 1976; Hussain and Khan, 1986
Oomyzus scapous (Thompson) (=*Tetrastichus scapus* Thompson; =*Tetrastichus coccinellidae* Kurdjumov)	Coccinellid aphidophagous	India	Hussain and Khan, 1986; Boucek, 1988; Narendran et al., 2001

Oomyzus cerococci (khan and Shafee)	Coccinellidae indicus (Muskell)	India: Karnataka, Bangalore	Khan and Shafee, 1981; Hussain and Khan, 1986; Graham et al., 1991; Narendran et al., 2001
Oomyzus lirimyzae Narendran	*Liriomyza* sp. Agromyzidae	India: Maharashtra, Rahuri	Narendran et al., 2001
Parachrysocharis javensis Girault	*Pyrilla perpusilla* (Walker) Hempitera: Lophopidae	India: Uttar Pradesh Gola Gorakhnath	Girault, 1913; Khan and Shafee, 1980
Quadratichus ovulorum (Ferriere)	*Epilachna vigintioctopunctata* (Fabricius)	India: Karnataka	Ferriere, 1930
Sigmophora polyseta (Saraswat)	Unidentified	India	Saraswat, 1975
Tachinobia repanda Boucek	*Papilio lagai* Dufrane (Lep: Papilionidae) *Bessa remota* (Aldrich) Diptera: Tachinidae *Sarcophaga banksi* Diptera: Sarcophagidae	India: Tami Nadu	Hussain and Khan, 1986
Tamarixia radiata (Waterston) (*Tetrastichus radiatus* Waterston)	*Diaphorina citri* K.(= *Euphalerus citri* (Kuwana) Hemiptera: Psyllidae	India: Uttar Pradesh, Agra and Fatehpur	Hussain and Khan, 1986
Tamarixia sheebae Narendran	Leaf galls of *Terminalia arjuna.* Hemiptera: Psyllidae, Triozidae	India: Kerala	Narendran et al., 2005; Kumar et al., 2020
Tetrastichus agarwali Shafee et al.	*Atherigona soccata*Rondani Diptera: Muscidae	India: Delhi; Uttarakhand	Hussain and Khan, 1986
T. ajmerensis Khan and Shafee	*Coccidohystrix insolitus* (Green) Hemiptera: Pseudococcidae	India: Rajasthan	Khan and Shafee, 1981; Hussain and Khan, 1986
Tetrastichus aligarhensis Khan and Shafee	*Pulvinariia sp.*	India: Uttar Pradesh	Khan and Shafee, 1981; Hussain and Khan, 1986
Tetrastichus annulicomis Khan and Shafee	*Coccidohystrix insolitus* (Green) Hemiptera: Pseudococcidae	India: Rajasthan	Khan and Shafee, 1981; Hussain and Khan, 1986
Tetrastichus anomalococci (Khan and Shafee) (=*Syntomosphyrum anomalococci* Khan andShafee)	*Anomalococcus crematogastri* (Green) Hom: Coccidae	India: Tamil Nadu	Khan and Shafee, 1980a; Hussain and Khan, 1986
Tetrastichus atomelli Khan	*Tropicomyza atömella* (Malloch) Diptera: Agromyzidae	India: Uttarakhand	Khan, 1983; Hussain and Khan, 1986
Tetrastichus coimbatorensis Rohwer	*Cecidomyiid* sp.	India: Tamil Nadu	Rohwer, 1921; Hussain and Khan, 1986
Tetrastichus colemani Crawford	*Aspidomorpha miliaris* (Fabricius)	India: Karnataka	Crawford, 1912: Hussain and Khan, 1986
*Tetrastichus delhiensis*Shafee et al.	Coleoptera: Chrysomelidae *Atherigona soccata* Rondani Diptera: Muscidae	India: Delhi	Hussain and Khan, 1986

(*Continued*)

TABLE 8.1 (Continued)
Host-Parasitoid Index of Eulophid Parasitoids in India

Parasitoids	Host	Distribution	Reference
Tetrastichus dhireni Saraswat	–	India: West Bengal(Probably)	Saraswat, 1978; Hussain and Khan, 1986
Tetrastichus fasciatus Ashmead	*Contarinia Caudata* Felt Diptera: Cecidomyiidae	India:Chumurajhagar and Chintaman	Ashmead, 1894; Gahan, 1922; Hussain and Khan, 1986
Tetrastichus flavidus Khan andShafee	Unidentified coccid	India: Andhra Pradesh	Khan and Shafee, 1981b; Hussain and Khan, 1986
Tetrastichus gardneri Ferriere	Pentatomid eggs	India: Uttarakhand	Ferriere, 1931a; Hussain and Khan, 1986
Tetrastichus indicus Khan and Shafee	Unidentified (Mealy Bug)	India: Tamil Nadu	Khan and Shafee, 1981; Hussain and Khan, 1986
Tetrastichus inferens Yoshimoto	*Seasamia inferens*Argyria Sp.	India	Yoshimoto, 1971. Mani and Kurian, 1953.
Terrostichus hawardi (= *T. ayyari* Rohwer; *Aprosioceius israel* Mani)	*Anadevidia peponis* (Fabricius) Lepidoptera: Noctuidae *Scirpophaga incertulas* (Walker) Lepidoptera: Pyralidae *Opisina arenosella* Walker Lepidoptera: Oecophoridae	India: Orissa	Ali and Subramanian, 1972; Narayan and Jairaj, 1973; Hussain and Khan, 1986
Tetrastichus isaaci Rohwer	Unidentified Gall	India: Tamil Nadu	Rohwer, 1921; Hussain and Khan, 1986
Tetrastichus kodaikanalensis Saraswat	–	India: Tamil Nadu	Saraswat, 1975; Hussain and Khan, 1986 Kaul and Saraswat, 1974;
Tetrastichus krishnaiahi Saraswat	–	India: Andhra Pradesh	Hussain and Khan, 1986
Tetrastichus kumaonensis Saraswat	–	India: Uttar Pradesh	Saraswat, 1975; Hussain and Khan, 1986
Tetrastichus lasiopterae Bhatnagar	*Lasioptera falcata* Felt Diptera: Cecidomyiidae	India: Kerala	Bhatnagar, 1952; Hussain and Khan, 1986
Tetrastichus longicauda (Kieffer) (= Hyperteleslongicauda Kieffer)	Cecidomyiid of Artemisa galls	India: Tamil Nadu	Kieffer, 1905; Pruthiand Mani, 1940
Tetrastichus malabarensis Saraswa	–	India: Kerala	Saraswat, 1975; Hussain and Khan, 1986
Tetrastichus nilamburensis Saraswat	–	India: Kerala	Saraswat, 1975; Hussain and Khan , 1986
Tetrastichus nyemitawas Rohwer	*Atherigona* sp. *Atherigona socata* Rondani Diptera: Muscidae	India: Madhya Pradesh; Maharashtra; Tamil Nadu; Rajasthan	Rohwer, 1921; Kundu and Kishore, 1973 Raodeo et al., 1972: Rawat andSahu, 1969; TaleyandThakare, 1979; Hussain Khan, 1986

Tetrastichus okawus Rohwer	Cumbu Grains	India: Tamil Nadu	Rohwer, 1921: Hussain and Khan, 1986
Tetrastichus ophiusae Crawford	*Achaea janata* (Linnaeus) Lepidoptera: Noctuidae	India: Karnataka	Crawford, 1912; Cherian and Basheer, 1947; Hussain and Khan, 1986
Tetrastichus ootyensis Saraswat	Unidentified	India: Tamil Nadu	Saraswat, 1975: Hussain and Khan, 1986
Tetrastichus plecopterae Kurian	*Plecoptera reflexa* Guenee Lepidoptera: Noctuidae	India: Uttarakhand	Hussain and Khan, 1986
Tetrastichus polyseta Saraswat	–	India: Tamil Nadu	Saraswat, 1975: Hussain and Khan, 1986
Tetrastichus psyllidis Khan andShafee	Psyllid Hom: Psyllidae	India: Uttar Pradesh, Aligarh	Khan and Shafee, 1981b; Hussain and Khan, 1986
Tetrastichus purpures(Cameron) (*Hadrothrix purpurea* Cameron; = *Tetrastichus immsii* Mahdihassan)	*Aonidiella orientalis* (Newstead)Acerialacca (Kerr) *Coccophagus ischirchii* Mahdihassan *Tachardiaephagus tachardiae* (Ashmead)	India: Bihar; Karnataka; West Bengal: Uttarakhand	Cameron, 1913: Mahdihassan, 1923; Ayyar and Margabandhu, 1934a.
Tetrastichus pyrillae Crawford	*Pyrilla aberrans* (Kirby) *Pyrilla perpusilla* (Walker) Hemiptera: Lophopidae	India: Bihar; Maharashtra	Narayanan and Kundan Lal, 1953: Ayyar and Margabandhu, 1934a: Hussain and Khan, 1986
Tetrastichus quadriseta Saraswat	–	India: Tamil Nadu	Saraswat, 1975; Hussain and Khan, 1986
Tetrastichus rhipophorathripscidis Narayanan et al.	*Mallothrips indicus* Ayyar *Rhipiphorothrips cruentatus* Hood Thysanoptera: Thripidae	India: Delhi	Narayanan et al., 1960; Hussain and Khan, 1986
Tetrastichus bicolor Saraswat	–	India: Tamil Nadu	Saraswat, 1975; Hussain and Khan, 1986
Tetrastichus satpurensis Saraswat	–	India: Madhya Pradesh	Saraswat, 1975; Hussain and Khan, 1986
Tetrastichus schnoebii Ferriere	*Spodoptera mauritia* Boisduvae Lepidoptera: Noctuidae *Scirpophaga incertulus* (Walker) Lepidoptera: Pyralidae	India: Andra Pradesh	Ferriere, 1931; Ayyarand Margabandhu, 1934a; Rai and Gowda, 1977; Rao and Ali, 1976
Tetrastichus sexmaculatus Kurian	*Menochilus sexmaculatus* (Fabricius) Coleoptera: Coccinellidae	India: Karnataka	Kurian, 1953; Hussain and Khan, 1986
Tetrastichus schencottensis Saraswat	–	India: Tamil Nadu	Saraswat, 1975; Hussain and Khan, 1986
Tetrastichus sokolowskii Kurdjumov	*Plutella xylostella* (Linnaeus) Lep: Yponomeutidae	India: (Palaerctic)	Cherian and Basheer, 1947; Hussain and Khan, 1986
Tetrastichus spirabilis Waterston	*Hypsiphla robuta* (Moore) Lep: Pyralidae	India: Uttarakhand	Hussain and Khan, 1986
Tetrastichus travancovensis Saraswat	–	India: Kerala	Saraswat, 1975; Hussain and Khan, 1986

(*Continued*)

TABLE 8.1 (Continued)
Host-Parasitoid Index of Eulophid Parasitoids in India

Parasitoids	Host	Distribution	Reference
Tetrastichus tritrichia Saraswat	–	India: Kerala	Saraswat, 1975; Hussain and Khan, 1986
Tetrastichus udaipurensis Khan and Shafee	*Pyrilla* sp.	India: Orissa; Rajasthan	Khan and Shafee, 1980; Hussain and Khan, 1986
Tetrastichus uniarticulata Saraswat	Unidentified	India: Kerala	Saraswat, 1975; Hussain and Khan, 1986
Tetrastichus mohani Khan and Sushil	Unidentified leaf miner Diptera: Agromyzidae	India: Uttarakhand, Pantnagar	Khan and Sushil, 1992
Tetrastichus mangiferae Khan and Sushil	Unidentified Coccids Hemiptera: Coccidae	India: Uttarakhand, Pantnagar	Khan and Sushil, 1992
Tetrastichus obliqua Khan and Sushil	*Spilosoma obliqua* Lepidoptera: Noctuidae	India: Uttarakhand, Pantnagar	Khan and Sushil, 1992
Tetrastichus chakrataensis Khan and Sushil	Unidentified leaf miner Lepidoptera		Khan and Sushil, 1992
Tetrastichus partellus Khan and Sushil	*Chilo partellus* Lepidoptera: Crambidae		Khan and Sushil, 1992
Tetrastichs spp.	*Epilachna vigintioctopunctata* Coleoptera: Coccinellidae	India: West Bengal	Reddy and Mandal, 2014
Tetrastichus thetisae Gupta	*Curetis thetis* Drury Lepidoptera: Lycaenidae	India	Gupta et al., 2015
Tetrastichus paraponyxyi Santhosh	*Paraponyx stagnalis* (Zeller) Lepidoptera: Crambidae	India: Kerala	Santhosh and Karthikeyan, 2017

phases, *viz.* host plant cultivation, mass rearing of host insect (*D. citri*) and mass rearing of parasitoids, which are briefly discussed below.

a. Cultivation of hostplants

Cantonese lemon, *Citrus limonia* or Orange Jasmine *Murrayapaniculata* are commonly used as hosts for the mass multiplication of Citrus psyllid *D. citri*. Seedlings of *M. paniculata* planted in 1.5 litre plastic bags are maintained in the nursery. 1% NPK (20: 20: 20) is sprayed once in two weeks as foliar fertilizer. Seedlings are sprayed with 10% detergent solution and then rinsed with water to prevent attacks by other sucking pests, like aphids. Irrigation has to be done twice a week and kept in 50% shade. Apical buds are removed to stimulate sprouting. At 25°C, new flushes are grown up to 1 cm (optimal size) in a week. Now the seedlings are ready to be used for the rearing of *D. citri*.

b. Rearing of host insect, *D. citri* in greenhouse

Seedlings with approximately 1 cm length of new flushes are transferred to cages for rearing *D. citri*. The rearing cages for *D. citri* must be cubical with ~50 cm dimension on each side. The sides of the cages are covered with 0.87 mm mesh to prevent psyllids from escaping. The cages are kept in the greenhouse, where the temperature is maintained at 28°C and RH 70–80%. A plastic tray was kept inside the cage to accommodate 5–6 seedlings. About 300 adult psyllids of about 12 days after eclosion were released in each cage and left for mating and oviposition for about 7 days. After oviposition, the adult psyllids were removed and seedlings with eggs are maintained so that the eggs hatch, develop into nymphs and then to the adult stage. Adult psyllids emerged in about 12 days from oviposition. In all adult psyllids developing in all cages, 30% are used to produce adults for the next batch of mass rearing and 70% are used to produce nymphs to be used as hosts for rearing *T. radiata*.

c. Mass Rearing of the Parasitoid

Parasitoids were reared in an air-conditioned room with temperature maintained at 25°C, RH 70–80% and photoperiod 14:10 (Light: dark) hours. Rearing cages for parasitoids are made of aluminium frame with top and front fitted with polystyrene for good light passage. An artificial light source of about 20 W is fitted above the cages. In each cage, two plastic trays were placed to accommodate 12 seedlings. After placing the seedling with nymphs (fourth and fifth instar) on the tray in the cages, adult parasitoids of *T. radiata* were released in 1:10 ratio (i.e. one parasitoid for 10 fourth or fifth instar nymphs of *D. citri*). On release, nymphs of *D. citri* were parasitized. Parasitoids develop from the nymphs and adults started to emerge after 12 days. Ten days after parasitism, the seedling containing mummies (parasitized nymphs) were cut and placed in emergence boxes. The emergent boxes were made up of black fibre board with an opening at the top, where a transparent glass bottle is attached. When the parasitoids emerged, they were attracted to the light from the bottle and thus collected. Parasitoids from the bottle were then transferred to glass tubes of one-inch diameter (200 adults/ tube). A drop of honey and pollen mixture was added to the tube and kept at 18°C until field release. Of all the parasitoids that emerged, 17% were used for the next batch of mass rearing and the remaining were used for the field release.

8.6.2 Mass Multiplication of *Tetrastichushowardii*

*Tetrastichushowardii*is a promising biocontrol agent for various lepidopteran pests, primarily of sugarcane internode borer, *Chilosachariphagus indicus*. *T. Howardii* has been mass multiplied on the larvae of *Spodoptera litura* (Sankar et al., 2014). The host insect, *S. litura* young stage larvae from 1st to 3rd instars have to be fed with castor leaves in a group and the later stages 4th and 5th instars must be reared separately on the artificial diet prepared by using besan flour, yeast and multivitamins substituting agar as a wetting agent to the media. The fully grown pupae of *S. litura*are used to mass produce the parasitoid. Several studies were conducted to find out various factitious hosts. Lepidopteran insects, such as cotton boll worm, *Helicoverpaarmigera,* and Pink boll worm, *Sesamiainferens* are reared on the artificial diet and eri silkworm, *Philosamiaricini* and Mulberry silkworm, *Bombyx mori* are reared on castor and mulberry leaves and used for the mass multiplication of *T. howardii.* Various attempts have also been made to rear it using pupae of cockroach, house fly, Uzi fly, *Chilopartellus*, *Chiloinfuscatellus* and *Chilosacchariphagus indica.* Among all, pupae of *Bombyx mori* and *Philosomiaricini* had shown high efficiency as hosts.

8.6.3 Mass Multiplication of *Tetrastichusisraeli* and *Trichospiluspupivora*

Five pupae of lepidopterous insects are taken in a 15 × 2.5 cm test tube. Thirty mated females are released inside the test tube with pupae. The parasitoids lay eggs inside the pupa and multiply further. After 15 days of release, adults start emerging. 20 adults/tree are released in the field for controlling *Opisinaarenosella* in the coconut (Mooventhan et al., 2018).

8.7 SUCCESSFUL BIOCONTROL AGENTS

Among all parasitic hymenopterans, eulophids have proven to be effective biocontrol agents for controlling pest populations. Various successful biocontrol programmes have been undertaken throughout the world including India using eulophid parasitoids.

Paper industries in India were facing heavy losses due to the devastating attacks of Eucalyptus gall wasp, *Leptocybe invasa.* In the year 2008, two eulophid parasitoids, viz. *Quadrastichus mendeli* and *Selitrichodes kryseri,* which had shown better control of the pest in Israel were imported from Australia into India by NBAIR. *Q. mendeli* was identified as best suited for Indian conditions and it is mass-produced and released in the field. Within two years of release, a drastic reduction in pest infestation was seen and the gall wasp was brought down below the pest status (Shylesha et al., 2018). Another parasitoid, *Aprostocetus gala,* is a potential parasitoid of erythrina gall wasp (*Quadrastichuserythrinae*) in India (NBAIR, 2014).

The citrus psyllid, *Diaphorina citri,* besides being a sap-sucking pest, acts as a vector for the gram-negative bacterium, which causes 'Citrus Greening' disease. Huanlongbinh or Citrus Greening is responsible for the decline of citrus cultivation across the world. *Tamarixia radiata* killed 90% of the presented nymphs through parasitism and host feeding in laboratory experiments (Aubert, 1991). Excellent control of Asian citrus psylla, *Diaphorina citri,* was achieved by the release of eulophid parasitoid, *Tamarixia radiata* into Reunion Island in 1978. Release of *T. radiata* wiped out the citrus greening from Reunion Island. Releases of eulophid larval parasitoid, *Pediobius foveolatus,* against Mexican bean beetle in major soybean areas of Maryland during the summer of 1973 resulted in high parasitism, ranging from 80 to 100%. The parasitoid spreads naturally to a distance exceeding 100 km from the released sites. Currently, the pest is successfully managed by annual inoculative releases.

Phymastichus offea (Hymenoptera: Eulophidae) is a successfully established biocontrol agent against coffee berry borer. The parasitoid released in Columbia in the year 1999–2000 was recovered in South India (Singh, 2004). *Telenomus remus* is an effective egg parasitoid of *Spodoptera litura. Pediobius foveolatus* has been showing high efficiency in the biocontrol of grub of the Hadda beetle, *Epilachna* spp.

Trichospiluspupivora and *Tetrastichus israeli* are highly promising biocontrol agents parasitizing the pupae of *Opisina arenosella,* a devastating pest of coconut in tropical India. *Tetrastichus howardii* is an effective parasitoid of various lepidopterous pests, such as *Helicoverpaarmigera, Spodoptera litura, Chilosachariphagus indicus, Chilopartellus,* etc. *Pediobius furvus* and *Tetrastichusinferens* showed an effective control of sugarcane tissue borers in Uganda and Thailand, respectively. *Pediobius imbrues* provided a natural parasitisation of 2–10% against the coconut slug caterpillar, *Macroplectranararia* Moore. A field efficacy study by Rao et al. (2017) showed 41.41% control of *M. nararia* by *P. imbrues*. Eulophid parasitoids, *Aprostocetus* spp., are the most important population regulating the natural enemy of the sorghum midge, *S. sorghicola* (Sharma and Davies, 1988; Nwanze et al., 1998).

Some other promising successful biocontrol agents are coconut leafminer *Promecothec acumingi* biocontrol project employing *Pediobius parvulus* and *Dimmockia javana* in South East Asia, *Pediobiusfurvus* for successful control of *Sesamia calarmistis* on sugarcane in Madagascar, *Trichospilus diatraeae* for sugarcane stem borer, *Chilo sacchariphagus*; *Tetrastichus pyrillae* and *T. ayyari* for *Pyrilla perpusilla* on sugarcane, *Tetrastichus sokolowskit* for Diamond black moth, *Plutella xylostella*, *Chrysocharis laricinellae* and *Elachertus argissa* for Larch casebearer *Coleopheralaricella* in the forest, *Diglyphus* spp. against *Liriomyza* spp., *Oomyzus sokolovskii* against diamond-back moth (*Plutellax ylostella*) and *Tetrastichus* spp. against Rice Yellow Stem Borer (*Scirpophaga incertulas*) (Khan et al., 2005).

8.8 CONCLUSIONS

In the era of concern for the global degradation of the environment, a lot of emphasis is given to eco-friendly means of pest management, which could potentially replace hazardous chemical pesticides. Among the various modes of pest management, biocontrol using suitable parasitoids and predators is the most promising and definitive measure. Eulophidae has shown tremendous records of pest control at various points of time during the last few decades. Considering their diversity, as discussed under various headings of this chapter, they are enormous in nature Yet, thorough and well-designed investigations on host-based studies, ecology, biology and behaviour are required to utilise them effectively in the biocontrol programmes. Standardisation and research on host-parasitoid relationship, methods of mass rearing, field release techniques are the need of the hour.

8.9 POINTS TO REMEMBER

1. Eulophidae is a hymenopteran parasitoid family of wasps, the largest family in the superfamily Chalcidoidea, having diversity across the world of about 6000 species grouped under 324 genera.
2. Eulophids are small to medium-sized chalcids, which can be easily recognized at once by the four segmented tarsi, antennae generally up to four funicular segments, and straight spur of the foretibia, and non-pedunculate wings, colour varies from metallic to black to yellow.
3. Widely accepted subfamily-level classification of the family is given by Gauthier et al. (2000). In which the family is divided into four subfamilies, Entedoninae, Euderinae, Eulophinae and Tetrastichinae, all of which are cosmopolitan in distribution
4. Eulophines are solitary or gregarious, idiobiont ectoparasitoids of the larvae, or sometimes the pupae of leaf-miners or similarly concealed hosts, such as leaf-folders, casebearers, gall-makers and stem-borers.
5. Entedonines are principally solitary or gregarious primary or secondary endoparasites of concealed eggs, larvae or pupae. Many of the genera have well-defined host ranges. For example, most *Entedon* species attack beetles.
6. Tetrastichines are usually primary endoparasitoids of the eggs, larvae or pupae of Diptera, Hymenoptera or Lepidoptera, through a marked association with small gallicolous hosts. Some even attack eriophyid mites or gall-forming nematodes.

7. Eulophid wasps have a diverse biology. They are found in agroecosystems as parasitoids, hyperparasitoids, predators and pests of crops. One important pest is the Eucalyptus gall wasp, *Leptocybe invasa.*
8. Eulophidae has a wide range of hosts ranging from Diptera, Mantodea, Lepidoptera, Coleoptera, etc. Many eulophids are recorded as leafminer parasitoids. Eulophids are mostly larval parasitoids, but some are egg and pupal parasitoids also.
9. Eulpohidparasitoids, such as *Pediobius foeviolatus*, *Trichospilus pupivora*, *Tamarixia radiata*, *Tetrastichus howardii*, *Quadrastichus mendeli,* are commercially used as biocontrol agents.
10. Several viable eulophid species are yet to be explored and mass production techniques for important biocontrol agents are to be standardised to facilitate biocontrol programmes.

REFERENCES

Ahmad, M.J., Khursheed, I. & Azim, M.N. (2013). A new species of *Diglyphus* walker (chalcidoidea: Eulophidae) on mustard leaf miner *Chromatomya horticola* (Goureau) from Kashmir, India. *Journal of Entomological Research*, 37(2): 163–166.

Ali, K. & Subramaniam, T.R. (1972). On the occurrence of *Tetrastichusisraeli* M & K (Eulophidae: Hymenoptera) as a pupal parasite of *Nephantis serinopa Meyr* (Cryptophasidae: Lepidoptera) in Tamilnadu. *Madras Agricultural Journal*, 54: 654.

Arifa, N. & Khan, M.A. (1991). Some new Chalcid Parasitoids (Hymenoptera: Eulophidae) Recorded from India. *Journal of Bombay Natural History Society*, 88 (3): 428–438.

Ashmead, W.H. (1894). Report on the parasitic Cynipidae, part of the Braconidae, the Ichneumonidae, the Proctotrypidae, and part of the Chalcidinae. Part II. *Journal of Linnaean Society (Zoology)*, 25: 108–188.

Ashmead, W.H. (1888). Description of some unknown parasitic Hymenoptera in the collection of the Kansas State Agricultural College, received from Prof. E.A. Popenoe. *Bulletin of the Kansas State Agricultural College*, 3: 1–VIII.

Ashmead. (1904). Reclassification of the chalcid flies of the superfamily Chalcidoidea, with description of new species in the Carnegie Museum, collected in South America by Herbert H. Smith. *Memoirs of the Carnegie Museum*, 1: 225–555.

Aubert, B. (1991). Citrus greening disease, a serious limiting factor for citriculture in Asia and Africa. *Proceedings of the International Society of Citriculture*, 1: 817–820.

Ayyar, T.V.R. (1925). Notes on the status of some parasitic Hymenoptera in South India. *Journal of Bombay Natural History Society*, 30: 487–491.

Ayyar, T.V.R. (1927). The parasitic Hymenoptera of economic importance in South India. *Bulletin of Entomological Research*, 18: 73–78.

Ayyar, T.V.R. & Margabandhu, V. (1934a). Further records of Indo-Ceylonese chalcid flies. *Journal of Bombay Natural History Society*, 37: 193–196.

Batra, R.C. & Sandhu, G.S. (1981). Differential population of citrus leaf-miner and its. parasites on some commercial citrus cultivars. *Journal of Research, Punjab agricultural University*, 18: 170–176.

Bhatnagar, S.P. (1952). Description of new and records of known Chalcidoidea from India. *Indian Journal of Agricultural Science*, 21: 155–178.

Boucek, Z. (1958). Revision der europaischeTetracampidae (Hym. Chalcidoidea) miteinemKatalog der Arten der Welt. Revise evropskychTetracampidae s Katalogendruhucelecho Sveta. *Acta EntomologicaMuseiNationalisPragae*, 32: 41–90.

Boucek, Z. (1959a). A study of central European Eulophidae, I: Eulophidae (Hymenoptera) *Acta EntomologicaMuseiNationalisPragae*, 33: 117–170.

Boucek, Z. (1959b). A study of central European Eulophidae II: *Diaulinopsis* and *Cirrospilus* (Hym.). *Acta EntomologicaMuseiNationalisPragae*, 33: 171–194.

Boucek, Z. (1974). On the Chalcidoidea (Hymenoptera) described by C. Rondani. *Redia.*, 55: 241–285.

Boucek, Z. (1976). Taxonomic studies of some Eulophidae (Hymenoptera) of economic interest mainly from *Africa Entomophaga*, 21: 401–414.

Boucek, Z. (1979). Description of a new eupelmid parasite (Hymenoptera: Chalcidoidea) of cockroaches in India. *Bulletin of Entomological Research*, 69: 93–96.

Boucek, Z. (1986). Taxonomic study of chalcidoid wasps (Hymenoptera) associated with gall midges (Diptera: Cecidomyiidae) on mango trees. *Bulletin of Entomological Research*, 76(3): 393–407.
Boucek, Z. (1988). Australasian Chalcidoidea (Hymenoptera) CAB International Wallingford, U.K. Page 584–758.
Budhachandra, T., Khan, M.A., Sunita, P., & Kalmesh, M. (2013). A new species of genus *Pediobius* Walker (Hymenoptera: Eulophidae) with new records of Eulophid parasitoids from Uttarakhand, India. *Pantnagar Journal of Research*, 11(1), 29–34.
Burks, B.D. (1943). The North American parasitic wasps of the genus Tetrastichus - A contribution to Biological control of insect pest. *Proceedings of the United States Natural History Museum*, 93: 505–608.
Cameron, P. (1912). On a collection of parasitic Hymenoptera (chiefly bred), made by Mr. Walker W. Froggalt, F.L.S., in New South Wales, with descriptions of new genera and species. Part. 3. *Proceedings of the Linnean Society. N.S.W*, 37: 172–216.
Cameron, P. (1913). *Hymenoptera of the George town Museum Par V-Timehri*, 3(3): 105–137.
Chatterjee P.N. (1943). Description of *Rhogesplecopterae* sp. nov. from India with notes on its biology and mosphology. *Indian Journal Entomology*, 5: 213–221.
Chatterjee, P.N. (1941). Notes on some parasites of Shisham defoliators at Allahabad and Dehra-Dun, India. *Indian Journal Entomology*, 3: 157–172.
Chatterjee, P.N. (1944). On the biology and morphology of *Euplectrusparvulus* Fern hymenop: Eulophidae. *Indian Journal Entomology*, 4: 95–101.
Cherian, M.C. & Basheer, M. (1947). *Indian Journal Entomology*, 9: 141.
Cherian, M.C. & Pillai. (1941). *Indian Journal Entomology*, 4: 236–237.
Cherian, M.C. & Margabundhu, V. (1942). A new species of *Trichospilus* (Hymenoptera: Chalcidoidea) from South India. *Indian Journal Entomology*, 4: 101–103.
Crawford, J.C. (1911). Description of new Hymenoptera1. *Proceedings of the United States Natural History Museum*, 39: 617–623.
Crawford, J.C. (1914). New Philippine Hymenoptera. *The Philippine Journal of Science*. 9: 457–464.
Crawford, J.C., (1909). New Chalcidoidea (Hymenoptera). *Proceedings of Entomological Society of Washington*, 13: 233–234.
Crawford, J.C., (1912). Description of new Hymenoptera. No. 5. *Proceedings of the United States Natural History Museum*, 43: 185.
Dharmaraju and Pradhan. (1977). Factors affecting the realisation of the biotic potential in *Perisicrola nephantides*Muesebeck and *Trichospilus pupivora* Ferriere, parasites of the coconut leaf caterpillar, *Nephantisserinopa* Meyrick. *Indian Journal Entomology*, 38: 370.
Domenichini, G. (1953). Studio sullamortologia dell addomedeg li Hymenoptera Chalcidoidea. *Boll. Zool. Agr. Bachic*, 19: 193–298.
Dubey, O.P. (1974). Some-new and little known chalcidoid parasites (Hymenoptera) of leaf-mining Agromyzidae (Diptera) from India. *Oriental Insects*, 8: 413–418.
Dutt, G.R. (1921). *Rept. Proc. 4th Ent. Mtgs. Pusa*, 1920, 4: 287–290.
Ekka, R.K., Gupta, A. & Shaw, S.S. (2020). A study on the distribution pattern of genus *Elasmus*westwood (Hymenoptera: Eulophidae) from Chhattisgarh. *Journal of Parmacognosy and phytochemistry*, 9(2): 299–302.
Erdos, J. (1944). Species hungarical generis EntedonDalm. Kalosca. 63 pp.
Everson, G.W. (2014). Hymenoptera. In: *Encyclopedia of toxicology*, Ed. Wexler, P., pp. 984–985. Elsevier.
Ferriere, C. (1930). Notes on Asiatic Chalcidoidea. *Bulletin of Entomological Research*, 21: 353–360.
Ferriere, C. (1931a). Notes on African Chalcidoidea. *Bulletin of Entomological Research*, 22: 127–135.
Ferriere, C. (1931b). New chalcidoid egg-parasites from South Asia. *Bulletin of Entomological Research*, 22: 279–295.
Ferriere, C. (1936). The parasites of the coffee leaf miners (*Leucoptera* spp.) in Africa. *Bulletin of Entomological Research*, 27: 477–491.
Ferriere, C. (1938). Description of some African Eulophidae (Hym., Chalci.). *Bulletin of Entomological Research*, 29: 141–147.
Ferriere, C. (1940a). Chalcid flies attacking noxious beetles in India and New Guinea. *Bulletin of Entomological Research*, 30: 163–168.
Ferriere, C. (1940b). On some parasites and hyperparasites of *Artonaca toxantha* Hamps. *Bulletin of Entomological Research.*, 31: 131–139.
Ferriere, C. (1941). New species of Euplectrini (Hym. Chalcidoidea) from. Europe, Africa and Asia. *Bulletin of Entomological Research*, 32: 17–48.

Ferriere, C. (1953). Les parasites de 'LithocolletisPlatani' on Italie. *Boll. 1st. Ent. Univ. Bologna*, 19: 395–404.

Ferriere, C. (1960). Un novveaeTetrastichinae parasite d Anthomyidea au Japan (Hymenopteres: Eulophidae). *Acta hymenopterologica, Tokyo*, 1: 105–108.

Forster. (1856). HymenopterologischeStudien., II. Heft. pp. 152.

Gadallah, N.S., Yefremova, Z.A., Yegorenkova, E.N., Soliman, A.M., El–Ghiet, U.M., Edmardash, Y.A. & Edmardash, Y.A. (2015). A review of the family Eulophidae (Hymenoptera: Chalcidoidea) of Egypt, with thirty three new records. *Zootaxa*, 4058(1), 66–80.

Gahan, A.B. (1922). Report on a small collection of parasitic Hymenoptera from Java and Sumatra. *Treubia*, 3: 47–52.

Gauthier, N., Lasalle, J., Quicke, D.L. & Godfray, H.C.J. (2000). Phylogeny of Eulophidae (Hymenoptera: Chalcidoidea), with a reclassification of Eulophinae and the recognition that Elasmidae are derived eulophids. *Systematic Entomology*, 25(4), 521–539.

Girault. (1913). Australian Hymenoptera Chalcidoidea-IV. The family Eulophidae with description of new genera and species. *Memoirs of the Queensland Museum*, 2: 140–296.

Godfray, H.C.J. & Godfray, H.C.J. (1994). *Parasitoids: behavioral and evolutionary ecology* (Vol. 67). Princeton University Press. pp. 472.

Gokulpure, R.S. (1972). Notes on the hosts and parasites of *Phytomyzaatricornis* (Meigen) (Diptera: Agromyidae). *Indian Journal of Agricultural Sciences*, 42(7): 630–638.

Graham, M. W. R. de v. & LaSalle, J. (1991). New synonym in European Tetrastichinae (Hymenoptera: Eulophidae) including designation of some neuotypes, lectotypes and new combinations. *Entomologist's Gazette*, 42 (2): 89–96.

Graham, M.W.R. de, V. (1959). Keys to the British genera and species of ElachertinaeEulophidae, Entedontinae and Euderinae (Hym., Chalcidoidea). *Trans. Soc. Br. Ent.* 13: 169–204.

Gupta, A., Gawas, S.M. & Bhambure, R. (2015a). On the parasitoid complex of butterflies with descriptions of two new species of parasitic wasps (Hymenoptera: Eulophidae) from Goa, India. *Systematic parasitology*, 92(3), 223–240.

Gupta, A., Yeshwanth, H.M., Hansson, C. & Venkatesan, T. (2015b). *Neochrysocharis* nr. *diastatae* (Howard) (Hymenoptera: Eulophidae) parasitic on eggs of Letana Walker (Orthoptera: Tettigoniidae) in India: first record of host association. *Journal of Biological Control*, 29(3), 121–124.

Hansson, C. (1990). A Taxonomic study on the Palearctic species of *Chrysonotomyia* Ashmead and *Neochrysocharis*Kurdjumov (Hymenoptera: Eulophidae). *Entomologica Scandinavica*, 21(1): 29–52.

Hansson, C. (1994). Re-evolution of the genus *Closterocerus*Weatwood (Hymenoptera: Eulophidae). *Entomologica Scandinavica*, 25: 1–25.

Hansson, C. (1995). Revision of the Nearctic species of *Neochrysocharis*Kurdjumov (Hymenoptera: Eulophidae). *Entomologica Scandinavica*, 26(1): 27–46.

Hansson, C. (1996). The status of the genera Asecodes Forster, *lonympha* Graham and *Teleopterussilvestri* (Hymenoptera: Eulophidae) with a review of Nearctic species. *Entomologica Scandinavica*, 27 (2): 159–167.

Hayat, M. (1985). Chalcidoidea of India and adjacent countries. *Oriental Insects*, 19: 246–252.

Hayat, M. & Shahi, M.H. (2004). Taxonomic Notes on Indian Eulophidae (Hymenoptera: Chalcidoidea)—1. On the Types of Some Tetrastichinae. *Oriental Insects*, 38(1), 303–314.

Howard, L.O. & Ashmead, W.H. (1896). On some reared parasitic hymenopterous insects from Ceylon. *Proceedings of the United States National Museum*, 18: 633–648.

Hussain, T. & Khan, M.Y. (1986). Family Eulophidae. *Oriental Insects*, 20: 211–245.

Ikeda, E. (1996a). Revision of the Japanese species of *Chrysocharis* (Hymenoptera: Eulophidae). II. *Japanese Journal of Entomology*, 64 (2): 275–287.

Ikeda, E. (1996b). Revision of the Japanese species of *Chrysocharis* (Hymenoptera: Eulophidae). III. *Japanese Journal of Entomology*, 64 (3): 551–569.

Ikeda, E. (1999). A revision of the world species of *Quadrastichodella*Girault, with description of four new species (Hymenoptera: Eulophidae). *InsectaMatsumurana*, 55: 13–35.

Jamali, M.M. (2020). New Record of *lonympha* Graham (Chalcidoidea: Eulophidae) from India with Description of a New Species. *Journal of the Entomological Research Society*, 22(2), 131–135.

Jamali, M.M., Zeya, S.B. & Fatima, F. (2017). A new species of the genus *Anumanniola* Narendran (Chalcidoidea: Eulophidae) from India. *Entomon*, 42(3), 251–254.

Jamali, M.M., Zeya, S.B. & Veenakumari, K. (2018). A new species of the genus *Pediobius* Walker (Eulophidae: Entedoninae) parasitizing spider from India. *Halteres*, 9, 111–115.

Kamarudin, N.H. et al. (1996). Hymenopterans Parasitoids associated with the bagworms *Metisa plana* and *Mahasenacorbetti* (Lepidoptera: Psychidae) on oil palms in Peninsular Malaysia. *Bulletin of Entomological Research*, 86(4): 423–439.

Kamijo, K. (1976). Notes on Ashmead's and Crawfors's types of Eulophidae (Hymenoptera: Chalcidoidea) from Japan. *Kontyu, Tokyo.*, 44(4): 482–495.

Kamijo, K. (1977a). Descriptions of a new species of *Chrysocharis* Forster from Japan with notes on two species (Hymenoptera: Eulophidae). *AKITU N. Ser.*, 13: 1–6.

Kamijo, K. (1977b). Notes on Ashmead's and Crawford's types of *Pediobius* Walker (Hymenoptera, Eulophidae) from Japan, with description on a new species. *Kontyu Tokyo*, 45(1): 12–22.

Kamijo, K. (1977c). Five new species of *Cotterellia* (Hymenoptera, Eulophidae) from Japan. *Kontyu Tokyo*, 45(2): 253–261.

Kamijo, K. (1979). Eulophidae (Hymenoptera) from Korea with descriptions of two new species. *Annales Museihistorico-naturalishungarici*, 71: 251–264.

Kamijo, K. (1996). A new genus and species of *Eulophina* (Hymenoptera: Eulophidae) reared from spider eggs sac with notes on its allied genus *Eulomorpha* Dodd. *Japanese Journal of Entomology*, 64(3): 482–488.

Kaul, B.K. & Saraswat, G.G. (1974). Collection of chalcids (Hymenoptera) from India. *Oriental Insects*, 8: 185–194.

Kazmi, S.I. & Kumar, G. (2013). The species of *Elasmus* Westwood (Hymenoptera: Eulophidae: Eulophinae) from West Bengal, India. *Prommalia*, 1(1), 79–90.

Kazmi, S.I. & Sheela, S. (2017). Record of *Elasmus* Westwood (Hymenoptera: Chalcidoidea: Eulophidae: Eulophinae: Elasmini) of Punjab, India. *Munis Entomology and Zoology*, 12(1), 280–284.

Kazmi, S.I. & Sheela, S. (2018). Annotated checklist of Indian (Hymenoptera: Chalcidoidea: Eulophidae) with a new synonymy. *Journal of the Entomological Research Society*, 20(1), 73–85.

Kerrich, G.J. (1970b). On the type material and synonymy of *Elasmusanticles* Walker and *Pediobiusimbreus* Walker (Hymenoptera: Chalcidoidea). *Oriental Insects*, 4: 89–92.

Kerrich, G.J. (1973). A revision of the tropical and sub-tropical species of the Eulophid genus *Pediobius* Walker (Hymenoptera: Chalcidoidea). *Bulletin British Musuem of Natural History*, 29: 115–199.

Kerrich, G.J. (1974a). Systematic studies on Eulophidae of economic significance (Hymenoptera: Chalcidoidea). *Bulletin of Entomological Research*, 63: 629–639.

Kerrich, G.J. (1969). Systematic studies on eulophid parasites (Hym: Chalcidoidea), mostly of coffee leaf miners in Africa. *Bulletin of Entomological Research*, 59: 195–228.

Kerrich, G.J. (1974b). Systematic studies on Eulophidae of economic significance (Hymenoptera: Chalcidoidea). *Bulletin of Entomological Research*, 63: 629–639.

Khan, F.R. (2020). Records of *Elasmus* species (Hymenoptera: Chalcidoidea: Eulophidae) from different States of India with thirteen new state records. *Records of the Zoological Survey of India*, 120(2), 123–131.

Khan, F.R. & Hayat, M. (2011). Record of some species of *Elasmus* from India (Hymenoptera: Chalcidoidea: Eulophidae). *Bionotes*, 13(1), 7–10.

Khan, M.A. & Sushil, S.N. (1998). A new genus Guptaiella gen. nov. of Eulophidae(Hymenptera: Chalcidoidea). *Shashpa*, 5(1): 1–4.

Khan, M.A. & Sushil, S. (1992). Parasitic wasps of the genus Tetrastichus Haliday (Hymenoptera: Eulophidae) from Northern India. *Journal of the Bombay Natural History Society*, 89(3): 329–347.

Khan, M.A., Agnihotri, M. & Sushil, S.N. (2005). Taxonomic studies of eulophid parasitoids (Hymenoptera: Chalcidoidea) of India. *Pantnagar Journal of Research*, 2(1): 110.

Khan, M.A., Samraj, D.A. & Khan, N. (1986). *Tetrastichusdavidi* sp. nov.(Hymenoptera: Eulophidae) a primary parasitoid associated with Eariasvittella (Fabricius) (Lepidoptera: Noctuidae) from India. *Journal of the Bombay Natural History Society*, 83(2): 409–411.

Khan, M.Y. & Shafee, S.A. (1980a). Taxonomic studies on some Indian eulophid parasites (Hymenoptera: Chalcidoidea). *Journal of Bombay natural History Society*, 76: 324–334.

Khan, M.Y. & Shafee, S.A. (1980b). Three new species of the genus *Neochrysocharis*Kurdyu (Eulophidae: Entedontinae) from India. *Journal of Bombay natural History Society*, 76(3): 486–490.

Khan, M.Y. & Shafee, S.A. (1981a). On some new species of *Tetrastichus*Haliday (Hymenoptera: Eulophidae). *Journal of Bombay natural History Society*, 78(2): 337–343.

Khan, M.Y. & Shafee, S.A. (1981b). Two new species of Tetrastichinae Foerster (Hymenoptera: Eulophidae) from India. *Journal of Bombay natural History Society*, 78(2): 344–347.

Khan, M.Y. & Shafee, S.A. (1981c). New species of *Chrysonotomyia* Ashmead (Hymenoptera: Eulophidae) from India. *Journal of Bombay natural History Society*, 78(2): 348–351.

Khan, M.Y. & Shafee, S.A. (1982). Species of the genus *Pediobius* Walker (Eulophidae:Entedontinae) from India. *Journal of Bombay natural History Society*, 79: 370–374.

Khan, M.A. (1982). A new species of *Tetrastichus*Haliday (Hymenoptera: Eulophidae) from high altitude of India. *Journal of Bombay natural History Society*, 79(2): 375–378.

Khan, M.A. (1983). A new species of *Tetrastichus* (Hymenoptera: Eulophidae) from India. *Journal of Bombay natural History Society*, 80(1): 185–188.

Khan, M.A. (1985a). Eulophid parasites (Hymenoptera: Eulophidae) of Agromyzidae in India. *Journal of Bombay natural History Society*, 82(1): 149–159.

Khan, M.A. (1985b). Two new species of *Chrysocharis* Forster (Hymenoptera: Eulophidae) from high altitude of India. *Journal of Bombay natural History Society*, 82(2): 376–380.

Khan, M.A. (1992). A new species of the genus *Stenomesius* Westwood (Hymenoptera: Eulophidae) from India. *Bull. Lab. Ent. Agr. Filippo Silvestri*, 49: 23–30.

Khan, M.A. (1993). *Mohaniella indica* gen and sp. nov. (Hymenoptera: Eulophidae) described from india. *Boll. Lab. Entomol. Agr. Filippo Silvestri*, 50: 3–7.

Khan, M.A. (1995). Parasitic wasps of the genus *Dicladocerus* (Hymenoptera: Eulophidae) from Northern India. *Journal of Insect Science*, 9(2): 102–111.

Khan, M.A. (1996). Parasitic wasps of the genus *Pediobius* Walker (Hymenoptera: Eulophidae) from India. *Journal of Insect Science*, 9(2): 102–111.

Khan, M.A. & Singh, R.S.J. (1994). A new species of *Steriomesius* Westwood (Chalcidoidea: Eulophidae) from Nainital hills. *Shashpa*. 1(1): 11–16.

Kieffer, J.J. (1905). New Eucharinae and Chalcidinae. *Deutsche entomologischeZeitschrift*, 49: 240–265.

Klian, M.A. & Sushil, S.N. (1992). Parasitic wasps of the genus *Tetrastichus*Haliday in.Northern India. *Journal of Bombay natural History Society*, 89: 329–347.

Kosheleva, O.V. (2019). Additions to the fauna and distribution of parasitoids from the family Eulophidae (Hymenoptera: Chalcidoidea) of Russia and some adjacent territories. In *ТрудыРусскогоэнтомологическогообщества*, 90 (pp. 22–32).

Krishna Ayyar, P.N. 1942. *Indian J. Ent.*, 4: 225–232.

Kumar, V., Kumar, S. & Rawat, S. (2020). A species of the genus *Tamarixia* (Hymenoptera: Eulophidae) parasitic on gall forming triozid (Homoptera: Psylloidea: Triozidae) infesting the leaves of *Terminalia arjuna* in Uttarakhand (India). *CIBTech Journal of Zoology*, 9: 93–99.

Kumar, V., Pant, P., Bisht, V., Bhat, S. & Kumar, S. (2018). Two reared species of the genus *Aprostocetus* (Hymenoptera: Eulophidae), parasitic on gall forming psylloid*Pauropsyllaficicola* Kieffer infesting *Ficus auriculata* Lour. in Uttarakhand, India. *Journal of Entomology and Zoology Studies*, 6(5), 906–910.

Kundu, G.G. & Kishore, P. (1973). *Indian Journal of Entomology*., 34: 80–81.

Kurian, C. (1953). Description of new and record of some known parasitic Hymenoptera from India. *Agra University Journal Research. (Sci.)*, 2: 113–124.

Kurian, C.(1954). Description of some new chalcid parasites (Parasitic Hymenoptera) from India. *Agra University Journal Research. (Sci.)*, 3: 119–134.

LaSalle, J. (1986). Notes on Tetrastichinae types in the Zoological Institute; Leningrad (Hym: Eulophidae). *Proceedings of the Entomological Society of Washington*, 88(3): 599–603.

LaSalle, J. (1989). Notes on the genus *Zagrammosoma* (Hymenoptera: Eulophidae) with description of a new species. *Proceedings of the Entomological Society of Washington*, 91: 230–236.

LaSalle, J. (1990a). A new genus and species of Tetrastichinae (Hymenoptera: Eulophidae) parasitic on the coffee berry borer *Hypothenemushampei* (Ferrasi) (Coleoptera: Scolytidae). *Bulletin of Entomological Research*, 80(1): 7–10.

LaSalle, J. (1990b). Tetrastichinal (Hymenoptera: Eulophidae) associated with spider egg sacs. *Journal of Natural History*, 24: 377–389.

LaSalle, J. (1994). North American genera of Tetrastichinae. (Hymenoptera: Eulophidae). *Journal of Natural History*, 28: 109–236.

LaSalle, J. & Schauf, M.E. (1990). New subfamily placement for some North American Eulophidae (Hymenoptera: Chalcidoidea). *Proceedings of the Entomological Society of Washington*, 92: 282–285.

LaSalle, J. & Schauf, M.E. (1992). Preliminary studies on neotropical Eulophidae (Hymenoptera: Chalcidoidea): Ashmead. Cameron. Howard and Walker species. *Contributions of the American Entomological Society*, 27: 47.

LaSalle, J. & Schauf, M.E. (1994). Systematics of the tribe Euderomphalini (Hymenoptera: Eulophidae): parasitoids of whiteflies (Homoptera: Aleyrodidaae). *Systematic Entomology*, 19(3): 235–25.

LaSalle, J.O.H.N. (1993). Parasitic Hymenoptera, biological control and biodiversity. *Hymenoptera and biodiversity*, 197–215.

LaSalle, J. & Pena, J.E. (1997). A new species of *Galeopsomyia* (Hymenoptera: Eulophidae: Tetrastichinae): a fortuitous parasitoid of the citrus leafminer, Phyllocnistiscitrella (Lepidoptera: Gracillariidae). *Florida Entomologist*, 80: 461–470.

Mahadihassan, S. (1923). Classification of lac insects from physiological stand point. *J. Sci. Maharaja's Coll. Vizinagaram*, 1: 47–99.

Mahendran, B., Agnihotri, M. & Pillai, K. (2014). A New Species of Parasitoid Wasp Euderus (Eulophidae: Chalcidoidea: Hymenoptera) from Uttarakhand, India on Dainha Shoot Weevil. *Madras Agricultural Journal*, 101.

Mani, M.S. (1938). *Catalogue of Indian Insects*. Part 23: 1–170.

Mani, M.S. (1939). Description of new.and records of some known Chalcidoid and other hymenopterous parasites from India. *Indian Journal of Entomology*, 1: 69–99.

Mani, M.S. (1941). Parasitic Hymenoptera from India. *Indian Journal of Entomology*, 3: 25–36.

Mani, M.S. (1971). Some Chalcidoid parasites (Hymenoptera) of leaf-mining Agromyzidae (Diptera) from India. *Journal of Natural History*, 5: 591–598.

Mani, M.S. & Kurian, C. (1953). Descriptionsand records of chalcids (Parasitic Hymenoptera) from India. *Indian Journal of Entomology*, 15: 1–21.

Mani. M.S. (1941). Parasitic Hymenoptera from India. *Indian Journal of Entomology*, 3: 25–36.

Masi, L. (1907). Contribuzioniallaconoscenzadeichalcididiitaliani. *Boll. Lab. Zoo. gen. agr. Scuolo Agric. Portia*, 1: 231–295.

Masi, L. (1917). Chalcididae of the Seychelles Islands. *Novit. Zool.*, 24: 121–230.

Mathur, R.N. (1942). *Indian Forest Rec. (N.S.)*, 7 (11): 1–65.

Miller, C.D.F.. (1962). Some nearctic species of Chalcid genus *Enasopa*Delucchi (Eulophidae: Entedontinae). *Canadian Entomologist*, 94: 1039–1052.

Miller, C.D.F. (1964). A new species of GrotiusomyiaGirault (Hymenoptera: Eulophidae). *Canadian Entomologist*, 96: 1021–1023.

Mooventhan, P., Murali Baskaran, R.K., Kumar, J. & Kaushal, P. (2020). Technical Bulletin on Current Status and Guidelines for Safe Use of Pesticides in Agriculture. ICAR-National Institute of Biotic Stress Management, Raipur, Chhattisgarh, India. pp. 32.

Mooventhan, P., Murali Baskaran, R.K., Sharma, K.C., Sivalingam, P.N., Sridhar, P., Mohan, S. & Kaushal, P. (2018). Eco-friendly pest management technologies. ICAR-National Institute of Biotic Stress Management, Raipur, Chhattisgarh, India. pp. 32.

Narayan, K. & Jairaj, S. (1973). *Madras Agriculture Journal*, 60: 640–641.

Narayanan, E.S. & Kundanlal. (1953). *Indian Journal of Entomology*, 15: 173–179.

Narayanan, E.S. (1960). Two new species of chalcidoid parasites from India. *Proceedings of Indian Academy of Science*, 52: 119–123.

Narayanan, E.S. & Subba Rao, B.R. (1960). New species of Encyrtid and Braconid parasites. *Indian Journal of Entomology*, 22: 75–79.

Narayanan, E.S., Subba Rao, B.R. & Ramachandra Rao, M. (1960). Some new species of chalcids from India. *Proceedings of the National Institue of Science, India*, 26 (B): 168–175.

Narendran, T.C. & Krishna Kumar, N.K. (1995). Systematics of *Ceratoneuraindi*Girault (Hymenoptera: Eulophidae) a pest of sweet pepper and brinjal in India. *Journal Tropical Agriculture*, 33(1): 47–49.

Narendran, T.C. (2007). Indian chalcidoidparasitoids of the Tetrastichinae (Hymenoptera: Eulophidae). *Occasional Paper-Records of the Zoological Survey of India*, 272. record number: 20083193101.

Narendran, T.C. & Bindu, K. (2009). A new species of *Elachertus* (Hymenoptera: Eulophidae) from Pepper leaf galls of Liothripskarnyi (Thysanoptera: Phaleothripidae) from India. *Oriental Insects*, 43(1): 45–48.

Narendran, T.C. & Hayat, M. (2007). A new species and a key to species of *Euderomphale*Girault (Hymenoptera: Eulophidae) from India. *Zoos' Print journal*, 22(10): 2862–2863.

Narendran, T.C. & Sheela, S. (1993). Descriptions of an interesting new genus and a new species of Epichrysomalinae (Hymenoptera: Agaonidae) from India. *Journal of the Zoological Society of Kerala*, 3(1): 7–12.

Narendran, T.C. & Sinu, P.A. (2003). A new genus and a new species of Eulophidae (Hymenoptera: Chalcidoidea) from India. *Zoos' print journal*, 18(3): 1031–1033.

Narendran, T.C., David, B.V. & Selvaraj, P. (2005). A new species of *Aprostocetus* Westwood (Hymenoptera: Eulophidae) parasitic on Melanagromyzaobtusa (Malloch)(Diptera: Agromyzidae) from India. *Entomon*, 30(3): 221–225.

Narendran, T.C., Fousi, K., Rajmohana, K. & Chandrika, M. (2002a). A new species of Eulophidae (Hymenoptera: Chalcidoidea) parasitoid on the slug caterpillar pest, *Contheylarotunda* Hampson (Lepidoptera: Cochliidae) of coconut in Kerala. *Journal of Applied Zoological Researches*, 13(1): 31–34.

Narendran, T.C., Girishkumar, P., Santhosh, S., Sheeba, M. & Jilcy, M.C. (2007). A taxonomic revision of *Pediobomyia*Girault (Hymenoptera: Eulophidae). *Journal of Entomological Research*, 31(2): 173.

Narendran, T.C., Kumar, P.G., Santhosh, S. & Jilcy, M.C. (2006). A revision of *Neotrichoporoides*Girault (Hymenoptera: Eulophidae) from India. *Oriental Insects*, 40(1): 1–21.

Narendran, T.C., Razak, N. & Sureshan, P.M. (2013). A taxonomic review of *Chrysocharis* Foerster (Hymenoptera: Eulophidae) of Indian subcontinent. *Records of the zoological Survey of India*, 113(2): 13–34.

Narendran, T.C., Santhosh, S. & Kumar, P.G. (2005). A new species of eulophidae (Hymenoptera: Chalcidoidea) from Kerala, India. *Phytoparasitica*, 33(5): 495–498.

Narendran, T.C., Sheeba, M., Santhosh, S., Jilcy, M.C. & Peter, A. (2007b). A Review of the species of *Deutereulophus* Schulz (Hymenoptera: Eulophidae) Of Oriental Region. *Uttar Pradesh Journal of Zoology*, 27, 61–67.

Narendran, T.C., Galande, S.M. & Mote, U.N. (2001). Two new species of Eulophidae (Hymenoptera: Chalcidoidea) from Maharashtra, India. *Uttar Pradesh Journal of Zoology*. 21(2): 153–157.

NBAIR., (2014). Directors report, Accessed at https://www.nbair.res.in/sites/default/files/2018-12/directorreport.pdf. Accessed on: 10.03.21.

Nees Ab Esenbeck, C.G. (1834). HymenopterorumIchneumonibusaffiniummonographical, genera europaea et species illustrantes. 2. Stuttgart & Tubingen. 448 pp.

Nikol'skaya, M.N. (1952). The chalcid fauna of the USSR (Chalcidoidea). *Opred. Faune Zool. Inst. Alcad. Nauk. SSR*, 44: 575.

Noyes, J.S. (2003). Universal Chalcidoidea database. Available from: http://www.nhm.ac.uk/research-curation/projects/chalcidoids/ (accessed 30 Dec 2020).

Nwanze, K.F., Reddy, Y.V.R., Nwilene, F.E., Kausalya, K.G. & Reddy, D.D.R. (1998). Tritrophic interactions in sorghum, midge (*Stenodiplosissorghicola*) and its parasitoid (*Aprostocetus* spp.). *Crop Protection*, 17(2): 165–169.

Olivier, M. (1791). Encycl method. Ins. 5. Panckoucke, Paris. pp. 753.

Pant, P., Bhat, V.K.S. & Kumar, S. (2019). A new host record of an eulophineparasitoid of the genus Elasmus (Hymenoptera: Eulophidae) from Karnataka, India. *Pantnagar Journal of Research*, 17: 3.

Parra, J.R.P., Alves, G.R., Diniz, A.J.F. & Vieira, J.M. (2016). Tamarixia radiata (Hymenoptera: Eulophidae)× Diaphorinacitri (Hemiptera: Liviidae): mass rearing and potential use of the parasitoid in Brazil. *Journal of Integrated Pest Management*, 7(1): 5.

Parshuram, M.S. & Agnihotri, M. (2017). A new species of Chrysonotomyia Ashmead (Hymenoptera: Eulophidae: Entedoninae) from Uttarakhand, India. *Journal of Entomology and Zoology Studies*, 5(3): 323–326.

Peck, O. (1963). A catalogue of the Nearctic Chalcidoidea (Insecta:Hymenoptera). *Can. Ent. Suppl.*, 30: 1092.

Pruthi, H.S. & Mani, M.S. (1940). Biological notes on Indian parasitic Chalcidoidea. *Misc. Bull. LC. A. R.*, 30: 1–44.

Rai & Gowda. (1977). *International Rice Communication Newsletter*, 26: 35.

Ramanagouda, S.H., Harish, K., Vastrad, A.S. & Shashidhar, V. (2010). Biology of the eucalyptus gall wasp, Leptocybeinvasa Fisher & La Salle. *Pest Management and Economic Zoology*, 18(1/2): 66–69.

Rao & Ali. 1976. *Indian Journal of Entomology*, 38: 191–193.

Rao, N.B.V., Padma, E. & Ramanandam, G. (2017). *Pediobius imbrues* (Hymenoptera: Eulophidae), a promising parasitoid of coconut slug caterpillar, *Macroplectranararia* Moore (Lepidoptera: Limacodidae) in Andhra Pradesh. *Pest Management in Horticultural Ecosystems*, 23(2): 176–178.

Rao, N.C., Rao, G.K. & Roshan, D.R. (2019). Field efficacy of the parasitoid, *Pediobius imbrues* (Hymenoptera: Eulophidae) on coconut slug caterpillar, *Macroplectranararia* Moore (Lepidoptera: Limacodidae). *Journal of Biological Control*, 33(4): 382–384.

Raodeo, A.K., Tikar, D.J. & Chundurwar, R.D. (1972). *Current Science*, 41: 430–431.

Ratzeburg, J.T.C. (1852). Die Ichneumones der Fortinsecten in forstlicher and entomologischerBeziehung. 3, xviii & 272 pp., 3 tbs. Berlin.

Rawat, R.R. & Sahu, H.R. (1969). *Indian Journal of Entomology*, 30: 319.

Reddy, S.S. & Mandal, S.K. (2014). Identification of natural enemies of the Hadda beetle, *Epilachnavigintioctopunctata* (Fabr.): Parasitoids and studies on life cycle duration of larval-pupal parasitoid*Pediobiusfoveolatus* (Crawford). *The Ecoscan*, 9: 757–762.

Risbec, J. (1951). Les Chalcidoid d' A.O.F. *Mem. Inst. Fr. Afr. Noire.*, 13 (7): 409.

Risbec, J. (1952). Contribution a l' etude des Chalcidoides de Madagascar - *Mémoires de l'InstitutScientifique de Madagascar*, 2: 1–449.

Rohwer, S.A. (1921). Descriptions of new chalcidoid flies from Coimbatore, South India. *Annals and Magazines of Natural History*, 9(7): 123–135.

Rondani, C. (1872). Sopra alcunivespariiparassiti. *Note BollettinodellaSocietàentomologicaitaliana.*, 4: 201–208.

Sankar, M., Jaigeetha, S. & Rao, M.S. (2014). Exploitation of biocontrol agents, *Trichogrammachilonis* and *Tetrastichushowardi* on yield improvement in sugarcane at EID Parry (India) Ltd., Sugar Mill Command Areas. *International Journal Of Innovative Research & Development*, 3(8): 314–318.

Santhosh, S. & Karthikeyan, K. (2017). A new species of *Tetrastichus*Haliday (Hymenoptera: Eulophidae) parasitizing the rice case worm, Parapoynxstagnalis (Zeller)(Lepidoptera: Crambidae) from India. *Oriental Insects*, 51(2): 160–165.

Santhosh, S., Babu, S.R. & Meena, P.K. (2020). Two New Host Records of *Euplectrus*euplexiaeRohwer (Hymenoptera: Eulophidae) from Southern Rajasthan, India. *Biosystematica*, 11(1&2): 13–17.

Saraswat, G.G. (1975). On some Tetrastichus (Hymenoptera: Chalcidoidea) from India. *Memoirs of the School Entomology, St. John's College, Agra*, 4: 1–34.

Saraswat, G.G. (1978). On a small collection of chalcids (Hymenoptera: Chalcidoidea) from India. *Memoirs of the School Entomology, St. John's College, Agra*, 5: 99–110.

Schauff, M.E. & LaSalle, J. (1993). Nomenclatural Notes on genera of North American Eulophidae (Hymenoptera: Chalcidoidea). *Proceedings of theEntomological Society of Washington*, 95(3): 488–503.

Schauff, M.E. LaSalle, J. & Wijesekara, G.A. (1998). The genera of chalcid parasitoids (Hymenoptera: Chalcidoidea) of citrus lesf miner *Phyllocnistiscitrella* Stainton (Lepidoptera: Gracillaridae). *Journal of Natural History*, 32(7): 1001–1056.

Schauff, M.E. (1985a). Taxonomic study of the Nearctic species of *Elachertusspinola* (Hymenoptera: Eulophidae). *Proceedings of theEntomological Society of Washington*, 87: 84–58.

Schauff, M.E. (1985b). Revision of the Nearctic species of *Hypossopus*Girault (HymEulophidae). *Journal of the New York Entomological Society* 93 (3): 1096–1108.

Schauff, M.E. (1987). Microdonophagus a new entedontine genus (Hym: Eulo.) from Panama. *Proceedings of theEntomological Society of Washington*, 88 (1): 167–173.

Schauff, M.E. (1988). The species of *Entedon* (Hymenoptora: Eulophidae) in America, North of Mexico. *Journal of the New York Entomological Society*, 96: 30–62.

Schauff, M.E. (1991). The Holarctic genera of Entedoninae (Hymenoptera: Eulophidae). *Contributions of the American Entomological Institute*. 26: 109.

Shafee, S.A. & Rizvi, S. (1984). *Terebratella indica*sp.n. representing a new genus of Tetrastichinae (Hym: Eulophidae) from Muzzafarpur, India. *Journal of the Swiss Entomological Society*, 57(4): 377–378.

Shafee, S.A. & Rizvi, S. (1985). Three new species of Pediobius Walker (Hym.: Eulo) from South India. *Journal ofBombay Natural History Society*, 82: 162–165.

Shafee, S.A. et al. (1983a). Descriptions of two new species of TetrastichusHaliday (Hym: Eulo.) from India. *Journal ofBombay Natural History Society*, 80 (2): 393–398.

Shafee, S.A. et al. (1983b). Two new species of Eulophidae (Hym: Chalcidoidea) from Andaman Island. *Journal ofBombay Natural History Society*, 80 (3): 618–620.

Sharkey, M.J. (2007). Phylogeny and classification of Hymenoptera. *Zootaxa*, 1668(1), 521–548.

Sharma, H.C. & Davies, J.C. (1988). Insect and other animal pests of millets. *International Crops Research Institute for the Semi-Arid Tropics, Patancheru, Andhra Pradesh, India*. 165 pp.

Shree, A. & Singh, S. (2015). Diversity of family Eulophidae (Hymenoptera: Chalcidoidea) in forest, horticulture, and habitats of Doon valley of Uttarakhand, India. *Indian Forester*, 141(8): 873–876.

Shylesha, A.N., Kumar, P.S., Sreedevi, K. & Ballal, C.R. (2018). Biocontrol rescues the paper industry: management of Eucalyptus gall wasp *Leptocybeinvasa. ICAR- National Bureau of Agricultural Insect Resources*. Accessed at https://nbair.res.in/sites/default/files/2018-12/EWG%20brochure_0.pdf. Accessed on: 10.03.21.

Silvestri, F. (1910). Introduzione in italia d'un imenotteroindiano per combottere la moscadellearanie. *Bollettino del Laboratorio di zoologiagenerale e agrariadella R. Scuola superiored'agricoltura in Portici.*, 4: 228–244.

Singh, R.S.J. & Khan, M.A. (1994a). Description of a new species of *Closterocerus* Westwood (Chalcidoidea: Eulophidae) a parasitoid associated with sisam leaf miner. *Journal of Insect Science* 9(2): 158–159.

Singh, R.S.J. & Khan, M.A. (1994b). *Mestocharella* sp., a new record of Eulophid from Northern India. *Journal of Bombay Natural History*, 92: 85–95.

Singh, R.S.J. & Khan, M.A. (1995). A new species of genus *Mestocharella*Girault (Hymenoptera: Eulophidae) from Pantnagar, India. *Shashpa*, 2(1): 23–26.

Singh, R.S.J. & Khan, M.A. (1996). Description of a new species of *Closterocerus* Westwood (Chalcidoidea: Eulophidae) a parasitoid associated with sisam leaf miner. *Journal of Insect Science*, 9(2): 158–159.

Singh, R.S.J. & Khan, M.A. (1997). A new genus and a new species of Eulophinae (Chalcidoidea: Eulophidae) from India. *Shashpa*, 4(1): 9–11.

Singh, R.S.J. & Khan, M.A. (1998). A new species of *Omphale*Haliday (Chalcidoidea: Eulophidae) parasitizing eggs of coccinellid beetle from Pantnagar, India. *Shashpa*, 5(2): 123–126.

Singh, S. (2005). Description of a new species of Euderus (Hymenoptera: Chalcidoidea: Eulophidae), an-eggparasitoid of *Alcidodesludificator* (Coleoptera: Curculionidae) a pest of *Gmelina arborea. Entomon* 30(4): 321–326.

Singh, S. (2017). Two new species of Aprostocetus (Hymenoptera: Eulophidae) parasitizing leaf gall forming Psylloids (Hemiptera: Triozidae) on two economically important trees in India. In *Annales Zoologici* (Vol. 67, No. 4, pp. 795–810). Museum and Institute of Zoology, Polish Academy of Sciences.

Singh, S. (2018). A new species of Aprostocetus Westwood (Hymenoptera: Eulophidae), parasitizing mango leaf gall midge (Diptera: Cecidomyiidae), from India. *Journal of Asia-Pacific Entomology*, 21(2), 553–559.

Singh, S.P. (2004). *Some success stories in classical biological control of agricultural pests in India* (No. C044. 006). Asia-Pacific Association of Agricultural Research Institutions (APAARI); FAO. Accessed at: http://www.apaari.org/wp-content/uploads/2009/05/ss_2004_02.pdf. Accessed on 10.03.21.

Subba Rao, B.R. (1957). Some new species of Indian Eulophidae. *Indian Journal of Entomology*, 19: 50–53.

Subba Rao, B.R. & Ramamani, S. (1965). Biology of *Cirrospiloideusphyllocnistoides* (Narayanan) and description of a new species, *Scotolinxquadristriate* (Hymenoptera: Eulophidae) as parasites of *Phyllocnistiscitrella* Stainton. *Indian Journal of Entomology*, 27(4): 408–413.

Subba Rao, B.R. & Sharma, A.K. (1966). Species of parasites reared from *Stomopteryxnerteria* (Meyrick). (Lepidoptera: Gelechidae). *Indian Journal of Entomology*, 28: 299–304.

Subba Rao, B.R., Baldev, P., Ram, A., Singh, R.P. & Srivastava, M.L. (1967). Further notes on Hyperapostica (Gyllenhal) and its natural enemies. *Indian Journal of Entomology*, 29: 370–379.

Surekha, K. & Narendran, T. (1993). Description of a new species of *Pediobius* Walker (Hymenoptera: Eulophidae) from India. *Geobios New Reports*, 12: 3–4.

Surekha, K. & Narendran, T.C. (1992). Taxonomy of four new species of Eulophidae (Hymenoptera: Chalcidoidea) from India. *Journal of Ecobiology*, 4(4): 265–270.

Surekha, K., La Sallie, J., Kumar, S.V.V. & Murphy, S.T. (1996). A new species of Sympiesis (Hymenoptera: Eulophidae) parasitic in the teak defoliator *Hyblaeapurera* (Lepidoptera: Hyblaeidae) in India. *Bulletin of Entomological Research.*, 86: 73–76.

Sureshan, P.M. & Narendran, T.C. (2001). A new species of LadnaBoucek (Hymenoptera: Chalcidoidea: Eulophidae) from India. *Oriental Insects*, 35: 203–206.

Sushil, S.N. & Khan, M.A. (1997). A new genus *Shardiella* gen. n. of Eulophidae (Hymenoptera: Chalcidoidea) from Northern India. *Shashpa*, 4(2): 87–90.

Taley, Y.M. & Thakare, K.R. (1979). *Indian Journal of Agricultural Sciences*, 49: 344–354.

Thomson, C.G. (1878). Hymenoptera scandinaviae (Vol. 1). typisexpressit Fr. Berling. pp286.

Triapitsyn, S. & Headrick, D.H. (1995). A review of the *Nearctic* spp. of the thrips attacking genus *Ceranisus* Walker (Hym: Eulo). *Transactions of American Entomological Society*, 121(4): 227–248.

Ujiye, T. & Adachi, I. (1995). Parasitoids of the citrus leafminer, Phyllocnistiscitrella Stainton (Lepidoptera: Phyllocnistidae) in Japan and Taiwan. *Bulletin of the Fruit Tree Research Station (Japan).* ISSN: 0916-5851.

Vanitha, K. (2015). A report on the occurrence of eulophid parasitoids on the Cashew Leaf Miner Acrocercopssyngramma Meyrick (Insecta: Lepidoptera: Gracillariidae). *Journal of Threatened Taxa*, 7(12), 7933–7936.

Vireck, H.L. (1916). Chalcidoidea connectient. *State Geological Natural History Survey Bulletin*, 22: 443–528.

Walker, F. (1839). MonographiaChalciditum. I. London. pp. 333.

Walker, F. (1872). Hormoceridae, Sphegigasteridae, Pteromalidae, Elasmidae, Elachistidae, Eulophidae, Entedonidae, Tetrastichidae and Trichogrammidae. Part 6. *Notes on Chalcidiae*, part 6: 89–105.

Westwood, J.O. (1840). An introduction to the modern classification of insects: founded on the natural habits and corresponding organisation of the different families (Vol. 2). Longman, Orme, Brown, Green, and Longmans.

Wijesekara, G.A.W. & Schauff, M.E. (1994). Revision of the tribe Euplectrina of Sri Lanka (Hymenoptera: Eulophidae). *Oriental Insects*, 28: 1–48.

Wijesekara, G.A.W. & Schauff, M.E. (1997). Two new genera and three new species of Euplectrinae (Hymenoptera: Eulophidae) from the new world. *Proceedings of Entomological Society of Washington*, 99(1): 101–109.

Yoshimoto, C.M. (1971). Revision of the genus Euderus of North of America Mexico (Hymenoptera: Eulophidae). *Can. Ent.*, 103: 541–578.

Yousuf, M., Singh, S., Ikram, M. & Singh, R.B. (2017). An overview on outbreak of Eucalyptus gall wasp, *Leptocybeinvasa* (Hymenoptera: Eulophidae) in Northern India. *Journal of Entomology and Zoology Studies*, 5(5): 496–501.

Zheng, X.L., Li, J., Yang, Z.D., Xian, Z.H., Wei, J.G., Lu, W. & Wang, X.P. (2014). A review of invasive biology, prevalence and management of *Leptocybeinvasa* Fisher & La Salle (Hymenoptera: Eulophidae: Tetrastichinae). *African Entomology*, 22(1): 68–79.

9 Trichogrammatid Parasitoids

Omprakash Navik, Richa Varshney, Y. Lalitha, and Sushil Kumar Jalali
ICAR-National Bureau of Agricultural Insect Resources, Bengaluru, India

CONTENTS

LEARNING OBJECTIVES

1. To know the systematics, history and classification of trichogrammatids.
2. To learn the size and distribution of trichogrammatids.
3. To know the diversity, host range and parasitism of trichogrammatids.
4. To know biology, ecology and behaviour of trichogrammatids.
5. To understand the importance of geographical strains of trichogrammatids.
6. To learn about improvements in trichogrammatid parasitoids.
7. To learn mass production and release strategies in biological control.

9.1 INTRODUCTION

Trichogrammatidae is a widespread family of Chalcidoidea, comprises of tiny parasitic wasps. Almost all trichogrammatids are primary, solitary or gregarious endoparasitoids of the eggs of other insects. Trichogrammatids parasitize the eggs of insects belonging to orders Lepidoptera,

DOI: 10.1201/9781003354239-9

Coleoptera, Hemiptera, Thysanoptera, Diptera, Hymenoptera and Neuroptera either in aquatic and terrestrial habitats (Querino et al., 2010). Trichogrammatids are very small to minute parasitic wasps, ranging in size from 0.2 to 1.5 mm (Pinto and Stouthamer, 1994). The members of this family are easily recognized by the three-segmented tarsi. The minute size, fragility and the problems related to the collection and curation of such insects are responsible for the great extent of their unexplored taxonomy and biology (Pinto and Stouthamer, 1994; Pinto, 2006). The lack of methods to preserve them permanently had hindered progress in the taxonomic study, therefore this group did not receive enough attention from taxonomists (Pinto, 2006; Noyes, 2003). However during the last forty years, steady progress has been made in collecting and identifying trichogrammatids from different parts of the world (Pinto, 2006; Hayat, 2008; 2009: Yousuf and Shafee, 1988; Querino and Zucchi, 2019).

Worldwide, more than 1000 species of trichogrammatids belonging to 96 genera are described from terrestrial and aquatic habitats (Noyes, 2019). Several trichogrammatids have been enormously used in biological control programs of various pests, particularly against the Lepidopteran pests, resulting in enormous economic importance (Jalali et al., 2016a, b). Trichogrammatid parasitoids, especially *Trichogramma* Westwood and its close relative *Trichogrammatoidea* Girault, have been studied extensively and intensively over the world for their use in biological control (Pinto and Stouthamer, 1994). The genus *Trichogramma* Westwood is the largest genus with about 235 described species, and *Trichogrammatoidea* Girault is a smaller taxon of 29 known species worldwide (Pinto, 2006; Querino and Zucchi, 2019). The ease of mass production of these two genera on factious hosts has resulted in large scale area coverage under biological control of agricultural and forest pests (Smith, 1996). Further, efforts were made to enhance the efficacy of trichogrammatid parasitoids through improved breeding and production systems, and also through inducing various traits against stresses that improve the performance of parasitoids in different agroecosystems. Therefore, this chapter reviews the work done on these important egg parasitoids across the world.

9.2 SYSTEMATICS

The Trichogrammatidae, a widespread family of Chalcidoidea, represents more than 1000 species and 96 genera worldwide. The trichogrammatids are a group of large minute parasitic wasps, which attack eggs of various insects, of which many are of economic importance. The members of Trichogrammatidae are one of the smallest egg parasitoids among the insect groups, ranging in length from 0.2 to 1.5 mm. The three-segmented tarsi separate the family from the other Chalcidoids (Pinto and Stouthamer, 1994). The trichogrammatid members are relatively uniform in morphology, compact body structure, reduced antennal segments and their development as egg parasitoids of various insect groups (Querino et al., 2010). The three-segmented tarsi, earlier considered to be a unique character of trichogrammatids, however, it has been found in a few species of other families, like Mymaridae (Huber and Beardsley, 2000), Eulophidae (Delvare and LaSalle, 2000), Aphelinidae (Kim and Triapitsyn, 2004). The placement of Trichogrammatidae in the superfamily Chalcidoidea remains in doubt, but the family considered to be closest to Aphelinidae and Eulophidae. The reduction in both antennal and tarsal segments and a short, straight tibial spur have suggested the affinity to the Eulophidae. Based on the male other morphological characteristics, such as genitalia and tarsal segmentation, indicate Aphelinidae likely a sister-group of family Trichogrammatidae (Heraty et al., 1997, Gibson et al., 1999, Pinto, 2006; Owen et al., 2007). Furthermore, the presence of straight tibial spur suggested closeness to *Cales* Howard (Hayat, 1994, 1998), while the placement of *Cales* at the family level remained questionable, which traditionally placed in Aphelinidae (Woolley, 1997). Recent molecular studies suggested that *Eretmocerus* (Aphelinidae) showed a distant relationship and placed close to Trichogrammatidae (Campbell et al., 2000).

Recently, Owen et al. (2007) proposed the phylogenetic study of the family Trichogrammatidae based on the molecular analysis data in the context of morphological characteristics. Based on this study, a new classification of the Trichogrammatidae proposed, and tribes are defined narrowly than earlier studies. It includes one tribe, Trichogrammatini, within the subfamily Trichogrammatinae and removed tribe Paracentrobiini, and placed it within subfamily Oligositinae with three tribes, Paracentrobiini, Chaetostrichini and Oligositini.

9.3 HISTORY AND CLASSIFICATION

Arsene Girault had done much of earlier work on Trichogrammatidae and described 52 genera (Girault, 1912), of which 37 are considered as valid by Doutt and Viggiani (1968). Nowicki (1935, 1936) studied the trichogrammatids of the Palearctic region, whereas Risbec (1955; 1957) dealt with trichogrammatids from Africa. Doutt and Viggiani (1968) have done more comprehensive work on the taxonomy of Trichogrammatidae, which included 64 genera and 347 species and provided a key to world genera. This work was based on the examination of original types of most of the species with illustrations. Other taxonomic works include a list of world genera and species (Yousuf and Shafee, 1986a, b). The catalogue of species from UAS (Burks, 1979), the Oriental Region (Hayat and Viggiani, 1984), Central Europe (Peck et al., 1964) and Latin America (De Santis, 1979, 1980, 1981), a list of genera and species (Lin, 1994), and a key to Nearctic genera (Pinto, 1997b) were documented. Pinto (2006) reviewed the trichogrammatid genera of New World, and Owen et al. (2007) studied the phylogeny of the family as the most recent attempts of classification.

The earlier classification of the family was done by Girault (1912; 1918), Kryger (1918) and Blood (1923). Viggiani (1971) found considerable variation in the morphology of male genitalia and divided the family into two subfamilies. Each of these includes two tribes (Oligositinae: Oligositini, Chaetostrichini; Trichogrammatinae: Trichogrammatini, Paracentrobiini). The subfamily Trichogrammatinae and Oligositinae was separated based on the presence or absence of aedeagal apodemes in the genitalia. The genitalia structure was most generalized in tribe Trichogrammatini and reduced in tribe Oligositini, while tribes Chaetostrichini and Paracentrobiini represent intermediate conditions (Viggiani, 1971; 1984). However, recently the family Trichogrammatidae is divided into the subfamilies Oligositinae (monophyletic) and Trichogrammatinae (paraphyletic) (Pinto, 2006; Owen et al., 2007). In Oligositinae, the genitalia with well-developed parameres and volsellae, whereas Trichogrammatinae has the most generalized genitalia with an aedeagus and genital capsule, as two separate components in male genitalia. The subfamily Trichogrammatinae includes only the tribe Trichogrammatini and Oligositinae divided into three monophyletic tribes, Oligositini, Chaetostrichini, and Paracentrobiini (Oligositinae), in which genitalia consists of a single component because of varying degree of aedeagal and genital capsule consolidation (Owen et al., 2007; Querino et al., 2010).

9.4 SIZE AND DISTRIBUTION

Most of the trichogrammatid genera are small and 74 genera include less than 10 species and 34 of these are monotypic. Only 22 genera have more than 10 species, and 13 of these have more than 20 species (Fig. 9.1; Noyes, 2019). *Trichogramma* is the largest genus, followed by *Oligosita* Walker with 237 and 96 described species, respectively, in the world. These two largest genera contribute more than 33% diversity of the family. The most numbers of described species of *Trichogramma* may be the result of its role in the biocontrol of insect pests (Pinto and Stouthamer, 1994). The importance of family is due to the *Trichogramma*, whose species are generalist eggs parasitoids and parasitize the various lepidopteran pests in a different ecosystems (Querino et al., 2010). Other relatively larger genera include *Oligosita*, *Pseudoligosita* Girault, *Ufens* Girault and *Paracentrobia* Howard. The revision of the genus *Ufens* has increased the species richness of the genera (Owen, 2011).

Several smaller genera of Trichogrammatidae are also widely distributed in different habitats (Pinto, 2006). Of the total number of described species, 358 alone are found in New World, followed

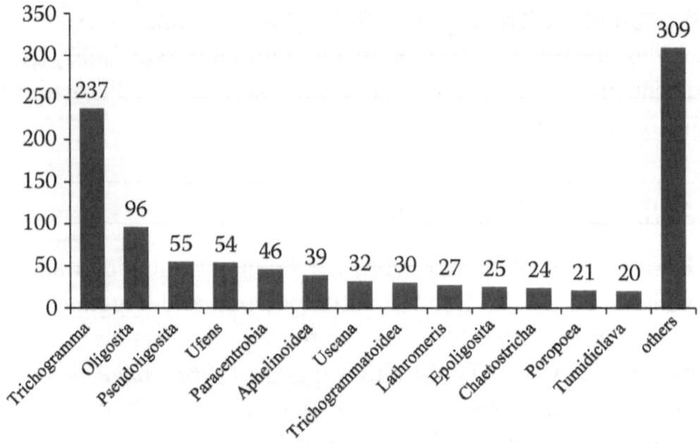

FIGURE 9.1 Number of species in 13 largest genera (>20 species) of Trichogrammatidae.

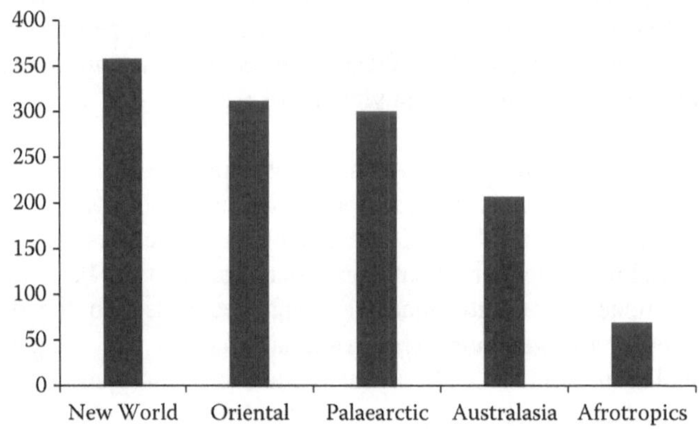

FIGURE 9.2 Number of species representing the different biogeographical region of world.

by Oriental (312) and Palearctic region (300), Australasia (207) and Afrotropical region (69) in the biogeographical regions (Fig. 9.2; Noyes, 2019). In the entire biogeographical region, *Trichogramma* is the largest genera with more number of described species. Most of the larger genera are cosmopolitan, except for *Trichogrammatoidea*, which is poorly represented in the New World and Palearctic regions.

9.5 DIVERSITY AND HOST RANGE

Trichogrammatids are primary endoparasitoids of insect eggs, either in solitary or gregarious form. Member of the family associated are varied and, a recent study suggested that the family initially parasitized members of Coleoptera and extended host range to other insect orders, such as Hemiptera and Lepidoptera (Owen et al., 2007). Ten insect orders have been known as the host of Trichogrammatidae (Pinto and Stouthamer, 1994; Querino et al., 2010). The highest numbers of Trichogrammatidae genera have associated with the members of Hemiptera, followed by Coleoptera and Lepidoptera (Fig. 9.3; Noyes, 2019). Worldwide, 32 genera of Trichogrammatidae are known to be associated with Hemiptera, 28 with Coleoptera and 8 with Lepidoptera, and the hosts of 38 genera are unknown. Most of the Trichogrammatidae genera have a broad range of hosts and many of them are recorded on more than one insect order. *Oligosita* has been recorded from Hemiptera, Orthoptera, Coleoptera, whereas *Trichogramma* usually found associated with Lepidoptera. However, *Trichogramma* also recorded from Diptera, Coleoptera, Hemiptera,

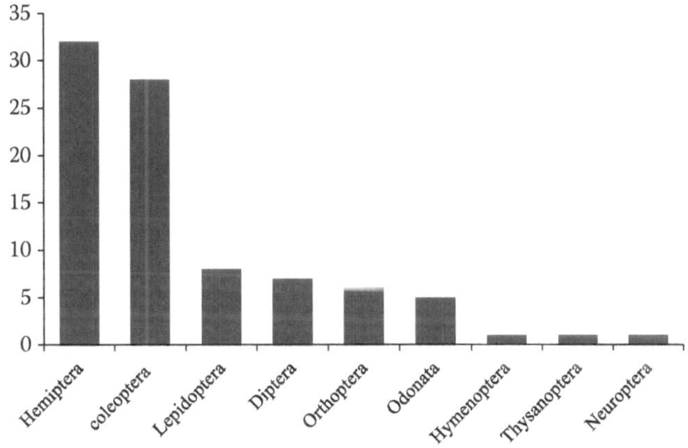

FIGURE 9.3 Number of Trichogrammatidae genera associated with insect orders.

Hymenoptera, and Neuroptera (Pinto, 1997a). This suggests that members of trichogrammatids have greater fidelity to microhabitat than the associated taxon of the hosts. Thus, the members of *Lathromeroidea* Girault have been associated with eggs of Hemiptera and Odonata that occur in aquatic environments. A strong influence of microhabitat than taxon of the host has been found in certain species of *Trichogramma* that parasitized the eggs of Lepidoptera, Hymenoptera and Neuroptera placed on the same plant (Pinto and Stouthamer, 1994). Some small genera of trichogrammatids are host specific and found associated with a few host insects, such as *Poropoea* Förster and *Ophioneurus* Ratzeburg collected from the eggs of Curculionoidea. Similarly, *Uscana* Girault is recorded only from Coleoptera, *Megaphragma* with narrow host range restricted to Thysanoptera and *Hydrophylita* Ghesquiere is recorded only from Odonata (Querino et al., 2010).

The natural occurrence of trichogrammatids and their parasitism is well known from the agroecosystem. The species predominant in the agroecosystem are mostly associated with their hosts and occur frequently, as *T. pretiosum* with 240 records that is usually not found in natural habitat (Pinto, 1999). However, little is known about trichogrammatids from the forest or natural reserves (Zucchi et al., 2010). A few genera of trichogrammatids extended their host range and parasitized the eggs of aquatic hosts, such as Notonectidae, Dytiscidae or Odonata, although the eggs are underneath the surface of the water. Globally, 22 species of trichogrammatids parasitize the eggs of aquatic insects and mainly belong to genera *Centrobiopsis* Girault, *Hydrophylita*, *Lathromeroidea*, *Prestwichia* Lubbock and *Pseudoligosita* Girault (Bennett, 2008; Querino and Hamada, 2009). The aquatic species *Prestwichia aquatica* Lubbock parasitize the aquatic beetle, bugs and damselflies (Irina et al., 2003). While *Hydrophylita lestesi* Costa Lima has been observed parasitizing the eggs of *Lestes* sp. (Lestidae) on leaves of *Hedychium coronarium* J. König (Querino and Hamada, 2009). An aquatic species *Hydrophylita* (*Lutzimicron*) *emporos* Shih & Polaszek parasitizes the eggs of the damselfly *Psolodesmus mandarinus mandarinus* McLachlan (Zygoptera: Calopterygidae) underwater (Shih et al., 2013). Besides, a few reports indicate *Lathromeroidea* sp. as a parasitoid of gerrid eggs (Henriquez and Spence, 1993) and *Pseudoligosita longifrangiata* (Viggiani) parasitizing eggs of damselflies (Querino and Hamada, 2009).

The parasitism level of trichogrammatids is not widely studied, even though the level of natural parasitism of trichogrammatids, particularly *Trichogramma* species, are high in the agroecosystem (Pinto, 1999; Souza et al., 2016; Jalali et al., 2018). The parasitism level of trichogrammatids primarily influenced by host and habitats, and varied in the different ecosystems (Table 9.1). Most of *Trichogramma* species parasitize lepidopteran eggs, particularly Noctuidae, which includes major agriculturally important pests (Zucchi et al., 2010; Jalali, 2013; Jalali et al., 2016b). Nevertheless, some species of *Trichogramma* also attack the non-economically important

TABLE 9.1
Parasitism of Trichogrammatid Species in Different Ecosystem

Crops	Species	Host insect	Parasitism rate (%)	Country	References
Rice	*Trichogramma japonicum*	*Scirpophaga incertulas*	1.6- 85.6	India	Sheeba and Prasad (2002)
	T. japonicum	*Scirpophaga incertulas*	9	China	Tang et al. (2017)
	T. chilonis	*S. incertulas*	15	China	Tang et al. (2017)
	Oligosita tachikawai	*Nilaparvata lugens*	10.6	India	Bentur and Kalode (1985)
	Oligosita yasumatsui *O. nephotetticum*	*N. lugens*	1–65%	Indonesia	Claridge et al. (1999)
	Oligosita naias	*N. lugens*	8.0	India	Shankar and Baskaran (1986)
	Oligosita spp. *Paracentrobia* spp.	*N. lugens*	44.3–78.6 0.2–5.5	Philippines	Sann et al. (2018)
	Oligosita spp. *Paracentrobia* spp.	*Nephotettix* spp.	2.6–7.6 6.4–70.3	Philippines	Sann et al. (2018)
	Oligosita naias	*Sogatella furcifera*	68	Malaysia	Watanabe et al. (1992)
	Oligosita aesopi	*S. furcifera.*	5	Malaysia	Watanabe et al. (1992)
Sorghum	*T. chilonis*	*Chilo partellus*	20–60	India	Taley and Thakare (1979)
	T. atopovirilia	*Spodoptera frugiperda*	2.47	Brazil	Jaraleño-Teniente et al. (2020)
Maize	*T. pretiosum* & *T. atopovirilia*	*S. frugiperda*	0.9–2.2	Brazil	Beserra et al. (2002)
	T. pretiosum & *T. atopovirilia*	*S. frugiperda*	0.72–2.36	Brazil	Jaraleño-Teniente et al. (2020)
	T. chilonis	*S. frugiperda*	20.9	Kenya	Sisay et al. (2019)
	T. atopovirilia	*S. frugiperda*	2.8–3.75	Brazil	Jaraleño-Teniente et al. (2020)
	T. pretiosum & *T. atopovirilia*	*S. frugiperda*	1.46–1.97	Brazil	Dequech et al. (2013)
	Aphelinoidea semifuscipennis, Oligosita clarimaculosa, Paracentrobia tapajosae, Pseudoligosita longifrangiata	*Dalbulus maidis*	22.6–60.9	Mexico	Moya-Raygoza et al. (2012; 2014)

	Oligosita desantisi, Oligosita giraulti, Paracentrobia tapajosae. Pseudoligosita longifrangiata Zagella nanula	*Dalbulus maidis*	3.4–44.8	Argentina	Moya-Raygoza et al. (2012; 2014)
	Aphelinoidea semifuscipennis, Burksiella platensis, Zagella nanula, Paracentrobia tapajosae, Paracentrobia sp. *Oligosita desantisi, Oligosita giraulti & Pseudoligosita longifrangiata*	*Dalbulus maidis*	0–56.7%	Argentina	Albarracin et al. (2017)
	T. chilonis	*S. frugiperda*	15.8–23.8%	India	Navik et al. (2021); Shylesha et al. (2018)
Johnson grass	*Zagella delicata*	*Homalodisca coagulata*	0.8–57.2	Argentina	Logarzo et al. (2004)
Sunn hemp	*Trichogrammatoidea armigera Tr. bactrae*	*Lampides boeticus*	13.80	India	Navik et al. (2019a)
Pigeon pea	*Trichogramma* spp.	*H. armigera*	0.1–8.3	India	Bhatnagar et al. (1982)
Cowpea	*Uscana* sp.	*Callosobruchus maculatus*	60%	Niger	Alzouma (1987)
	Uscana sp.	*Bruchidius atrolineatus*	21%	Niger	Alzouma (1987)
Lucerne	*T. chilonis*	*H. armigera*	84.2–98.2	India	Yadav et al. (1985)
Cotton	*T. achaeae*	*E. vittella*	27.2	India	Sekhon and Varma (1985)
	T. chilonis	*E. vittella*	9.5–15.8	India	Telang et al. (2004); Mohapatra (2006)
	T. chilonis	*H. armigera*	22.1	India	Mohapatra (2006)
	T. chilotraeae	*E. vittella*	33.3	India	Manjunath (1974)
	T. chilotraeae	*H. armigera*	83.3	India	Manjunath (1974)
Sugarcane	*T. pretiosum & T. galloi*	*Diatraea saccharalis*	100.0	Brazil	Souza et al. (2016)
	T. japonicum	*Chilo sacchariphagus indicus*	7–55	India	Venugopal et al. (1968)
Cabbage	*T. chilonis*	*Spodoptera litura*	31.8	India	Srivastava and Kushawaha (1995)
	T. chilonis	*Plutella xylostella*	7.69–42	India	Navik et al. (2019a)
Cauliflower	*T. chilonis*	*P. xylostella*	4	India	Yadav et al. (2001)
Mustard	*T. chilonis*	*P. xylostella*	77–94.8	India	Yadav et al. (2001)

(*Continued*)

TABLE 9.1 (Continued)
Parasitism of Trichogrammatid Species in Different Ecosystem

Crops	Species	Host insect	Parasitism rate (%)	Country	References
Tomato	*T. achaeae*	*H. armigera*	28.5–36.0	India	Navik et al. (2019a)
	Tr. achaeae	*T. absoluta*	2.5	Portugal	Oliveira et al. (2017)
	T. achaeae *Tr. bactrae*	*T. absoluta*	5.35	India	Navik et al. (2019a)
	T. pretiosum	*Neoleucinodes elegantalis*	13–19	Brazil	Oliveira et al. (2020)
	T. chilonis	*H. armigera*	16.21	India	Navik et al. (2019a)
	Trichogramma sp.	*T. absoluta*	8	Spain	Gabarra et al. (2014)
	Trichogramma sp.	*T. absoluta*	–	Italy	Zappala et al. (2012)
Brinjal	*T. chilonis*	*Acherontia styx*	79–93	India	Rao et al. (1980)
	T. achaeae *T. chilonis*	*A. styx*	54–64	India	Krishnamoorthy et al. (1999)
Okra	*T. chilonis*	*H. armigera*	8	India	Thontadarya et al. (1978)
Sweet potato	*T. chilonis* *T. achaeae* *T. agriae*	*Agrius convolvuli*	63.8	India	Nagarkatti (1973)
Potato	*T. chilonis*	*H. armigera*	84.2–98.2	India	Yadav (1980)
Apple	*T. embryophagum* & *T. cacoeciae*	*Cydia pomonella*	10.9- 22.4	India	Lalitha et al. (2015)
Citrus	*T. chilonis*	*Othreis materna*	21–50	India	Dodia et al. (1986)
Pomegranate	*T. chilonis*	*Deudorix epijarbas*	2.7	India	Gupta and Sood (2001)
Carica papaya	*T. pretiosum* & *T. manicobai* *Trichogramma* sp.	*Protambulyx strigilis*	22.9	Brazil	Souza et al. (2016)
Coconut	*T. chilonis*	*Opisina arenosella*	1–5	India	Jalali et al. (2005b)
Cassava	*T. pretiosum,* *T. manicobai* *T. marandobai*	*Erinnyis ello*	43.5	Brazil	Souza et al. (2016)
Passion fruit	*T. pretiosum*	*Agraulis vanillae*	29.0	Brazil	Souza et al. (2016)
Avocado	*Megaphragma mymaripenne*	*Heliothrips haemorrhoidalis*	3–51	USA	Hessein and McMurtry (1988)

Calotropis	*T. danausicida*	*Danaus* sp.	57.4	India	Navik et al. (2019a)
Calotropis procera	*T. pretiosum & T. manicobai*	*Danaus* sp.	35.3	Brazil	Souza et al. (2016)
Crotalaria	*Tr. bactrae*	*Lampides boeticus*	21.05	India	Navik et al. (2019a)
Tabernae montana	*T. chilonis*	Sphinx moth	60	India	Navik et al. (2019a)
Jasmin	*T. chilonis*	Sphinx moth	26.92	India	Navik et al. (2019a)
Olive	*T. cordubensis*	*Prays oleae*	1.1–16.6	Egypt	Hegazi et al. (2005)
	T. pretiosum	*Prays oleae*	1.1–7.2	Egypt	Hegazi et al. (2005)
	T. cacoeciae	Prays oleae	3.8–5.1	Egypt	Hegazi et al. (2005)
	T. cordubensis	*Palpita unionalis*	0.3–15.3	Egypt	Hegazi et al. (2005)
	T. pretiosum	*Palpita unionalis*	0.3–10.3	Egypt	Hegazi et al. (2005)
Tonina fluviatilis	*Pseudoligosita longifrangiata*	*Argia insipida*	10.6	Brazil	Querino and Hamada (2009)

butterflies in the forest habitats, such as *Trichogramma bruni* and *T. acacioi* (Querino and Zucchi, 2002; Zucchi, 2010). Many species occur in several habitats, like natural reserves, forests and crop habitats, while many of *Trichogramma* species only occur in the forest ecosystem. Querino and Zucchi (2003a, b, c) reported the occurrence of *T. acuminatum*, *T. alloeovirilia*, *T. bertii*, *T. esalqueanum*, *T. maxacalii*, *T. parrai*, *T. psyllium*, *T. tupiense* and *T. zucchii* from the forests in Brazil and other species that occur in agroecosystem could parasitize the forest pests as well. In India, *T. chilonis*, *T. achaeae*, *Trichogrammatoidea bactrae* and *Tr. armigera* dominated the agroecosystem, whereas, *T. danausicida* is found in undisturbed habitats (Navik et al., 2019a). In addition, the distribution of trichogrammatids varies with the altitude. In New World, occurrence of only 12 species of *Trichogramma* are recorded at altitudes more than 2,000 m, and the occurrence of *T. fasciatum* recorded above 3,000 m in Costa Rica (Zucchi et al., 2010).

9.6 BIOLOGY

The immature stages of egg parasitoids develop within host eggs. Most of the trichogrammatid genera have hymenopteriform eggs (Pak and Oatman, 1982; Dahlan and Gordh, 1996; Jarjees and Merritt, 2002; Boivin, 2010), with an exception of genus *Poropoe*, where stalked eggs are found (Clausen, 1940). Generally, *Trichogramma* eggs are 100–140 μm in length and 30–50 μm in width at the time of oviposition (Tanaka, 1985; Manweiler, 1986; Saakian-Baranova, 1990; Dahlan and Gordh, 1996; Jarjees and Merritt, 2002), however *T. chilonis* lay comparatively smaller eggs than other species (Tanaka, 1985). The egg of genus *Brachyufens* measures 193 μm in length and 41 μm in width at the widest point with spindle shape, transparent egg, and had a pedicel (Jacas et al., 2009). Trichgrammatids eggs are hydropic, and their size increase as they absorb nutrition from host egg after oviposition (Jarjees and Merritt, 2002). Females of *T. chilonis* lay eggs in or near the peripheral yolk of the host egg (Tanaka, 1985), whereas *T. dendrolimi* females lay in the host egg yolk (Takada et al., 2000). After laying, the egg and embryo increase gradually in 22–24 hours within the host eggs before eclosion (Dahlan and Gordh, 1996; Jarjees and Merritt, 2002). After 18–19 hours of oviposition, the developing embryo surrounded by a chorion and an inner membrane, which is probably an embryonic cuticle called 'embryonic exuvia'. Between egg envelopes and embryonic exuvia, free cells with large nuclei can be observed (Volkoff et al., 1995). Initially, trichogrammatid eggs laid within host are yolkless, however, as the development proceeds, the embryo gets cellularized, formation of stomodaeum and proctodaeum appears and later in hours gastrulation occurred. Jarjees and Merritt (2002) reported the embryonic development of *T. australicum* and found midgut formation and stomodaeaum developed after 12–16 hours of oviposition and the component of the foregut, anal vesicle, and most of the posterior part of the hindgut, are well-developed after 20 hours of oviposition. Similarly, Volkoff et al. (1995) reported the development of *T. cacoeciae* and found that stomodaeaum, proctodeum and midgut formation competes 18–19 hours after oviposition.

The late-staged trichogrammatid embryo (after 20–27 hours of oviposition) shows vigorous movement inside the embryonic exuvia at the anterior region in *T. australicum* (Jarjees et al., 1998) and the posterior end in *T. cacoeciae* (Volkoff et al., 1995). This vigorous movement eventually ruptures the exuvia and release larva into the host egg. Soon after hatching, no active feeding takes place by neonate, termed as "free-living embryo" (Volkoff et al., 1995, Consoli et al., 1999b). These non-active neonates of *T. australicum* and *T. cacoeciae* are vermiform. However, as the active larva start feeding on host yolk, the size increases and gradually changes to pyriform. A full matured become sacciform ingest full egg host content and shows no further changes in its shape. As the larva grows, the mandible changes from yellow to brownish. The appearance of scattered urate bodies indicates the onset of the prepupal stage. The prepupal stage of *T. australicum* appears after 49–51 hours and lasts for 50–52 hours (Jarjees and Merritt, 2002), whereas, in *T. cacoeciae*, it appears after 66–67 hours after oviposition (Volkoff et al., 1995). The compound eyes and ocelli turn red, and sclerotization of the mandible and oviposition occurs during the pupal stage. The adults of *T. australicum* and *T. cacoeciae* emerged after 190–192 and 241–242 hours, respectively, after oviposition (Jarjees and Merritt, 2002; Volkoff et al., 1995).

Most trichogrammatid egg parasitoids larvae undergo hypermetamorphosis (Hagen, 1964), except genus *Trichogramma*, where larval development is simplified (Boivin, 2009). The number of instars reported within Trichogrammatidae remained controversial due to minute immature stages and their confined development within the host egg (Jarjees and Merritt, 2002). In most Trichogrammatidae genera, one to five larval instars are reported based on morphological features, like larval size and shape, mandible size and presence of exuviae (Jarjees and Merritt, 2002). The genera *Ophioneurus* (Bakkendorf, 1934) and *Brachyufens* (Jacas et al., 2009) had two larval instars, whereas only one larval instar is reported in genera *Chaetostricha, Prestwichia, Trichogramma* (Bakkendorf, 1934) and *Oligosita* (Taylor 1937). However, one to five larval instars are reported in the genus *Poropoea* (Silvestri, 1916; Bakkendorf, 1934). In genus *Trichogrammatoidea*, two to three larval instars have been reported, especially two instars in *Trichogrammatoidea bactrae* Nagaraja (Hutchison et al., 1990), and three instars in *Tr. armigera* Nagaraja (Manjunath, 1972) and *Tr. simmondsi* Nagaraja (Zongo et al., 1993). Several authors reported a different number of larval instars in the species of *Trichogramma*, which varies from one to five. Takada et al. (2000) observed two larval instars in *T. dendrolimi*, while three instars are reported in *T. buesi* Voegele (Abbas, 1989), *T. evanescens, T. embryophagum* Hartig (Flanders, 1937), *T. cacoeciae* (Kochetova, 1969), *T. pintoi, T. semblidis, T. platneri* Nagarkatti (Aurivillius) (Saakian-Baranova, 1990), *T. maidis* (Hawlitzky and Boulay, 1982), *T. minutum* (Manweiler, 1986), and *T. brassicae* Bezdenko (Wu et al., 2000). However, Pak and Oatman (1982) reported four instars in *T. brevicapillum* Pinto and Platner. Saakian-Baranova (1990) reported thee larval instars in *Trichogramma australicum* Girault, *T. brassicae* and *T. cacoeciae*, however later study revealed only one larval instar in these species (Volkoff et al., 1995, Dahlan and Gordh, 1996; 1997; Wu et al., 2000, Jarjees and Merritt, 2002). The recent studies of Volkoff et al. (1995) and Jarjees and Merritt (2002) reported only one larval stage present in *Trichogramma*, as there was no evidence of larval ecdysis and exuvia found between embryonic exuvia and the pupation. Further, the changes in first instar larvae due to ingestion of food and changes in the curvature of mandibles reported, however no evidence of moulting was reported the larval development (Volkoff et al., 1995). Similarly, the embryonic exuvium remained attached to the developing larvae till the prepupal stage and no increase in the size of mandibles in developing till larva supports the presence of only one instar in *Trichogramma* (Volkoff et al., 1995; Jarjees and Merritt, 2002). This finding supports that only one larval instar present in *T. cacoeciae* and *T. australicum*, however, this statement cannot be generalized for whole species of *Trichogramma*. For most of *Trichogramma* species, the series of development from embryo to adult emergence is completed in 7–14 days (Schöller et al., 2001; 2006).

9.7 ECOLOGY AND BEHAVIOUR

Host acceptance is considered to be an important criterion in the selection of suitable biocontrol agent. Though the species of *Trichogramma* are polyphagous (Smith, 1996), several studies demonstrated the suitability of host species for *Trichogramma,* such as host egg volume (Bai et al., 1992; Pak, 1988), chorion thickness (Pak et al., 1990), nutritional contents (Barrett and Schmidt, 1991), age (Garcia, 2000; Monje et al., 1999) and host immune system (Abdel-Latief et al., 2008; Reed et al., 2007). These characteristics influence parasitism rates as well as the number of wasps per host egg (Honda and Luck, 2001), body size and fecundity (Greenberg et al., 1998; Kazmer and Luck, 1991), longevity (Kuhlmann and Mills, 1999), and sex ratio (Corrigan and Laing, 1994; Kuhlmann and Mills, 1999) of *Trichogramma* offspring. Among eggs of six noctuid Lepidoptera: *Thysanoplusia orichalcea* Fabricius, *Xestia c-nigrum* (L.) *Peridroma saucia* (Hübner), *Phlogophora meticulosa* (L.), *Noctua pronuba* (L.), and *N. atlantica* (Warren), eggs of *P. saucia* followed by *P. meticulosa* were the least suitable hosts for *T. cordubensis* Vargas and Cabello, due to significantly longer preimaginal development and, significantly lower parasitism as well the adult emergence (Roriz et al., 2006). Schafer and Herz (2020) demonstrated that *T. pintoi*, *T. nerudai* and *T. cacoeciae* attained a similar level of parasitism on *Tuta absoluta* Meyrick eggs on potted tomato plants as *T. achaeae.*

Temperature is one of the most paramount environmental factors that influence the insect physiology and behaviour (Ratte, 1985), survival, lifetime fecundity and other biological parameters and regulating seasonal or daily cycles and indirectly influence various aspects of insect biology (Pratissoli et al., 2000; Andrade et al., 2011). Understanding the thermal requirements of *Trichogramma* spp. is essential for planning its mass rearing procedures in the laboratory (Maceda et al., 2003; Parra et al., 2010a) for the specific region and to determine its potential as a biocontrol agent for a given pest (Pratissoli et al., 2000). However, the temperature effects on the biological parameters and thermal requirements of *Trichogramma* depends on species/strains studied (Schöller et al., 2001; Foerster and Foerster, 2009) and the host used (Özder et al., 2010). Pino et al. (2020) demonstrated that *T. achaeae* developed successfully from egg to adult in the temperature range of 15°C to 30°C. Furthermore, at higher temperature (35°C and above) though parasitization was observed, it prevented adult emergence (Pino et al., 2020; Ayvaz et al., 2008; Schöller et al., 2001; Cabello and Vargas, 1988). There are several reports on the adverse effect of temperature on adult longevity, fecundity, life duration, parasite emergence, and sex ratio of *Trichogramma* species (Pino et al., 2020; Coelho et al., 2016; Maceda et al., 2003; Foerster et al., 2014; Chihrane et al., 1997; Chihrane and Lauge, 1996; Carriere and Boivin, 1997). Firake and Khan (2014) observed that though *T. chilonis* and *T. poliae* Nagaraja could survive and reproduce at temperatures up to 40°C, their releases for pest control was effective between 23 ± 1°C and 26 ± 1°C.

Parasitoids are a useful model to understand the climate change effects on fitness because their physiological and behavioural responses can easily be measured under laboratory conditions (Denis and Pierre, 2012). However, the impact of climate change on most used egg parasitoids is poorly studied (Dyer and Coley, 2002; Jeffs and Lewis, 2013). *Trichogramma* species widely distributed across different latitudes vary in terms of the minimum developmental thresholds (T_0) and the thermal constant (K) (the amount of heat that each species requires to complete its lifecycle expressed in degree-days, DD) (Pizzol et al., 2010). Wu et al. (2016) in a meta-analysis study reported that minimum developmental thresholds for *T. aurosum* Sugonjaev and Sorokina, *Trichogramma buesi* Voegele, *T. dendrolimi* Matsumura and *T. zahiri* Polaszek acquired from Russia, Germany, China and Bangladesh are positively correlated with an increase of absolute latitude. On the other hand, the thermal constant was negatively correlated with absolute latitude indicating that the *Trichogramma* species originate at higher latitude had a lower thermal requirement to complete their life cycle.

Volatile cues play an important role in interspecies communications for insect pests and natural enemies, to find, feed and reproduce in an agroecosystem (El-Wakeil et al., 2017; Rodriguez et al., 2012; Vet and Dicke, 1992). Egg parasitoids are observed to locate lepidopteran host by exploiting pheromones, compounds released from scales of the wings of ovipositing females (Shu and Jones, 1989, Colazza et al., 1997) and plant volatiles in response to egg-laying by the host (Nina et al., 2014). The study indicated that augmentation of Trichogrammatids in tomato crop ecosystem amalgamated with allelochemical cues released by tomato variety K-21 can prove to be an effective biocontrol procedure against economically important insect pests of tomato (Singh et al., 2017). They also observed a higher foraging response by trichogrammatids under the influence of volatile cues derived from tomato variety K-21 of vegetative stage growing in field condition (VSF). Study shows that *T. cordubensis* was found to be attracted to volatiles from tomato plants with *T. absoluta* eggs 24 h after oviposition whereas *T. achaeae* did not discriminate between egg-infested tomato plants (Milonas et al., 2019).

Functional response studies are very essential to study host searching ability and effectiveness of a parasitoid before deploying them in biocontrol programme (Fernandez-Arhex and Corley, 2003; Luo et al., 2014; Poncio et al., 2016). These qualities are crucial to achieving effective and successful control of target pest (Fernandez-Arhex and Corley, 2003). The functional response is an essential element of the dynamics of host-parasitoid association and is an important determinant of the stability of the system (Oaten and Murdoch, 1975). A type II functional response is common in trichogrammatids and has been reported for many species of *Trichogramma* to different pests. For example *T. pretiosum* to *T. absoluta* (Faria et al., 2000), *T. achaeae, T. pretiosum, T. chilonis* and *T. pieridis* to *T. absoluta* (Manohar et al., 2020); *Trichogramma brassicae* to *Sitotroga cerealella* (Olivier) (Lepidoptera: Gelechiidae) and *Ephestia kuehniella* (Zeller) (Lepidoptera: Pyralidae) (Karimian, 1998; Moezipour et al., 2008; Farrokhi et al., 2010), *T. chilonis* to *Chilo sacchariphagus* (Kapur) (Lepidoptera: Crambidae); *T. ostriniae* to *Ostrinia nubilalis* (Hubner) (Crambidae: Lepidoptera) (Wang and Ferro, 1998). However, a few studies have exhibited a type III functional response in *Trichogramma* spp. to their hosts (Wang and Ferro, 1998; Fathipour et al., 2003; Jones et al., 2006; Moezipour et al., 2008).

When two natural enemies share the same resources they might engage in intraguild predation (IGP)/competition. IGP between predators and parasitoids favour predator and asymmetrical (Polis et al., 1989, Brodeur and Rosenheim, 2000). Furthermore, the intensity of IGP depends on the rates of predation on parasitized and non-parasitized prey (Rosenheim, 1998). It has been observed that *Trichogramma* spp. after parasitizing the host eggs induce modifications, such as changes in the texture, shape or colouration and the composition or quantity of external semiochemical cues to avoid IGP (Vinson, 1994). Sometimes to manage insect pests, two or more natural enemies have to be deployed. Understanding the competitive interactions between the two parasitoids is very important to achieve maximum host suppression. Ksentini and Herz (2019) found that releases of *T. cacoeciae* may disturb the natural control of codling moth made by the endemic parasitoid, *Ascogaster quadridentata* Wesmael. For managing recent invasive fall armyworm (FAW), *Spodoptera frugiperda* (J.E. Smith) two egg parasitoids, *viz. T. chilonis* and *Telenomus remus* Nixon, were observed parasitizing FAW eggs (Navik et al., 2021). Laboratory studies showed that though *T. remus* is a potential parasitoid of FAW eggs, *T. chilonis* is equally competent and these two parasitoids are complementary to each other when released simultaneously against FAW eggs (Varshney, 2021 unpublished data).

9.8 GEOGRAPHICAL STRAINS OF TRICHOGRAMMATIDS

The efficiency of trichogrammatid parasitoids in controlling the pests depends not only on their ability to parasitize the host eggs but also on their adaptability to the geographical regions and the prevailing climatic conditions of the region where they planned to be released (Pizzol et al., 2010). Thus, the effectiveness of parasitoid varies based on the strains and climatic conditions (Smith, 1996; Klug and Meyhofer, 2009; Tabone et al., 2010). These parasitoids are composed of various populations that have biological traits adapted to the specific climatic conditions of different geographical regions in which they naturally occur and survive (Pizzol et al., 2010; Fatouros et al., 2014). In the field conditions, several factors, such as plant morphology, crop architecture, prey distribution and environmental conditions influence the parasitizing efficiency and behaviour of parasitoids and ultimately impact the efficacy of biocontrol (Andow and Prokrym, 1990; Rutledge et al., 2003; Tabone et al., 2006; Carrillo et al., 2008; Desneux and Ramirez-Romero, 2009). Further, the parasitism of trichogrammatid parasitoid greatly varied with the habitats, plant species, plant structures, spacing, odour and chemistry plays an important role in determining the efficiency of trichogrammatids as biocontrol agents (Gingras et al., 2002; Romeis et al., 2005). Chailleux et al. (2012) assessed 29 European geographical strains recovered from various hosts, host plants and geographical locations for their parasitism potential and found the parasitism rate varied among the strains of *Trichogramma*. The parasitism potential of geographical strains of parasitoids linked directly with the species under evaluation and origin of habitat, host plants family, and the original host insect also influence the parasitism of parasitoids. Two strains of *Trichogramma cacoeciae* Marchal, which originated from France (Alsace) and Tunisia (Degache) showed differences in parasitism level under different temperature regimes suggested that the French strain parasitized and emerged better at lower temperature against pests on the grapevine and various orchards in Western Europe (Wiackowski and Wiackowska, 1966; Pizzol, 2004), whereas Tunisian strain performed better at higher temperature in palm plantations. This showed that *Trichogramma* parasitoids obtained from various geographical locations inherited the biological traits for the potential adaptation under specific abiotic conditions (eg. temperature), occurring in their respective countries of origin (Pizzol et al., 2010). Similarly, *T. cacoeciae* found effective against the apple codling moth, *Cydia pomonella* L., in apple orchards in the temperate region of India (Lalitha et al., 2015). Further, the influence climatic condition and host plant were assessed for 17 trichogrammatids strains include two indigenous strains from France under greenhouse conditions against *Plutella xylostella* (L). The level of parasitism differed between the strains and strains within the species. The parasitism level of *T. ostriniae* originate from Japan has four time higher parasitism than the Moldavian strain on the eggs of *P. xylostella*. Further, the effect of cauliflower plants has a positive effect on eight strains, a negative effect on four strains and no effect on five strains (Tabone et al., 2010). A significant geographical variation in biological, biochemical and morphological characteristics of *Trichogramma evanescens* Westwood observed when collected from different regions of Eurasia (Ram, 1995). Therefore, the biological differences observed in the morphological, physiological, ecological or behavioural features among species or populations of parasitoids due to the geographical variations, environmental and biological interactions, which reflected in the adaptations to the local environment. However, differences in adaptiveness in geographical strains of parasitoids, such as climatic tolerance, specificity toward host and evasion of host-immunity response which reduced their effectiveness as biocontrol control agents (Messenger and Van den Bosch, 1971).

The indigenous geographical strains generally perform better in the native conditions. The collection of various indigenous strains from the different geographical regions and screening them to further select a potential strain against the pest infesting crops in a particular region is one of the important selection criteria for parasitoids in the success of the biocontrol program (Hassan, 1989; Hassan and Guo, 1991; Wührer and Hassan, 1993; Bai et al., 1995). Further, the strains collected from different geographical regions differ in reproductive capacity. This is evidenced from the field efficacy of several native geographical strains of *T. minutum* parasitized the eggs of spruce budworm *C. fumiferana* (Smith and Hubbes, 1986). Herz and Hassan (2006) observed the difference in strains of *T. cordubensis* originated from Egypt and Portugal collected from olive and tomato crops, respectively. The strain of *T. cordubensis* collected from olive was more efficient in host finding and searching capacity, survival in arid conditions and parasitized significantly higher percentage of olive moth and jasmine moth eggs as compared to the strain of *T. cordubensis* collected from Portugal in tomato (Herz and Hassan, 2006). It is emphasized that preference may be given to native strains or species that are collected from the local environment (van Lenteren et al., 2003).

9.9 IMPROVEMENT OF TRICHOGRAMMATIDS

In nature, if an effective wild strain of trichogrammatids cannot be found, it might be possible to increase and improve the effectiveness of laboratory strains already under rearing in the laboratory. Improvement in parasitoids aims to increase the activity of released parasitoid at the field level and this can be achieved through genetic improvements of a strain by selective breeding (Jalali and Singh, 1993; Bertin et al., 2017). Generally, the native species or strains are selected for the improvement and subsequent releases on the ecological basis that native strains are better adapted to the proposed climate, habitat, and host conditions (Smith, 1996). However, the selection of the appropriate trichogrammatids for improvement is a complex process because there is considerable interspecies variation among the species (Frankham, 2008; Tayeh et al., 2012). The selection criteria of *Trichogramma* strain may be focused on quantitative traits, which need an improvement to increase the efficacy of parasitoids because lack of knowledge on traits appears to be a major limiting factor. Moreover, if the natural selection act against the artificially improved strains with selected traits or outbreeding with a natural population occurs, this may lead to a gradual revert of artificial strain to a wild strain after release in the field over time (Hoy, 1976; Nayak et al., 2018). The genetic improvement of parasitoids is promising because it provides an opportunity to maintain desired traits and quality by using continuous selection during parasitoid production (Jalali et al., 2006; Kumar et al., 2008).

In recent time, the genetic improvement of various species of trichogrammatids was carried out through selection and fortified for various traits against the various physical and chemical stresses encountered in the targeted crop systems (Venkatesan and Jalali, 2013). The genetic improvement of parasitoids, such as trichogrammatids, is required to overcome the problems associated with abiotic factors, susceptibility to insecticides and adaptability to new cropping patterns and systems because trichogrammatids are more habitat-specific than the host (Bertin et al., 2017). The potential *Trichogramma* strains possess the high searching ability, higher biotic potential and female biases sex ratio is often regarded as a better strain for inundative releases at the field level (Smith,1996; Prabhulinga et al., 2016). However, the potential strain combined with tolerant stresses, such as temperature and insecticides, would be more efficient against target crop pests in regions where such a stressed ecosystem is prevailing (Jalali et al., 2006). Therefore, the development of tolerant or resistant strains to various selected traits by adopting an approach of genetic improvement was undertaken to evolve tolerant strains of parasitoids worldwide (Jalali et al., 2006; Benvenuto et al., 2012; Jalali et al., 2016b). In India, the most commonly used egg parasitoid *T. chilonis* was selected for the development of tolerance against multiple insecticides and high-temperature tolerant strain (MIHTTs), high quotient strain (HQs) and temperature tolerant strains (TTs). Additionally, the high searching strain and temperature tolerant strain of *T. japonicum* was also developed through conventional and selective breeding techniques (Jalali et al., 2003; Lalitha et al., 2015).

The insecticide-tolerant strains have been advocated and widely used in integrated pest management worldwide (Ballal et al., 2009; Jalali et al., 2016a; Navik et al., 2019b), because of selection such trichogrammatids are potent and enhance the efficacy in pest management in the field level (Venkatesan and Jalali, 2013). Moreover, the exposure of trichogrammatid parasitoid to chemical insecticides reduced the parasitism and adult emergence by causing perturbation of nerve transmissions induced by the insecticides (Delpuech and Meyet, 2003; Bastos et al., 2006). However, the insecticide tolerant (endosulfan-tolerant) strain of *T. chilonis* was developed and found effective in suppressing the pest population under endosulfan sprayed crop ecosystems (Jalali et al., 2006). Besides, a higher parasitism rate (54–91%) recorded with the release of insecticide-tolerant strain during screenhouse evaluation (Sithanantham et al., 2013). Subsequently, a multiple insecticides tolerant strain of *T. chilonis,* tolerant to endosulfan, monocrotophos and fenvalerate, was developed and field evaluated against *H. armigera* on cotton showed higher parasitism of 8.3–43%. Further, the *T. chilonis* strain exposed to newer insecticides groups, such as indoxacarb, spinosad and tebufenoside, was found to increase the tolerance against the frequently used insecticides by farmers, and this strain enables to suppress the insect pest's infestations in tomato (Jalali et al., 2016) and cabbage with increased yield and reduction of insecticidal spray (Navik et al., 2019b).

The survival and parasitizing capacity of *T. chilonis* decreases with an increase in the temperature higher than 32°C. Therefore, it is necessary to enhance temperature tolerant traits so that parasitoids can survive at higher temperature. These adaptive traits may increase the efficacy and performance of trichogrammatids under a regime of higher temperature (Jalali and Singh, 1992; Venkatesan and Jalali, 2013). Trichogrammatid parasitoids effectively perform at temperature 20–32°C. The heat hardening process induces the tolerance level of *Trichogramma carverae* Oatman & Pinto when exposed to 1–2 hours at 33 or 35°C and able to survive at 40°C. However, this developmental hardening reduced the longevity of adult males and females and also reduced the parasitism rate (Scott et al., 1997). Moreover, heat hardening is a well-known means of acclimatization in which invertebrates exposed to high but sublethal temperature protects against subsequent heat-induced death and enhance adult fitness under elevated temperature at field level. Hence, the high-temperature-tolerant strain of trichogrammatids may perform better under a heat-stressed environment. In an evaluation study, the temperature-tolerant strain of *T. chilonis* showed a more than 55% parasitism rate compared to no parasitism by laboratory strain. This indicates that laboratory strain is capable to survive at 32–40°C, but unable to parasitize the eggs of targeted pests (Jalali and Singh, 1992). Therefore, temperature-tolerant *Trichogramma* developed through adaptation to higher temperature may be released to control the Lepidoptera pests in a various agroecosystems. Further, the field-tested genetically improved strain of *T. chilonis* have combined tolerance to newer insecticides and the high temperature was evaluated against pest of cotton, rice and tomato. In cotton, tolerant strain significantly reduced the bollworm larvae (0.2 larvae/boll) with increase yield compared to susceptible strain. Similarly, the tolerant strain of *T. chilonis* parasitized 70% of eggs of stem borer as compared to susceptible strain (25%) in rice. Field release of the tolerant strain of *T. chilonis* produced 65–95% egg parasitism of *H. armigera*, while susceptible strain could parasitize 8–12% host eggs in tomato. Further, this tolerant strain parasitized 31% egg of *P. xylostella* infesting cabbage compared to 5% by the susceptible strain of *T. chilonis* (Lalitha et al., 2015).

The improved strains of trichogrammatid parasitoids, such as multiple insecticide tolerant strain (MITs), a temperature tolerant strain (TTs) and a high quotient strain (HQs) were efficient against the targeted pest in the sugarcane and cotton ecosystem. The MITs strain of *T. chilonis* could significantly reduce pest populations, its damage and increased yield. Further, the TTs strain of *T. chilonis* and *T. japonicum* reduced the population of *S. incertulas* and *C. medinalis* and resulted in 19.59 to 51.2% parasitism. The TTs and HQs strains of *T. chilonis* were also performed efficiently to reduce the early shoot and internode borers' incidence in sugarcane with increasing yield (Lalitha et al., 2015).

The extensive research on the manipulation of parasitoids to improve the searching ability by using contact or volatile kairomone substances to the habitat where hosts present are at low densities might increase the effectiveness of host findings (Gross, 1981; Bakthavatsalam et al., 2013; Chailleux et al., 2019). Several host insect kairomones may stimulate the searching behaviour and increase the retention time of wasps in a target area. This kind of response suppresses the dispersal of the wasps from a crop habitat (Gross et al., 1984; Rodriguez et al., 2012). The various kinds of kairomones found in the scales of lepidopteran hosts or host plant they feed on have increased the host searching behaviour of female of *Trichogramma* species, which subsequently increased parasitism rate (Nordlund and Greenberg, 2002). Several chemical compounds, such as octacosane, pentacosane, hexacosane, tricosane, hexatriacontane, and docosane, are considered as favourable hydrocarbons derived or emitted from the host were the main source of attraction and subsequently increased the activity of *Trichogramma* spp. (Paramasivan et al., 2004; Yadav et al., 2005; 2008; Rani et al., 2007). The parasitism of *Trichogramma* spp. have been increased from 13% to 22%, when extract from the moth scales or synthetic kairomones sprayed on the plants (Lewis et al., 1975a, b). The application of moth scales or the synthetic kairomones impregnated particles of diatomaceous earth on plants at field level showed results only at high host densities. However, at low to intermediate host densities, parasitism rates were enhanced if moth scales or the synthetic kairomone were applied around the host eggs (Lewis et al., 1979; Gross, 1981). Similarly, application of dust formulation or rubber septa impregnated with-n-tricosane has increased the parasitism rate of *T. chilonis* in cotton, while it was not evident in the case of *T. japonicum* in rice (Bakthavatsalam et al., 2013).

Trichogrammatids search their host in a complex and dynamic environment and mostly relying on various chemical cues emitted by the adult host, its products, egg itself, host plants and by the interaction between egg and plant surface (Romeis et al., 2005; Fatouros et al., 2008; Salerno et al., 2009; Colazza et al., 2007). Therefore, more recently the importance of the rearing system "plant + host eggs" has been widely adopted to enhance the parasitizing capacity of *Trichogramma* parasitoids (Bueno et al., 2009; Tabone et al., 2010; Cascone et al., 2015). Further, the strains of *Trichogramma* collected from the original host parasitize more effectively the same host eggs or relatively similar host eggs from the same groups in fields (Chailleux et al., 2012). This behaviour of trichogrammatid parasitoids is usually determined by the preadult experience of eggs that used as rearing substrate. Several studies have shown that laboratory hosts such as *E. kuehniella, S. cerealella* and *Corcyra cephalonica* (Stainton) alter the rate of parasitism (Makee, 2005; Kolliker-Ott et al., 2003; Grenier et al., 2001; Brotodjojo and Walter, 2006). Thus, to enhance the parasitizing efficiency, two main approaches are the selective rearing on the target host and priming of unselected antagonists by the target host before release (Cascone et al., 2015). The successful result obtained in the case of mass production of *T. brassicae* on *E. kuehniella* (flour moth) when combined with parasitoid rearing for one generation on the target host, the European corn borer, *O. nubilalis* (Hoffmann et al., 2001). The rearing system "host + plant" is a pivotal factor that influences the parasitoid performance in the field when comparing the routine rearing system (Tabone et al., 2010). Similarly, the parasitizing efficacy of *T. achaeae* against *T. absoluta* enhanced when reared for a generation on tomato borer, which previously reared on laboratory host *E. kuehniella*. The parameters related to the *T. achaeae* efficiency has significantly improved and parasitizing efficiency increased to 69% when parasitoids were exposed for a generation on target host eggs as compared routinely reread on *E. kuehniella* eggs where only 25% of *T. absoluta* eggs were parasitized (Cascone et al., 2015). However, these induced adaptations may disappear in a continuous rearing system. Besides, the variation in temperature during rearing and release may also affect parasitoid performance. Therefore, field efficacy of parasitoid might be improved through acclimation to a relevant temperature by adapting the rearing of parasitoids at least one generation on the target host before field releases (Lessard and Boivin, 2012; Cascone et al., 2015).

9.10 MASS PRODUCTION AND BIOLOGICAL CONTROL

The production of selected parasitoid at a large scale is crucial for cost efficiency, which enables to cover of the large-scale area through inundative releases (Parra et al., 2010a; Zhang et al., 2015). The production system involves several stages, such as drying and cleaning of the host eggs, egg cards preparation, inoculation of parasitoid, and selection and collection of parasitized host eggs (Liu et al., 1991; Song et al., 1994; Wang et al., 1999). The process involved in mass rearing has been mechanized or automated in many parts of the world to produce a cost-effective system by using economical laboratory hosts for the production of selected *Trichogramma* species (Zang et al., 2020). Initially, *Trichogramma* was reared on Angoumois grain moth *Sitotroga cerealella* Olivier (Flanders, 1929). Afterwards, many small host eggs of lepidopteran insects were also used for *Trichogramma* production worldwide based on the availability of the rearing medium and the cost economics (Smith, 1996). Rice moth *C. cephalonica* has been most efficiently used as factitious hosts for many species of *Trichogramma* in many Asian countries including India, China and the Philippines. *Sitotroga cerealella* has been utilized as host eggs in Australia, Russia, East Europe; Bulgaria; Columbia, Germany, Peru, North and South American countries (Pitcher et al., 2002; Luck and Forster, 2003; Reznik et al., 2009). The Mediterranean flour moth, *Ephestia kuehniella* Zeller, the most commonly used host egg for the large scale production of *Trichogramma* in Latin America (Brazil), France, Switzerland and many countries of Western Europe (Bertin et al., 2017; Parra, 2010b). However, in China, the most successful mass production methods involved use of large size hosts eggs such as *Antheraea pernyi* Guérin-Méneville and *Samia cynthia ricini* Boisduval (Li et al., 2019; Wang et al., 2014). However, later the production of *Trichogramma* on *S. cynthia ricini* eggs was superseded by one relying on *A. pernyi* eggs (Wang, 2001). The host *A. pernyi* is reared in open conditions on oak trees in the northeastern forests of China, whereas *S. cynthia ricini* reared indoors. Using the large size host eggs of *A. pernyi* and *S. cynthia ricini* produces up to 260 and 60 *Trichogramma* per eggs, respectively (Wang et al., 2014). Moreover, large host eggs produce too many parasitoids, which sometimes reduce longevity, fitness and body size and increased the sex ratio of the parasitoids (Gao, 1992; Liu et al., 2017; Zhang et al., 2010). Therefore, the optimization of female: host egg ratios and optimal exposure time were taken into consideration for the production of *Trichogramma* on large size host eggs (Henry et al., 2008; Bertin et al., 2017). This helped in the production of *T. dendrolimi* and *T. chilonis* at a commercial scale using large host eggs in China (Hassan et al., 2004; Zhang et al., 2015). In Indian conditions, the *C. cephalonica* is the best host, which is accepted by most *Trichogramma* species for their mass rearing (Jalali et al., 2007; Nagaraja, 2013). While, *S. cerealella* and *C. cephalonica* are the most suitable host for mass rearing of *T. ostriniae* and *T. japonicum*, respectively in China and India (Chen et al., 2000; Jia et al., 2002). In addition to lepidopteran eggs, *Trichogramma* has been mass-reared on artificial eggs developed in China. Earlier studies showed that insect hemolymph, animal serum, and egg yolk were the most important components for developing artificial media (Xie et al., 1982; Consoli et al., 1999a). The rearing of *Trichogramma* on artificial eggs produced 90% parasitism and 75% adult emergence (Han et al., 1994; Zhang et al., 1993). This full mechanized production system with the possibility to produce *Trichogramma* on large scale throughout the year did not prove cost-effective and discontinued to search for a cost-effective alternative to insect hemolymph with trehalose (Lü et al., 2013).

9.11 RELEASE STRATEGIES OF TRICHOGRAMMATIDS

The inundative release method of *Trichogramma* species adapted in various biological control programmes. Parasitoids are released in large numbers, which act as a biopesticide spray (Smith, 1996). Large-scale inundative releases in biological control programs have been used successfully against the various lepidopteran pests infesting various crops at different release rates (Table 9.2). A release of single species (*T. dendrolimi* at 225,000 wasps/ha) to control the pest, like Asian corn borer (ACB) *Ostrina furnacalis* (Guenée) over 15 release at 5–7 days interval practiced in China (Zhang et al., 2015). In Brazil, three species, notably *T. pretiosum*, *T. galloi* and *T. atopovirilia* are released to control the pests. Among *T. pretiosum* has been used on nearly 250,000 ha to control lepidopteran pests in soybean, cotton, corn, citrus, beans, millet, and others crops (Parra et al., 2015). In India, *T. chilonis* released at large numbers with the rate of 50,000 to 250,000 ha to control the pests of rice, maize, fruits and vegetable crops. Besides, *T. japonicum* has been released at the rate of 50,000 ha to suppress the pests of rice and sugarcane (Jalali et al., 2016). In Europe and the Americas, the most studied and used trichogrammatids in biocontrol are *T. evanescens, T. dendrolimi, T. pretiosum, T. brassicae (=muidis),* and *T. nubilale*. However, the first three species have dominated in large releases due to their plasticity in habitat and host selection (Smith, 1996).

A mixed-species release also explored to increase the efficacy and effectiveness of the biological control. In China, mixed releases of *T. japonicum* and *T. dendrolimi* controlled the *C. suppressalis* in rice. As the production cost of *T. dendrolimi* on *A. pernyi* is only 1/75 of that *T. japonicum* reared on *C. cephalonica*; therefore, release of mixed-species method adopted by identifying the optimal ratio between the two species. The mixed-species release proved effective and cheaper than the release of sole species for pest management (Zang et al., 2020). In other methods, a mixture of different developmental stages of *Trichogramma* may be released, thereby staggering emergence, particularly if only one or a few releases can be made in the fields. This approach ensures the continuous availability of females actively searching throughout the host for oviposition. However, practically this approach is limited in practice due to parasitized eggs exposed to predation or extreme environmental conditions (Smith, 1996).

Several approaches are used for *Trichogramma* releases at the field level. Traditionally, parasitized egg cards of *Trichogramma* are used to release in the fields just before adult emergence (Parra et al., 2015). Release of adults, a sprinkling of parasitized eggs mixed with vermiculite and Semolina resulted in 61 to 77% parasitism in laboratory cages (Jalali et al., 2005a). Nowadays, the parasitoids are released in the field either as an adult or at the late pupal stage that are very close to their emergence (Pinto et al., 2003) or parasitized eggs encased in cardboard capsules. These capsules developed based on crop characteristics and environmental conditions and these allow the parasitoids to emerge in a protected environment, avoiding exposure of immatures to environmental stressors (rain and sunshine) and impeding attack by predatory insects, especially ants

TABLE 9.2
Trichogrammatids Used for Inundative Releases Against Insect Pest at Field Level

Crop	Species	Target pest	Number of Releases	Release rate (per ha)	Country	References
Rice	*Trichogramma japonicum*	*Scirpophaga incertulas*	7–9	50,000–100,000	India	Lalitha et al. (2015)
	T. chilonis	*Cnaphalocrocis mainsails*	4–9	100,000	India	Bentur et al. (1994)
	T. chilonis & *T. japonicum*	*Scirpophaga incertulas*	–	50,000–200,000	China	Tang et al. (2017)
Sorghum	*T. chilonis*	*Chilo partellus*	6	100,000	India	Singh (1999)
Maize	*T. chilonis*	*Chilo partellus*	6	75,000–100,000	India	Singh and Jalali (1994) Singh (1999)
	Trichogramma ostriniae	*Ostrinia nubilalis*	-	75,000	USA	Wright et al. (2002); Hoffmann et al. (2006)
	Trichogramma brassicae	*O. nubilalis*	2	375,000	Italy	Razinger et al. (2015)
	T. brassicae	*O. nubilalis*	2	375,000	France	Razinger et al. (2015)
	T. brassicae	*O. nubilalis*	2	375,000	Slovenia	Razinger et al. (2015)
	T. pretiosum	*Helicoverpa zea*	–	100,000	Brazil	Parra et al. (2015)
	T. evanescens	*O. nubilalis*	2	75,000	Turkey	Öztemiz and Kornosor, 2007
	T. pretiosum	*Spodoptera frugiperda*	3	100,000	Brazil	Figueiredo et al. (2015)
	T. chilonis	*H. armigera*	–	150,000	China	Wang et al. (2002)
	T. pretiosum	*Helicoverpa zea*	2	150,000–200,000	USA	Manandhar and Wright (2015)
	T. pretiosum	*S. frugiperda*	4	100,000	India	Varshney et al. (2021)
Sweet corn	*T. brassicae*	*O. nubilalis*	10	300,000	USA	Mertz et al. (1995)
Pearl millet,	*Trichogrammatoidea armigera*	*Heliocheilus albipunctella*	1	150000	Niger	Karimoune et al. (2020)
Sugarcane	*T. chilonis*	*Chilo auricilius*, *Chilo sacchariphagus indicus* & *C. infuscatellus*	12	50,000	India	Lalitha et al. (2015)
	T. japonicum	*Scirpophaga excerptalis*	4–12	50,000	India	Lalitha et al. (2015)
	T. galloi	*Diatraea saccharalis*	3	50,000–200,00	Brazil	Parra et al. (2015)
Cotton	*T. chilonis*	*Helicoverpa armigera*	13	150,000	India	Virk et al. (2004)
	T. chilonis	*Earias* spp.	3–8	150,000	India	Singh and Jalali (1994)
	Trichogrammatoidea bactrae	*Pectinophora gossypiella*	8	100,000	India	Lalitha et al. (2015)

(*Continued*)

TABLE 9.2 (Continued)
Trichogrammatids Used for Inundative Releases Against Insect Pest at Field Level

Crop	Species	Target pest	Number of Releases	Release rate (per ha)	Country	References
Sugarcane	*T. chilonis*	*Chilo infuscatellus*	8–9	50,000	India	Sharma et al. (2020)
	T. chilonis	*C. sacchariphagus indicus*	6–13	75,000–100,000	India	Manisegaran (2004); Geetha (2009; 2010)
	T. japonicum	*Scirpophaga excerptalis*	8	50,000	India	Sharma et al. (2020)
	T. chilonis	*Chilo auricilius*	10–12	50,000	India	Shenhmar et al. (2003); Sharma et al. (2020)
		Chilo tumidicostalis	6–8	50,000	India	Srikanth et al. (2016);
Soybean	*T. pretiosum*	*Chrysodeixis includes* *H. armigera*	–	250,000	Brazil	Parra (2014)
Castor	*T. achaeae*	*Achaea janata* *Conogethes punctiferalis*	–	100000	India	Lalitha et al. (2015)
Yardlong beans	*T. evanescens*	*Maruca vitrata*	4	150,000	Philippines	Ulrichs and Mewis (2004)
Cabbage	*T. chilonis*	*Plutella xylostella*	3–6	100,000	India	Navik et al. (2019b)
	Tr. bactrae	*P. xylostella*	6	50,000	India	Krishnamoorthy (2003)
	T. brassicae	*P. xylostella*	3–6	100,000	India	Lalitha et al. (2015)
Cauliflower	*T. chilonis*	*P. xylostella*	6	100,000	India	Lalitha et al. (2015)
	T. peiridis	*P. xylostella*	6	100,000	India	Lalitha et al. (2015)
	T. brassicae	*Pieris brassicae*	3	100,000	India	Lalitha et al. (2015)
Tomato	*T. chilonis*	*H. armigera*	5–6	100,000	India	Jalali et al. (2016a)
	T. pretiosum	*T. absoluta* *H. zea* *N. elegantalis*	–	400,000	Brazil	Parra et al. (2015)
	T. pretiosum	*H. armigera*	4–5	50,000	India	Gupta and Babu (1998)
	T. achaeae	*T. absoluta*	9–12	140 adults/m^2	Portugal	Oliveira et al. (2017)
	T. euproctidis & *T. achaeae*	*T. absoluta*	7	25–75 adults/m^2	Egypt	El-Arnaouty et al. (2014)
Brinajl	*T. chilonis*	*Leucinodes orbonalis*	5–6	50,000	India	Lalitha et al. (2015)
Potato	*T. atopovirilia*	*Phthorimaea operculella*	–	400,000	Brazil	Parra et al. (2015)

Okra	*T. chilonis*	*Earias* spp.	4–6	50,000	India	Lalitha et al. (2015)
Apple	*T. embryophagum* & *T. cacoeciae*	*Cydia pomonella*	2–4	2000–5000 adults/tree	India	Lalitha et al. (2015)
	T. platneri	*C. pomonella*	2	3000 adults/tree	Canada	Cossentine and Jensen (2000)
Ber	*T. chilonis*	*Meridarchis scyrodes*	5	100,000	India	Mani et al. (2005)
Pomegranate	*T. chilonis*	*Deudorix isocrates*	4–6	100,000–250,000	India	Karuppuchamy et al. (2001); Lalitha et al. (2015).
Coconut	*T. chilonis* *T. embryophagum*	*Opisina arenosella*	–	1000–4000 adults/palm	India	Jalali et al. (2005b)
Avocado	*Trichogrammatoidea annulata* & *T. atopovirilia*	*Stenoma catenifer*	–	30 wasp/eggs	Brazil	Parra et al. (2015)

(Parra et al., 2015). In China, a box-shaped capsule was made based on maize leaf phenology to release parasitized eggs of *A. pernyi* for control of Asian corn borer (Zhang et al., 2015), whereas, ball-shaped capsules used to release the parasitoid in rice fields produced from biodegradable materials that provide an automatic floating posture and enable parasitoid exit holes to remain above the water surface (Gibson, 2019). The manual release of *Trichogramma* is expensive and not practical in large area coverage and during peak agricultural activities. Use of unmanned aerial vehicles like agricultural drones are deployed to expand the area coverage and suitable for complex farmlands with numerous separate plots and small farms (Wang et al., 2018). The release of *Trichogramma pintoi* by using agricultural drones controlled mangrove pests with 92% parasitism in a 500-ha area (Luo et al., 2018). Besides, *T. pretiosum* was released by using motorcycles for pest control of tomato, soybean, cotton and bean fields in Brazil (Parra et al., 2015).

9.12 CONCLUSIONS

Trichogrammatids are the most diverse egg parasitoids group and are widely distributed throughout the world in terrestrial and aquatic ecosystems. The minute size of the trichogrammatids makes them difficult to collect and curating from a different ecosystem; therefore, the taxonomic studies remained much difficult as compared to other parasitoid groups. However, in the recent past, the use of an integrated approach by using morphological and molecular data helped to understand the taxonomy and systematics of trichogrammatids. Trichogrammatids possess a wide host range and have the ability to adapt, develop and survive in different ecological conditions. Among the trichogrammatids, species of *Trichogramma* and *Trichogrammatoidea* constitute important component under integrated pest management. *Trichogramma* is the most utilized egg parasitoids across the globe due to its ease of mass production and release against the target pests. The adaptability of trichogrammatids to the various crop ecosystems, wide host range, climatic conditions and their production at a large scale in commercial laboratories has increased the use of these egg parasitoids in pest management. Further, genetic improvement of local or potential strains of *Trichogramma* for the tolerance to various abiotic stresses has increased the efficacy of parasitoids at field level under stress conditions. Additionally, the development of an improved rearing system for parasitoids and manipulation of their behaviour by using volatile kairomones has increased the host finding capacity even at low pest densities. Thus, trichogrammatid parasitoids showed their potential and usefulness for the management of crop pests worldwide.

9.13 POINTS TO REMEMBER

1. More than 1000 species of trichogrammatid egg parasitoids are known across the globe from terrestrial and aquatic ecosystems.
2. Trichogrammatids are primary endoparasitoids of insect eggs either in solitary or gregarious form.
3. The highest numbers of Trichogrammatidae genera have been associated with members of Hemiptera, followed by Coleoptera and Lepidoptera.
4. Globally, 22 species of trichogrammatids parasitize the eggs of aquatic insects and mainly belong to genera *Centrobiopsis* Girault, *Hydrophylita* Ghesquiere, *Lathromeroidea* Girault, *Prestwichia* Lubbock and *Pseudoligosita*.
5. *Trichogramma* is the most speciose genus with 235 described species of the family Trichogrammatidae.
6. The members of *Trichogramma* are generalist eggs parasitoids and parasitize the various lepidopteran pests in different ecosystems.
7. Temperature is one of the most important environmental factors that influence the physiology and behaviour of *Trichogramma* spp., which is essential for planning its mass rearing procedures in the laboratory.

8. More emphasis should be given to indigenous strains or species that are collected from the same region for pest control.
9. The improvements of trichogrammatids are important through selection and fortification for various traits against the various physical and chemical stresses, which encounters targeted crop systems.
10. Some of the most utilized species are *Trichogramma achaeae*, *T. brassicae*, *T. chilonis*, *T. evanescens*, *T. japonicum*, *T. nubilalis*, *T. pretiosum*, *Trichogrammatoidea armigera* and *Tr. bactrae* in agri-horticulture systems in the world.

ACKNOWLEDGEMENTS

The authors are thankful to the Director, ICAR-National Bureau of Agricultural Insect Resources, Bengaluru, India, for the constant support for research and writing the review papers.

REFERENCES

Abbas, M.S.T. (1989). Studies on *Trichogramma buesi* as a biocontrol agent against *Pieris rapae* in Egypt. *Entomophaga*, 34: 447–451.

Abdel-Latief, M. & Hilker, M. (2008). Innate immunity: Eggs of *Manduca sexta* are able to respond to parasitism by *Trichogramma evanescens*. *Insect Biochemistry and Molecular Biology*, 38: 136–145.

Albarracin, L.E., Triapitsyn, S.V. & Virla, E.G. (2017). Egg Parasitoid Complex of the Corn Leafhopper, *Dalbulus maidis* (DeLong) (Hemiptera: Cicadellidae), in Argentina. *Neotropical Entomology*, 46(6): 666–677.

Alzouma, I. (1987). Reproduction et developpement de *Bruchidius atrolineatus* Pic (Coleoptera: Bruchidae) aux depens des cultures de *Vigna unguiculata* L. *Walp (Leguminosae: Papilionacea) dans un agro-systeme sahelien au Niger.* These presentee a I Universite Francois-Rabelais de Tours, France. 274.

Andow D.A. & Prokrym, D.R. (1990). Plant structural complexity and host finding by a parasitoid. *Oecologia*, 82: 162–165.

Andrade, G.S., Pratissoli, D. Dalvi, L.P., Desneux, N. & dos Santos, H.J.G. (2011). Performance of four *Trichogramma* species (Hymenoptera: Trichogrammatidae) as biocontrol agents of *Heliothis virescens* (Lepidoptera: Noctuidae) under various temperature regimes. *Journal of Pest Science*, 84: 313–320.

Ayvaz, A., Karasu, E., Karaborklu, J. & Tuncbilek, A.S. (2008). Effects of cold storage, rearing temperature, parasitoid age and irradiation on the performance of *Trichogramma evanescens* Westwood (Hymenoptera: Trichogrammatidae). *Journal of Stored Product Research*, 44: 232–240.

Bai, B., Çobanoĝlu, S. & Smith, S.M. (1995). Assessment of *Trichogramma* species for biological control of forest lepidopteran defoliators. *Entomologia Experimentalis et Applicata*, 75(2): 135–143.

Bai, B., Luck, R.F., Forster, L., Stephens, B. & Janssen, J.A.M. (1992). The effect of host size on quality attributes of the egg parasitoid, *Trichogramma pretiosum*. *Entomologia Experimentalis et Applicata*, 64: 37–48.

Bakkendorf, O. (1934). Biological investigations on some Danish Hymenopterous egg-parasites, especially in Homopterous and Heteropterous eggs, with taxonomic remarks and descriptions of new species. *Entomologiske Meddelelser*, 19: 1–135.

Bakthavatsalam, N., Tandon, P.L. & Bhagat, D. (2013). Trichogrammatids: Behavioural ecology. In: *Biological control of insect Pests using egg parasitoids*, ed. S. Sithantham, et al., pp. 77–102. India: Springer.

Ballal, C.R., Srinivasan, R. & Jalali, S.K. (2009). Evaluation of an endosulfan tolerant strain of *Trichogramma chilonis* on cotton. *BioControl*, 54(6): 723–732.

Barrett, M. & Schmidt, J.M. (1991). A comparison between the amino acid composition of an egg parasitoid wasp and some of its hosts. *Entomologia Experimentalis et Applicata*, 59: 29–41.

Bastos, C.S., de Almeida, R.P. & Suinaga F.A. (2006). Selectivity of pesticides used on cotton (*Gossypium hirsutum*) to *Trichogramma pretiosum* reared on two laboratory hosts. *Pest Management Science*, 62: 91–98.

Bennett, A.M.R. (2008). Global diversity of hymenopterans (Hymenoptera: Insecta) in freshwater. *Hydrobiologia*, 595(1): 529–534.

Bentur, J.S. & Kalode, M.B. (1985). Natural enemies of rice leaf- and plant hoppers in Andhra Pradesh. *Entomon*, 10(4): 271–274.

Bentur, J.S., Kalode, M.B., Rajendran, B. & Patil, V.S. (1994). Field evaluation of the egg parasitoid. *Trichogramma japonicum* Ash. (Hym: Trichogrammatidae) against the rice leaf roller, *Cnaphalocrocis medinalis* (Guen.) (Lep: Pyralidae) in India. *Journal of Applied Entomology*, 117: 257–261.

Benvenuto, C., Tabone, E., Vercken, E., Sorbier, N., Colombel, E., Warot, S., Xevier, S. & Ris, N. (2012). Intraspecific variability in the parasitoid wasp *Trichogramma chilonis*: can we predict the outcome of hybridization? *Evolutionary Applications*, 5(5): 498–510.

Bertin, A., Pavinato, V.A.C. & Parra, J.R.P. (2017). Fitness-related changes in laboratory populations of the egg parasitoid *Trichogramma galloi* and the implications of rearing on factitious hosts. *BioControl*, 62: 435–444.

Beserra, E.B., Dias, C.T.D.S. & Parra, J.R.P. (2002). Distribution and natural parasitism of *Spodoptera frugiperda* (Lepidoptera: Noctuidae) eggs at different phenological stages of corn. *Florida Entomologist*, 85: 588–593.

Bhatnagar, V.S., Lateef, S.S., Sithanantham, S., Pawar, C.S. & Reed, W. (1982). *Heliothis* research at ICRISAT. In: Proceedings of international workshop on *Heliothis* management, ICRISAT, Patancheru, 15–20 Nov 1981, p. 385–395.

Blood, B.N. (1923). Notes on Trichogrammatinae taken around Bristol. *Annual Report and Proceedings of the Bristol Naturalists' Society*, 5(5): 254.

Boivin, G. (2010). Reproduction and immature development of egg parasitoids. In: *Egg parasitoids in agroecosystems with emphasis on Trichogramma* ed. Consoli, F.L. et al., pp. 1–23. Dordrecht: Springer.

Brodeur, J. & Rosenheim, J.A. (2000). Intraguild interactions in aphid parasitoids. *Entomologia Experimentalis et Applicata*, 97: 93–108.

Brotodjojo, R.R. & Walter, G.H. (2006). Oviposition and reproductive performance of a generalist parasitoid (*Trichogramma pretiosum*) exposed to host species that differ in their physical characteristics. *Biological Control*, 39(3): 300–312.

Bueno, R.C., Parra, J.R., Bueno, A.F. & Haddad, M.L. (2009). Performance of trichogrammatids as bio-control agents of *Pseudoplusia includens* Walker (Lepidoptera: Noctuidae). *Neotropical Entomology*, 38(3): 389–394.

Burks, B.D. (1979). Chalcidoidea (part) and Cynipoidea. In: *Catalog of Hymenoptera in America North of Mexico* ed. K.V. Krombein et al., 1: 768–1107.

Cabello, T. & Vargas, P. (1988). The effect of temperature on the bionomic of *Trichogramma cordubensis* Vargas and Cabello (Hym.: Trichogrammatidae). In: *Trichogramma and other egg parasites*, eds.Voegele, J. et al., pp. 155–164. Paris, France: I.N.R.A. Pub.

Campbell, B., Heraty, J.M., Rasplus, J., Chan, K., Steffen-Campbell, J. & Babcock, C. (2000). Molecular systematics of the Chalcidoidea using 28S-D2 rDNA. *Hymenoptera: Evolution, Biodiversity, Biological Control*. ed. Austin, A. & Dowton, M., pp. 59–73. Melbourne: ANIC Publishing.

Carriere, Y. & Boivin, G. (1997). Evolution of thermal sensitivity of parasitization capacity in egg parasitoids. *Evolution*, 51: 2028–2032.

Carrillo, D., Pena, J.E. & Capinera, J.L. (2008). Effect of host plants on successful parasitism by *Haeckeliania sperata* (Hymenoptera: Trichogrammatidae) on *Diaprepes abbreviatus* (Coleoptera: Curculionidae) eggs. *Environmental Entomology*, 37: 1565–1572.

Cascone, P., Carpenito, S., Slotsbo, S., Iodice, L., Sørensen, J.G., Holmstrup, M. & Guerrieri, E. (2015). Improving the efficiency of *Trichogramma achaeae* to control *Tuta absoluta*. *BioControl*, 60(6): 761–771.

Chailleux, A., Desneux, N., Seguret, J., Maignet, P., Khanh, H.D.T. & Tabone, E. (2012). Assessing European egg parasitoids as a mean of controlling the invasive South American tomato pinworm *Tuta absoluta*. *PLoS One* 7: e48068.

Chailleux, A., Stirnemann, A., Leyes, J. & Deletre, E. (2019). Manipulating natural enemy behavior to improve biological control: attractants and repellents of a weaver ant. *Entomologia Generalis*, 38(3): 191–210.

Chen, H.Y., Wang, S.Y. & Chen, C.F. (2000). Quality control of *Trichogramma ostriniae* (Hymenoptera: Trichogrammatidae) mass reared with rice moth *Corcyra cephalonica* eggs. *Natural Enemies of Insects*, 22: 145–150.

Chihrane, J. & Lauge, G. (1996). Loss of parasitization efficiency of *Trichogramma brassicae* (Hymenoptera: Trichogrammatidae) under high temperature conditions. *Biological Control*, 87: 85–89.

Chihrane, J., Derrien, A. & Lauge, G. (1997). Locomotor activity of *Trichogramma brassicae* (Hymenoptera) under the influence of high temperature shocks. *Journal of Insect Behavior*, 10: 203–211.

Claridge, M.F., Morgan, J.C., Steenkiste, A.E., Iman, M. & Damayanti, D. (1999). Seasonal patterns of egg parasitism and natural biological control of rice brown planthopper in Indonesia. *Agricultural and Forest Entomology*, 1(4): 297–304.

Clausen, C.P. (1940). *Entomophagous insects*. McGraw-Hill, New York.

Coelho, A., Rugman-Jones, P.F., Reigada, C., Stouthamer, R. & Parra, J.R.P. (2016). Laboratory performance predicts the success of field releases in inbred lines of the egg parasitoid *Trichogramma pretiosum* (Hymenoptera: Trichogrammatidae). *PLoS ONE*, 11: e0146153.

Colazza, S., Aquila, G., De Pasquale, C., Peri, E. & Millar, J.G. (2007). The egg parasitoid *Trissolcus basalis* uses n-nonadecane, a cuticular hydrocarbon from its stink bug host *Nezara viridula*, to discriminate between female and male hosts. *Journal of Chemical Ecology*, 33(7): 1405–1420.

Colazza, S., Rosi, M. & Clemente, A. (1997). Response of the egg parasitoid *Telenomus busseolae* to sex pheromone of *Sesamia nonagrioides*. *Journal of Chemical Ecology*, 23: 2437–2444.

Consoli, F.L., Kitajima, E.W. & Parra, J.R.P. (1999a). Sensilla on the antenna and ovipositor of the parasitic wasps *Trichogramma galloi* Zucchi and *T. pretiosum* Riley (Hym., Trichogrammatidae). *Microscopy Research and Technique*, 45: 313–324.

Consoli, F.L., Rossi, M.M. & Parra, J.R.P. (1999b). Developmental time and characteristics of the immature stages of *Trichogramma galloi* and *T. pretiosum* (Hymenoptera: Trichogrammatidae). *Revista Brasileira de Entomologia*, 43: 271–275.

Corrigan, J.E. & Laing, J.E. (1994). Effects of the rearing host species and the host species attacked on performance by *Trichogramma minutum* Riley (Hymenoptera, Trichogrammatidae). *Environmental Entomology*, 23: 755–760.

Cossentine, J.E. & Jensen, L.B.M. (2000). Releases of *Trichogramma platneri* (Hymenoptera: Trichogrammatidae) in apple orchards under a sterile codling moth release program. *Biological Control*, 18(3): 179–186.

Dahlan, A.N. & Gordh, G. (1996). Development of *Trichogramma australicum* Girault (Hymenoptera: Trichogrammatidae) on *Helicoverpa armigera* (Hübner) eggs (Lepidoptera: Noctuidae). *Australian Journal of Entomology*, 35: 337–344.

Dahlan, A.N. & Gordh, G. (1997). Development of *Trichogramma australicum* (Hym.: Trichogrammatidae) at low and high population density in artificial diet. *Entomophaga* 42: 525–536.

De Santis, L. (1979). *Catálogo de los himenopteros calcidoideos de América do Sur de los Estados Unidos*. La Plata: Comisión de Investigaciones Científicas de la Provincia de Buenos Aires. p. 488.

De Santis, L. (1980). *Catálogo de los himenopteros brasileños de la serie parasitica incluindo Bethyloidea*. UFPR, Curitiba. p. 395.

De Santis, L. (1981). Catálogo de Los himenopteros calcidoideos de América do sur de los Estados Unidos – Primer Suplemento. *Revista Peruana de Entolomogía*, 24: 1–38.

Delpuech, J.M. & Meyet, J. (2003). Reduction in the sex ratio of the progeny of a parasitoid wasp (*Trichogramma brassicae*) surviving the insecticide chlorpyrifos. *Archives of Environmental Contamination and Toxicology*, 45: 203–208.

Delvare, G. & LaSalle, J. (2000). *Trisecodes* gen. n. (Hymenoptera, Eulophidae, Entedoninae) the first eulophid with three tarsal segments. *Journal of Hymenoptera Research*, 9: 305–312.

Denis, D., Pierre, J.S., van Baaren, J. & van Alphen, J.J.M. (2012). Physiological adaptations to climate change in pro-ovigenic parasitoids. *Journal of Theoretical Biology*, 309: 67–77.

Dequech, S.T.B., Camera, C., Sturza, V.S., Ribeiro, L.D., Querino, R.B. & Poncio, S. (2013). Population fluctuation of *Spodoptera frugiperda* eggs and natural parasitism by *Trichogramma* in maize. *Acta Scientiarum Agronomy*, 35: 295–300.

Desneux, N. & Ramirez-Romero, R. (2009). Plant characteristics mediated by growing conditions can impact parasitoid's ability to attack host aphids in winter canola. *Journal of Pest Science*, 82: 335–342.

Dodia, J.F., Yadav, D.N. & Patel, R.C. (1986). Management of fruit sucking moth, *Othreis fullonica* L. (Lepidoptera: Noctuidae) in citrus orchards at Anand (Gujarat). *GAU Research Journal*, 11: 72–73.

Doutt, R.L. & Viggiani, G. (1968). The classification of the Trichogrammatidae (Hymenoptera: Chalcidoidea). *Proceedings of the California Academy of Sciences*, 35: 477–586.

Dyer, L.A. & Coley, P.D. (2002). Tritrophic interactions in tropical versus temperate communities. In: *Multitrophic Level Interactions*, eds. Tscharntke, T. & Hawkins, A., pp. 67–88. New York: Cambridge.

El-Arnaouty, S.A., Pizzol, J., Galal, H.H., Kortam, M.N., Afifi, A.I., Beyssat, V., Desneux, N., Biondi, A. & Heikal I.H. (2014). Assessment of Two *Trichogramma* Species for the Control of *Tuta absoluta* in North African Tomato Greenhouses. *African Entomology*, 22(4), 801–809.

El-Wakeil, N., Saleh, M., Gaafar, N. & Elbehery, H. (2017). Conservation biological control practices. In: *Biological Control of Pest and Vector Insects*, ed. Shields, V.D.C., pp. 41–69. Croatia: InTech Publisher.

Faria, C.A., Angela, J.B.T. & Farias, M.I. (2000). Functional response of *Trichogramma pretiosum* (Hymenopetra: Trichogrammatidae) to *Tuta absoluta* (Meyrick) (Lepidoptera: Gelechiidae) eggs: effect of host age. *Neotropical Entomology*, 29: 85–93.

Farrokhi, S., Ashouri, A., Shirazi, J., Allahyari, H. & Huigens, M.E. (2010). A comparative study on the functional response of *Wolbachia* infected and uninfected forms of the parasitoid wasp *Trichogramma brassicae*. *Journal of Insect Science*, 10: 1–11.

Fathipour, Y., Haghani, M., Attaran, M., Talebi, A.A., Moharramipour, S. (2003). Functional response of *Trichogramma embryophagum* (Hym.: Trichogrammatidae) on two laboratory hosts. *Journal of Entomological Society of Iran*, 23: 41–45.

Fatouros, N.E., Dicke, M., Mumm, R., Meiners, T. & Hilker, M. (2008). Foraging behavior of egg parasitoids exploiting chemical information. *Behavioral Ecology*, 19(3): 677–689

Fatouros, N.E., Pineda, A., Huigens, M.E., Broekgaarden, C., Shimwela, M.M., Candia, I.A.F., Verbaarschot. P. & Bukovinszky, T. (2014). Synergistic effects of direct and indirect defences on herbivore egg survival in a wild crucifer. *Proceedings of the Royal Society B: Biological Sciences*, 281: 20141254.

Fernandez-Arhex, V. & Corley, J.C. (2003). The functional response of parasitoids and its implications for biological control. *Biocontrol Science & Technology*, 13: 403–413.

Figueiredo, M.C.D., Cruz, I., da Silva, R.B. & Foster, J.E. (2015). Biological control with *Trichogramma pretiosum* increases organic maize productivity by 19.4%. *Agronomy for Sustainable Development*, 35(3): 1175–1183.

Firake, D.M. & Khan, M.A. (2014). Alternating temperatures affect the performance of *Trichogramma* species. *Journal of Insect Science*, 14(1): 41.

Flanders, S.E. (1929). The mass production of *Trichogramma minutum* Riley and observations on the natural and artificial parasitism of the codling moth egg. *Proceedings of 4th International Congress of Entomol Transcript*, 2: 110–130.

Flanders, S.E. (1937). Notes on life history and anatomy of *Trichogramma*. *Ann. Entomol. Sot. Amer*. 30: 304–308.

Foerster, M.R. & Foerster, L.A. (2009). Effects of temperature on the immature development and emergence of five species of *Trichogramma*. *Biocontrol*, 54: 445–450.

Foerster, M.R., Marchioro, C.A. & Foerster, L.A. (2014). Temperature dependent parasitism, survival, and longevity of five species of *Trichogramma* Westwood (Hymenoptera: Trichogrammatidae) associated with *Anticarsia gemmatalis* Hübner (Lepidoptera: Noctuidae). *Neotropical Entomology*, 43: 176–182.

Frankham, R. (2008). Genetic adaptation to captivity in species conservation programs. *Molecular Ecology*, 17: 325–333.

Gabarra, R., Arno, J., Lara, L., Maria, J.V., Antoni, R., Francisco, B., Urbaneja, A., Tellez, M., Molla, O. & Jordi, R. (2014). Native parasitoids associated with *Tuta absoluta* in the tomato production areas of the Spanish Mediterranean Coast. *BioControl*, 59: 45–54.

Garcia, P. (2000). Biologia de *Trichogramma cordubensis* Vargas & Cabello (Hym., Trichogrammatidae) numaperspectiva de controlobiológico. Ph.D. Thesis, Departamento de Biologia, Universidade dos Açores, pp. 1–238.

Geetha, N. (2009). Effect of timing, increase in frequency of release and dose of *Trichogramma chilonis* Ishii for the management of the internode borer *Chilo sacchariphagus indicus* (Kapur), in sugarcane. *Pest Management and Economic Zoology*, 17(1): 17–23.

Geetha, N. (2010). Compatibility of pheromones and *Trichogramma chilonis* Ishii for the management of inter node borer *Chilo sacchariphagus indicus* (Kapur) in sugarcane. *Journal of Insect Science*, 23(3): 301–307.

Gibson, G.A.P., Heraty, J.M. & Woolley, J.B. (1999). Phylogenetics and classification of Chalcidoidea and Mymarommatoidea – a review of current concepts (Hymenoptera, Apocrita). *Zoologica Scripta*, 28: 87–124.

Gibson, G.A.P. (2019). *Institute of Biological Control, Jilin Agricultural University, Changchun, China. Hamuli*, 10: 2–3.

Gingras, D., Dutilleul, P. &. Boivin, G. (2002). Modeling the impact of plant structure on host-finding behavior of parasitoids. *Oecologia*, 130: 396–402.

Girault, A.A. (1912). The chalcidoid family Trichogrammatidae. I Tables of subfamilies and genera and revised catalogue. *Bulletin of Wisconsin National History Society*, 10: 81–100.

Girault, A.A. (1918). *North American Hymenoptera Trichogrammatidae*. Girault, A.A. (1918), North American Hymenoptera Trichogrammatidae Private Publication, Sydney. p. 11.

Greenberg, S.M., Nordlund, D.A. & Wu, Z. (1998). Influence of rearing host on adult size and oviposition behaviour of mass produced female *Trichogramma minutum* Riley and *Trichogramma pretiosum* Riley (Hymenoptera: Trichogrammatidae). *Biological Control*, 11: 43–48.

Grenier, S., Basso, C. & Pintureau, B. (2001). Effects of the host species and the number of parasitoids per host on the size of some *Trichogramma* species (Hymenoptera: Trichogrammatidae). *Biocontrol Science and Technology*, 11(1): 21–26.

Gross, H.R., Lewis, W.J., Beevers, M. & Norlund, D.A. (1984). *Trichogramma pretiosum* (Hymenoptera: Trichogrammatidae): Effects of augmented densities and distributions of *Heliothis zea* (Lepidoptera: Noctuidae) host eggs and kairomones on field performance. *Environmental Entomology*, 13: 981–985.

Gross, H.R., Harrell, E.A., Lewis, W.J. & Nordlund, D.A. (1981). *Trichogramma* Spp.: concurrent ground application of parasitized eggs, supplemental *Heliothis zea* host eggs, and host-seeking stimuli. *Journal of Economic Entomology*, 74: 227–229.

Guo, M. (1992). Study on parasitising behaviour of *Trichogramma*. I. Superparasitism and progeny allocation. *Natural Enemy Insects*, 14: 6–12.

Gupta, P.R. & Babu, B.R.R.M. (1998). Management of *Helicoverpa armigera* on tomato with *Trichogramma pretiosum* and *Bacillus thuringiensis* var. *Kurstaki*. In: Advances in IPM for horticultural crops (Ed.Reddy P.P. et al.). Proceeding of the first national symposium on pest management in horticultural crops: environmental implications and thrusts, Bangalore, 15–17 Oct 1997. p. 75–80.

Gupta, P.R. & Sood, A. (2001). Role of egg parasitoids for the control of pomegranate fruit-borer in mid hills of Himachal Pradesh. In: Biological control – contributed papers (Ed.Darshan, S. et al.). Proceedings of the symposium on biological based pest management for quality crop protection in the current millennium, Punjab Agricultural University, Ludhiana, 18–19 July 2001. p. 57.

Hagen, K.S. (1964). Developmental stages of parasites. In: *Biological control of insect pests and weeds*, ed. DeBach, P., pp. 168–246. London: Chapman & Hall.

Han, S.C., Chen, Q.X. & Li, L.Y. (1994). A study on the oviposition synergists for in vitro rearing *Trichogramma* spp. *Acta Entomologica Sinica*, 1: 333–338.

Hassan, S.A. & Guo, M.F. (1991). A simple method to control the quality of mass reared egg parasites of the genus *Trichogramma*. In: Proceedings of the 5th International Workshop on Quality Control of Mass Reared Arthropods (Ed. Bigler, F.). IOBC Wageningen, The Netherlands. p. 127–137.

Hassan, S.A. (1989). Selection of suitable *Trichogramma* strains to control the codling moth *Cydia pomonella* and the two summer fruit tortrix moths *Adoxophyes orana*, *Pandemis heparana* (Lep.: Tortricidae). *Entomophaga*, 34(1): 19–27.

Hassan, S.A., Liscsinszky, H. & Zhang, G. (2004). The oak-silkworm egg *da Silva, R.B. & Foster, J.E. (2015). Biological* (Lepidoptera: Anthelidae) as a mass-rearing host for parasitoids of the genus *Trichogramma* (Hymenoptera: Trichogrammatidae). *Biocontrol Science and Technology*, 14: 269–279.

Hawlitzky, N. & Boulay, C. (1982). Régimes alimentaires et développement chez *Trichogramma maidis* Pintureau et Voegele (Hym. Trichogrammatidae) dans l' oeuf *d' Anagasta kuehniella* Zeller (Lep. *Pyralidae*). *Colloques l' INRA*, 9: 101–106.

Hayat, M. & Viggiani, G. (1984). A preliminary catalogue of the Oriental Trichogrammatidae (Hymenoptera, Chalcidoidea). *Bollettino del Laboratorio di Entomologia Agraria Filippo Silvestri*, 41: 23–52.

Hayat, M. (1994). Notes on some genera of the Aphelinidae (Hymenoptera: Chalcidoidea), with comments on the classification of the family. *Orient Insects*, 28: 81–96.

Hayat, M. (1998). Aphelinidae of India (Hymenoptera: Chalcidoidea): a taxonomic revision. Memoirs on Entomology, *International*, 13: 1–416.

Hayat, M. (2008). Taxonomic notes on the Indian Trichogrammatidae (Hymenoptera: Chalcidoidea), with redescriptions and records of some species, *Oriental Insects*, 42(1): 1–32.

Hayat, M. (2009). Records and descriptions of Trichogrammatidae from India (Hymenoptera: Chalcidoidea). *Oiental Insects*, 43: 201–227.

Hegazi, E.M., Herz, A., Hassan, S., Agamy, E., Khafagi, W., Shweil, S., Zaitoun, A., Mostafa, S., Hafez, M., El-Shazly, A., El-Said, S., Abo-Abdala, L., Khamis, N. & El-Kemny, S. (2005). Naturally occurring *Trichogramma* species in olive farms in Egypt. *Insect Sciecne*, 12, 185–192.

Henriquez, N.P. & Spence, J.R. (1993). Studies of *Lathromeroidea* sp. (Hymenoptera, Trichogrammatidae), a parasitoid of gerrid eggs. *Candian Entomologist*, 125: 693–702.

Henry, L.M., Roitberg, B.D. & Gillespie, D.R. (2008.) Host-range evolution in Aphidius parasitoids: fidelity, virulence and fitness trade-offs on an ancestral host. *Evolution*, 62: 689–699.

Heraty, J.M., Woolley, J.B. & Darling, D.C. (1997). Phylogenetic implications of the mesofurca in Chalcidoidea (Hymenoptera), with emphasis on Aphelinidae. *Systematic Entomology*, 22: 45–65.

Herz, A. & Hassan, S.A. (2006). Are indigenous strains of *Trichogramma* sp. (Hym., Trichogrammatidae) better candidates for biological control of Lepidopterous pests of the olive tree? *Biocontrol Science and Technology*, 16(8): 841–857.

Hessein, N.A. & McMurtry, J.A. (1988). Observations on *Megaphragma mymaripenne* Timberlake (Hymenoptera: Trichogrammatidae), an egg parasite of *Heliothrips haemorrhoidalis* Bouché (Thysanoptera: Thripidae). *Pan Pacific Entomologist*, 64: 250–254.

Hoffmann, M.P., Ode, P.R., Walker, D.L., Gardner, J., van Nouhuys, S. & Shelton, A.M. (2001). Performance of *Trichogramma ostriniae* (Hymenoptera: Trichogrammatidae) reared on factitious hosts, including the target host, *Ostrinia nubilalis* (Lepidoptera: Crambidae). *Biological Control*, 21(1): 1–10.

Hoffmann, M.P., Pitcher, S.A., Cheever, S.A., Gardner, J., Losey, J.E., Kuhar, T.P., & Youngman, R.R. (2006). Efficacy of inoculative releases of *Trichogramma ostriniae* (Hymenoptera: Trichogrammatidae) against European corn borer *Ostrinia nubilalis* (Lepidoptera: Crambidae) in field corn. *Biological Control*, 36(3): 345–349.

Honda, J.Y. & Luck, R.F. (2001). Interactions between host attributes and wasp size: a laboratory evaluation of *Trichogramma platneri* as an augmentative biological control agent for two avocado pests. *Entomologia Experimentalis et Applicata*, 100: 1–13.

Hoy, M.A. (1976). Genetic improvement of insects: fact or fantasy? *Environmental Entomology*, 5: 833–839.

Huber, J.T. & Beardsley, J.W. (2000). A new genus of fairyfly, *Kikiki*, from the Hawaiian Islands (Hymenoptera: Mymaridae). *Proceedings of the Hawaiian Entomological Society*, 34: 65–70.

Hutchison, W.D., Moratorio, M. & Martin, J.M. (1990). Morphology and biology of *Trichogrammatoidea bactrae* (Hymenoptera: Trichogrammatidae), imported from Australia as a parasitoid of pink bollworm (Lepidoptera: Gelechiidae) eggs. *Annals of the Entomological Society of America*, 83: 46–54.

Irina, S., Fursov, V., Dmitri, G. & Leonid, F. (2003). Swimming in the diving wasp *Prestwichia aquatica* (Hymenoptera: Trichogrammatidae). *Entomologia Generalis*, 26 (4): 225–234.

Jacas, J.A., Peña, J.E. & Duncan, R.E. (2009). Morphology and development of the immature stages of *Brachyufens osborni* (Hymenoptera: Trichogrammatidae), an egg parasitoid of broad-nosed weevil species (Coleoptera: Curculionidae). *Annals of the Entomological Society of America*, 102(1): 112–118.

Jalali, S.K. & Singh, S.P. (1992). Differential response of four *Trichogramma* species to low temperatures for short term storage. *Entomophaga* 37: 159–165.

Jalali, S.K. & Singh, S.P. (1993). Superior strain selection of the egg parasitoid *Trichogramma chilonis* Ishii biological parameters. *Journal of Biological Control* 7: 57–60.

Jalali, S.K. (2013). Natural occurrence, host range and distribution of trichogrammatid egg parasitoids. In: *Biological control of insect pests using egg parasitoids*, ed. Sithanantham, S. et al., p. 67–76. New Delhi: Springer.

Jalali, S.K., Mohanraj, P., & Lakshmi, B.L. (2016b). Trichogrammatids. In. *Ecofriendly pest management for food security*, ed. Omkar, pp. 139–181. CRC Press.

Jalali, S.K., Murthy, K.S., Venkatesan, T., Lalitha, Y. & Devi, P.S. (2005a). Preliminary testing of some new release methods of egg parasitoid *Trichogramma* spp. *Journal of Biological Control*, 19(2): 99–103.

Jalali, S.K., Navik, O., Murthy, K.S., Venkatesan, T. & Lalitha, Y. (2018). Predilection for host egg and host plant by Trichogrammatid species collected from different crops. *Annals of Plant protection Sciences*, 26: 1–5.

Jalali, S.K., Singh S.P., Venkatesan, T., Murthy K.S. & Lalitha Y. (2006). Development of endosulfan tolerant strain of egg parasitoid *Trichogramma chilonis* Ishii (Hymenoptera: Trichogrammatidae). *Indian Journal of Experimental Biology*, 44: 584–590.

Jalali, S.K., Singh, S.P. & Venkatesan, T. (2003) Natural enemies of cotton bollworms in sprayed farmers' field in different states of India. *National Journal of Plant Improvement*, 5: 31–33.

Jalali, S.K., Venkatesan, T., Murthy, K.S. & Ojha, R. (2016a). Management of *Helicoverpa armigera* (Hübner) on tomato using insecticide resistance egg parasitoid, *Trichogramma chilonis* Ishii in farmers' field. *Indian Journal of Horticulture*, 73(4): 611–614.

Jalali, S.K., Venkatesan, T., Murthy, K.S., Bhaskaran, T.V. & Lalitha, Y. (2005b). Preliminary attempt at biological control of coconut leaf eating caterpillar, *Opisina arenosella* Walker using tow *Trichogramma* species. *Cord*, 21(2): 1–6.

Jalali, S.K., Venkatesan, T., Murthy, K.S., Rabindra, R.J. & Lalitha, Y. (2007). Vacuum packaging of *Corcyra cephalonica* (Stainton) eggs to enhance shelf life for parasitization by the egg parasitoid *Trichogramma chilonis*. *Biological Control*, 41(1), 64–67.

Jaraleño-Teniente, J., Lomeli-Flores, J.R., Rodríguez-Leyva, E., Bujanos-Muñiz, R. & Rodríguez-Rodríguez, S.E. (2020). Egg parasitoids survey of *Spodoptera frugiperda* (Smith) (Lepidoptera: Noctuidae) in maize and sorghum in Central Mexico. *Insects*, 11(3): 157.

Jarjees, E., Merritt, D.J. & Gordh, G. (1998). Anatomy of the mouthparts and digestive tract during feeding in larvae of the parasitoid wasp *Trichogramma australicum* Girault (Hymenoptera: Trichogrammatidae). *International Journal of Insect Morphology and Embryology*, 27: 103–110.

Jarjees, E.A. & Merritt, D.J. (2002). Development of *Trichogramma australicum* Girault (Hymenoptera: Trichogrammatidae) in *Helicoverpa* (Lepidoptera: Noctuidae) host eggs. *Australian Journal of Entomology*, 41: 310–315.

Jeffs, T. & Lewis, O.T. (2013). Effects of climate warming on host-parasitoid interactions. *Ecological Entomology*, 38: 209–218.

Jia, N.X., Huang, Y.G., Li, Q.Z. & Li. Y. (2002). Studies on reproducing *Trichogramma* by using *Sitotroga cerealella* Oliver eggs. *Journal of Jilin Agricultutal University*, 24: 58–60.

Jones, R.F., Rochat, J., Goebel, F.R. & Tabone, E. (2006). Functional response of *Trichogramma chilonis* to *Galleria mellonella* and *Chilo sacchariphagus* eggs. *Entomologia Experimentalis et Applicata*, 118: 229–236.

Karimia, Z. (1998). Bioecology of *Trichogramma brassicae* in paddy fields of Guilan province. M.Sc. Thesis. Faculty of Agriculture. University of Guilan, Iran. pp. 99.

Karimoune, L., Ba, M.N., Baoua, I.B. & Muniappan, R. (2020). Field performance of the parasitoid wasp, *Trichogrammatoidea armigera* (Hymenoptera: Trichogrammatidae) following releases against the millet head miner, *Heliocheilus albipunctella* (Lepidoptera: Noctuidae) in the Sahel. *BioControl*, 65: 389–399.

Karuppuchamy, P., Balasubramanian, G. & Sundarababu, P.C. (2001). Use of neem products and biocontrol agents for the management pomegranate fruit borer, *Deudorix isocrates* (Fabr.). *Madras Agricultural Journal*, 88: 334–336.

Kazmer, D.J. & Luck, W. (1991). Female body size, fitness and biological control quality: field experiments with *Trichogramma pretiosum*. In: *Trichogramma and Other Eggs Parasitoids.* 3rd International Symposium (Eds. E. Wajnberg and S. B. Vinson), Les Colloques de INRA,, Paris, p. 37–40.

Kim, J.W. & Triapitsyn, S.V. (2004). A new species of *Pteroptrix* (Hymenoptera: Aphelinidae) from Argentina, the first known aphelinid with three-segmented tarsi. *Entomological News*, 114: 10–17.

Klug, T. & Meyhofer, R. (2009). Performance of two *Trichogramma brassicae* strains under greenhouse and field conditions for biocontrol of the silver Y moth in spinach cultures. *Journal of Pest Science*, 82: 73–79.

Kochetova, N.I. (1969). Adaptations to parasitism in some egg parasites of the genus *Trichogramma* (Hymenoptera. Trichogrammatidae). *Zool. Zh.* 48: 1816–1822.

Kolliker-Ott, U.M., Bigler, F. & Hoffmann, A.A. (2003). Does mass rearing of field collected *Trichogramma brassicae* wasps influence acceptance of European corn borer eggs? *Entomologia Experimentalis et Applicata*, 109(3): 197–203.

Krishnamoorthy, A. (2003). IPM package for tomato with safe pesticide residue. IIHR Extension folder (NATP) – 03.

Krishnamoorthy, A., Kumar, N.K. & Mani, M. (1999). Record of egg Parasitoid on *Acherontia styx* infesting brinjal. *Pest Management in Horticultural Eco-systems* 5(1): 57–58.

Kryger, J.P. (1918). The European Trichogramminae. *Entomologiske Meddelelser*, 12: 257- 354.

Ksentini, I. & Herz, A. (2019). Studies on interactions between parasitoids: the case of the idiobiont *Trichogramma cacoeciae* and the koinobiont *Ascogaster quadridentata* on *Cydia pomonella* eggs. *Bulletin of Insectology*, 72(2): 207–218.

Kuhlmann, U. & Mills, N.J. (1999). Comparative analysis of the reproductive attributes of three commercial-produced *Trichogramma* species (Hymenoptera: Trichogrammatidae). *Biocontrol Science & Technology*, 9: 335–346.

Kumar, G.A., Jalali, S.K., Venkatesan, T., Nagesh, M., and Lalitha, Y. (2008). Genetic improvement of egg parasitoid *Trichogramma chilonis* for combined tolerance to multiple insecticides and high temperature. *Journal of Biological control*, 22(2): 347–356.

Lalitha, Y., Jalali, S.K., Ballal, C.R. & Mohanraj, P. (2015). *Current status of Trichogramma in India. In Biological control of pest using Trichogramma: Current status and perspective.* (Etd. Vinsion et al.) Northwest A & F University press. pp. 186–250.

Lessard, E., & Boivin, G. (2012). Effect of low temperature on emergence, fecundity, longevity and host-feeding by *Trichogramma brassicae*. *BioControl*, 58(3): 319–329.

Lewis, W.J., Beevers, M., Nordlund, D.A., Gross, H.R. & Hagen, K.S. (1979). Kairomones and their use for management of entomophagous insects. IX. Investigations of.various kairomone-treatment patterns for *Trichogramma* spp. *Journal of Chemical Ecology*, 5: 673–680.

Lewis, W.J., Jones, R.L., Nordlund, D.A. & Gross, H.R. (1975b). Kairomones and their use for management of entomophagous insects. II. Mechanisms causing increase in rate of parasitization by *Trichogramma* spp. *Journal of Chemical Ecology*, 1: 349–360.

Lewis, W.J., Jones, R.L., Nordlund, D.A. & Sparks, A.N. (1975a). Kairomones and their use for management of entomophagous insects. I. Evaluation for increasing rates of parasitization by *Trichogramma* spp. in the field. *Journal of Chemical Ecology*, 1: 343–347.

Li, T.H., Tian, C.Y., Zang L.S., Hou, Y.Y., Ruan, C.C., Yang, X., Monticelli, L. & Desneux, N. (2019). Multiparasitism with *Trichogramma dendrolimi* on egg of Chinese oak silkworm, *Antheraea pernyi*, enhances emergence of *Trichogramma ostriniae*. *Journal of Pest Science*, 92: 707–713.

Lin, N. (1994). *Systematic studies of Chinese Trichogrammatidae*. Fuzhou, Fujian. pp. 365.

Liu, H., Fang, X., Liu, J., Huang, S.S. & Ouyang, G. (2017). Superparasitism of *Trichogramma chilonis* (Hymenoptera: Trichogrammatidae) by different inoculation proportion and its influences on offspring. *Chinese Journal of Plant Protection*, 37: 5–10.

Liu, Z.C., Yang, H.W., Wang, C.X., Liu, J.F., Liang, Y.F. & Zeng, J.C. (1991). A machine for obtaining clean infertile eggs of oak silkworm used in *Trichogramma* production. *Chinese Journal of Biological Control*, 7: 38–40.

Logarzo, G.A., Virla, E.G., Triapitsyn, S.V. & Jones, W.A. (2004). Biology of *Zagella delicata* (Hymenoptera: Trichogrammatidae), an egg parasitoid of the sharpshooter *Tapajosa rubromarginata* (Hemiptera: Clypeorrhyncha: Cicadellidae) in Argentina. *Florida Entomologist* 87(4): 511–516.

Lü, X., Han, S.C., Li, L.Y., Grenier, S. & De Clercq, P. (2013). The potential of trehalose to replace insect hemolymph in artificial media for *Trichogramma dendrolimi* (Hymenoptera: Trichogrammatidae). *Insect Science*, 20: 629–636.

Luck, R.F. & Forster, L.D. (2003). Quality of augmentative biological control agents: a historical perspective and lessons learned from evaluating *Tricogramma*. In: *Quality control and production of biological control agents: Theory and testing procedures*, ed. van Lenteren, J.C., pp. 231–246. Wallingford, CT: CAB International.

Luo, J., Chang, M.S., Li, D.W., Mou, X.W. & Cheng H.Y. et al. (2018). Technique on control of mangrove pests by spreading *Trichogramma* using UAV. *Guangxi Forestry Science*, 47: 473–475.

Luo, S.P., Li, H.M., Lu, Y.H., Zhang, F., Haye, T., Kuhlmann, U. & Wu, K. (2014). Functional response and mutual interference of *Peristenus spretus* (Hymenoptera: Braconidae), a parasitoid of *Apolygus lucorum* (Heteroptera: Miridae). *Biocontrol Science & Technology*, 24: 247–256.

Maceda, A., Hohmann, C.L. & Santos, H.R. (2003). Temperature effects on *Trichogramma pretiosum* Riley and *Trichogrammatoidea annulata* De Santis. *Brazilian Archives of Biology and Technology*, 46: 27–32.

Makee, H. (2005). Factors influencing the parasitism of codling moth eggs by *Trichogramma cacoeciae* March. and *T. principium* Sug. et Sor. (Hymen. Trichogrammatidae). *Journal of Pest Science*, 78(1): 31–39.

Manandhar, R. & Wright, M.G. (2015). Enhancing biological control of corn earworm, *Helicoverpa zea* and thrips through habitat management and inundative release of *Trichogramma pretiosum* in corn cropping systems. *Biological Control*, 89: 84–90.

Mani, M., Krishnamoorthy A. & Gopalakrishnan, C. (2005). Biological control of lepidopterous pests of horticultural crops in India: a review. *Agriculture Reviews*, 26: 39–49.

Manisegaran, S. (2004). Revalidation of *Trichogramma chilonis* for the control of internode borer, *Chilo sacchariphagus indicus* in sugarcane. *Indian Journal of Entomology*, 66: 24–26.

Manjunath, T.M. (1972). Biological studies on *Trichogrammatoidea armigera* Nagaraja, a new dimorphic egg parasite of *Heliothis armigera* (Hübner) in India. *Entomophaga*, 17: 131–147.

Manjunath, T.M. (1974). Evaluation of the effect of two insecticidal schedules on important pests of Hybrid – 4 cotton and their natural enemies. M.Sc. (Agri.) thesis submitted to Gujarat Agricultural University, Anand, pp. 66.

Manohar, T.N., Sharma, P.L., Verma, S.C., Sharma, K.C. & Chandel, R.S. (2020). Functional response of indigenous *Trichogramma* spp. to invasive tomato leafminer, *Tuta absoluta* (Meyrick) under laboratory conditions. *International Journal of Tropical Insect Science*, 40: 101–107.

Manweiler, S.A. (1986). Developmental and ecological comparisons of *Trichogramma minutum* and *Trichogramma platneri* (Hymenoptera: Trichogrammatidae). *Pan Pacific Entomologist*, 62: 128–139.

Mertz, B.P., Fleischer, S.J., Calvin, D.D. & Ridgway, R.L. (1995). Field Assessment of *Trichogramma brassicae* (Hymenoptera: Trichogrammatidae) and *Bacillus thuringiensis* for Control of *Ostrinia nubilalis* (Lepidoptera: Pyralidae) in Sweet Corn. *Journal of Economic Entomology*, 88(6): 1616–1625.

Messenger, P.S. & van den Bosch, R. (1971). The adaptability of introduced biological control agents. In: *Biological Control*, ed. Huffaker, C.B., pp. 68–92. Boston, MA: Springer.

Milonas, P.G., Anastasaki, E. & Partsinevelos, G. (2019). Oviposition-Induced Volatiles Affect Electrophysiological and Behavioral Responses of Egg Parasitoids. *Insects*, 10: 437.

Moezipour, M., Kafil, M. & Allahyari H. (2008). Functional response of *Trichogramma brassicae* at different temperatures and relative humidities. *Bulletin of Insectology*, 61: 245–250.

Mohapatra, L.N. (2006). Biological control agents and their role in regulation of cotton bollworms population. *Journal of Cotton Research Development*, 20: 255–258.

Monje, J.C., Zebitz, C.P.W. & Ohnesorge, B. (1999). Host and host age preference of *Trichogramma galloi* and *T. pretiosum* (Hymenoptera: Trichogrammatidae) reared on different host. *Journal of Economic Entomology*, 92: 97–103.

Moya-Raygoza, G., Albarracin, E.L. & Virla, E.G. (2012). Diversity of Egg Parasitoids Attacking *Dalbulus maidis* (Hemiptera: Cicadellidae) populations at low and high elevation sites in Mexico and Argentina. *Florida Entomologist*, 95(1): 105–112.

Moya-Raygoza, G., Renteria, I.C., Albarracin, E.L. & Virla, E.G. (2014). Egg parasitoids of the leafhoppers *Dalbulus maidis* and *Dalbulus elimatus* (Hemiptera: Cicadellidae) in two Maize habitats. *Florida Entomologist*, 97: 309–312.

Nagaraja, H. (2013). Mass production of Trichogrammatid parasitoids. In: *Biological control of insect pests using egg parasitoids*, ed. Sithanantham, S. et al., pp. 175–189. New Delhi: Springer.

Nagarkatti, S. (1973). A note on egg parasitism of *Agrius convolvuli* (L.) in Karnataka. *Technical Bulletin of Commonwealth Institute of Biological Control*, 16: 23–27.

Navik, O., Jalali, S.K. & Lalitha, Y. (2019a). Natural parasitism by trichogrammatids (Hymenoptera: Trichogrammatidae) on lepidopteran eggs under diverse cropping system. *Journal of Biological Control*, 33: 279–284.

Navik, O., Ramya, R.S., Varshney, R., Jalali, S.K., Shivalingaswamy, T.M., Rangeshwaran, R., Lalitha, Y., Patil, J. & Ballal, C.R. (2019b). Integrating biocontrol agents with farmer's practice: impact on diamondback moth, *Plutella xylostella* (Linnaeus) (Lepidoptera: Plutellidae) and cabbage yield. *Egypt Journal of Biological Pest Control*, 29: 35. https://doi.org/10.1186/s41938-019-0140-x

Navik, O., Shylesha, A.N., Patil, J., Venkatesan, T., Lalitha, Y. & Ashika, T.R. (2021). Damage, distribution and natural enemies of invasive fall armyworm *Spodoptera frugiperda* (J.E. Smith) under rainfed maize in Karnataka, India. *Crop Protection*. https://doi.org/10.1016/j.cropro.2021.105536

Nayak, S.B. Seni, A. & Vinay, B.K. (2018). Genetic improvement of egg parasitoid *Trichogramma chilonis* (Hymenoptera: Trichogrammatidae) by artificial selection. *International Journal of Chemical Studies*, 6(2): 2623–2626.

Nina, E., Fatouros, A., Pineda, M.E., Huigens, C., Broekgaarden, M.M., Shimwela, I.A. (2014). Synergistic effects of direct and indirect defences on herbivore egg survival in awild crucifer. *Journal of Proceedings of the Royal Society of London*, B281: 20141254.

Nordlund, D.A. & Greenberg, S.M. (2002). Comparison of the influence of external treatments on oviposition by *Trichogramma pretiosum* and *Trichogramma minutum* (Hymenoptera: Trichogrammatidae) into stretched plastic artificial eggs. *Subtropical Plant Science*, 54: 49–53.

Nowicki, S. (1935). Descriptions of new genera and species of the family Trichogrammatidae from the Palearctic region, with notes. *I. Z. Angewandte Entomologie*, 21: 556–596.

Nowicki, S. (1936). Descriptions of new genera and species of the family Trichogrammatidae from the Palearctic region, with notes. II. Z. *Angewandte Entomologie*, 23: 114–148.

Noyes, J.S. (2003) Universal Chalcidoidea Database. Disponível em http://www.nhm.ac.uk/entomology/chalcidoids/index.html (Accessed on 20 Jan 2006).

Noyes, J.S. (2019). Universal Chalcidoidea Database. World Wide Web electronic publication. http://www.nhm.ac.uk/chalcidoids (Accessed on 11 November 2020).

Oaten, A. & Murdoch, W.W. (1975). Functional response and stability in predator-prey systems. *The American Naturalist*, 109: 289–298.

Oliveira, L., Durão, A.C., Fontes, J., Roja, I.S. & Tavares, J. (2017). Potential of *Trichogramma achaeae* (Hymenoptera: Trichogrammatidae) in biological control of *Tuta absoluta* (Lepidoptera: Gelechiidae) in azorean greenhouse tomato crops. *Journal of Economic Entomology*, 110(5): 2010–2015.

Oliveira, R.C.M., Pastori, P.L., Coutinho, C.R., Juvenal, S.O. & Aguiar, C.V.S. (2020). Natural parasitism of *Trichogramma pretiosum* (Hymenoptera: Trichogrammatidae) in *Neoleucinodes elegantalis* (Lepidoptera: Crambidae) eggs on tomato (Solanales: Solanaceae) in the Northeast region, Brazil. *Brazilian Journal of Biology*, 80(2): 474–475.

Owen, A.K. (2011) Revision of *Ufens* Girault, 1911 (Hymenoptera: Trichogrammatidae). *University of California Publications in Entomology*, 131: 1–176.

Owen, A.K., George, J., Pinto J.D. & Heraty, J.M. (2007). A molecular phylogeny of the Trichogrammatidae (Hymenoptera: Chalcidoidea), with an evaluation of the utility of their male genitalia for higher level classification. *Systematic Entomology*, 32: 227–251.

Özder, N. & Kara, G. (2010). Comparative biology and life tables of *Trichogramma cacoeciae*, *T. brassicae* and *T. evanescens* (Hymenoptera: Trichogrammatidae) with *Ephestia kuehniella* and *Cadra cautella* (Lepidoptera: Pyralidae) as hosts at three constant temperatures. *Biocontrol Science & Technology*, 20: 245–255.

Öztemiz, S. & Kornosor, S. (2007).The effects of different irrigation systems on the inundative release of *Trichogramma evanescens* Westwood (Hymenoptera: Trichogrammatidae) against *Ostrinia nubilalis* Hubner (Lepidoptera: Pyralidae) in the second crop maize. *Turkish Journal of Agriculture and Forestry*, 31: 23–30.

Pak, G., Van Dalen, A., Kaashoek, N. & Dijkman, H. (1990). Host egg chorion structure influencing host suitability for the egg parasitoid *Trichogramma* Westwood. *Journal of Insect Physiology*, 36: 869–875.

Pak, G.A. & Oatman, E.R. (1982). Biology of *Trichogramma brevicapillum. Entomologia Experimentalis et Applicata*, 32: 61–67.

Pak, G.A. (1988). *Selection of Trichogramma for Inundative Biological Control*. Wageningen, Netherlands. pp. 220.

Paramasivan, A., Paul, A.V.N. & Dureja, P. (2004). Kairomones of *Trichogramma platneri (Hymenoptera: Trichogrammatidae* (Swinhoe) and their impact on the egg parasitoid *Trichogramma chilonis* Ishii. *Indian Journal of Entomology* 66: 78–84.

Parra, J.R.P. (2010a). Egg parasitoids commercialization in the New World. In: *Egg Parasitoids in Agroecosystems with emphasis on Trichogramma*, ed. Consoli, F.L. et al., pp. 373–388. Dordrecht: Springer.

Parra, J.R.P. (2010b). Mass rearing of egg parasitoids for biological control programs. In: *Egg parasitoids in agroecosystems with emphasis on Trichogramma*, ed. Consoli, F.L. et al., pp. 267–292. Dordrecht: Springer.

Parra, J.R.P. (2014). Biological control in Brazil an Overview. *Scientia Agricola*, 71: 420 -429.

Parra, J.R.P., Zucchi, R.A., Coelho, A., Gerernias, L.D. & Consoli, F.L. (2015). *Trichogramma as a tool for IPM in Brazil* (Ed. Vinsion et al.). Northwest A & F University press. pp. 472–496.

Peck, O., Boucek, Z. & Hoffe A. (1964). Keys to the Chalcidoidea of Czechoslovakia (Insecta: Hymenoptera). *The Memoirs of the Entomological Society of Canada*, 34: 170.

Pino, D.M., Gallego, J.R., Hernández Suárez, E., & Cabello, T. (2020). Effect of Temperature on Life History and Parasitization Behavior of *Trichogramma achaeae* Nagaraja and Nagarkatti (Hym.: Trichogrammatidae). *Insects*, 11(8): 482.

Pinto, A.S., Parra, J.R.P., Oliveira, H.N. & Arrigoni, E.B. (2003). Comparação de técnicas de liberação e parasitismo de *Trichogramma galloi* Zucchi (Hymenoptera: Trichogrammatidae) para o controle de *Diatraea saccharalis* (Fabricius) (Lepidoptera: Crambidae). *Neotropical Entomology*, 32: 311–318.

Pinto, J.D. & Stouthamer, R. (1994). Systematics of the Trichogrammatidae with emphasis on *Trichogramma*. In: *Biological control with egg parasitoids* (Ed. Wajnberg, E. & Hassan, S.A.). pp. 1–36. Wallingford, UK: CAB Int.

Pinto, J.D. (1997a). Taxonomia de Trichogrammatidae (Hymenoptera) com ênfase nos gêneros que parasitam Lepidoptera, In: *Trichogramma e o controle biológico aplicado*, ed. Parra, J.R.P. & Zucchi, R.A., pp. 13–40. Piracicaba: FEALQ.

Pinto, J.D. (1997b). Trichogrammatidae. In: *Annotated keys to the genera of Nearctic Chalcidoidea (Hymenoptera)*, ed. Gibson, G.A.P. et al., pp. 726–752. Ottawa: NRC Research Press.

Pinto, J.D. (1999). Systematics of the North American species of *Trichogramma* Westwood (Hymenoptera: Trichogrammatidae). *Memoirs Entomological Society of Washington*, 22: 1–287.

Pinto, J.D. (2006). A review of the New World genera of Trichogrammatidae (Hymenoptera). *Journal of Hymenoptera Research*, 15: 38–163.

Pitcher, S.A., Hoffmann, M.P., Gardner, J., Wright, M.G. & Kuhar, T.P. (2002). Cold storage of *Trichogramma ostriniae* reared on *Sitotroga cerealella* eggs. *BioControl*, 47: 525–535.

Pizzol, J. (2004). Etudes bioe´cologiques de *Trichogramma cacoeciae* Marchal, parasitoı¨de oophage de l'eude´mis de la vigne, en vue de son utilisation en lutte biologique. Diplome d'Inge´nieur Diplome´ par l'Etat, option Agriculture, ENSAM Montpellier, 80 pp

Pizzol, J., Pintureau, B., Khoualdia, O. & Desneux, N. (2010). Temperature-dependent differences in biological traits between two strains of *Trichogramma cacoeciae* (Hymenoptera: Trichogrammatidae). *Journal of Pest Science*, 83(4): 447–452.

Polis, G.A., Myers, C.A. & Holt, R.D. (1989). The ecology and evolution of intraguild predation: potential competitors that eat each other. *Annual Review of Ecology and Systematics*, 20: 297–330.

Poncio, S., Montoya, P., Cancino, J., Nava, D.E. (2016). Determining the functional response and mutual interference of *Utetes anastrephae* (Hymenoptera: Braconidae) on *Anastrepha oblique* (Diptera: Tephritidae) larvae for mass rearing purposes. *Annals of the Entomological Society of America*, 109: 518–525.

Prabhulinga, T., Jalali, S.K., Kumar, K.P. & Doddabasappa, B. (2016). Biological characteristics of Arrhenotokous and Thelytokous *Trichogramma pretiosum* Riley. In *Arthropod diversity and conservation in the Tropics and Sub-tropics*, ed. Chakravarthy, A. & Sridhara, S., pp. 327–333. Singapore: Springer.

Pratissoli, D. & Parra, J.R.P. (2000). Fertility life table of *Trichogramma pretiosum* (Hym., Trichogrammatidae) in eggs of *Tuta absoluta* and *Phthorimaea operculella* (Lep., Gelechiidae) at different temperatures. *Journal of Applied Entomology*, 124: 339–342.

Querino, R.B. & Hamada, N. (2009). An aquatic microhymenopterous egg-parasitoid of *Argia insipida* Hagen in Selys (Odonata: Coenagrionidae) and biological observations in the Central Amazon, Brazil. *Neotropical Entomology*, 38: 346–351..

Querino, R.B. & Zucchi, R.A. (2003a). Six new species of *Trichogramma* Westwood (Hymenoptera: Trichogrammatidae) from a Brasilian forest reserve. *Zootaxa*, 134: 1–11.

Querino, R.B. & Zucchi, R.A. (2003b). New species of *Trichogramma* Westwood (Hymenoptera: Trichogrammatidae) associated with lepidopterous eggs in Brasil. *Zootaxa*, 163: 1–10.

Querino, R.B. & Zucchi, R.A. (2003c). Caracterização morfológica de dez espécies de *Trichogramma* (Hymenoptera: Trichogrammatidae) registradas na América do Sul. *Neotropic Entomol*, 32: 597–613.

Querino, R. B. & Zucchi, R.A. (2002). Intraspecific variation in *Trichogramma bruni* Nagaraja, 1983 (Hymenoptera: Trichogrammatidae) associated with different hosts. *Brazilian Journal of Biology*, 62: 665–679.

Querino, R.B., Zucchi, R.A. & Pinto, J.D. (2010). Systematics of the Trichogrammatidae (Hymenoptera: Chalcidoidea) with a focus on the genera attacking Lepidoptera. In: *Egg parasitoids in agroecosystems with emphasis on Trichogramma*. Springer Science, Dordrecht. p. 191–218.

Querino, R.B. & Zucchi, R.A. (2019). Annotated checklist and illustrated key to the species of *Trichogramma* Westwood (Hymenoptera: Trichogrammatidae) from South America. *Zootaxa*, 4656(2): 201–231.

Ram, P. (1995). Studies on strains of *Trichogramma evanescens* Westwood from different regions of Eurasia. *Biocontrol Science and Technology*, 5(3): 329–338.

Rani, P.U., Kumari, S.I., Sriramakrishna, T. & Sudhakar, T.R. (2007). Kairomones extracted from rice yellow stem borer and their influence on egg parasitization by *Trichogramma japonicum* Ashmead. *Journal of Chemical Ecology*, 33: 59–73.

Rao, J.G., Rao, P.K. & Murthy, M.S. (1980). Record of *Trichogramma australicum* (Hymenoptera: Trichogrammatidae) as a parasite of *Acherontia styx* Westwood (Lepidoptera: Sphingidae). *Entomon*, 5(1): 83.

Ratte, H.T. (1985). *Temperature and insect development. Environmental Physiology and Biochemistry of Insects* (Ed.Hoffmann, K.H.). Springer, Berlin, Germany. pp. 33–66.

Razinger, J., Vasileiadis, V.P., Giraud, M., van Dijk, W., Modic, S., Sattin, M., & Urek, G. (2015). On-farm evaluation of inundative biological control of *Ostrinia nubilalis* (Lepidoptera: Crambidae) by *Trichogramma brassicae* (Hymenoptera: Trichogrammatidae) in three European maize-producing regions. *Pest Management Science*, 72(2): 246–254.

Reed, D.A., Luhring, K.A., Stafford, C.A., Hansen, A.K., Millar, J.G., Hanks, L.M. & Paine, T.D. (2007). Host defensive response against an egg parasitoid involves cellular encapsulation and melanization. *Biological Control*, 41: 214–222.

Reznik, S.Y., Voinovich, N.D. & Karpova, S.G. (2009). Daily rhythms in parasitization of the Angoumois grain moth *Sitotroga cerealella* Oliv. (Lepidoptera, Gelechiidae) eggs by the egg parasitoid *Trichogramma principium* Sug. et Sor. (Hymenoptera, Trichogrammatidae) females. *Entomological Review*, 89: 5–15.

Risbec, J. (1955). I. Chalcidoides, D'A. O.F. II. Les Microgasterinae D'A. O.F. *Mem. Inst. Fr. Afr. N.*, 13: 297–299.

Risbec, J. (1957). Mymaridae et Trichogrammatidae des malagaches. *Bulletin du Muséum National d'Histoire Naturelle*, 2 27(4): 311–313.

Rodriguez, S.C., Blaauw, B.R. & Isaacs, R. (2012). Manipulation of natural enemies in agroecosystems: habitat and semiochemicals for sustainable insect pest control. In: *Integrated Pest Management and Pest Control-Current and Future Tactics*. ed. Soloneski, S., pp. 89–126. Croatia: InTech Publisher.

Romeis, J., Babendreier, D., Wackers F.L. & Shanower, T.G. (2005). Habitat and plant specificity of *Trichogramma* egg parasitoids–underlying mechanisms and implications. *Basic and Applied Ecology*, 6: 215–236.

Roriz, V., Oliveira, L. & Garcia, P. (2006). Host suitability and preference studies of *Trichogramma cordubensis* (Hymenoptera: Trichogrammatidae). *Biological Control*, 36: 331–336.

Rosenheim, J.A. (1998). Higher-order predators and the regulation of insect herbivore populations. *Annual Review of Entomology*, 43: 421–447.

Rutledge, C.E., Robinson, A.P. & Eigenbrode, S.D. (2003). Effects of a simple plant morphological mutation on the arthropod community and the impacts of predators on a principal insect herbivore. *Oecologia*, 135: 39–50.

Saakian-Baranova, A.A. (1990). Morphological study of preimaginal stages of six species of *Trichogramma* Westwood (Hymenoptera: Trichogrammatidae). *Entomologicheskoe Obozrenie*, 2: 257–263.

Salerno, G., Frati, F., Conti, E., De Pasquale, C., Peri, E. & Colazza, S. (2009). A finely tuned strategy adopted by an egg parasitoid to exploit chemical traces from host adults. *Journal of Experimental Biology*, 212(12): 1825–1831.

Sann, C., Theodorou, P., Heong, K.L., Villareal, S., Settele, J., Vidal, S., & Westphal, C. (2018). Hopper parasitoids do not significantly benefit from non-crop habitats in rice production landscapes. *Agriculture, Ecosystems & Environment*, 254: 224–232.

Schafer, L. & Herz, A. (2020). Suitability of European *Trichogramma* species as biocontrol agents against the tomato leaf miner *Tuta absoluta*. *Insects*, 11(6): 357.

Schöller, M. & Hassan, S.A. (2001). Comparative biology and life tables of *Trichogramma evanescens* and *T. cacoeciae* with *Ephestia elutella* as host at four constant temperatures. *Entomologia Experimentalis et Applicata*, 98: 35–40.

Schöller, M.E., Flinn, P.W., Grieshop, M.J. & Žd'árková, E. (2006). Biological control of stored-product pests. In: *Insect Management for food storage and processing*, ed. Heaps, J.W., pp. 67–87. AACC International Press.

Scott, M., Berrigan, D. & Hoffmann, A.A. (1997) Cost and benefits of acclimation to elevated temperature in Trichogramma carverae. *Entomologia Experimentalis et Applicata*, 85, 211–219.

Sekhon, B.S. & Varma, G.C. (1985). Cotton crop protection in Punjab. *Has it entered crisis phase? Annals of biology*, 1(2): 123–131.

Shankar, G. & Baskaran, P. (1986). Relative abundance of three species of egg parasitoids of rice brown planthopper (BPH). *International Rice Research Newsletter*, 11(4): 35.

Sharma, S., Shera, P.S., Kaur, R. & Sangha, K.S. (2020). Evaluation of augmentative biological control strategy against major borer insect pests of sugarcane—a large-scale field appraisal. *Egypt Journal of Biological Pest Control*, 30: 127.

Sheeba, S.P.K. & Prasad, G.S. (2002). Seasonal parasitism of *Trichogramma japonicum* in South Andamans. *Insect Environ*, 8: 139–140.

Shenhmar, M. & Brar, K.S. (2003). Role of augmentative release of *Trichogramma japonicum* Ashmead in the enhancement of parasitism and control of *Scirpophaga excerptalis* (Walker) on sugarcane. In: *Biological control of pests*, ed. S. Ignacimuthu, S. & Jayaraj, S., pp. 16–19. New Delhi: Phoenix Publishing House Pvt. Ltd.

Shih, Y.T., Ko, C.C., Pan, K.T., Lin, S.C. & Polaszek, A. (2013) *Hydrophylita* (*Lutzimicron*) *emporos* Shih & Polaszek (Hymenoptera: Trichogrammatidae) from Taiwan, Parasitising Eggs, and Phoretic on Adults, of the Damselfly *Psolodesmus mandarinus mandarinus* (Zygoptera: Calopterygidae). *PLoS ONE*, 8(7): e69331.

Shu, S. & Jones, R.L. (1989). Kinetic effects of kairomone in moth scales of the European corn borer on *Trichogramma nubilale* Ertleand Davis (Hymenoptera: Trichogrammatidae). *The Journal of Insect Behavior*, 2: 123–131.

Shylesha, A.N., Jalali, S.K., Gupta, A., Varshney, R., Venkatesan, T., Shetty, P., Ojha, R., Ganiger, P.C., Navik, O., Subaharan, K., Bakthavatsalam, N., Ballal, C.R. (2018). Studies on new invasive pest *Spodoptera frugiperda* (J.E. Smith) (Lepidoptera: Noctuidae) and its natural enemies. *J. Biol. Contr.*, 32: 1–7.

Silvestri, F. (1916). Contribuzione alla conoscenza del genero *Poropoea* Forster (Hymenoptera: Chalcididae). *Boll. Lab. Zool. Gen. Agr. Portici*, 11: 120–135.

Singh, S.P. & Jalali, S.K. (1992). Biological suppression of *Chilo auricilius* on sugarcane in India. *Trichogramma News*, 6: 25.

Singh, S.P. & Jalali, S.K. (1994). Trichogrammatids. Technical Bulletin, No. 7. Project Directorate of Biological Control, Bangalore, p 93.

Singh, S.P. (1999). Biological control in India. *Indian Journal of Plant Protection*, 27(1&2): 126–138.

Singh, U.P., Singh, S., Kumar, R. & Kumar, A. (2017). Study on foraging response of Trichogrammatids governed by volatile cues. *Tropical Ecology*, 58(4): 741–750.

Sisay, B., Simiyu, J., Mendesil, E., Likhayo, P., Ayalew, G., Mohamed, S. & Tefera, T. (2019). Fall Armyworm, *Spodoptera frugiperda* Infestations in East Africa: Assessment of Damage and Parasitism. *Insects*, 10(7): 195.

Sithanantham, S., Geetha, N., Baitha, A. & Jalali, S.K. (2013). Utility of *Trichogramma* for biocontrol of sugarcane borers. In: *Biological Control of Insect Pests Using Egg Parasitoids*, ed. Sithanantham, S. et al., pp. 271–300. India: Springer.

Smith, S.M. (1996). Biological control with *Trichogramma*: advances, successes and potential of their use. *Annual Review of Entomology*, 41: 375–406.

Smith, S.M., & Hubbes, M. (1986). Strains of the egg parasitoid *Trichogramma minutum* Riley. *Journal of Applied Entomology*, 102(1-5): 81–93.

Song, R.C., Chi, Y.M., Wang, H.P., Lu, G.Y., Wang, Q.X. & Zhang, Q.B. (1994). A study on setting equipment of manufactured reproduction of *Trichogramma*. *Trans. CSAE*, 10: 48–52.

Souza, A.R., Giustolin, T.A., Querino, R.B. & Alvarenga, C.D. (2016). Natural parasitism of Lepidopteran eggs by *Trichogramma* species (Hymenoptera: Trichogrammatidae) in agricultural crops in Minas Gerais, Brazil. *The Florida Entomologist*, 99(2): 221–225.

Srikanth, J., Easwaramoorthy, S. & Jalali, S.K. (2016) A 100 years of biological control of sugarcane pests in India: review and perspective. *CAB Review*, 11: 1–32.

Srivastava, B. & Kushwaha, K.S. (1995). Sequential appearance and frequency of the parasites of *Spodoptera litura* (Fabr.) on cauliflower and cabbage at Udaipur, Rajasthan. *Advances in Agricultural Research in India*, 3: 130–140.

Tabone, E., Bardon, C., Pintureau, B. & Alauzet, C. (2006). Importance of host oviposition pattern and plant size for the selection of *Trichogramma* strains to control the diamondback moth. *Entomologia Experimentalis et Applicata*, 119: 47–51.

Tabone, T., Bardon, C., Desneux, N. & Wajnberg, E. (2010). Comparative assessment of parasitism of different *Trichogramma* spp. on *Plutella xylostella* L. on greenhouse cauliflower. *Journal of Pest Science*, 83: 251–256.

Takada, Y., Kawamura, S. & Tanaka T. (2000). Biological characterisics: growth and development of the egg parasitoid *Trichogramma dendrolimi* (Hymenoptera: Trichogrammatidae) on the cabbage armyworm *Mamestra brassicae* (Lepidoptera: Noctuidae). *Applied Entomology and Zoology*, 35: 369–379.

Taley, Y.M. & Thakare K.R. (1979). Biology of seven new hymenopterous parasitoids of *Atherigona soccata* (Rondani). *Indian Journal of Agricultural Sciences*, 49: 344–354

Tanaka, M. (1985). Early embryonic development of the parasitic wasp, *Trichogramma chilonis* (Hymenoptera, Trichogrammatidae). In: *Recent advances in insect embryology in Japan*, eds. Ando, H. & Miya, K., pp. 171–179. Tsukuba, Japan: ISEBU.

Tang, R., Babendreier, D., Zhang, F., Kang, M., Song, K. & Hou, M.L. (2017). Assessment of *Trichogramma japonicum* and *T. chilonis* as potential biological control agents of yellow stem orer in Rice. *Insects*, 8: 19.

Tayeh, A., Estoup, A., Laugier, G., Loiseau, A., Turgeon, J., Toepfer S. & Facon, B. (2012). Evolution in biocontrol strains: insight from the harlequin ladybird *Harmonia axyridis*. *Evolutionary Applications*, 5: 481–488.

Taylor, T.H.C. (1937). The biological control of an insect in Fiji. An account of the coconut leaf-mining beetle and its parasite complex, 55–63. London, United Kingdom: The Imperial Institute of Entomology.

Telang, S.M., Rathod K.S. Rathod, R.M. (2004). Parasitization by different parasites of *E. vitella* in okra. *Journal of Soil & Crops*, 14(2): 335–339

Thontadarya, T.S., Hanumanna, M. & Rao, K.J. (1978). Record of *Trichogramma australicum* Girault on the eggs of *Heliothis armigera* (Hubner) infesting bhendi (*Abelmoschus esculentus* (H) Moench). *Current Research*, 7(12): 206.

Ulrichs, C. & Mewis, I. (2004). Evaluation of the efficacy of *Trichogramma evanescens* Westwood (Hym., Trichogrammatidae) inundative releases for the control of *Maruca vitrata* F. (Lep., Pyralidae). *Journal of Applied Entomology*, 128(6): 426–431.

van Lenteren, J.C. (2003). *Quality control and Production of Biological Control Agents: Theory and Testing Procedures*. Wallingford, UK: CABI Publishing.

Varshney, R., Poornesha, B., Raghavendra, A., Lalitha, Y., Apoorva, V., Ramanujam, B., Rangeshwaran, R., Subaharan, K., Shylesha, A.N., Bakthavatsalam, N., Chaudhary, M. & Pandit, V. (2021). Biocontrol-based management of fall armyworm, *Spodoptera frugiperda* (J.E Smith) (Lepidoptera: Noctuidae) on Indian Maize. *Journal of Plant Disease and Protection*, 128: 87–95.

Venkatesan, T. & Jalali, S.K. (2013). Trichogrammatids: adaptation to stresses. In: *Biological Control of Insect Pests Using Egg Parasitoids*, ed. Sithanantham, S. et al., pp. 105–125. India: Springer.

Venugopal, S., Subramanium, H. & Natrajan, R. (1968). Notes on *Trichogramma japonicum* Ashm., an egg parasite of internode borer (*Proceras indicus* Kapur) of sugarcane in Madras State. *Indian Journal of Agricultural Sciences*, 38: 874–875.

Vet, L.E. & Dicke, M. (1992). Ecology of infochemical use by natural enemies in a tritrophic context. *Annual Review of Entomology*, 37: 141–172.

Viggiani, G. (1971). Ricerche sugil Hymenoptera Chalcidoidea XXVIII. Studie morfologico comparativo dell'armature genitale esterna maschile dei Trichogrammatidae. *Boll. Lab. Entomol. Agraria. Filippo Silvestri Portici*, 29: 181–222.

Viggiani, G. (1984). Further contribution to the knowledge of the male genitalia in the Trichogrammatidae (Hym. Chalcidoidea). *Bollettino del Laboratorio di Entomologia Agraria 'Filippo Silvestri' di Portici*, 41: 173–182

Vinson, S.B. (1994). Physiological interactions between egg parasitoids and their hosts, In: *Biological control with egg parasitoids*. eds. Wajnberg, E. & Hassan, S.A., pp. 245–271. Oxford, United Kingdom: CAB International.

Virk, J.S., Brar, K.S. & Sohi, A.S. (2004). Role of trap crops in increasing parasitism efficiency of *T. chilonis* Ishii in cotton. *Journal of Biological Control*, 18: 61–64.

Volkoff, A.N., Daumal, J., Barry, P., François, M.C., Hawlitzky, N. & Rossi, M.M. (1995). Development of *Trichogramma cacoeciae* Marchal (Hymenoptera: Trichogrammatidae): time table and evidence for a single larval instar. *International Journal of Insect Morphology and Embryology*, 24: 459–466.

Wang, B. & Ferro, D.N. (1998). Functional response of *Trichogramma ostriniae* (Hymenoptera: Trichogrammatidae) to *Ostrinia nubilalis* (Lepidoptera: Pyralidae) under laboratory and field conditions. *Environmental Entomology*, 27: 752–758.

Wang, C.L., Mao, G., Gao, Y., Li, L.J. & Cui, D.J. (1999). An instrument for selecting healthy parasitized host eggs. *Chinese Journal of Biological Control*, 15: 139–140.

Wang, G., Lan, Y., Qi, H., Chen, P., Hewitt, A. & Han, Y. (2018). Field evaluation of an unmanned aerial vehicle (UAV) sprayer: effect of spray volume on deposition and the control of pests and disease in wheat. *Pest Management Science*, 75: 1546–1555.

Wang, S.Q. (2001). Research progress in *Trichogramma* mass rearing by using artificial host eggs. *Plant Protection Technology and Extension*, 21: 40–41.

Wang, Z.Y., He, K.L., Wen, L.P., Zhang, G.Y. & Zheng, L. (2002). Intercrop effect on parasitization of *Helicoverpa armigera* (Hübner) eggs by *Trichogramma chilonis* Ishii. In: International conference on environmentally sustainable agriculture for dry areas for the 3rd Millennium Proceedings, Shijiazhuang, Dobing Enterprises, September 15–19. p. 301–304.

Wang, Z.Y., He, K.L., Zhang, F., Lu, X. & Babendreier, D. (2014). Mass-rearing and release of *Trichogramma* for biological control of insect pests of corn in China. *Biological Control*, 68: 136–144.

Watanabe, T., Wada, T., Mohd, N., Noor, B. & Salleh N. (1992) Parasitic activities of egg parasitoids on the rice planthoppers, *Nilaparvata lugens* (Stal.) and *Sogatella furcifera* (Horv´ath) (Homoptera: Delphacidae), in the Muda Area of Peninsular Malaysia. *Applied Entomology and Zoology*, 27: 205–211.

Wiackowski, S.K. & Wiackowska, I. (1966). Biological, microbiological and chemical control of the plum moth *Laspeyresia funebrana* Tr. (Lep. Tortricidae) in Poland. *Entomophaga*, 11: 261–267.

Woolley, J.B. (1997). Aphelinidae. In: *Annotated keys to the genera of Nearctic Chalcidoidea (Hymenoptera)*, ed. Gibson, G.A.P. et al., pp. 134–150. Ottawa, Ontario: NRC Research Press.

Wright, M.G., Kuhar, T.P., Hoffmann, M.P. & Chenus, S.A. (2002). Effect of inoculative releases of *Trichogramma ostriniae* on populations of *Ostrinia nubilalis* and damage to Sweet Corn and Field Corn. *Biological Control*, 23(2): 149–155.

Wu, L.-H., Hoffmann, A.A. & Linda, J.T. (2016). Potential impact of climate change on parasitism efficiency of egg parasitoids: A meta-analysis of *Trichogramma* under variable climate conditions. *Agriculture, Ecosystems and Environment*, 231: 143–155.

Wu, Z.X., Cohen, A.C. & Nordlund, D.A. (2000). The feeding behavior of *Trichogramma brassicae*: new evidence for selective ingestion of solid food. *Entomologia Experimentalis et Applicata*, 96: 1–8.

Wührer, B.G., & Hassan, S.A. (1993). Selection of effective species/strains of *Trichogramma* (Hym., Trichogrammatidae) to control the diamondback moth *Plutella xylostella* L. (Lep., Plutellidae). *Journal of Applied Entomology*, 116: 80–89.

Xie, Z.N., Wu, P.Y. & Deng, X.L. (1982). Research of artificial diet of *Trichogramma*. 1. Amino acid analysis of hemolymph of *Philosamia cynthia ricini* larvae and pupae. *Natural Enemies Insects*, 4: 5–11.

Yadav, B., Paul, A.V.N. & Gautam, R.K. (2008). Kairomonal effect of different combinations of synthetic kairomones on the egg parasitoids *Trichogramma exiguum*, Pinto and Platner and *Trichogramma japonicum* (Ashmead). *Indian Journal of Entomology*, 70: 68–70.

Yadav, B., Paul, A.V.N., Gopal, M. & Gautam, R.K. (2005). Kairomonal effect of some saturated fatty acids on the egg parasitoids *Trichogramma exiguum*, Pinto and Platner and *Trichogramma japonicum* (Ashmead). *Indian Journal of Entomology*, 67: 7–11.

Yadav, D.N., Anand, J. & Devi, P.K. (2001). *Trichogramma chilonis* on *Plutella xylostella* (Lepidoptera: Plutellidae) in Gujarat. *Indian Journal of Agricultural Science*, 71: 69–70.

Yadav, D.N. (1980). Studies on the natural enemies of *Heliothis armigera* (Hubner) and its biological control using an egg-parasite, *Trichogramma australicum* Girault (Hymenoptera: Trichogrammatidae). *Gujarat Agricultural University Research Journal*, 6(1): 62–63.

Yadav, D.N., Patel, R.C. & Patel, D.S. (1985). Impact of inundative releases of *Trichogramma chilonis* Ishii against *Heliothis armigera* (Hbn.) in Gujarat (India). *Journal of Entomological Research*, 9: 153–159.

Yousuf, M. & Shafee, S.A. (1986a). Catalogue of genus-group names of world Trichogrammatidae (Hymenoptera). *Indian Journal of Systematic Entomology*, 3: 13–27.

Yousuf, M. & Shafee, S.A. (1986b) Check list of species and bibliography of the world Trichogrammatidae (Hymenoptera). *Indian Journal of Systematic Entomology*, 3: 29–82.

Yousuf, M. & Shafee, S.A. (1988) *Taxonomy of Indian Trichogrammatidae (Hymenoptera: Chalcidoidea Indian Journal of Systematic Entomology*, 4(2): 55–200.

Zang, L.S., Wang, S., Zhang, F. & Desneux, N. (2020). Biological Control with *Trichogramma* in China: History, Present Status and Perspectives. *Annual Review of Entomology*, 66: 463–484.

Zappala, L., Bernardo, U., Biondi, A., Cocco, A., Deliperi, S., Delrio, G., Giorgini, M., Pedata, P., Rapisarda, C., Tropea, G.G. & Siscaro, G. (2012) Recruitment of native parasitoids by the exotic pest *Tuta absoluta* in Southern Italy. *Bulletin of Insectology*, 65: 51–61.

Zhang, J.J., Ruan, C.C., Zang, L.S., Shao, X.W. & Shi, S.S. (2015).Technological improvements for mass production of *Trichogramma* and current status of their applications for biological control on agricultural pests in China. *Chinese Journal of Biological Control*, 31: 638–646.

Zhang, L.W. (1993). Attractive material for the oviposition of *Trichogramma dendrolimi* on artificial "eggs." *Natural Enemies Insects*, 15: 101–105.

Zhang, Y.F., Song, Q.T., Zhang, F. & Li, Y.X. (2010). Comparison of superparasitism and suitability to superparasitism between *Trichogramma ostriniae* and *T. dendrolimi*. *Chinese Journal of Biological Control*, 26: 377–384.

Zongo, J.O., Vincent, C. & Stewart, R.K. (1993). Biology of *Trichogrammatoidea simmondsi* (Hym.: Trichogrammatidae) on sorghum shoot fly, *Atherigona soccata* (Dipt.: Muscidae) eggs. *Entomophaga*, 38: 207–212.

Zucchi, R.A., Querino, R.B., & Monteiro, R.C. (2010). Diversity and hosts of *Trichogramma* in the New World, with emphasis in South America. In: *Egg parasitoids in agroecosystems with emphasis on Trichogramma*, ed. Consoli, F.L. et al., pp. 219–236. Dordrecht: Springer.

10 Scelionid Parasitoids

K Rajmohana and Sunita Patra
Zoological Survey of India, M-Block, New Alipore, West Bengal, India

CONTENTS

LEARNING OBJECTIVES

1. To get introduced to the scelionid wasps and aspects about their diversity.
2. To learn their diagnostic characteristics and biological specialties.
3. To understand their traits as efficient natural enemies and as potential biocontrol agents.
4. To get to know instances of successful biocontrol programs using scelionid wasps in India.

10.1 INTRODUCTION

10.1.1 Systematic Position

Kingdom	:	Animalia
Phylum	:	Arthropoda
Class	:	Insecta
Order	:	Hymenoptera
Superfamily	:	Platygastroidea
Family	:	Scelionidae

DOI: 10.1201/9781003354239-10

Scelionidae (Superfamily- Platygastroidea; Order- Hymenoptera) are idiobiont endoparasitoids of insects and spider eggs (van Noort, 2021). Platygastroidea is the third largest superfamily after Ichneumonoidea and Chalcidoidea (Austin et al., 2005). Scelionids are true egg parasitoids, completing their development within their host egg itself (Mills, 1994), eventually killing the host embryo. Compared to larval and pupal parasitoids, egg parasitoids, like scelionids, are extremely beneficial biocontrol agents, since they kill their hosts before hatching and there exists no question of crop damage due to emerging caterpillars.

A historical review of the classification of family relying on the results of a phylogenetic study of the platygastrid wasps, based on sequences from the 18S rRNA, 28S rRNA and CO1 genes by Murphy et al. (2007), Sharkey (2007) had formally synonymised Scelionidae under family Platygastridae. As a result, both Scelionidae and Platygastridae were treated as a single family, Platygastridae. However, this classification did not get much acceptance among the international experts in the group, since the analysis did not include an optimum number of valid genera. Hence the classification proposed by Masner (1993) is being followed even at present. However, an updated classification based on robust phylogenetic data is much awaited (van Noort, 2021). The last hard copy of a species catalogue of Scelionidae was provided by Johnson in 1992, which was later supplemented by a few highly informative and userfriendly web databases.

The basis of taxonomic studies on Scelionidae of India is the review work by Mani and Sharma (1982). This annotated critical catalogue summarised all hitherto available information on the group. Identification keys to subfamilies, genera and species were provided along with species diagnosis and distribution. The primary types of several species of Scelionidae mentioned in the above work, which were described through the prestigious PL 480 project of Prof. M.S. Mani in the early 1970s are now housed in the National Museum of Natural History Smithsonian Institution, U.S.A. World revisions are available only in very few scelionid taxa, namely *Paratelenomus* Dodd (Johnson, 1996), *Psix* Kozlov and Le, (Johnson and Masner, 1985), *Nixonia* Masner (Johnson and Masner, 2006), *Fusicornia* Risbec (Taekul et al., 2008), *Platyscelio* Kieffer (Taekul et al., 2010) (Figure 10.6), *Heptascelio* Kieffer (Johnson et al., 2008), *Odontacolus* Kieffer (Vaelrio et al., 2013), *Oxyscelio* Kieffer (Burks et al., 2013), *Scelio pulchripennis* group (Yoder et al., 2009), *Protelenomus* Kieffer (Veenakumari et. al., 2019) and *Xenomerus* Walker (Miko et al., 2010). The need for large-scale revisions to the taxonomy of Indian Scelionidae cannot be overruled.

Though scelionids are potential biocontrol agents, the lack of a solid systematic base curtailed their exploitation to the fullest. Several species remain yet to be discovered and described (Orr, 1988). More than 200 species were described new to science in the last 10 years in India under Scelionidae, including Abhilash and Rajmohana (2014), Anjana and Rajmohana (2015), Anjana et al. (2016a, 2016b, 2017), Rajmohana (2007a, 2007b, 2011, 2014), Rajmohana and Narendran (2007a and 2007b), Rajmohana and Talukdar (2010), Rajmohana and Bijoy (2011), Rajmohana and Peter (2012), Rajmohana and Nisha (2013), Rajmohana and Peter (2013), Rajmohana and Nisha (2013), Rajmohana and Anto (2014), Rajmohana et al. (2013a, 2013b, 2017) and Veenakumari et al. (2011, 2014, 2016, 2017). Veenakumari and Prashanth (2017) and Veenakumari et al. (2014,2017–18), described twenty species of *Cremastobaeus* Ashmead new to India and 40 species of *Leptacis,* respectively. Several new genera - *Neoduta* Rajmohana and Peter (2012), *Neoceratobaeus* Rajmohana (2014), *Chakra* Rajmohana and Veenakumari (in Rajmohana and Veenakumari, 2014), *Anokha* Rajmohana and Veenakumari (in Rajmohana et al., 2017) and *Indiscelio* Veenakumari, Popovici and Talamas (Veenakumari et al., 2018) under subfamily Scelioninae and *Dvivarnus* Rajmohana and Veenakumari (Veenakumari et al., 2011) in subfamily Telasinae were described in the last decade.

Molecular studies are still awaited in a good majority of Indian Scelionidae. Mt CO1 barcodes are available only for a few species of Indian Scelionidae under the subfamily Telenominae (Rajmohana and Nisha, 2013; Rajmohana and Anto, 2014; and Rajmohana et al., 2013, 2019).

10.2 CHARACTERISTICS OF SCELIONIDAE

Scelionid wasps can be easily distinguished from other families of parasitic Hymenoptera, through well-represented morphological characters. Species taxonomy is female-based. Species identification keys for males are lacking in most genera, hence associating the sexes of the species accurately, remains challenging, unless the samples are biologically reared.

The sister family Platygastridae can be told apart, at once, since they never possess more than 10 antennal segments. Further in both sexes, the antennae are elbowed after scape and also after fourth or fifth segment and never with more than 10 segments in females (Rajmohana and Patra, 2019).

10.2.1 Diagnosistic Characters of Scelionidae

a. Body size ranges from 0.6 mm to 12 mm.
b. Body mostly black and nonmetallic, though some are yellow or brown.
c. Sculpture on the body is an important taxonomic attribute.
d. Sexual dimorphism is seen mostly in the nature of the antenna or in the number of antennal segments. The male antenna has more antennomeres and the metasoma has more tergites and sternites than in females. The fifth antennal segment in males is modified.
e. Usually antenna is with 11 or 12 segments, at times with 6 and 14 segments too.
f. Wings are with very less venation. Only submarginal, marginal, stigmal and postmarginal veins, are generally present on forewings. The submarginal vein in the hindwing is complete in most genera.
g. Metasomal segments can be subequal or with variable length. If metasomal segment 2 is the longest, then the forewing has complete venation, with the submarginal vein reaching the anterior margin.
h. The metasomal segment 7 in females can be either external or internal, with cerci or sensory plates, and may be extruded with the ovipositor during oviposition or remain attached to tergum 6.

The specialised sensilla on the female antenna and the ovipositor system in females are unique to Platygastroidea, including Scelionidae. The claval segments of the female antenna sensilla, on their ventral side, have one or two unique 'multiporous gustatory sensilla', which aid in the recognition of host-associated cues. Two types of ovipositor systems are observed. In the *Ceratobeaus* type, the ovipositor extension is by the contraction of muscles attached to hypodemes, along with a change in hydrostatic pressure. In *Scelio* type, the more apomorphic arrangement is seen, where the ovipositor is extended at the end of a telescopic membranous tube, by hydrostatic pressure alone (Austin et al., 2005).

To exploit cryptic habitats and for specialised host utilisation, scelionids have remarkable adaptations.Wing reduction in scelionids includes aptery, brachyptery or microptery. These adaptations aid them in searching for their host in the soil. Such features along with a streamlined body are common in genera, like *Idris, Ceratobeaus* and *Beaus*, the spider egg parasitoids. Wing reduction helps the female wasps to penetrate in the silken egg sacs of their host spiders. Reduced wings, frontal prominence, and protruded or elongated mandibles are sand burrowing adaptations for *Encyrtoscelio* Dodd to reach the eggs of *Cydnus* bugs in soil. These are habitat dependent morphological modifications towards functional specialisation in groups having cryptobiotic niches (Austin et al., 2005).

Scelionids have a modified structure of the abdomen in relation to ovipositor extension and retraction. Several genera like *Calliscelio, Baryconus* and *Duta,* have telescopic ovipositors, which are longer than their body length, aiding them to reach their hidden host eggs. To house the internally retracted ovipositor, some species have developed horns or humps, or tubular extensions dorsally on their first abdominal segment.

10.3 DIVERSITY OF SCELIONIDAE

Scelionidae is classified into 3 subfamilies- Telenominae, Teleasinae and Scelioninae (Masner, 1976, 1993), of which Scelioniae is the most diverse and the oldest (Masner, 1993). Under Telenominae, about 20 genera and 904 species are considered to be valid. The bulk of the species, fall under *Telenomus* Haliday (652 valid species) (Taekul et al., 2014).

The most recent tabulation of the diversity of Scelionidae in India is 65 valid genera with more than 400 species. More than 70% of the species come under the subfamily Scelioninae, under 51 genera. Subfamily Telenominae is known by 8 genera and more than 60 species, while subfamily Teleasinae has 6 genera, with more than 40 valid species.

The diversity of Scelionidae in India is largely unexplored. *Telenomus* Haliday and *Trissolcus* Ashmead, *Scelio* Latreille, *Calliscelio,* Ashmead, *Gryon* Haliday, *Cremastobaeus* Kieffer, *Certaobaeus* Ashmead, *Baeus* Haliday, *Idris* Förster, and *Trimorus* Förster are some of the species-rich genera of Scelionidae in India.

10.4 BIOLOGY

The ground plan biology of Scelionds places them as egg parasitoids of insects and spiders. They are one among the very few taxa, which exclusively use eggs as their hosts and are true egg parasitoids. With their hypodermic-like ovipositor, the female scelionid deposits a single egg or rarely multiple eggs into the egg of the host. The developing larva consumes the contents of the host egg, pupates and finally the adult wasp emerges out. The diverse array of host taxa includes the major insect orders, like Lepidoptera, Orthoptera, Coleoptera, Hemiptera, Diptera Mantodea and Neuroptera, other than the spider taxa, Araneae. Several hosts of Scelionidae are economically important major pests in agriculture [lepiodpteran stem borerers (*Spodoptera* spp.), locusts (*Locusta migratoria*), lablab bugs (*Megacopta cribraria*), southern green stink bug (*Nezara viridula*)], forestry [the gypsy moth (*Lymantria dispar*)] while a few are relevant in the human and animal health sector [kissing bugs (*Triatoma*, *Rhodnius*), and horse flies (*Tabanus* spp.) Hessian fly (*Mayetiola destructor*)] (Austin et al., 2005). Scelionid wasps are well known for their host specificity at tribal level (Table 10.1).

Scelionid eggs are small, contain no yolk and are surrounded by a very thin chorion. After meiosis, the oocyte nucleus produces an egg pronucleus and polar bodies giving rise to the embryo and polar body cells, respectively. The polar bodies produce an extraembryonic

TABLE 10.1
Scelionid Tribes and their Host Specificity (Masner, 1976; Galloway and Austin, 1984; Austin et al., 2005)

Sl. No.	Tribe	Host specificity
1	Baeini	Araneae (Araneomorphe)
2	Baryconini	Tettigonidae
3	Calliscelionini	Tettignoidae, Gryllidae
4	Embidobiini	Embioptera, Araneae
5	Gryonini	Heteroptera
6	Idrini	Araneae (Araneomorphe)
7	Mantibariini	Mantodea
8	Platyscelionini	Tettigonidae
9	Scelionini	Acrididae
10	Thoronini	Aquatic Heteroptera (mostly Gerridae and Nepidae)

membrane. Once chorion ruptures, the embryo develops into a larva, while the extraembryonic membrane dissociates into teratocytes (Strand, 2014). Some studies in Telenominae have shown that teratocytes form the source of the secretory products to digest host eggs extra orally (Strand, 1986).

Rearing records of the subfamily Telenominae indicate that their hosts are restricted to Lepidoptera, Hemiptera, Diptera or Neuroptera. Genus *Trissolcus* with about 150 described species are exclusive egg parasitoids of Hemiptera, at least a total of 327 host-parasitoid records are available associating *Trissolcus* hemipteran eggs. Members of genus *Telenomus* with around 628 species are primarily parasitoids of Lepidoptera, though they also attack Hemiptera (Pentatomidae and Scutellaridae), Diptera and Neuroptera. Of the 484 host-parasitoid records known, more than 50% of associations are between *Telenomus* and Lepidoptera (Fatouros et al., 2020). The genus *Phanuromyia* Dodd and the *crassiclava* group of *Telenomus* Haliday, are recorded as parasitoids of planthopper eggs, the fulgorids in particular. *Eumicrosoma* attacking the eggs of seed bugs (Lygaeoidea: Blissidae), *Paratelenomus* as egg parasitoids of turtle bugs (Pentatomoidea: Plataspidae), *Phanuromyia* attacking auchenorrhynchous Hemiptera (Issidae, Fulgoridae) and *Psix* parasitising Coreoidea and Pentatomoidea are examples of host specificity at generic level (Taekul et al., 2014).

Unlike in Chalcid families, the scelionid species do not parasitise across different orders (Taekul et al., 2014). Definitive host associations of scelionids in India are scanty (Rajmohana and Sunita, 2019). Table 10.2 summarises the reported host associations of scelionids in India.

Gryon is a very large genus of Scelioninae, the vast majority attacks heteropteran eggs. *Gryon clavigrallae*, the parasitoid attacking *Clavigralla gibbosa* and *C. scutellaris* (Hemiptera: Coreidae), the important insect pests of pigeonpea, *Cajanus cajan* (Fabaceae), in India and Africa, attacked host eggs of all ages, though they preferred eggs lesser than 4 days old. Superparasitism though rare, happens when unparasitised eggs were unavailable (Romeis et al., 2000). Martela et al., (2019) give baseline information for laboratory rearing of *Gryon gonikopalense* including designing host specificity tests. *G. gonikopalense* male and female parasitoids emerged from host eggs in 25 and 27 days, respectively. Fed with honey and water, under laboratory conditions, the parasitoid lived for nearly 66 days, however, their life span was reduced to 55 days, when the individuals had reproductive activity. All females started parasitism soon after emergence. In its entire lifetime, a single female produced 20 to 94 offspring.

The age of host eggs affects the oviposition of the parasitoids. As per a study on *Telenomus podisi,* the time required for mating, host-searching, and oviposition are the main factors limiting the number of generations (Yeargan, 1980). *Telenomus busseolae*, the egg parasitoid of *Sesamia* spp., attacking cereals, is a highly efficient parasitoid, for its ability to successfully develop in late host stages and high egg retention capacity (Chabi, 1997). In *Telenomus. busseolae*, the females oviposit within hours after emergence, and if sufficient hosts are available, they lay more than half of their eggs within the first 6 hours.

Though polyembrony is common in some egg parasitoids, like encyrtids and trichogrammatids, this phenomenon is not yet reported in Scelionidae (Austin et al., 2005).

Phoresy is a remarkable hitchhiking feature, an obligatory adaptation in some egg parasitoids, for accessing eggs immediately after egg laying, before they become unsuitable as hosts. The parasitoids get transported to the egg deposition site, hitchhiking the host, mostly the females. This has been reported in a few species in some genera of Scelionidae. Species of *Sceliocerdo* and *Synoditella* are phoretic on acridid hosts, *Mantibaria* on mantids; *Thoronella* and *Calotelea* on Odonata; *Epigryon*, *Protelenomus*, *Paratelenomus* (Figure 10.4) and *Telenomus* on Heteroptera and *Telenomus* on Lepidoptera (Clausen, 1976; Masner, 1995; Austin et al., 2005; Veenakumari et al., 2012; Rajmohana et al., 2019; Fatouros et al., 2020).

TABLE 10.2
Host association of Scelionidae in India

	Subfamily	Parasitoid	Host	Ref.
1	Scelionidae	*Ceratobaeus dunensis* Mukerjee	Ex eggs -*Cheriacanthium* sp. (Araneomorphae: Miturgidae)	Shweta and Rajmohana, 2014
2	Scelionidae	*Gryon diadematis* Mineo	Ex. eggs on Heteroptera on Teak leaf	Mineo, 1983
3	Scelionidae	*Gryon fulviventre* (Crawford)	Ex eggs of *Antestiopsis cruciate* (Hemiptera: Pentatomidae) on Jasmine; Ex eggs of *Nezara viridula* (Hemiptera: Pentatomidae); Ex eggs of *Clavigralla gibbosa* (Hemiptera: Coreidae)	Mani and Sharma, 1982
4	Scelionidae	*Gryon ingens* Veenakumari and Rajmohana	Ex. Reduvid eggs	Veenakumari, 2016
5	Scelionidae	*Gryon orestes* (Dodd)	Ex. Coreid Eggs	Personal observation
6	Scelionidae	*Platyscelio pulchricornis* Kieffer	Ex. Orthopteran egg on paddy leaf	Voucher specimen examined at NBAIR, Bangalore
7	Scelionidae	*Sceliocerdo viatrix* (Brues)	Phoretic on *Neorthacris acuticeps* (Bolìvar) (Orthoptera: Acrididae)	Veenakumari et al., 2012
8	Scelionidae	*Eumicrosoma cumaeum* (Nixon)	Ex. eggs of *Macropes excavatus* Distant = *Cavelerius excavatus* (Distant) (Heteroptera: Lygaeidae)on Sugarcane	Nixon, 1938
9	Scelionidae	*Eumicrosoma phaeax* (Nixon)	Ex. eggs of *Macropes excavatus* Distant = *Cavelerius excavatus* (Distant) (Heteroptera: Lygaeidae) on Sugarcane	Nixon, 1938
10	Scelionidae	*Paratelenomus saccharalis* (Dodd)	Phoretic on *Megacopta cribraria* (Hemiptera: Plataspidae:) on *Dolichos lablab*	Rajmohana and Narendran, 2001
11	Scelionidae	*Protelenomus flavicornis* Kieffer Coreidae)	Phoretic on *Anoplocnemis phasiana* (Heteroptera: Coreidae)	Veenakumar and Prashanth, 2015
12	Scelionidae	*Psix robustus* Rajmohana	Ex. Heteropteran Eggs	Rajmohana and Talukdar, 2010
13	Scelionidae	*Psix sunithae* Singh, Johnson and Ramamurthy	Ex. eggs of Coreidae (Heteroptera) on *Nux vomica* leaves	Singh et al., 2012
14	Scelionidae	*Telenomus beneficiens* (Zehntner)	Ex. eggs of *Chilo infuscatellus* (Lepidoptera: Crambidae) , *Chilo sacchariphagus* (Lepidoptera: Crambidae) *and Scirpophaga nivella* (Lepidoptera: Crambidae) on sugarcane	Narasimham et al., 1997
15	Scelionidae	*Telenomus colemani* (Crawford)	Ex. Eggs of *Clostera cupreata* (Lepidoptera: Notodontidae) on poplar	Ahmad et al., 1997
16	Scelionidae	*Telenomus cuspis* Rajmohana and Srikumar	Ex. eggs of *Helopeltis antonii* (Hemiptera: Miridae) on guava neem and tea	Rajmohana et al., 2013
17	Scelionidae	*Telenomus dignoides* Nixon	Ex. egg mass of *Scirpophaga nivella* (Lepidoptera: Crambidae) on sugarcane, *Erianthus munja*, *Chilo saccariphagus indicus* (Lepidoptera: Crambidae) on sugarcane.	Nixon, 1937
18	Scelionidae	*Telenomus dignus* (Gahan)	Ex. eggs of *Scirpophaga nivella* (Lepidoptera: Crambidae) on sugarcane, *Scirpophaga incertulas.*	Gahan, 1925
19	Scelionidae	*Telenomus dilatus* Rajmohana and Anto	Ex. eggs of *Papilio* sp. (Lepidoptera: Papilionidae)	Rajmohana and Anto, 2014
20	Scelionidae	*Telenomus euproctiscidis* (Wilcox)	Ex eggs of *Euproctis lunata* (Lepidoptera: Erebidae)	Mani and Sharma, 1982
21	Scelionidae	*Telenomus globosus* Bin and Johnson	Ex eggs of *Chilo sacchariphagus* (Lepidoptera: Crambidae) on sugarcane	Bin and Johnson, 1982

TABLE 10.2 (Continued)
Host association of Scelionidae in India

	Subfamily	Parasitoid	Host	Ref.
22	Scelionidae	*Telenomus manolus* Nixon	Ex eggs of *Amsacta albistriga* (Lepidoptera: Arctiidae) on Groundnut	Sundaramurthy et al., 1976
23	Scelionidae	*Telenomus molorchus* Nixon	Ex eggs of *Spilarctia oblique* (Lepidoptera: Arctiidae) on Pigeon pea	Joshi et al., 1983
24	Scelionidae	*Telenomus oryzae* Rajmohana and Nisha	Ex. eggs of *Scotinophara* sp. (Heteroptera: Scutellaridae) on *Oryza sativa*	Rajmohana and Nisha, 2013
25	Scelionidae	*Telenomus otones* Nixon	Ex eggs of *Tarucus Theophrastus* (Lepidoptera: Lycaenidae) on lac insects	Nixon, 1940
26	Scelionidae	*Telenomus pegasus* Nixon	Ex eggs of *Tarucus theophrastus* (Lepidoptera: Lycaenidae) on lac insects	Nixon, 1940
27	Scelionidae	*Telenomus proditor* Nixon	Hosts: Ex. Eggs of *Eupterote undata* (Lepidoptera: Eupterotidae) on *Gmelina arborea*	Nixon, 1937
28	Scelionidae	*Telenomus remus* Nixon	Ex. eggs of *Spodoptera litura* (Lepidoptera: Noctuidae)	Narasimham et al., 1997
29	Scelionidae	*Telenomus rowani* Gahan	Ex. eggs of *Scirpophaga nivella* (Lepidoptera: Crambidae) on paddy and *S. incertulas* (Lepidoptera: Crambidae) on Sugarcane	Narasimham et al., 1997
30	Scelionidae	*Telenomus samueli* (Mani)	Ex. eggs of *Bagrada cruciferarum* (Hemiptera: Pentatomidae)	Mani and Sharma, 1982
31	Scelionidae	*Telenomus seychellensis* Kieffer	Ex eggs of *Antestiopsis cruciata* (Hemiptera: Pentatomidae) on Jasmine; Ex eggs of *Cantheconidea furcellata* (Hemiptera: Pentatomidae)	Mani and Sharma, 1982
32	Scelionidae	*Telenomus talaus* Nixon	Ex. eggs of *Papilio* sp. (Lepidoptera: Papilionidae)	Nixon, 1937
33	Scelionidae	*Trissolcus barrowi* (Dodd)	Ex. eggs of Sphingid moth (Lepidoptera)	Mani and Sharma, 1982
34	Scelionidae	*Trissolcus hyalinipennis* Rajmohana and Narendran	Ex. eggs of *Bagrada cruciferarum* Kirkaldy (Hemiptera: Pentatomidae)	Subba Rao and Chacko, 1961; Rajmohana and Narendran, 2007b
35	Scelionidae	*Trissolcus jatrophae* Rajmohana and Narendran	Ex. eggs of *Scutellara nobilis* Fabricius on *Jatropha* (Hemiptera: Scutelleridae)	Rajmohana and Narendran, 2011
36	Scelionidae	*Trissolcus nigrus* Rajmohana and Narendran	Host: Ex Eggs of *Plautia fimbriata* (Hemiptera: Pentatomidae)	Narayanan and Kaur 1959; Rajmohana and Narendran, 2007
37	Scelionidae	*Trissolus orontes* (Nixon)	Ex eggs of an unidentified insect on Tiliaceae and Ex eggs of *Utetheisa pulchella* (Lepidoptera: Arctiidae)	Nixon, 1935

10.5 EFFECTIVE SCELIONID BIOCONTROL AGENTS

The most important Scelionid genera relevant in the field of biocontrol are *Telenomus* Haliday (Figure 10.1, Figure 10.2), *Trissolcus* Ashmead (Figure 10.3) *Gryon* Haliday. The fall armyworm, *Spodoptera frugiperda* (Lepidoptera: Noctuidae) an alien invasive pest species from the tropical and subtropical Americas was first reported from India in 2018 (Sharanabasappa et al., 2018). Since then, the pest has spread to several states including Kerala, Karnataka, Tamil Nadu, Maharashtra, Gujarat, Andhra Pradesh, Odisha, Bihar, Jharkhand and West Bengal (EPPO, 2018, 2019). This highly destructive pest is known to attack over 350 plants, of which 40% are economically important crops (Montezano, 2018), cereal crops like rice, wheat and maize, are the most affected. *Telenomus remus* Nixon attacking the eggs of *S. frugiperda* is a promising

FIGURE 10.1 *Telenomus oryzae* emerging from Rice Black Bug eggs on rice.

FIGURE 10.2 *Telenomus oryzae* female.

biological control agent. *Telenomus remus* is known to parasitise eggs of five different species of *Spodoptera*, they can parasitise overlapping layers of egg masses too. *T. remus* females produce more than 250 eggs in its reproductive lifespan. *T. remus* is also reported from the eggs of *S. litura* and *S. exigua* (Hübner) (Chou, 1987, Tang et al., 2010). They were was released in corn field areas, as part of IPM programs in Venezuela, where the parasitism rates reached up to 90%, demonstrating the high potential of this biocontrol agent of several species of *Spodoptera*.

The scelionids attack heteropteran families, like Coreidae, Alydidae (Figure 10.5), Pentatomidae and have been known to be efficient candidates for classical biocontrol programs. The egg parasitoid, *Gryon gonikopalense* Sharma, which is widely distributed in India and Pakistan, is being evaluated as a candidate for classical biocontrol against the *Bagrada* bug, *B. hilaris* in California (Martel et al., 2019). The bug is a serious pest of crucieferous vegetables. The parasitoid–host association of *Trissolcus basalis* (Wollaston) and *Nezara viridula*, the notorious green stinkbug serve as a

FIGURE 10.3 *Trissolcus jatrophae* female.

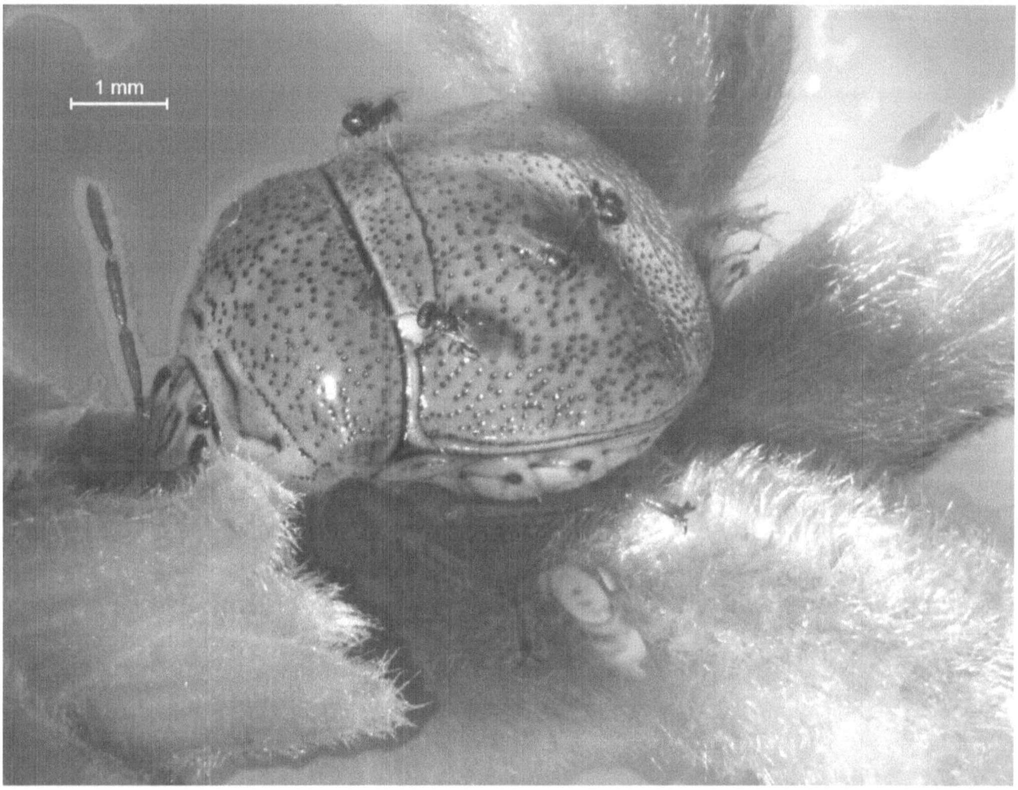

FIGURE 10.4 *Paratelenomus anu* phoretic on *Megacopta cribraria.*

FIGURE 10.5 *Gryon* female emerging from *Leptocorisa* eggs.

FIGURE 10.6 *Platytelenomus pulchricornis* Kieffer.

much favoured study model system in entomological research (Austin et al., 2005). Ganjisaffar et al. (2018) report the discovery of *Trissolcus hyalinipennis,* an Old World egg parasitoid described erstwhile from India from the eggs of *Bagrada hilaris,* the invasive stinkbug in North America. This can be an example of fortuitous biocontrol, which is defined as accidental immigration and establishment of an exotic natural enemy, a phenomenon of great significance in the biocontrol scenario.

Against *Helopeltis* spp. (Hemiptera: Miridae), a polyphagous pest attacking cashew, cocoa, neem, guava, tea, black pepper, cinchona, etc., *Telenomus cuspis*, the newly described native egg parasitoid (Rajmohana et al., 2013b) showed a good potential for biocontrol in India (Srikumar et al., 2014). For the egg parasitoid, *Telenomus laeviceps*, a promising biocontrol candidate of the cabbage moth, *Mamestra brassicae* (Noctuidae), provision of selected flowers as nectar resources, had an impact on the survival and parasitisation performance in the field (Barloggio et al., 2018).

In Colombia, releases that integrate *Trichogramma* with both Bt and *Apanteles* or *Telenomus* have reduced the use of insecticides by 50% on tomato and cassava (Masry and Wakeil, 2020). Studies on the host searching behaviour of *Trissolcus japonicus* (Ashmead), an egg parasitoid and the primary candidate for the biocontrol of *Halyomorpha halys* (Stål) (Hemiptera: Pentatomidae), the brown marmorated stinkbug proved that contact kairomones from different host species elicit varied responses in the host foraging behavior of the parasitoid (Boyle et al., 2020).

Members of genus *Scelio* are known to attack the eggs of grasshoppers. Acridids inflict severe damage to cereal and vegetable crops, fruit trees and forests worldwide.

Desirable traits of Scelionids as efficient natural enemies (Summarised from Orr, 1988)

a. Most scelionid species have good searching abilities.
b. The potential to inflict mortality on host populations is higher in scelionids compared to other egg parasitoids.
c. Species, like *Trissolcus basalis,* shows high ecological compatibility, a trait advantageous for exerting control throughout the ecological range of the pest. This favours the biocontrol programmes, since parasitoids are often more sensitive to environmental conditions than their hosts.
d. Ability to survive on an alternate host, when primary hosts are not available, is a feature expressed by several scelionids. Their wide host ranges contributed to the success of *Telenomus alsophilae* Viereck and *Trissolcus basalis*, as candidates in classical biocontrol programs.
e. Scelionids react in a density-dependent manner with their hosts, in most cases- responding in a positive manner to increase in host density.
f. Scelionids are well synchronised with field activities and the life cycles of their hosts. This has contributed to their success as natural enemies.
g. High dispersal ability, which is a positive trait for biocontrol agents, is seen in several scelionds. Some species are even phoretic on their hosts to reach new habitats, hitchhiking their hosts.
h. While rearing for biocontrol programs, scelionids can survive in laboratory with simple diets, like nectar and aphid honey dew, they do not require a specialised diets.
i. Scelionids generally lack hyperparasitoids or primary pathogens. Hence their effectiveness in the field is not compromised due to secondary parasitism. However, a couple of stray reports of hyperparasitoids on scelionids include *Trichogramma pretiosum* Riley as a facultative parasitoid of *Telenomus heliothidis* (Strand and Vinson, 1984) and the pteromalid *Acroclisoides sinicus* as an obligate hyperparasitoid of *Trissolcus japonicus*, *T. mitsukurii*, and also *Telenomus* sp. in pentatomid eggs (Giovannini, 2021).
j. *Telenomus remus* colonies reared on a factitious lepidopteran host, *Corcyra cephalonica* and on its natural host, *S. frugiperda*, did not lose their ability to parasitise host eggs until 19 generations in the laboratory (Naranjo-Geuvara, 2020).
k. Mass-releases of parasitoids in thefield, help in host control. Releasing *Trissolcus basalis* in mass, into soybean fields resulted in a significant increase in the percentage of parasitism of *Nezara viridula.*

The scelionids though are vigorous in the field, the difficulties encountered in mass rearing delimit their extensive utilisation as biocontrol agents. Bayram et al. (2005) investigated the effect of cold storage of both pupae and adults of *Telenomus busseolae*. Storage had a significant adverse effect on mean adult emergence. The survival percentage was less in the case of long storage times. As per Martela et al. (2019) *G. gonikopalense* female rejected the frozen host eggs for oviposition, though freezing of host eggs did not affect the development of parasitoid larvae. The suitability of frozen host eggs for rearing Scelionidae species is less studied.

10.6 SCELIONIDS IN BIOCONTROL PROGRAMS

Among the parasitoids used in classical biocontrol programs worldwide, 20 were scelionids and 12 of them released against 11 target pest species could establish successfully (Orr, 1988). In India, so far 166 exotic biocontrol agents have been introduced. While 33 could not be released in the field, 71 were recovered after release. Excellent control was provided by 6 of them, 7 yielded substantial control and 4 gave partial control (Singh, 2004).

10.6.1 Successful Biocontrol Programs in India, Using Scelionidae (Singh, 2004)

1. From New Guinea, *Telenomus proditor*? was imported into India to control *Achaea janata* on castor in Hyderabad; parasitoid recovery was 60–80% (Kulshreshtha et al., 1967).
2. *Telenomus remus* was imported to India from New Guinea in 1963 to control *Achaea janata* on castor in Andra Pradesh and *Spodoptera litura* on papaya, cabbage and tobacco nursery in Rajahmundry (Andhra Pradesh) and on lucerne and cauliflower in Gujarat.
3. *Telenomus alecto* imported from Colombia in 1966 and Trinidad, West Indies in 1982, released in Plassey (West Bengal) and was recovered from Plassey borer, *Chilo tumidicostalis*. Also released and recovered in Motipur (Bihar), while they established after release at Golagokarnath (Uttar Pradesh).
4. *Telenomus* sp. imported from Bolivia, released in sugarcane fields, was recovered from eggs of *Chilo* spp.
5. *Telenomus nephele* from Ghana was imported in 1972 to control Sugarcane Tissue Boreres.
6. *Telenomus nawai* was imported from New Guinea in 1963 to control the Castor Semilooper.

10.7 CONCLUSIONS

Scelionidae is a diverse, species-rich family of parasitic Hymenoptera with noteworthy economic and ecological significance. As egg parasitoids, they play a vital role in the regulation of the population of other arthropods, insects and spiders in particular. Being natural enemies of notorious pests of crops and forestry, they have the potential as efficient biocontrol agents, contributing to the success of Integrated Pest Management Programs. They are both taxonomically and biologically less known. Extensive and intensive taxonomic surveys are needed to unearth their native biodiversity. Habitat or ecosystem-based baseline data on their diversity is the need of the hour. An improved understanding and utilisation of these valuable biocontrol agents can be achieved by integrating the conventional morphological approach as well as molecular characterisation. They have the potential to assist us in our stride towards sustainable agriculture.

10.8 POINTS TO REMEMBER

1. Scelionids, in general, are solitary endoparasitoids and belong to the category of beneficial insect fauna.
2. Members of Scelionidae attack exclusively the egg stage of their natural enemies, hence they suppress the pests before they hatch and prevent crop damage.
3. Host range of scelionids includes members of Lepidoptera, Hemiptera, Coleoptera, Diptera, Neuroptera, Orthoptera, Mantodea and Odonata and also Araneomorph spiders.
4. Their morphological modification and adaptations towards functional specialisation enable them to exploit even cryptic habitats.
5. From biocontrol perspective, the most important genera of scelionds include *Telenomus*, *Trissolcus* and *Gryon*.

REFERENCES

Abhilash, P. & Rajmohana, K. (2014). A new species of *Gryon* Haliday (Hymenoptera: Platygastridae) from India. *Journal of Threatened Taxa*, 6(14): 6711–6714.

Ahmad, M., Ahmad, M.J. & Mishra, R.K. (1997). Indigenous micro-hymenopteran parasitoid speciesassociated with poplar defoliator, *Clostera cupreata* (Lepidoptera: Notodontidae). *Annals of Forestry*, 5(1): 26–34.

Anjana, G., Rajmohana, K. & Shweta, M. (2016a). New species of *Gastrotrypes* Brues (Hymenoptera: Platygastridae) from India along with key to world species. *Oriental Insects*, 50(1): 9–22.

Anjana, M. & Rajmohana K. (2015). Three new species of *Amblyaspis* Förster (Hymenoptera: Platygastridae) from India along with a key to Indian species. *Halteres*, 6: 113–120.

Anjana, M., Rajmohana, K. & Sinu, P.A. (2017). Two New Species of Iphitrachelus Haliday (Hymenoptera: Platygastridae) from India with A Key to Oriental Species. *Insect Diversity and Taxonomy, T. C. N. Com.* Vol. November 2017: 213–224.

Anjana, M., Rajmohana, K., Athira, A. & Ranjith, A. P. (2016b). New report and host association of *Trichacoides* Dodd (Hymenoptera: Platygastridae), with a key to species from India. *Journal of Insect Systematics*, 2(2): 81–88.

Austin, A.D., Johnson, N.F. & Dowton, M. (2005). Systematics, evolution, and biology of scelionid and platygastrid wasps. *Annual Review of Entomology*, 50: 553–582.

Barloggio, G., Tamm, L., Nagel, P. & Luka, H. (2018). Selective flowers to attract and enhance *Telenomus laeviceps* (Hymenoptera: Scelionidae): a released biocontrol agent of *Mamestra brassicae* (Lepidoptera: Noctuidae). *Bulletin of Entomological Research*, 1–9. 10.1017/s0007485318000287

Bayram, A., Ozcan, H. & Kornosor, S. (2005). Effect of cold storage on the performance of *Telenomus busseolae* Gahan (Hymenoptera: Scelionidae), an egg parasitoid of *Sesamia nonagrioides* (Lefebvre) (Lepidoptera: Noctuidae). *Biological Control*, 35: 68–77.

Bin, F. & Johnson, N.F. (1982). New species of *Telenomus* (Hymenoptera: Scelionidae), egg parasitoids of tropical pyralid pests (Lepidoptera: Pyralidae). *Redia*, 65: 229–252.

Boyle, M.S., Weber, D. C., Hough-Goldstein, J. & Hoelmer, K.A. (2020). Host *Kairomones* Influence Searching Behavior of *Trissolcus japonicus* (Hymenoptera: Scelionidae), a Parasitoid of *Halyomorpha halys* (Heteroptera: Pentatomidae). *Environmental Entomology*, 49(1): 15–20. 10.1093/ee/nvz155.

Burks, R.A., Masner, L., Johnson N.F. & Austin A.D. (2013). Systematics of the parasitic wasp genus *Oxyscelio* Kieffer (Hymenoptera, Platygastridae s.l.), Part I: Indo-Malayan and Palearctic fauna. *ZooKeys*, 292: 1–263. 10.3897/zookeys.292.3867

Chabi O.A., Schulthess, F.,Shanower, T.G. & Bosque-Pe´Rez N. A. (1997). Factors Influencing the Developmental Rates and Reproductive Potentials of *Telenomus busseolae* (Gahan) [Hym.: Scelionidae], an Egg Parasitoid of *Sesamia calamistis* Hampson (Lep.: Noctuidae). *Biological Control*, 8: 15–21.

Chou, L.Y. (1987). Note on *Telenomus remus* (Hymenoptera: Scelionidae). Bulletin of Society of Entomology, *National Chung-Hsing University*, 20: 15–20.

Clausen, C.P. (1976). Phoresy among entomophagous insects. *Annual Review Entomology*, 21: 343–367.

EPPO. (2018). EPPO Global Database, https://gd.eppo.int/ EPPO, Paris, France.

EPPO. (2019). *Spodoptera frugiperda* continues to spread in Asia. In: *EPPO Reporting Service*, (No. 2019/053), Paris, France: EPPO. https://gd.eppo.int/reporting/article-6483.

Fatouros, N. E., Cusumano, A., Bin, F., Polaszek, A. & van Lenteren, J. C. (2020). How to escape from insect egg parasitoids: a review of potential factors explaining parasitoid absence across the Insecta. *Proceedings of the Royal Society* B.2872020034420200344. 10.1098/rspb.2020.0344

Gahan, A.B. (1925). A second lot of parasitic Hymenoptera from the Philippines. *Philippine Journal of Science*, 27: 83–109.

Galloway, I. D., & Austin, A. D. (1984). Revision of the Scelioninae (Hymenoptera: Scelionidae) in Australia. *Australian Journal of Zoology*, Supplementary Series 99, pp.1–138.

Ganjisaffar, F., Talamas, E.J., Bon, M.C., Gonzalez, L., Brown, B.V. & Perring, T.M. (2018). *Trissolcus hyalinipennis* Rajmohana & Narendran (Hymenoptera, Scelionidae), a parasitoid of *Bagrada hilaris* (Burmeister) (Hemiptera, Pentatomidae), emerges in North America. *Journal of Hymenoptera Research*, 65: 111–130. 10.3897/jhr.65.25620

Giovannini, L., Sabbatini-Peverieri, G., Tillman, P.G., Hoelmer, K.A. & Roversi, P.F. (2021). Reproductive and Developmental Biology of *Acroclisoides sinicus*, a Hyperparasitoid of Scelionid Parasitoids. *Biology*, 10: 229. 10.3390/ biology10030229.

Johnson, N.F. (1992). Catalog of world Proctotrupoidea excluding Platygastridae. *Memoirs of the American Entomological Institute*, 51: 1–825. 10.5281/zenodo.23657

Johnson, N. F. (1996). Revision of world species of *Paratelenomus* (Hymenoptera: Scelionidae). *The Canadian Entomologist*, 128: 273–291.

Johnson, N. F. & Masner, L. (1985). Revision of the genus *Psix* Kozlov & Lê (Hymenoptera: Scelionidae). *Systematic Entomology*, 10: 33–58.

Johnson, N.F. & Masner, L. (2006). Revision of world species of the genus *Nixonia* Masner (Hymenoptera: Platygastroidea, Scelionidae). *American Museum Novitates*, 3518: 1–32. 10.1206/0003-0082(2006)3518[1:ROWSOT]2.0.CO;2

Johnson, N. F., Masner, L, Musetti, L, van Noort, S., Rajmohana, K., Darling, D. C., Guidottiand, A. E. & Polaszek, A. (2008). Revision of world species of the genus *Heptascelio* Kieffer (Hymenoptera: Platygastroidea,Platygastridae). *Zootaxa*, 1776: 1–51.

Joshi, R.C., Rao, R.K. & Rao, B.H.K. (1983). Occurrence of *Telenomus (Aholcus) molorchus* Nixon as an egg parasitoid of Bihar hairy caterpillar. *Indian Journal of Entomology*, 45(4): 499.

Kulshreshtha, J.P., Sanghi, P.K., Thobbi, V.V. & Kulkarni, L.G. (1967). Mass multiplication of *Telenomus sp.* (Hymenoptera: Scelionidae) an egg parasite imported in India for the biological control of the castor semilooper *Achaea janata* Linnaeus (Lepidoptera: Noctuidae) and some observations on its biology. *Indian Journal of Environment*, 29(3): 290–294.

Mani, M. S. &. Sharma, S. K. (1982). Proctotrupoidea (Hymenoptera) from India. A review. *Oriental Insects*, 16: 135–258.

Martela, D.G., Augéb, M., Talamasc, E., Rochea, M., Smitha, L. & Sforzaa, R.F.H. (2019). First laboratory evaluation of *Gryon gonikopalense* (Hymenoptera: Scelionidae), as potential biological control agent of *Bagrada hilaris* (Hemiptera: Pentatomidae). *Biological Control*, 135: 48–56.

Masner, L. (1976). Revisionary notes and keys to world genera of Scelionidae (Hymenoptera: Proctotrupoidea). *Memoirs of the Entomological Society of Canada*, 97: 1–87.

Masner, L. (1993). Superfamily Proctotrupoidea, Superfamily Platygastroidea, Superfamily Ceraphronoidea. In *Hymenoptera of the world: an identification guide to families* eds. Goulet, H. & Huber, J., pp. 537–569. Monograph: Agriculture Canada Research Branch, 1894E, 668 pp.

Masner, L. (1995). The proctotrupoid families. In *The Hymenoptera of Costa Rica*, ed. Hanson, P.E. & Gauld, I.D., pp. 209–265. Oxford: Oxford University Press.

Masry, S. H. D. & Wakeil, N.E. (2020). Egg parasitoid production and their role in controlling insect pests. *Cottage Industry of Biocontrol Agents and Their Applications*, pp. 3–47.

Miko, L., Masner, L. A. & Deans, R. (2010). World revision of *Xenomerus* Walker (Hymenoptera: Platygastroidea, Platygastridae). *Zootaxa*, 2708: 73.

Mills, N.J. (1994). Parasitoid guilds" defining the structure of the parasitoid communities of endopterygote insect hosts. *Environmental Entomology*, 23: 1066–1983.

Mineo, G. (1983). Studies on the Scelionidae (Hymenoptera: Proctotrupoidea) XV111. Revision of the genus *Gryon* Haliday (Ethiopian-Oriental regions) the charon group. *Phytophaga*, 1: 11–26.

Montezano, D.G., Sosa-Gómez, D.R., Roque-Specht, V.F., Sousa-Silva, J.C., Paula-Moraes, S.V., Peterson, J.A. & Hunt, T.E. (2018). Host plants of *Spodoptera frugiperda* (Lepidoptera: Noctuidae) in the Americas. *African Entomology*, 26(2): 286–300. 10.4001/003.026.0286.

Murphy, N.P., Carey, D., Castro, L.R., Downton, M. & Austin, A.D. (2007). Phylogeny of the platygastroid wasps (Hymenoptera) based on sequences from the 18SrRNA, 28S rRNA and cytochrome oxidase I genes: implications for the evolution of the ovipositor system and host relationships. *Biological Journal of the Linnaean Society*, 91: 653–669.

Naranjo-Guevara, N., dos Santos, Odorizzi, L.A., Barbosa, C. P., Cristina, N., e Castro, Corrêa, A.C.M. & Aparecido, F.O. (2020). Long-Term Mass Rearing Impacts Performance of the Egg Parasitoid *Telenomus remus* (Hymenoptera: Platygastridae). *Journal of Entomological Science*, 55(1): 69–86.

Narasimham, A.U., Bhumannavar, B.S., Ramani, S., Poorani, J. & Rajeshwari, S.K. (1997). A Catalogue of natural enemies and other insects in PDBC reference collection, Technical bulletin no. 17, Project Directorate of Biological Control, Bangalore.

Narayanan, E.S. & Kaur, R.B. (1959). A new species of *Microphanurus* Kieffer. *Proceedings of the Indian Academy of Science, (B)*, 49: 136–138.

Nixon, G.E.J. (1935). A revision of the African Telenominae (Hymenoptera: Proctotrupoidea: Scelionidae). *Proceedings of the Royal Society*, 5: 131–134.

Nixon, G.E.J. (1937). Some Asiatic Telenominae (Hymenoptera: Proctotrupoidea). *The Annals and Magazine of Natural History*, 10(20): 444–475.

Nixon, G.E.J. (1938). Asiatic species of *Microphanurus* (Hymenoptera: Proctotrupoideae). *The Annals and Magazine of Natural History*, 11(2): 122–139.

Nixon, G.E.J. (1940). New species of Proctotrupoidea. *The Annals and Magazine of Natural History*, 11(6): 497–512.

Orr, D.B. (1988). Scelionid wasps as biological control agents: a review. *Florida Entomologist*, 71: 506–528.

Rajmohana, K. (2007a). On a new species of *Duta* Nixon (Hymenoptera: Scelionidae) from India. *Records of Zoological Survey of India*, 107(4): 7–12.

Rajmohana, K. (2007b) Insecta: Scelionidae (Platygastroidea): Hymenoptera Fauna of Kudremukh National Park. *Conservation Area Series*, 32: 49–69.

Rajmohana, K, (2010). First report and a redescription of *Duta serraticeps* (Priesner) (Hymenoptera: Platygastridae) from India. *Biosystematica*, 4(2): 51–55.

Rajmohana, K. (2011). Descriptions of new taxa in Thoronini (Scelioninae: Platygastridae: Hymenoptera: Insecta) from India. Recent Advances in Biodiversity of India, published by Director. *Zoological Survey of India*. (pp. 267–276).

Rajmohana, K. (2014). A systematic inventory of Scelioninae and Teleasinae (Hymenoptera: Platygastridae) in the rice ecosystems of North central Kerala. *Memoirs of Zoological Survey of India*, 22: 72.

Rajmohana, K. & Narendran, T. C. (2001). Parasitoid complex of *Coptosoma cribrarium* (Fabricius) (Plataspididae: Hemiptera). *Insect Environment*, 6(4): 163.

Rajmohana, K. & Narendran, T.C. (2007a). A new species of *Paratelenomus* (Hymenoptera: Scelionidae) from India. *Zoos' Print Journal*, 22(1): 2522–2523.

Rajmohana, K. & Narendran, T. C. (2007b). A systematic note on the genus *Trissolcus* Ashmead (Hymenoptera: Scelionidae) with a key to species from India. *Records of the Zoological Survey of India*, 107(3): 101–103.

Rajmohana, K. & Talukdar. (2010). A new species of *Psix* Kozlov & Lê (Hymenoptera: Platygastridae) from India. *Biosystematica*, 4(2): 57–62.

Rajmohana, K & Bijoy. C. (2011). First report of phoresy of *Trissolcus* Ashmead (Hymenoptera: Platygastridae) from India. Skaphion. *Platygastroidea Planetary Biodiversity Inventory Project Newsletter*, 5: 49–50.

Rajmohana, K. & Peter, A. (2012). On a new genus and species of Scelioninae (Hymenoptera: Platygastridae) from India. *Biosystematica*, 6(1): 19–25.

Rajmohana, K. & Peter, A. (2013). Fauna of Bhadra Wildlife Sanctuary & Tiger Reserve. *Insecta: Platygastridae (Hymenoptera: Platygastroidea). Conservation Area Series*, 46: 11–23.

Rajmohana, K. & Nisha, M.S. (2013). *Telenomus oryzae* (Hymenoptera: Platygastridae), a new egg parasitoid of the rice black bug, *Scotinophara* (Heteroptera: Pentatomidae) from India. *Halteres*, 4: 79–86.

Rajmohana, K. & Anto, M. (2014). *Telenomus dilatus* sp. n. (Hymenoptera: Platygastridae) – an egg parasitoid of swallowtail butterflies from South India. *Halteres*, 5: 73–78.

Rajmohana, K. & Veenakumari. K. (2014). *Chakra*, a new genus of Scelioninae (Hymenoptera: Platygastridae) from India, along with description of a new species. *Zootaxa*, 3821(2): 285–290.

Rajmohana, K. & Patra, S. (2019). Scelionidae and Platygastridae (Hymenoptera: Platygastroidea) of India. *Indian Insects Diversity and Science*, CRC Press Taylor & Francis Group, pp. 159–172.

Rajmohana, K, Narendran, T.C. & Manoharan, T. (2011). A new species of *Trissolcus* (Hymenoptera: Platygastridae): egg parasitoid of *Scutellera nobilis* Fabricius (Hemiptera: Scutelleridae) on *Jatropha curcas* Linnaeus (Euphorbiacea) in India. *Hexapoda*, 18(2): 106–110.

Rajmohana, K., Abhilash, P. & Narendran, T.C. (2013a). First record of the male sex of the wide spread *Calliscelio elegans* (Perkins) (Hymenoptera: Platygastridae) along with some taxonomic notes on the species. *Biodiversity Data Journal*, 1–13. Pensoft Publishers. 10.3897/BDJ.1.e983.

Rajmohana, K., Srikumar, K.K., Bhat, P.S., Raviprasad, T.N. & Jalali, S.K. (2013). A new species of platygastrid *Telenomus cuspis* sp.nov. (Hymenoptera), egg parasitoid of tea mosquito bug from India, 339 with notes on its bionomics and mtCo1 data. *Oriental Insects*, 47: 226–232.

Rajmohana, K., Veenakumari, K., Bijoy, C., Prashanth, M., Sinu, P.A. & Ranjith, A.P. (2017). *Anokha* gen. n. (Hymenoptera: Platygastroidea: Scelionidae) and two new species from India. *Halteres*, 8: 77–84.

Rajmohana, K., Sachin, J.P., Elijah, J. T., Mukundan, S., Shamyasree, Jalali, S. K. & Ojha, R. (2019). *Paratelenomus anu* Rajmohana, Sachin & Talamas (Hymenoptera, Scelionidae): description and biology of a new species of phoretic egg parasitoid of Megacopta cribraria (Fab.) (Hemiptera, Plataspidae). *Journal of Hymenoptera Research*, 73(73): 103–123.

Romeis, J., Shanower, T.G. & Madhuri, K. (2000). Biology and field performance of *Gryon clavigrallae* (Hymenoptera: Scelionidae), an egg parasitoid of *Clavigralla spp.* (Hemiptera: Coreidae) in India. *Bulletin of Entomological Research*, 90(3), 253–263.

Sharanabasappa, Kalleshwaraswamy, C. M., Asokan, R., Mahadeva Swamy, H. M., Maruthi, M. S., Pavithra, H. B., Kavita Hedge, Shivaray Navi, Prabhu, S. T. & Goergen, G. (2018). First report of the Fall armyworm, Spodoptera *frugiperda* (J E Smith) (Lepidoptera: Noctuidae), an alien invasive pest on maize in India. *Pest Management in Horticultural Ecosystems*, 24(1): 23–29.

Sharkey, M. (2007). Phylogeny and classification of Hymenoptera. *Zootaxa*, 1668, 521–548.

Shweta, M., Rajmohana, K. & Bijoy, C. (2014). A biosystematic account on Baeini wasps (Hymenoptera: Platygastridae), the little known natural enemy complex of Spiders in India. *Proceedings of 26th Kerala Science Congress Pookode*, Wayanad: 28-31 Jan. 10(84): 3654–3662.

Singh, J. & Ramamurthy. (2012). New species of the genus *Psix* Kozlov & Lê (Hymenoptera: Platygastridae). *Zootaxa*, 3530: 35–42.

Singh, S.P. (2004). Some Success stories in Classical Biological control of Agricultural Pests in India. Bangkok, APAARI. [Retrieved from http://www.apaari.org/wpcontent/uploads/2009/05/ss_2004_02.pdf. (Accessed on 20th June, 2021)].

Srikumar, K., Bhat, P. S., Vanitha, K., Rajmohana, K. & Raviprasad, T. (2014). Phenology and Parasitization Behaviour of *Telenomus cuspis* (Hymenoptera: Platygastridae) an Egg Parasitoid of *Helopeltis antonii* (Hemiptera: Miridae) in Cashew. *Proceedings of the National Academy of Sciences, India Section B: Biological Sciences*. 10.1007/s40011-014-0365-0

Strand, M.R. (1986). The physiological interactions of parasitoids with their hosts and their influence on reproductive strategies. *In Insect Parasitoids*. eds. Waage, J & Greathead, D., pp. 97–136. Academic Press.

Strand, M.R. (2014). Teratocytes and their functions in parasitoids. *Current Opinion in Insect Science*, 6: 68–73. 10.1016/j.cois.2014.09.005.

Strand, M.R. & Vinson, S.B. (1984). Facultative hyperparasitism by the egg parasitoid *Trichogrammaa pretiosum* (Hymenoptera: Trichogrammatidae). *Annals of the Entomological Society of America*, 77: 679–686.

Subba Rao, B.R. & Chacko, M.J. (1961). Studies on *Allophanurus indicus* n. sp., an egg parasite of *Bagrada cruciferum* Kirkalady. *Beitraege zur Entomologie*, 11: 812–824.

Sundaramurthy, V.T., Jayaraj, S. & Swamiappan, M. (1976). *Telenomus manolus* Nixon, a potential egg parasite of groundnut red hairy caterpillar, *Amsacta albistriga* Walker. *Current Science*, 45(21): 775.

Taekul, C., Johnson, N. F., Masner, L, Rajmohana K. & Chen S. P. (2008). Revision of the world species of the genus *Fusicornia* Risbec (Hymenoptera: Platygastridae, Scelioninae). *Zootaxa*, 1966: 1–52.

Taekul, C., Johnson, N.F., Masner, L., Polaszek, A. & Rajmohana K. (2010). World species of the genus *Platyscelio* Kieffer (Hymenoptera, Platygastridae). *ZooKeys*, 50: 97–126.

Taekul, C., Valerio, A.A., Austin, A.D., Klompen, H. & Johnson, N.F. (2014). Molecular phylogeny of telenomine egg parasitoids (Hymenoptera: Platygastridae s.l.: Telenominae): evolution of host shifts and implications for classification. *Systematic Entomology*, 39: 24–35.

Tang, Y.L., Chen, K.W. & Xu, Z.F. (2010). Study on ontogenesis of *Telenomus remus* Nixon (Hymenoptera: Scelionidae). *Journal of Changjiang Vegetables*, 18: 1–3.

Valerio, A. A., Austin, A. D., Masner, L. & Johnson, N.F. (2013). Systematics of Old World *Odontacolus* Kieffer s.l. (Hymenoptera, Platygastridae s.l.): parasitoids of spider eggs. *ZooKeys*, 314: 1–151.

Van Noort, S. (2021). WaspWeb: Hymenoptera of the Afrotropical region. URL: www.waspweb.org (accessed on 20th June, 2021).

Veenakumari, K. & Prashanth, M. (2015). Redescription of *Protelenomus flavicornis* Kieffer (Platygastroidea: Platygastridae) from India. *Entomofauna, Zeitschrift für Entomologie*, 36: 305–312.

Veenakumari, K. & Prashanth, M. (2017). The genus *Cremastobaeus* Ashmead (Hymenotpera: Scelionidae: Cremastobaeini) from India. *Journal of Natural History*, 51: 1989–2056.

Veenakumari, K., Rajmohana, K & Prashanth, M. (2012). Studies on phoretic Scelioninae (Hymenoptera: Platygastridae) from India along with description of a new species of *Mantibaria* Kirby. *Linzer Biologische Beiträge*, 44(2): 1715–1725.

Veenakumari, K., Khan, F. R. & Prashanth, M. (2014). Three new species of Teleasinae (Hymenoptera: Platygastridae) from India. *Entomologist's Monthly Magazine*, 150: 227–239.

Veenakumari, K., Notton, D. & Polaszek, A. (2019). World Revision of The Genus *Protelenomus* kieffer (Hymenoptera:Scelionidae:Telenominae). *Annales Zoologici*, 69(2): 381.

Veenakumari, K., Rajmohana, K., Prashanth, M. & Abhilash, P. (2016). An unusual, new, sexually dimorphic species of *Gryon* Haliday (Hymenoptera: Scelionidae) from India. *Oriental Insects*, 10.1080/00305316.2016.1142482

Veenakumari, K., Buhl, P. N., Prashanth, M. & Khan, F.R. (2017-2018). Revision of Indian species of *Leptacis* Forster (Hymenoptera: Platygastroidea, Platygsatridae). *Entomologist's Monthly Magazine, (Four parts)*, 153: 205–231, 279-312; 154: 21–52.

Veenakumari, K., Popovici, O.A., Talamas, E.J. & Prashanth, M. (2018). *Indiscelio*: A new genus of Scelionidae (Platygastroidea) from India. *Journal Asia-Pacific Entomology*, 21: 571–577.

Veenakumari, K., Rajmohana, K., Manickavasagam, S. & Mohanraj, P. (2011). On a new genus of Teleasinae (Hymenoptera: Platygastridae) from India. *Biosystematica*, 5(2): 39–46.

Veenakumari, K., Talamas, E., Rajmohana, K. & Prashanth, M. (2017). Two new species of *Apteroscelio* Kieffer (Hymenoptera: Scelionidae) from India. *Zootaxa*, 4277(1): 137–143.

Yeargan, K.V. (1980). Effects of Temperature on Developmental Rate of *Telenomus podisi* (Hymenoptera: Scelionidae). *Annals of the Entomological Society of America*, 73(3): 339–342. 10.1093/aesa/73.3.339

Yoder, M. J., Valerio, A.A., Polaszek, A., Masner, L. & Johnson, N.F. (2009). Revision of *Scelio pulchripennis* group species (Hymenoptera: Platygastroidea, Platygastridae). *ZooKeys*, 20: 53–118.

11 Aphelinid Parasitoids

Mani Chellappan and M.T. Ranjith
Department of Agricultural Entomology, College of Agriculture, KAU, Thrissur, Kerala

CONTENTS

DOI: 10.1201/9781003354239-11

LEARNING OBJECTIVES

1. To know the important parasitoid species in Aphelinidae with their host insects and stages of parasitisation.
2. To have an idea of the diversity and distribution of aphelinid species.
3. To learn the biology and mass rearing of important aphelinid species.
4. To have an idea on the exploitation of aphelinid parasitoids in a successful bio-control programme.

11.1 INTRODUCTION

The Family Aphelinidae, belongs to the Superfamily Chalcidoidea of the order Hymenoptera, and consists of 36 genera and over a thousand species worldwide, making it a key source of biocontrol agents for economically important pest species (Noyes, 2003). However, in India, a total of 269 species belonging to 24 genera have been documented (Manickavasagam and Menakadevi, 2014). The Aphelinidaeis a medium-sized chalcidid family, that is sometimes overlooked due to their tiny size (0.5–1.5 mm). The antenna in this group has four or less funicular segments, the tarsi have four or five segments, the foretibial spur is large, curved, and bifid, the metasoma is broad, attached to the thorax (not petiolate), and the notauli are complete, straight, deep, and widely separated in the posterior region (generally more than the length of scutellum) (Stringer et al., 2012). The body colour varies from yellow, grey, brown, black, or a combination of these hues. Several authors consider this group to be of family rank, with morphological affinities to the Eulophidae and Trichogrammatidae and biological affinities to the Signiphoridae and Encyrtidae (Viggiani, 1984). The majority of aphelinids are primary ecto-or- endo parasitoids of stemorrhynchousHomoptera (Aphidoidea, Aleyrodoidea, and Coccoidea) but, a few are reported to develop on other hosts, such as eggs of Orthoptera and Lepidoptera and puparia of Diptera. However, some species exhibit hyperparasitism, which affects both sexes or only the male in some aphelinids. It is a complicated phenomenon that involves egg adaptations, larval morphology, and host interactions, among other things. A trophic pyramid is formed when a group of aphelinids is related to the same phytophagous host (Yasnosh, 1979; Viggiani, 1984; Polaszek, 1991; Hayat, 1998). In various parts of the world, several species of Aphilinids have been employed successfully in the biocontrol of insect pests, *viz.* aphids, scale insects, and whiteflies. The species belonging to the genera *Encarsia* Foerster (Polaszek et al., 2004), and *Eretmocerus* Haldeman (Rose and Zolnerowich, 1997) parasitise the whiteflies, Aleyrodidae, whereas the members of the genus *Aphelinus* Dalman mainly parasitise the aphids (Hayat, 1998). However, most of them develop by consuming scale- insects belonging to the families, *viz.* Coccidae, Diaspididae, Aclerdidae, Cerococcidae, Asterolecaniidae, Margarodidae, Tachardiidae, Kermococcidae, Eriococcidae, Lecanodiaspididae, Eriococcidae, Kermococcidae, Lecanodiaspididae and Psyllidae either as endo- or as ectoparasitoids (Viggiani, 1984; Hayat, 1998).

11.2 DISTRIBUTION AND DIVERSITY

Aphelinid parasitoids are insects with a wide range of shapes and colours (Janssen, 1961; Traboulsi, 1969; Heraty, 2016). There are 1168 species, 33 genera, and seven subfamilies in this family, which live in a range of environments all over the world (Noyes, 2016). The majority of the described species are found in the Palaearctic and Australian regions, while a few are found in other zoogeographical regions. Some genera' numbers and definitions have recently been updated (Hayat, 1998). However, the evolutionary relationships of families or subfamilies have not been examined in a cladistic framework. Although the monophyletic origin of Aphelinidae is still debated; Gibson (1989) believed Aphelinidae (excluding Eriaporinae) to be paraphyletic with Signiphoridae due to the anterior extension of the metasternum between the mesocoxae. Because of the close kinship between the Azotinae (Aphelinidae) and the Signiphoridae, Aphelinidae could be considered paraphyletic to Signiphoridae (Woolley, 1988). Thus, no single character can be used to define the Aphelinidae family without excluding other families of Chalcidoidea. Molecular data have also failed to support any higher-level groups in the family, with Aphelininae (excluding Eretmocerus), Coccophaginae, Calesinae, Azotinae, and Eriaphytinae being classified as separate and distantly related (Campbell et al., 2000; Munro et al., 2011) Aphelinidae is said to be cosmopolitan in distribution and its 10 genera are recorded from the entire six zoogeographical regions; they are *Ablerus, Aphelinus, Aphytis, Centrodora, Coccobius, Coccophagus, Encarsia, Eretmocerus, Marietta and Pteroptrix*. The Aphelinid genera are exclusively parasiticon whiteflies (aleyrodids). The genus *Encarsia* Foersterwith over 200 described species in the world contains most of the species that are considered as parasitoids of whiteflies (Begum et al., 2011). The diversity and distribution of various aphelinid species recorded in India are summarised in Table 11.1.

11.3 BIOLOGY

The majority of aphelinids found around the world are parasitic on sternorrhynchous Homoptera. While some feed on Aphidoidea, Aleyrodoidea, or Psylloidea, most of them feed on scale-insects (Coccoidea), either as internal parasitoids, ectoparasitoids (under the scale), or predators of their eggs. Males of a number of species grow as hyperparasitoids of coccoids through their main parasitoids viz., eulophids, aphelinids, or encyrtids. Many aphelinids are internal egg parasitoids of various Auchenorrhyncha, Reduvioidea, Lepidoptera, or Orthoptera, whereas only a few develop on Dryinidae (Hymenoptera), Cecidomyiidae or Chamaemyiidae larvae or pupae (Viggiani, 1984). Aphelinid species are primary endoparasitoids of aphids and make up a significant component of the aphelinid fauna in the United Kingdom (e.g. *Aphelinus* spp.).

Adult aphelinids feed on honeydew excreted by their hosts or fluids from oviposition wounds. Honeydew secretion is induced by Coccophagus species caressing their coccid hosts (Cendana, 1937). Predatory host feeding may provide proteinaceous nourishment for female aphelinids. When the parasite inserts its ovipositor into the host's body and then withdraws it after a series of movements, bodily fluids may exude through a tiny tube created by the aphelinid.

Rosen and DeBach (1979) provided a considerably more extensive explanation of predatory feeding on a host, including the construction of a tube from accessory gland secretions, maceration of the host's body with the ovipositor, and so on. In most cases, the host insect develops big, brown necrotic patches and dies within hours to days.

Most of the aphelinid species are biparental. Females who have been mated can lay both fertilised (diploid) and unfertilised (haploid) eggs, which mature into females and males, respectively. Unfertilised (haploid) eggs are produced by unmated females, which mature into males (arrhenotokous parthenogenesis).

Aphelinid eggs are commonly stalked. The ectophagous species do not oviposit on specific parts of the host's body. Female *Aphytis* oviposit exclusively on hosts with a free body beneath the scale cover; eggs are placed on the dorsal or ventral surface. The oviposition behaviour of endophagous species is more diverse. Some parasites appear to have no special preference, but others have specific requirements that may vary depending on whether or not the ovipositing female is mated (Viggiani, 1984). Ectophagous and some endophagous species have hymenopteriform first instar larvae with four pairs of spiracles, but endoparasitic species have an apneustic caudate primary larva. Endoparasitoids (e.g. Coccophagous) have larvae with neither spiracles nor a functional tracheal system (Rosen and DeBach, 1979). Ectoparasitic species (e.g. *Aphytis* spp.) have larvae with both spiracles and a viable tracheal system (Rosen and DeBach, 1979). Pupation may take

TABLE 11.1
The Checklist of Various Aphelinid Species Recorded in India

Sl. No	Species	Distribution
1	*Ablerus aleuroides* Husain and Agarwal, 1982	Uttar Pradesh (Husain and Agarwal, 1982).
2	*Ablerus aligarhensis* Khan and Shafee, 1976	Uttar Pradesh (Khan and Shafee, 1976).
3	*Ablerus aonidiellae* Hayat, 1974a	Rajasthan (Hayat, 1974a); Uttar Pradesh (Hayat, 1998).
4	*Ablerus bharathius* Subba Rao, 1984	Andhra Pradesh, Karnataka, Tamil Nadu and Uttar Pradesh (Hayat, 1998).
5	*Ablerus chionaspidis* Howard, 1914	Karnataka, Tamil Nadu and Uttar Pradesh (Hayat, 1986).
6	*Ableru sconnectens* Silvestri, 1927	Karnataka (Hayat and Anis, 2004); Tamil Nadu (Menakadevi and Manickavasagam, 2011).
7	*Ablerus delhiensis* Lal, 1938	Bihar, Haryana, Madhya Pradesh, Odisha and Uttar Pradesh (Hayat, 1998).
8	*Ablerus gargarae* Hayat, 1998	Tamil Nadu (Hayat, 1998).
9	*Ablerus inquirenda* Silvestri, 1927	Tamil Nadu (Hayat, 1998).
10	*Ablerus perspeciosus* Girault, 1916	Jammu & Kashmir and Uttar Pradesh (Hayat, 1998).
11	*Ablerus williamsi* Annecke and Insley, 1970)	Assam, Odisha (Hayat and Khan, 2010); Tamil Nadu, Kerala (Menakadevi and Manickavasagam, 2011).
12	*Ablerus xenex* Hayat, 2010	Sikkim (Hayat and Khan, 2010). Genus Aphelinus Dalman, 1820
13	*Aphelinus abdominalis* Dalman, 1820	Andhra Pradesh, Assam, Himachal Pradesh, Jammu & Kashmir, Karnataka, Kerala, Meghalaya, Punjab, Rajasthan, Tamil Nadu and West Bengal (Hayat, 1998).
14	*Aphelinus accuminatus* Hayat, 1998	Meghalaya (Hayat, 1998).
15	*Aphelinus albipodus* Hayat and Fatima, 1992	Andhra Pradesh, Jammu & Kashmir, Karnataka, Rajasthan and Tamil Nadu (Hayat and Fatima, 1992).
16	*Aphelinus ancer* Hayat, 1990	Karnataka (Hayat, 1990)
17	*Aphelinus argiope* Walker, 1839	Himachal Pradesh and Uttar Pradesh (Hayat, 1986).
18	*Aphelinus asychis* Walker, 1839	Himachal Pradesh, Jammu & Kashmir, Karnataka, Meghalaya, Uttar Pradesh and West Bengal (Hayat, 1986).
19	*Aphelinus basilicus* Hayat, 1998	Andhra Pradesh, Assam, Bihar, Goa, Daman & Diu and West Bengal (Hayat, 1998); Tamil Nadu (Menakadevi and Manickavasagam, 2011).
20	*Aphelinus desantisi* Hayat, 1972a	Nagaland (Tripathi, 1995); Punjab (Hayat, 1972a); Karnataka, Rajasthan and Uttar Pradesh (Hayat, 1986); Tamil Nadu (Menakadevi and Manickavasagam, 2011).
21	*Aphelinus gossypii* Timberlake, 1924	Bihar (Ahmad and Singh 1996); Himachal Pradesh (Thakur et al., 1988); Jammu & Kashmir and Uttar Pradesh (Hayat, 1986).
22	*Aphelinu shumilis* Mercet, 1927	Uttar Pradesh and West Bengal (Hayat, 1986).
23	*Aphelinus lankaensis* Hayat, 1994	Karnataka (Hayat, 1994).
24	*Aphelinus mali* Haldeman, 1851	Andhra Pradesh, Himachal Pradesh, Jammu & Kashmir, Kerala and Tamil Nadu (Hayat, 1998).
25	*Aphelinus megadontus* Hayat, 1998	Sikkim (Hayat, 1998).
26	*Aphelinus polaszeki* Hayat, 1998	Himachal Pradesh (Hayat, 1998).
27	*Aphelinus semiflavus* Howard, 1908	India (Hayat, 1998).
28	*Aphelinus sharpae* Hayat, 1998	Assam and Meghalaya (Hayat, 1998).
29	*Aphytis alami* Agarwal, 1964	Uttar Pradesh (Agarwal, 1964).
30	*Aphytis aligarhensis* Hayat, 1998	Uttar Pradesh (Hayat, 1998).
31	*Aphytis bangalorensis* Rosen and DeBach, 1986	Karnataka (Rosen and DeBach, 1986).
32	*Aphytis chrysomphali* Mercet, 1912a	India (Hayat, 1986).
33	*Aphytis diaspidis* Howard, 1881	Bihar, Jammu & Kashmir and Uttar Pradesh (Hayat, 1998).
34	*Aphytis fioriniae* Rosen and Rose, 1989	Assam (Rosen and Rose, 1989).
35	*Aphytis fisheri* DeBach, 1959	India (Hayat, 1986).
36	*Aphytis gordoni* DeBach and Rosen, 1976	Karnataka (Hayat, 1998).
37	*Aphytis hispanicus* Mercet, 1912b	Assam and Uttar Pradesh (Hayat, 1998).
38	*Aphytis holoxanthus* DeBach, 1960	Karnataka and Tamil Nadu (Hayat, 1986).

TABLE 11.1 (Continued)
The Checklist of Various Aphelinid Species Recorded in India

Sl. No	Species	Distribution
39	*Aphytis lepidosaphes* Compere, 1955	India (Hayat, 1986).
40	*Aphytis linganensis* Compere, 1955	Assam, Delhi, Karnataka, Kerala, Tamil Nadu and Uttar Pradesh (Hayat, 1998).
41	*Aphytis maculatus* Shafee, 1970	Tamil Nadu (Hayat, 1986).
42	*Aphytis maculicornis* Masi, 1911	India (Hayat, 1986).
43	*Aphytis manu* Hayat, 1998	Karnataka (Hayat, 1998)
44	*Aphytis melinus* DeBach, 1959	Delhi, Karnataka, Maharashtra, Pondicherry and Tamil Nadu (DeBach, 1959).
45	*Aphytis mytilaspidis* Le Barron, 1870	India (Hayat, 1986).
46	*Aphytis paramaculicornis* DeBach and Rosen, 1976	Jammu & Kashmir (Amin Masoodi andTrali, 1987).
47	*Aphytis peculiaris* Girault, 1932	Bihar and Delhi (Hayat, 1998).
48	*Aphytis philippiensis* DeBach and Rosen, 1976	Assam, Karnataka and Tamil Nadu (DeBach and Rosen, 1976, Fry, 1989, Hayat, 1986 and Rosen and DeBach, 1979).
49	*Aphytis proclia* Walker, 1839	Himachal Pradesh (Hayat, 1998) and Jammu and Kashmir (Amin Masoodi and Trali, 1987).
50	*Aphytis sankarani* Rosen and DeBach, 1986	Karnataka (Fry, 1989 & Rosen and DeBach, 1986).
51	*Aphytis theae* Cameron, 1891	Assam and Himachal Pradesh (Hayat, 1998).
52	*Botryoideclava bharatiya* Subba Rao, 1980	Maharashtra (Jadhav and Varma, 2001); Uttar Pradesh (Hayat, 1998); Andhra Pradesh (Manickavasagam and Menakadevi, 2012).
53	*Centrodora azimi* Hayat, 1973	Andhra Pradesh (Hayat, 1986).
54	*Centrodora bahara* Hayat, 2010	Uttar Pradesh (Hayat, 2010).
55	*Centrodora bifasciata* Hayat,1988	Assam (Hayat, 1988).
56	*Centrodora flemingiae* Hayat, 2012	Jharkhand
57	*Centrodora ghorpadei* Hayat, 1998	Karnataka (Hayat, 1998).
58	*Centrodora hamira* Hayat, 2010	Himachal Pradesh and Odisha (Hayat, 2010).
59	*Centrodora hexatricha* Erdos and Novicky, 1953	Assam (Hayat and Zeya, 1993).
60	*Centrodora idioceri* Ferriere, 1931	Assam, Bihar, Goa, Daman & Diu, Jammu & Kashmir, Tamil Nadu and Uttar Pradesh (Hayat, 1998).
61	*Centrodora lakma* Hayat, 2010	Uttar Pradesh (Hayat, 2010).
62	*Centrodora lineascapa* Hayat, 1998	Uttar Pradesh (Hayat, 1998), Kerala, Tamil Nadu (Menakadevi and Manickavasagam, 2011).
63	*Centrodora mala* Hayat, 2010	Odisha (Hayat, 2010).
64	*Centrodora mandiana* Hayat, 2010	Himachal Pradesh (Hayat, 2010).
65	*Centrodora oophaga* Girault, 1917	Tamil Nadu (Hayat, 1998).
66	*Centrodora ossira* Hayat, 2010	Odisha (Hayat, 2010).
67	*Centrodora pathania* Hayat, 2010	Punjab (Hayat, 2010).
68	*Centrodora tetratricha* Hayat, 2010	Uttar Pradesh (Hayat, 2010).
69	*Centrodora uttara* Hayat, 2010	Uttar Pradesh (Hayat, 2010).
70	*Coccobius amala* Hayat and Khan, 2012	Uttar Pradesh (Hayat and Khan, 2012).
71	*Coccobius aligarhensis* (Hayat, 1974a)	Uttar Pradesh (Hayat, 1998).
72	*Coccobius aspidiellae* Hayat, 2010	Kerala (Hayat and Khan, 2010).
73	*Coccobius assamensis* Hayat, 2010	Assam (Hayat and Khan, 2010).
74	*Coccobius ceroplastidis* Agarwal, 1964	Uttar Pradesh (Hayat, 1998).
75	*Coccobius comperei* Hayat, 1971	Karnataka and Kerala (Hayat, 1998).
76	*Coccobius confusus* Compere and Annecke, 1961)	India (Hayat, 1985).
77	*Coccobius cosibus* Hayat, 2010	Himachal Pradesh, Meghalaya and Odisha (Hayat and Khan, 2010), Tamil Nadu (Manickavasagan and Menakadevi, 2012)
78	*Coccobius debachi* Compere and Annecke, 1961	India (Hayat, 1985).
79	*Coccobius flaviceps* Girault, 1915	India (Hayat, 1971).
80	*Coccobius fulvus* Compere and Annecke, 1961	Karnataka (Hayat, 1986) Maharashtra (Agarwal, 1964 & Shafee et al., 1988).

(*Continued*)

TABLE 11.1 (Continued)
The Checklist of Various Aphelinid Species Recorded in India

Sl. No	Species	Distribution
81	*Coccobius himalayanus* Hayat, 2010 (in Hayat and Khan, 2010).	Himachal Pradesh (Hayat and Khan, 2010).
82	*Coccobius himpradianus* Hayat, 2010 (in Hayat and Khan, 2010).	Himachal Pradesh (Hayat and Khan, 2010).
83	*Coccobius mirus* Hayat, 1998	Andhra Pradesh, Assam and Uttar Pradesh (Hayat, 1998), Tamil Nadu (Menakadevi and Manickavsagam, 2011).
84	*Coccobius phoebus* Hayat, 2010 (in Hayat and Khan, 2010)	Himachal Pradesh, Karnataka, Odisha and Uttar Pradesh (Hayat and Khan, 2010).
85	*Coccobius reticulatus* Compere and Annecke, 1961	Andhra Pradesh, Delhi, Karnataka, Kerala, Maharashtra and Uttar Pradesh (Hayat, 1998); Tamil Nadu (Menakadevi and Manickavasagam, 2011).
86	*Coccobius subflavus* Annecke and Insley, 1970	India (Noyes, 2012).
87	*Coccobius udakamundus* Hayat, 1998	Tamil Nadu (Hayat, 1998).
88	*Coccobius zalama* Hayat and Khan, 2012	Uttar Pradesh
89	*Coccophagoides orientalis* Agarwal, 1964	Bihar, Rajasthan, Uttar Pradesh (Hayat, 1998); Tamil Nadu (Menakadevi and Manickavasagam, 2012).
90	*Coccophagoides utilis* Doutt, 1966	Bihar (Hayat, 1998).
91	*Coccophagus acanthosceles* Waterston, 1916	Assam (Hayat, 1986).
92	*Coccophagus anthracinus* Compere, 1925	Delhi (Hayat, 1986)
93	*Coccophagus assamensis* Hayat, 1993	Assam (Hayat, 1993).
94	*Coccophagus atratus* Compere, 1926	Delhi (Hayat, 1986).
95	*Coccophagus bivittatus* Compere, 1926	Haryana, Rajasthan, Tamil Nadu, Uttar Pradesh (Hayat, 1998).
96	*Coccophagus bogoriensis* Koningsberger, 1897	Andhra Pradesh, Assam, Karnataka, Kerala (Hayat, 1998).
97	*Coccophagus burksi* Hayat, 1971	Puducherry (Hayat, 1998); Tamil Nadu (Menakadevi and Manickavasagam, 2011).
98	*Coccophagus candidus* Hayat, 1993	Assam (Hayat, 1993).
99	*Coccophagus ceroplastae* Howard, 1895	widely distributed recorded from several Indian states (Hayat, 1998), Meghalaya (Hayat and Khan, 2010).
100	*Coccophagus chloropulvinariae* Hayat, 1974b	Assam (Hayat, 1993) Uttar Pradesh (Hayat, 1974b).
101	*Coccophagus clavatus* Husain and Agarwal, 1982	Andhra Pradesh (Husain and Agarwal, 1982).
102	*Coccophagus copernicus* Husain and Agarwal, 1982	Andhra Pradesh (Husain and Agarwal, 1982).
103	*Coccophagus cowper*, ?????????????1917	Andhra Pradesh, Goa, Kerala, Karnataka, Maharashtra, Puducherry and Tamil Nadu (Hayat, 1998).
104	*Coccophagus dius* Hayat, 1998	Karnataka (Hayat, 1998).
105	*Coccophagus flavicorpus* Husain and Agarwal, 1982	Andhra Pradesh (Husain and Agarwal, 1982).
106	*Coccophagu flavifrons* Howard, 1885	Delhi (Hayat, 1986).
107	*Coccophagus formicariae* Hayat, 1998	Assam (Hayat, 1998).
108	*Coccophagus fumadus* Hayat, 2010 (in Hayat and Khan, 2010).	Odisha (Hayat and Khan, 2010); Kerala (Menakadevi and Manickavasagan, 2011).
109	*Coccophagus gilvus* Hayat, 1971	Goa (Hayat, 1998); Tamil Nadu (Menakadevi and Manickavasagam, 2011).
110	*Coccophagus indochraceus* Hayat, 1998	Uttar Pradesh (Hayat, 1998).
111	*Coccophagus insignis* Hayat and Zeya, 1993	Assam (Hayat and Zeya, 1993).
112	*Coccophagus longiclavatus* Shafee, 1972	Assam, Bihar, Karnataka, Uttar Pradesh (Hayat, 1998).
113	*Coccophagus longicornis* Hayat, 1971	Punjab (Hayat, 1998); Kerala (Menakadevi and Manickavasagam, 2011).
114	*Coccophagus longifasciatus* Howard, 1907	Odisha (Hayat and Khan, 2010).
115	*Coccophagus longipedicellus* Shafee, 1972	Andaman & Nicobar Islands, Karnataka (Hayat, 1998); Tamil Nadu (Menakadevi and Manickavasagam, 2011).
116	*Coccophagus lutescens* Compere, 1931	Rajasthan (Hayat, 1998).
117	*Coccophagus lycimnia* Walker, 1839	Karnataka (CIBC, 1976).
118	*Coccophagus mangiferae* Dozier, 1932	Tamil Nadu, Uttar Pradesh (Hayat and Viggiani, 1984).
119	*Coccophagus meghaianus* Hayat, 2010 (in Hayat and Khan, 2010).	Meghalaya (Hayat and Khan, 2010).

TABLE 11.1 (Continued)
The Checklist of Various Aphelinid Species Recorded in India

Sl. No	Species	Distribution
120	*Coccophagus narendrani* Hayat and Zeya, 1993	Kerala (Hayat and Zeya, 1993).
121	*Coccophagus nigricorpus* Shafee, 1972	Andhra Pradesh (Hayat, 1998).
122	*Coccophagus ochraceus* Howard, 1895	Uttar Pradesh (Hayat, 1986).
123	*Coccophagus ophicus* Husain and Agarwal, 1982	Andhra Pradesh (Husain and Agarwal, 1982).
124	*Coccophagus parlobatae* Hayat, 2007	Karnataka (Hayat, 2007).
125	*Coccophagus perhispidus* Girault, 1926	Delhi (Hayat, 1986).
126	*Coccophagus prinslooi* Hayat, 1998	Karnataka (Hayat, 1998).
127	*Coccophagus pseudococci* Compere, 1933	Andaman & Nicobar Islands, Bihar, Delhi, Goa, Himachal Pradesh, Karnataka, Maharashtra, Punjab, Rajasthan, Tamil Nadu, Uttar Pradesh (Hayat, 1998).
128	*Coccophagus samarae* Hayat, 1998	Tamil Nadu (Hayat, 1998).
129	*Coccophagus sexvittatus* Hayat, 1974b	Madhya Pradesh (Hayat, 1998).
130	*Coccophagus shafeei* Hayat, 1974b	Goa, Karnataka (Hayat, 1998).
131	*Coccophagus shillongensis* Hayat and Singh, 1989	Meghalaya (Hayat and Singh, 1989).
132	*Coccophagus silvestrii* Compere, 1931	Uttar Pradesh (Hayat, 1998).
133	*Coccophagus subflavescens* Hayat, 1971	Uttar Pradesh (Hayat, 1998).
134	*Coccophagus sudhiri* Hayat, 1993	Assam (Hayat, 1993).
135	*Coccophagus tschirchii* Mahdihassan, 1923	Assam, Bihar, Jammu & Kashmir, Karnataka, Punjab, West Bengal (Hayat, 1998).
136	*Coccophagus zebratus* Howard, 1907	Karnataka (Hayat, 2012)
137	*Coccophagus zeyai* Hayat, 1998	Tamil Nadu (Hayat, 1998).
138	*Diaspiniphagus decorus* Hayat, 2011a	Uttar Pradesh (Hayat, 2011a).
139	*Encarsia acaudaleyrodis* Hayat, 1976	Delhi, Rajasthan, Tamil Nadu (Hayat, 1976); Uttar Pradesh (Hayat, 2011b)
140	*Encarsia albiscutellum* Girault, 1913	Uttar Pradesh (Hayat, 2011b).
141	*Encarsia aseta* Hayat and Polaszek, 1992	Maharashtra (Hayat, 1998); Uttar Pradesh (Hayat, 2011b).
142	*Encarsia asfar* Hayat, 2011b	Uttar Pradesh (Hayat, 2011b).
143	*Encarsia assamensis* Hayat, 2011b	Assam (Hayat, 2011b).
144	*Encarsia asterobemisiae* Viggiani and Mazzone, 1980	Himachal Pradesh (Hayat, 2011b).
145	*Encarsia aurantii* Howard, 1894	Andhra Pradesh, Karnataka (Hayat, 1998), Tamil Nadu (Menakadevi and Manickavasagam, 2011).
146	*Encarsia azimi* Hayat, 1986	Himachal Pradesh, Punjab, Tamil Nadu, Uttar Pradesh, Odisha (Hayat, 1998); Meghalaya, Assam, Uttarkhand, Karnataka (Hayat, 2011b).
147	*Encarsia bangalorensis* Hayat, 1989	Karnataka (Hayat, 1998).
148	*Encarsia bennetti* Hayat, 1984	Encarsiabennetti Hayat, 1984
149	*Encarsia bimaculata* Heratyand Polaszek, 2000	Distribution: Assam, Karnataka, Maharashtra, Odisha and West Bengal (Hayat, 2011b).
150	*Encarsia bolangera* Hayat, 2011b	Karnataka (Hayat, 2011b).
151	*Encarsia borkusiensis* Hayat, 2011b	Assam (Hayat, 2011b).
152	*Encarsia boswelli* Girault, 1915	Karnataka, Tamil Nadu (Polaszek and Hayat, 1992).
153	*Encarsia brevicalcar* Hayat, 2011b	Goa (Hayat, 2011b).
154	*Encarsia brevivalvula* Hayat, 1989	Uttar Pradesh (Hayat, 1998), Tamil Nadu (Menakadevi and Manickavasagam, 2011).
155	*Encarsia brevivena* Hayat, 1989	Karnataka (Hayat, 1989).
156	*Encarsia brimblecombei* Girault, 1933	Assam (Hayat, 1989).
157	*Encarsia cibcensis* Lopez-Avila, 1987	India (Schmidt et al., 2001).
158	*Encarsia citrina* Craw, 1891	Meghalaya, Karnataka (Hayat, 2011b).
159	*Encarsia clypealis* Silvestri, 1927	Andhra Pradesh, Assam, Karnataka, Maharashtra, Tamil Nadu, Uttar Pradesh (Hayat, 1998).
160	*Encarsia coimbatorensis* Hayat, 1989	Tamil Nadu (Hayat, 1998).
161	*Encarsia confusa* Hayat, 1989	Bihar and Uttar Pradesh (Hayat, 1998).
162	*Encarsia colima* Myartseva, 2005	Tamil Nadu (Menakadevi & Manickavsagam, 2011).

(Continued)

TABLE 11.1 (Continued)
The Checklist of Various Aphelinid Species Recorded in India

Sl. No	Species	Distribution
163	*Encarsia davidi* Viggiani and Mazzone, 1980	Rajasthan and Uttar Pradesh (Hayat, 2011b).
164	*Encarsia dialeurodis* Hayat, 1989	Uttar Pradesh (Hayat, 1998).
165	*Encarsia diaspidicola* Silvestri, 1909	Meghalaya (Hayat, 2011b).
166	*Encarsia dispersa* Polaszek, 2004 (In Polaszek et al., 2004)	Karnataka, Kerala and Lakshadweep Islands (Hayat, 2011b).
167	*Encarsia divergens* Silvestri, 1926	Assam (Hayat, 1998).
168	*Encarsia duorunga* Hayat, 1989	Uttar Pradesh (Hayat, 1998), Tamil Nadu (Menakadevi and Manickavasagam (2012); Assam, Meghalaya, Uttarakhand and West Bengal (Hayat, 2011b).
169	*Encarsia elegans* Masi, 1911	Bihar, Delhi and Uttar Pradesh (Hayat, 1998); Assam, Himachal Pradesh, Meghalaya and Uttarakhand (Hayat, 2011a).
170	*Encarsia elongata* Dozier, 1937	Assam (Hayat, 1998).
171	*Encarsia farmani* Hayat, 2011b	Uttarakhand (Hayat, 2011b).
172	*Encarsia flavoscutellum* Zehntner, 1900	Assam (Hayat, 1998).
173	*Encarsia guadeloupae* Viggiani, 1987	Bangaluru (Antony et al., 2006); Andhra Pradesh, Kerala (Hayat, 2011b); Tamil Nadu (Menakadevi and Manickavasagam, 2011).
174	*Encarsia gunturensis* Azim and Shafee, 1980	Andhra Pradesh, Tamil Nadu and Uttar Pradesh (Hayat, 1998).
175	*Encarsia hitam* Hayat, 2011b	Meghalaya, Uttarakhand, Uttar Pradesh (Hayat, 2011b); Tamil Nadu (Menakadevi and Manickavasagam, 2011).
176	*Encarsia hragila* Hayat, 2011b	Uttar Pradesh (Hayat, 2011b).
177	*Encarsia inaron* Walker, 1839	Karnataka, Uttar Pradesh (Hayat, 2011b)
178	*Encarsia indica* Shafee, 1973	Puducherry (Hayat, 1998).
179	*Encarsia inida* Hayat, 2011b	Sikkim (Hayat, 2011b).
180	*Encarsia interstrica* Huang and Polaszek, 1998	West Bengal (Hayat, 2011b).
181	*Encarsia isaaci* Mani, 1941	Andhra Pradesh, Odisha, Tamil Nadu and Uttar Pradesh (Hayat, 1998).
182	*Encarsia ixorae* Krishnan and Vasantharaj David, 1996	Tamil Nadu (Hayat, 1998).
183	*Encarsia jowaiana* Hayat, 2011b	Meghalaya (Hayat, 2011b).
184	*Encarsia khadijae* Hayat, 2011b	Odisha, Uttarakhand, Uttar Pradesh (Hayat, 2011b).
185	*Encarsia lahorensis* Howard, 1911	Bihar (Hayat, 1998).
186	*Encarsia levo* Hayat, 200	West Bengal (Hayat, 2006).
187	*Encarsia lipaleyrodis* Krishnan and Vasantharaj David, 1996	Tamil Nadu (Krishnan and Vasantharaj David, 1996).
188	*Encarsia liliyingae* Viggiani and Ren	Assam (Hayat, 1998).
189	*Encarsia longicauda* Hayat, 1989	Andhra Pradesh, Rajasthan and Tamil Nadu (Hayat, 1998) Orissa (Hayat, 2011b).
190	*Encarsia longifasciata* Subba Rao, 1984	Karnataka (Hayat, 1998).
191	*Encarsia lounsburyi* Berlese and Paoli, 1916	Sikkim (Hayat, 2011b).
192	*Encarsia lutea* Masi, 1909	Bihar and Uttar Pradesh (Hayat, 1998)
193	*Encarsia macroptera* Viggiani, 1985	Maharashtra, Punjab and Uttar Pradesh (Hayat, 1998).
194	*Encarsia meghalayana* Hayat, 2011b	Meghalaya (Hayat, 2011b).
195	*Encarsia merceti* Silvestri, 1926	Assam, Delhi (Hayat, 1998).
196	*Encarsia muliyali* Mani, 1941	Odisha, Uttar Pradesh (Hayat, 1998); Tamil Nadu (Menakadevi and Manickavasagam, 2011).
197	*Encarsia nanjingensis* Huang and Polaszek, 1998	Assam (Hayat, 1998).
198	*Encarsia narayanani* Agarwal, 1964	Karnataka, Maharashtra, Odisha and Uttar Pradesh (Hayat, 1998).
199	*Encarsianorani* Hayat, 1989	Uttar Pradesh (Hayat, 1998).
200	*Encarsia obtusiclava* Hayat, 1989	Karnataka, Uttar Pradesh (Hayat, 1998); Tamil Nadu (Menakadevi and Manickavasagam, 2011).
201	*Encarsia occultans* Hayat, 1998	Bihar, Odisha, Uttar Pradesh (Hayat, 1998).
202	*Encarsia orangae* Raina et al., 1996	Maharashtra (Raina et al., 1996).
203	*Encarsia ossira* Hayat, 2011b	Odisha (Hayat, 2011b).

TABLE 11.1 (Continued)
The Checklist of Various Aphelinid Species Recorded in India

Sl. No	Species	Distribution
204	*Encarsia perfl* ava Hayat, 1989	Uttar Pradesh (Hayat, 1998), Assam, Madhya Pradesh and Uttarakhand (Hayat, 2011b).
205	*Encarsia perniciosi* Tower, 1913	Jammu & Kashmir, Himachal Pradesh, Punjab, Uttar Pradesh (Hayat, 1998).
206	*Encarsia perplexa* Huang and Polaszek, 1998	Uttar Pradesh (Hayat, 2006).
207	*Encarsia pseudococci* Agarwal, 1964	Uttar Pradesh (Agarwal, 1964)
208	*Encarsia punicae* Hayat, 1998	Andhra Pradesh, Karnataka (Hayat, 1998); Uttar Pradesh (Hayat, 2011b).
209	*Encarsia pura* Hayat, 1998	Maharashtra (Hayat, 1998).
210	*Encarsia reeda* Hayat, 2011b	Assam, Meghalaya, Uttarakhand (Hayat, 2011b).
211	*Encarsia sankarani* Hayat, 1989	Assam (Hayat, 1989).
212	*Encarsia septentrionalis* Hayat, 1989	Bihar, Delhi, Uttar Pradesh (Hayat, 1989).
213	*Encarsia sikkimiana* Hayat, 2011a	Sikkim (Hayat, 2011b).
214	*Encarsia smithi* Silvestri, 1926	Bihar, Karnataka, Maharashtra (Hayat, 1998).
215	*Encarsia Sophia* Girault, 1915	India.
216	*Encarsia strenua* Silvestri, 1927	Uttar Pradesh and West Bengal (Noyes, 2012).
217	*Encarsia terebrator* Shafee, 1974	Andhra Pradesh (Hayat, 1998).
218	*Encarsia tinctoriae* Krishnan and VasantharajDavid, 1996	Tamil Nadu (Krishnan and Vasantharaj David, 1996).
219	*Encarsia tristis* Zehntner, 1896	Bihar, Delhi, Karnataka, Maharashtra, Punjab, Tamil Nadu, Uttar Pradesh (Hayat, 1998)
220	*Encarsia trivittata* Hayat, 1989	Uttar Pradesh (Hayat, 1989).
221	*Encarsia udaipuriensis* Shafee, 1973	Bihar, Odisha, Rajasthan, Tamil Nadu and Uttar Pradesh (Hayat, 1998); Meghalaya (Hayat, 2011b).
222	*Eretmocerus adustiscutum* Krishnan and Vasantharaj David, 1996	Karnataka, Kerala, Maharashtra, Tamil Nadu and Uttar Pradesh (Krishnan and Vasantharaj David, 1996).
223	*Eretmocerus bisetae* Hayat, 1998	Uttar Pradesh (Hayat, 1998).
224	*Eretmocerus breviclavus* Subba Rao, 1984	Karnataka (Subba Rao, 1984).
225	*Eretmocerus corni* Haldeman, 1851	Maharashtra (Hayat, 1986).
226	*Eretmocerus delhiensis* Mani, 1941	Bihar, Delhi, Rajasthan and Uttar Pradesh (Hayat, 1998).
227	*Eretmocerus dialeurolongae* Krishnan and Vasantharaj David, 1996	Tamil Nadu (Krishnan and Vasantharaj David, 1996).
228	*Eretmocerus flavus* Krishnan and Vasantharaj David, 1996	Tamil Nadu (Krishnan and Vasantharaj David, 1996).
229	*Eretmocerus gunturiensis* Hayat, 1972b	Andhra Pradesh and Maharashtra (Hayat, 1998).
230	*Eretmocerus haldemani* Howard, 1908	Uttar Pradesh (Hayat, 1986).
231	*Eretmocerus hayati* Zolnerowich and Rose, 1998	India (Noyes, 2012).
232	*Eretmoceru shydrabadensis* Husain and Agarwal, 1982	Andhra Pradesh (Husain and Agarwal, 1982).
233	*Eretmocerus indicus* Hayat, 1972	Uttar Pradesh (Hayat, 1998).
234	*Eretmocerus longiscapus* Hayat, 1998	Uttar Pradesh (Hayat, 1998).
235	*Eretmocerus mundus* Mercet, 1931	Andhra Pradesh, Bihar, Maharashtra, Tamil Nadu, Uttar Pradesh (Hayat, 1998).
236	*Eretmocerus paulistus* Hempel, 1904	Uttar Pradesh (Hayat, 1986).
237	*Eretmocerus rajasthanicus* Hayat, 1976	Rajasthan (Hayat, 1998); Tamil Nadu (Menakadevi and Manickavasagam, 2011).
238	*Eretmocerus sculpturatus* Hayat, 1998	Bihar (Hayat, 1998).
239	*Eretmocerus serius* Silvestri, 1927	Andhra Pradesh, Karnataka and Uttar Pradesh (Hayat, 1998).
240	*Eretmocerus trialeurodis* Hayat, 1998	Tamil Nadu (Hayat, 1998).
241	*Eriaphytis chackoi* Subba Rao, 1980	Karnataka (Hayat, 1998).
242	*Eriaphytis orientalis* Hayat, 1972c	Kerala, Rajasthan, Tamil Nadu and Uttar Pradesh (Hayat, 1998).
243	*Euryischia indica* Mani and Kurian, 1953	Andhra Pradesh, Bihar, Tamil Nadu, Uttar Pradesh (Hayat, 1998).
244	*Euryischomyiawashingtoni* Girault, 1914	Andhra Pradesh, Karnataka and Tamil Nadu (Hayat, 1998).
245	*Eutrichosomella indica* Singh and Srinivasa, 2010.	Western Ghats (Singh and Srinivasa, 2010).

(Continued)

TABLE 11.1 (Continued)
The Checklist of Various Aphelinid Species Recorded in India

Sl. No	Species	Distribution
246	*Eutrichosomella keralensis* Manickavasagam and Menakadevi, 2012	Kerala (Manickavasagam and Menakadevi, 2012).
247	*Idiococcobius encarsoides* Hayat, 2010	West Bengal (Hayat and Khan, 2010)
248	*Marietta carnesi* (Howard, 1910)	Jammu & Kashmir, Uttar Pradesh (Hayat, 1998).
249	*Marietta leopardina* Motschulsky, 1863	Throughout India including Andaman & Nicobar Islands (Hayat, 1998).
250	*Marietta picta* Andre, 1878	Uttar Pradesh (Hayat, 1998).
251	*Marlattiella maculata* Hayat, 1974a	Uttar Pradesh (Hayat, 1974a).
252	*Myiocnema comperei* Ashmead, 1900	India (Thompson, 1955).
253	*Paraphytis maculatus* Shafee, 1970	Tamil Nadu (Shafee, 1970).
254	*Paraphytis pecularis* Girault, 1932	Bihar and Delhi (Girault, 1932)
255	*Paraphytis transversa* Huang, 1994	Kerala (Hayat, 2012).
256	*Proaphelinoides anomalus* Hayat, 1985	Karnataka (Hayat, 1986).
257	*Proaphelinoides assamensis* Hayat, 2012	Assam (Hayat, 2012).
258	*Proaphelinoides chidambaramensis* Manickavasagam and Menakadevi, 2012 (in Menakadevi and Manickavasagam, 2012).	Tamil Nadu (Menakadevi and Manickavasagam, 2012).
259	*Promuscideaun fasciativentris* Girault, 1917	Andaman & Nicobar Islands, Andhra Pradesh, Assam, Bihar, Delhi, Jammu & Kashmir, Karnataka, Kerala, Madhya Pradesh, Odisha, Punjab, Tamil Nadu, Uttar Pradesh, West Bengal (Hayat, 1998).
260	*Protaphelinus nikolskajae* Yasnosh, 1963	Himachal Pradesh, Jammu & Kashmir (Hayat, 1998).
261	*Pteroptri xalbifemur* Girault, 1915	Karnataka (Hayat, 1998).
262	*Pteroptrix aster* Hayat, 2010	Uttar Pradesh (Hayat and Khan, 2010).
263	*Pteroptrix bemisiae* Mani	Bihar (Hayat, 1986).
264	*Pteroptrix chinensis* Howard, 1907	Jammu & Kashmir (Hayat, 1998).
265	*Pteroptrix koebelei* Howard, 1907	Karnataka (Hayat, 1998).
266	*Pteroptrix machiaveli* (Girault, 1915)	Karnataka (Hayat, 1998); Uttar Pradesh, Odisha, Meghalaya, Assam (Hayat and Khan, 2010); Tamil Nadu (Menakadevi and Manickavasagam, 2011).
267	*Pteroptrix matpada* Hayat, 2010 Hayat and Khan, 2010	Orissa (Hayat and Khan, 2010).
268	*Pteroprtix longiclavata* Shafee et al., 1988	Utttar Pradesh (Shafee et al., 1988)
269	*Pteroptrix zoza* Hayat, 2010	Uttar Pradesh (Hayat and Khan, 2010).

place inside or outside the body of the host. Some species pupate inside the living host in an air-filled pupation chamber. As in the Encyrtidae, there is some indication that the air in this chamber comes from the host's tracheal system (Clausen, 1940). Scale and aphid parasitoids emerge by cutting a hole in the host mummy's integument, but if the scale has a fragile coating, they push their way out from behind it. Adults of some of these species lack functional mandibles (Viggiani, 1984). Overwintering is usually done as a fully developed larva or pupa.

The ambient factors as well as the stages of the host insect determine the length of the developmental period of parasitoid from oviposition to adult emergence. In the laboratory, temperature influences parasitoid development in the appropriate host stage, and generally males have a shorter developmental period when compared to females. For example, the life cycle of *Aphytis chilensis* takes 45–61 days at 19°C, 24–30 days at 24°C and 17–21 days at 28°C. At 24°C, the female takes 15 days and the male takes 13–14 days in *Encarsia pergandiella* (Gerling, 1966). *Aphytis melinus* has the shortest life cycle, lasting 12–13 days at 26.7°C (Rosen and DeBach, 1979). The pupal stage normally takes the same time as all of the larval instars combined to grow.

The temperature threshold for activity varies depending on the stage of development and it is 6–7°C for the larva, 13°C for the pupa and 15–16°C for the adult of *Aphelinus mali* (Bonnemaison, 1965). The aphelinid parasitoids of multivoltine hosts have a very short life cycle (15–35 days) under field conditions, at least during spring and summer, and the number of generations per year varies with local climatic conditions and is unrelated to the host generation cycle. For instance, *Aphelinus mali* has 5–6 generations per year on *Eriosoma lanigerum* in northern Europe and as many as 14–15 generations on *Eriosoma lanigerum* in Israel and India (Bonnemaison, 1965). When compared to their host, the Aphelinid species linked with hosts that develop a restricted number of generations (2–4) each year; have the same number or more generations. *Encarsia lahorensis* has at least 4–6 generations, whereas *Dialeurodes citri* has just 2–3 generations (Viggiani and Mazzone, 1978). However, some aphelinids that live in univoltine hosts have many generations per year. In Japan, *Coccophagus hawaiiensis* has four generations on *Ceroplastes ruhens* compared to its host's single generation (Clausen, 1940). Aphelinids frequently overwinter in various phases of preimaginal development. Internal larval stages, which are highly protected by the host body, may survive even in extremely cold temperatures. *Aphytis, Coccophagus*, *Encarsia*, and other species have been known to overwinter in the adult stage.

11.4 HOSTS OF APHELINID PARASITOIDS

The aphelinid parasitoids are associated with those host insects that are morphologically very similar; they have an oval or semi-oval shape and are typically less than 5 mm long. The majority of them belong to the Sternorrynchan Homoptera (Aleyrodoidea, Aphidoidea, Coccoidea), with a few Auchenorrynchan species. However, systematically unrelated hosts, such as eggs of Orthoptera, Heteropteraand Lepidoptera, and pupae of Diptera may be utilised if they have a similar shape and external structure (Jasnosh, 1979; Viggiani, 1981). Aphelinidae consists of polyphagous, oligophagous, and monophagous species. *Encarsia citrina* is one of the most polyphagous species, whereas *Aphytis lepidosaphes* appears to be highly host-specific, with *Cornuaspis beki* as its only known host. The apparent exceptions in some aphelinids' host relationships necessitate a thorough examination of their morpho-typical adaptations (Viggiani, 1984).

The majority of aphelinid species develop in the nymphal stage of their host insect, whereas only a few are primary egg parasitoids. Each group of species (or even the species of genus) prefers a certain instar stage of their insect hosts. Several species oviposit directly on or into the host in which their larvae develop, whereas parasitoids that oviposit in younger hosts can arrest their development until it reaches the right stage or size. The primary parasitoids of Homoptera rarely finish their development in the first nymphal instar of their host, and the parasitoid larvae develop in the second or subsequent instars of their host in the rare cases when first instar is accepted for oviposition. Though *Encarsia formosa* parasitizes first and second instars of whiteflies, it prefers to feed on third and fourth instar nymphs for their development, and both the host and the parasitoid die before reaching adulthood (Neil et al., 1976).On the other hand, *E. pergandiella* prefers second and third instar nymphs for its development (Gerling, 1966). Small bodies and slender forewings with long fringes characterise species that develop in small hosts, i.e. *Encarsia pergandiella*, *E. citrina*., etc. (Viggiani, 1984). The aphelinid genera infesting various hosts and their stages are summarised in Table 11.2.

11.5 IMPORTANT PARASITOIDS

Aphelinids have been successfully employed in the biocontrol of homopterous insects, such as aphids, whiteflies, armoured scale insects, and soft scales in their natural habitat. The important species found under various genera in the family Aphelinidaeand their diagnostic characters are given below.

TABLE 11.2
Aphelinid Genera Parasitising Stages of Various Insect Hosts

Host	Stage parasitised	Aphelinid genera
Orthoptera		
1. Acrididae	Egg	*Centrodora*
2. Tettigoniidae		
Lepidoptera		
1. Lasiocampidae	Egg	*Ablerus, Azotus*
2. Noctuidae		*Encarsia (males)*
3. Notodontidae		*Cales*
4. Tortricidae		*Encarsia (males)*
Diptera		
1. Cecidomyiidae	Pupa	*Centrodora*
2. Chamaemyiida		*Ablerus. Euryischia,*
		Marietta
Hymenoptera		
1. Aphelinidae	Larva and pupa	*Ablerus. Azotus,*
		Coccophagus. Encarsia
2. Dryinidae		*Centrodora*
3. Encyrtidae		*Ablerus, Azotus*
4. Eulophidae		*Coccophagus, Marietta*
5. Scelionidae		*Physcus*
		Azotus
Heteroptera		*Oolathron*
1. Reduvioidea	Egg	
Homoptera		
1. Auchenorrhyncha	Egg	
2. Cercopidae		*Centrodora*
3. Cicadellidae		*Azotus, Centrodora*
4. Cicadidae		*Centrodora*
5. Fulgoridae		*Centrodora*
6. Issidae		*Azotus*
7. Membracidae		*Azotus, Centrodora*
8. Ricaniidae		*Centrodora*
Sternorrhyncha		
1. Psylloidea	Immature stages	*Encarsia (unpubl. data)*
2. Aleyrodoidea	Immature stages	*Ablerus, A zotus, Cales, Coccophagus, Encarsia, Eretmocerus, Euxanthellu*
Aphidoidea		
Aphididae	Immature and adult stages	*Aphelinus, Marietta*
Callaphididae		*Mesidia, Mesidiopsis*
Chaetophoridae		*Aphelinus, Marietta*
Pemphigidae		*Protaphelinus Aphelinus, Marietta*
Thelaxidae		
Coccoidea		*Aclerdephagus, Botryoideclava, Marietta*
Aclerdida		
Asterolecanidae		*Azotus .Eriaphytis*
		Prococcophagus

TABLE 11.2 (Continued)
Aphelinid Genera Parasitising Stages of Various Insect Hosts

Host	Stage parasitised	Aphelinid genera
Coccidae		*Aneristus. Azotus. Coccophagus. Encarsia. Euxanthellus. Lounsburyia. Marietta. Neocales. Prococcophagus. Promuscidea. Timberlakiella*
Diaspididae	Immature and adult stages	*Ablerus. Aphelosoma. Aphytis. Archenomus. Artas. Azotus. Bardylis. Coccophagoides. Hispaniella. Marietta. Marlattiella. Myocnema. Physcus. Pteroptrix*
Kermococcidae		*Myocnema*
Kerridae		*Coccophagus. Marietta*
Lecaniodiaspididae		*Marietta*
Margarodidae		*Eunotiscus*
Pseudococcidae		*Coccophagus. Eunotiscus. Euryischomyia Marietta. Myocnema. Promuscidea*
Stictococcidae		*Promuscidea*

11.5.1 Genus: *Aphelinus* Dalman

Diagnosis: Female with 6-segmented antennae, formula 1131, F1 and F2 short and transverse, or F1–3 quadrate to longer than broad. Distance between posterior pair of scutellar setae is usually equal to or greater than that between the anterior pair. The forewing is broad and densely setose; the submarginal vein usually with 4 or more setae; the linea calva is distinct, proximally bordered by one to several lines of setae. Mid-tibial spur is usually densely setose (Plate 11.1a).

Species: The sixteen important species recorded in India are *Aphelinus abdominalis (*Dalman), *A. acuminatus* Hayat, *A. albipodus*Hayat & Fatima, *A. ancer* Hayat, *A. argiope* Walker, *A. asychis* Walker, *A.basilicus* Hayat, *A. desantisi* Hayat, *A. gossypii* Timberlake, *A. humilis* Mercet, *A. lankaensis* Hayat, *A. mali* (Haldeman), *A. megadontus* Hayat, *A. polaszeki* Hayat, *A. semiflavus* Howard, and *A. sharpae* Hayat. They are almost exclusively internal parasitoids of aphids (Hemiptera: Aphidoidea) (Hayat, 1998; NBAIR, 2021).

11.5.2 Genus: *Aphytis* Howard

Diagnosis: Most of the species are lemon yellow or paler in colour with various degrees of infuscation; Antenna 6-segmented, formula 1131 or 1122; in some cases 5 or 4 segmented, formula 1121, 1111; Forewing variable, but with venation long, reaching clearly beyond half-length of wing; Linea calva distinct, with setae proximad of linea calva always longer than those on disc; submarginal vein usually with two setae; Posterior margin of propodeum almost always crenulate, which can be observed only at high magnification (Plate 11.1b).

Species: The twenty-two species found associated with scale insects are *Aphytis alami* Agarwal, *A. bangalorensis* Rosen & DeBach, *A. aligarhensis* Hayat, *A. chrysomphali* (Mercet), *A. diaspidis* (Howard), *A. fioriniae* Rosen & Rose, *A. fisheri*DeBach, *A. gordoni* DeBach&Rosen, *A. hispanicus* (Mercet), *A. holoxanthus* DeBach, *A. lepidosaphes* Compere, *A. lingnanensis* Compere, *A. maculicornis* (Masi), *A. manii* Hayat, *A.melinus*DeBach, *A. mytilaspidis* (Le Baron), *A. paramaculicornis* DeBach& Rosen, *A. philippinensis* DeBach& Rosen, *A. pinnaspidis* Rosen &DeBach, *A. proclia* (Walker), *A. sankarani* Rosen &DeBach, *A. theae* (Cameron). They are ectoparasitoids of diaspidid scales (Hemiptera: Diaspididae). They are the most widely used parasitoids in applied biocontrol worldwide. Species identification is often very difficult due to the presence of biological races/strains and the lack of reliable morphological characters (Hayat, 1998; NBAIR, 2021).

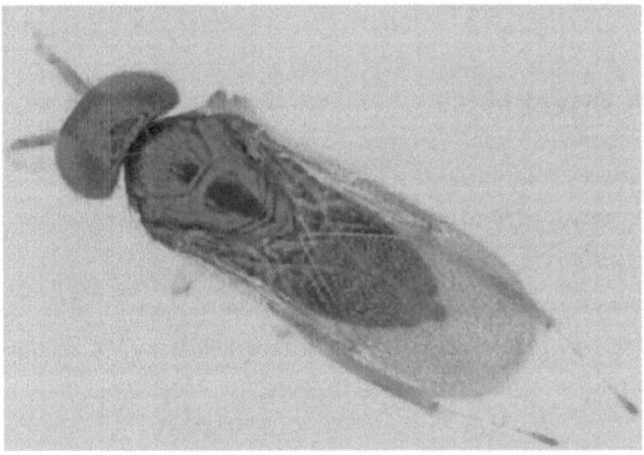

Photo courtesy: NBAIR

Plate 1a. *Aphelinus* Dalman 1981

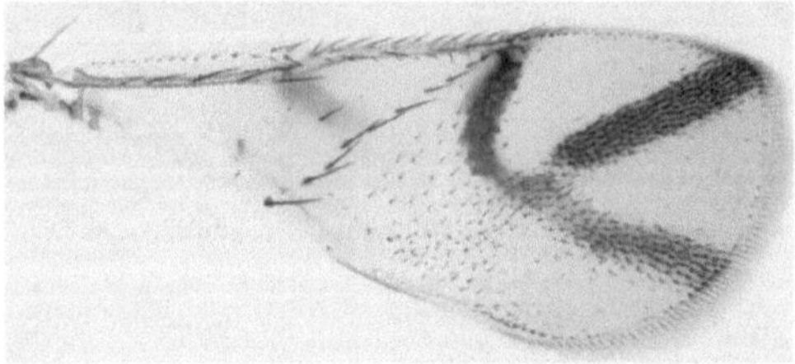

Photo courtesy: NBAIR

Plate 1a. Wing of *Aphelinus* Dalman 1981

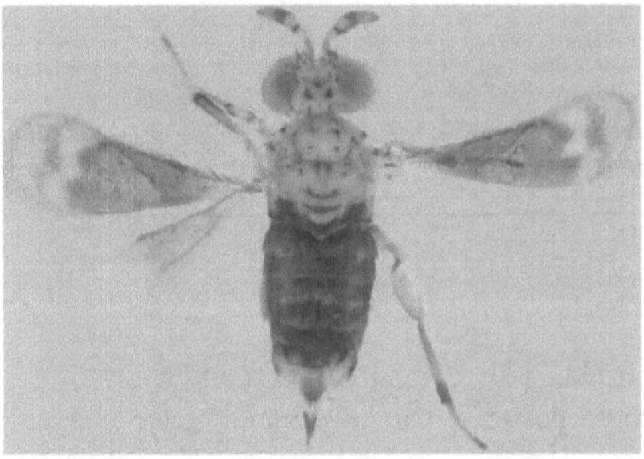

Photo courtesy: NBAIR

Plate 1b. *Aphytis* Howard 1900

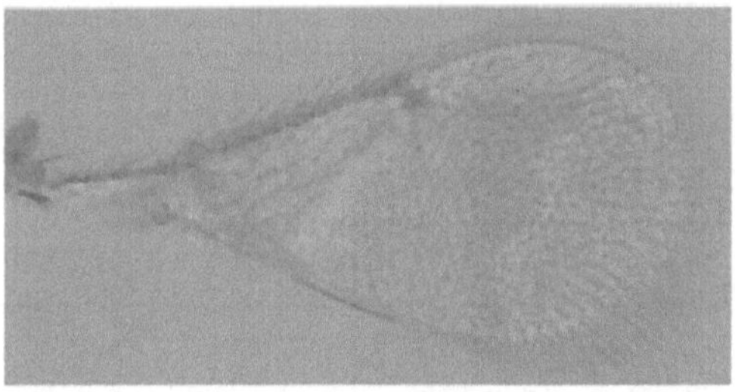

Photo courtesy: NBAIR

Plate 1b. Wing of *Aphytis* Howard 1900

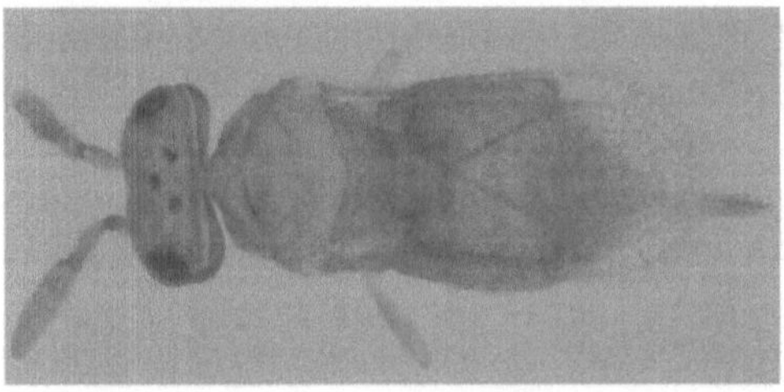

Photo courtesy: NBAIR

Plate 1c. *Botryoideclava* Subba Rao, 1980

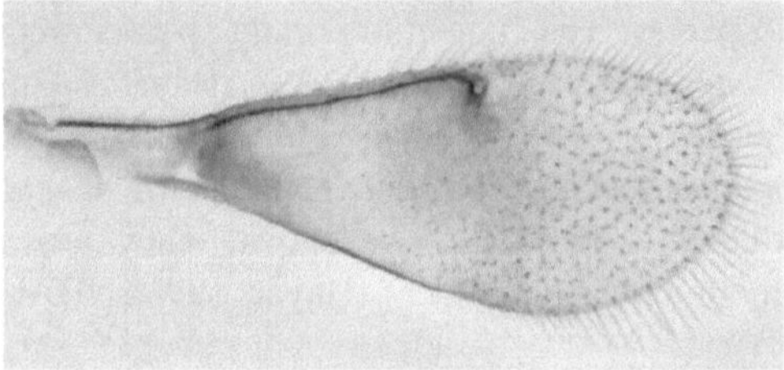

Photo courtesy: NBAIR

Plate 1c. *Botryoideclava* Subba Rao, 1980

Photo courtesy: NBAIR

Plate 1d. *Centrodora* Foerster 1878

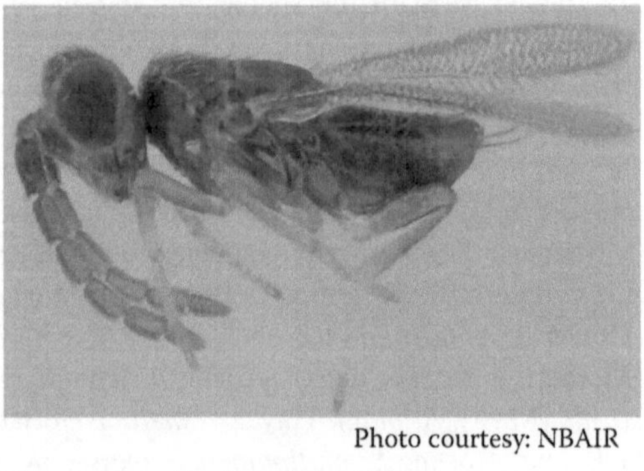

Photo courtesy: NBAIR

Plate 1e. *Coccobius* Ratzeburg, 1852

PLATE 11.1 Important genera of aphelinid parasitoids.

Photo courtesy: NBAIR

Plate 1f. *Coccophagoides* Girault 1915

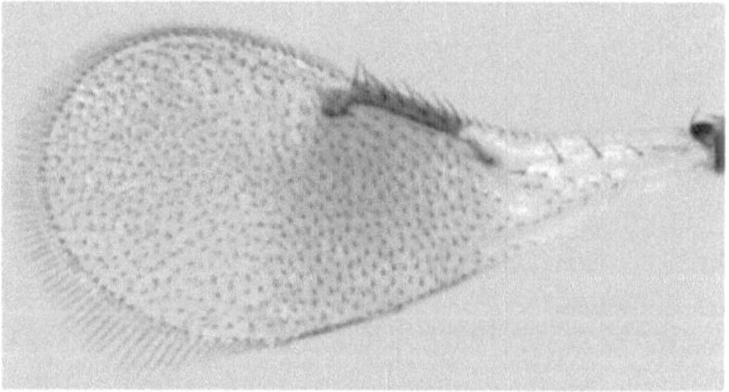

Photo courtesy: NBAIR

Plate 1f. Wing of *Coccophagoides* Girault 1915

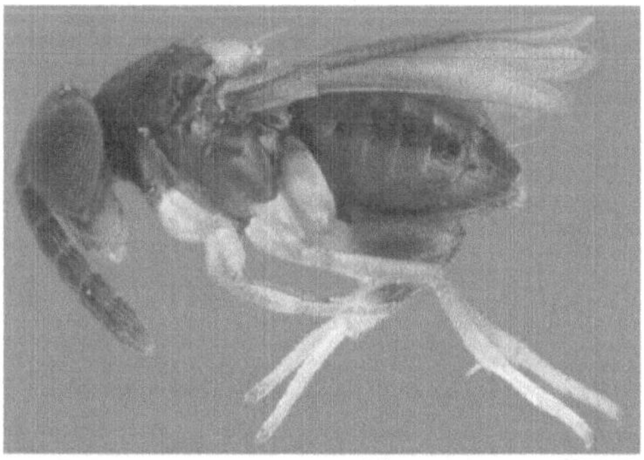

Photo courtesy: NBAIR

Plate 1g. *Coccophagus* Westwood 1833

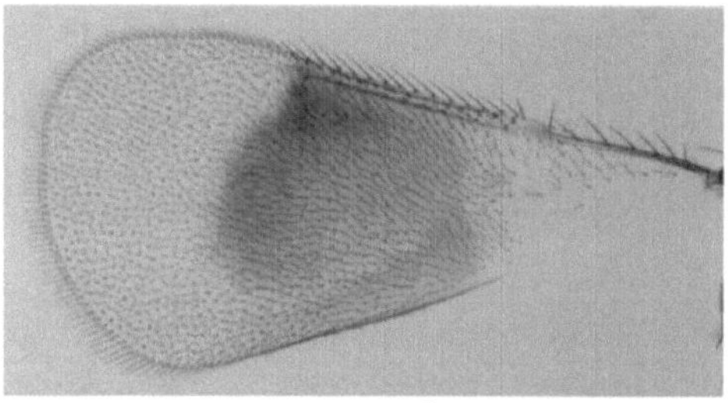

Photo courtesy: NBAIR

Plate 1g. Wing of *Coccophagus* Westwood 1833

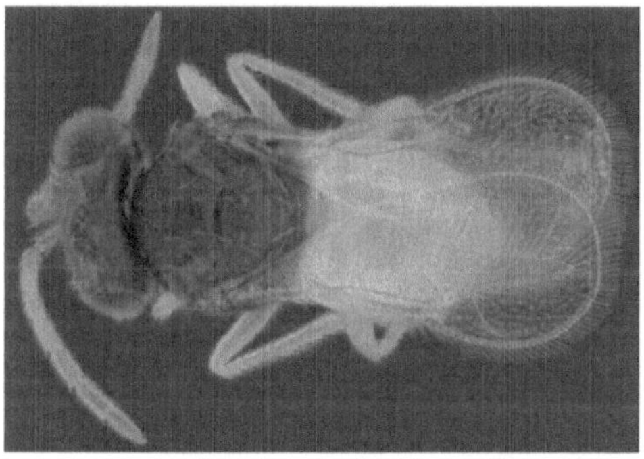

Photo courtesy: NBAIR

Plate 1h. *Encarsia* Foerster 1878

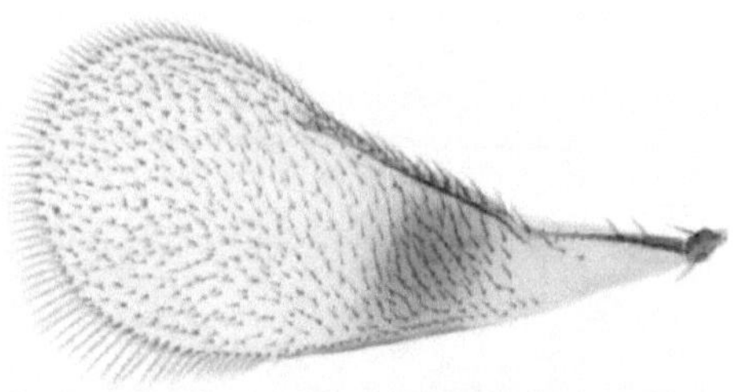

Photo courtesy: NBAIR

Plate 1h. Wing of *Encarsia* Foerster 1878

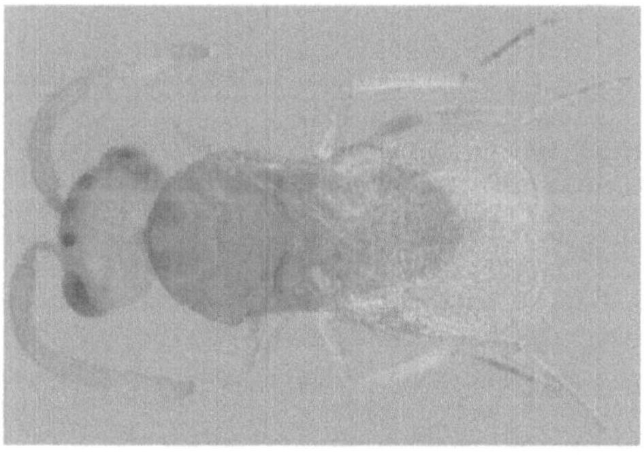

Photo courtesy: NBAIR

Plate 1i. *Eretmocerus* Haldeman 1850

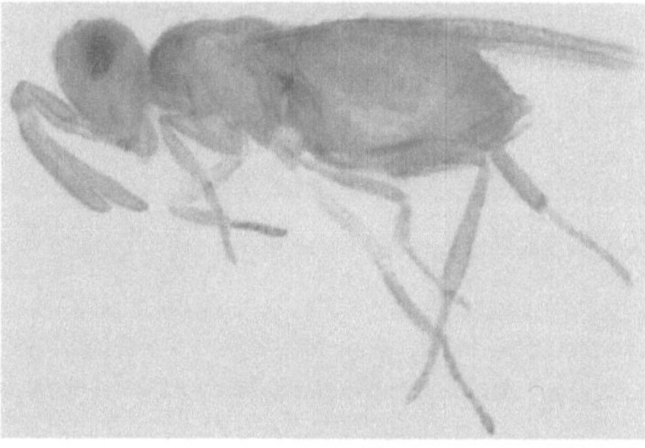

Photo courtesy: NBAIR

Plate 1i. *Eretmocerus* Haldeman 1850

PLATE 11.1 (*continued*)

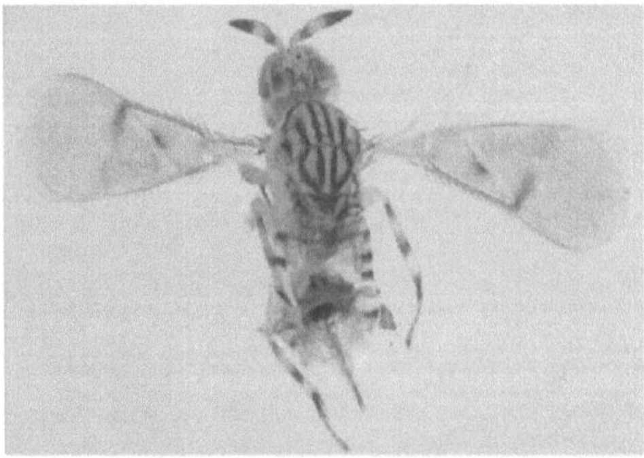

Photo courtesy: NBAIR

Plate 1j. *Paraphytis* Compere 1925

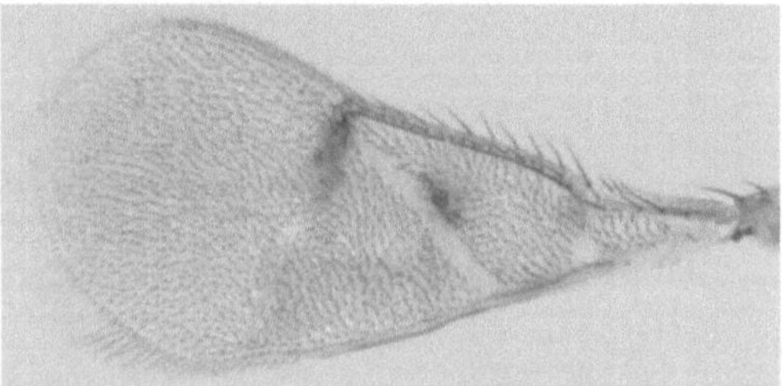

Photo courtesy: NBAIR

Plate 1j. Wing of *Paraphytis* Compere 1925

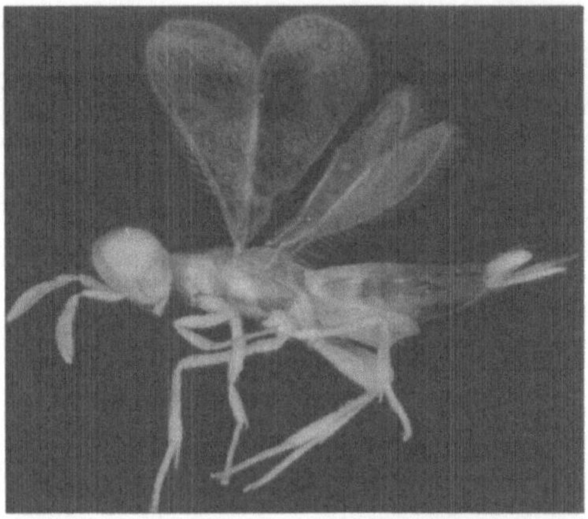

Photo courtesy: NBAIR

Plate 1k. *Proaphelinoides* Girault 1917

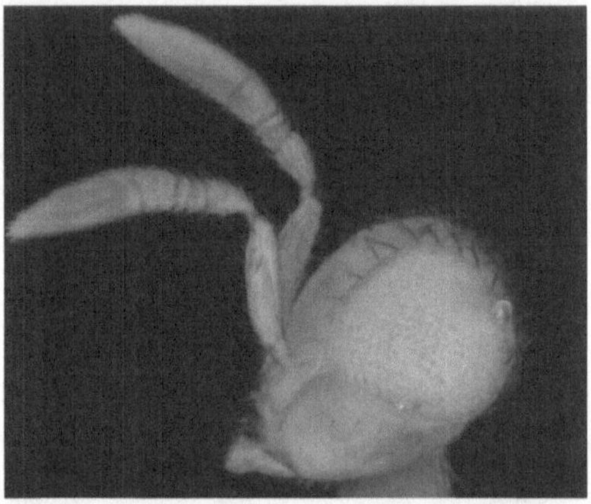

Photo courtesy: NBAIR

Plate 1k. Head of *Proaphelinoides* Girault 1917

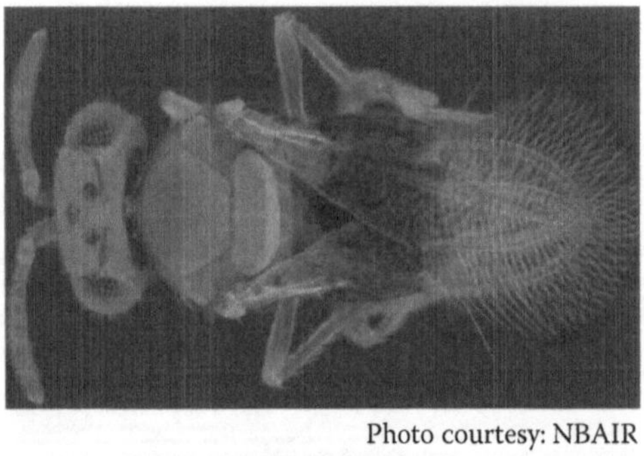

Photo courtesy: NBAIR

Plate 1l. *Pteroptrix* Westwood 1833

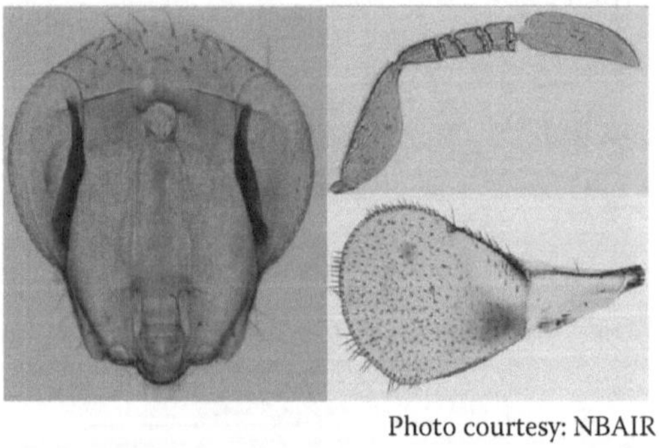

Photo courtesy: NBAIR

Plate 1m. Umairia Hayat 2014
(Head, Antenna and wing)

PLATE 11.1 *(continued)*

11.5.3 Genus: *Botryoideclava* Subba Rao

Diagnosis: They are elongated, somewhat depressed and flattened, ovipositor distinctly exserted; Head with well-developed temples; Antenna 6-segmented with 1122 formula, both funicle segments small, transverse; Clava distinct and large, 2-segmented, usually with numerous, conspicuous, elongate oval sensillae; Pronotum medially divided; Axillae small, slightly projecting forward;

Forewing with venation exceeding mid-length of the forewing, costal cell shorter than the marginal vein, costal cell without setae; submarginal vein with 2 setae; stigmal vein short; linea calva not well defined; Tarsal formula 5-5-5 (Plate 11.1c).

Species: They are distributed in Oriental regions (India and Thailand). The important species recorded in India are *Botryoideclava bharatiya* Subba Rao and *B. thailandica* Hayat. The host insects are *Aclerdata kahashii* Kuwana (Aclerdidae) *Melanaspis glomerata* (Green) (Diaspididae) infesting sugarcane. (Easwaramoorthy et al., 1996; Hayat, 1998; NBAIR, 2021)

11.5.4 Genus: *Centrodora* Foerster

Diagnosis: Their body is slim and elongated; ovipositor long, usually 2.0x or more of the mid-tibia. The forewing is usually long and narrow, with venation not extending 0.5x of wing length; costal cell at least as long as the marginal vein; the submarginal vein with at least 3 setae.

Species: The species found in India are *Centrodora azizi* Hayat, *C. bahara* Hayat, *C. bifasciata* Hayat, *C. flemingiae* Hayat, *C. ghorpadei* Hayat, *C. hamira* Hayat, *C. hexatricha* Erdös & Novicky, *C. idioceri* Ferrière, *C.lakma* Hayat, *C. lineascapa* Hayat, *C. mala* Hayat, *C. mandiana* Hayat, *C. oophaga* (Girault), *C. ossira* Hayat, *C. pathania* Hayat, *C. tetratricha* Hayat, *C. uttara* Hayat. A majority of species are zoophagous, parasitising eggs of Hemiptera, but at least two species are parasitoids of diaspidids. Some species parasitise dipterous larvae, while a few are polyphagous and/or hyperparasitoids (Hayat, 1998; NBAIR, 2021).

11.5.5 Genus: *Coccobius* Ratzeburg

Diagnosis: Their antenna is 7-segmented, anellus minute, funicle with 3 segments, club 2-segmented; Mid lobe of mesoscutum with numerous setae, each axilla hardly projecting with a single seta; Scutellum with a broadly rounded apex and usually with 4–6 setae; Forewing without a linea calva; submarginal vein with a row of four or more setae; marginal vein not short than a costal cell, post marginal vein nearly obsolete or at most very short (Plate 11.1e).

Species: The important species recorded from various parts of India are *Coccobius aligarhensis* (Hayat), *C. aspidiellae* Hayat, *C. assamensis* Hayat, *C. ceroplastidis* (Agarwal), *C. comperei* (Hayat), *C. confusus* (Compere &Annecke), *C. cosibus* Hayat, *C. debachi* (Compere &Annecke), *C. flaviceps* (Girault& Dodd), *C. fulvus* (Compere &Annecke), *C. himalayanus* Hayat, *C. himpradianus* Hayat, *C. mirus* Hayat, *C. phoebus* Hayat, *C. reticulatus* (Compere &Annecke), *C. subflavus* (Annecke&Insley), *C. udakamundus* Hayat, *C. zalama* Hayat & Khan. The females are primary parasitoids in diaspidid scale insects (Hemiptera: Diaspididae). Males, bred from diaspidids or other coccoid hosts, are generally hyperparasitoids on other primary parasitoids, including females of their own species (Hayat, 1998; NBAIR, 2021).

11.5.6 Genus: *Coccophagoides* Girault

Diagnosis: Theirantennae 8-segmented with antennal formula 1133 or 116, usually with an anellus; flagellum elongate, spindle-shaped, F6 conical; Forewing with costal cell distinctly longer than a marginal vein, parastigma well developed, submarginal vein with 3 or more setae; Axillae strongly projecting, each axillla with 1–2 setae; scutellum with 4–6 setae; Hypopygium in female prominent, extending nearly to apex of gaster or only to about the level of cercal plates; Tarsal formula 5-5-5 (Plate 11.1f).

Species: They are distributed in Australian, Afro-tropical, Oriental, Nearctic, and Palaearctic regions and the species recorded in India are *Coccophagoides orientalis* (Agarwal) and *Coccophagoides utilis* Doutt. They are parasitoids of diaspine scale insects (Hemiptera: Coccoidea: Diaspididae) (Hayat, 1998; NBAIR, 2021).

11.5.7 Genus: *Coccophagus* Westwood

Diagnosis: Antenna is 8-segmented with formula 1133, with a distinct anellus; Pronotum entire, not medially membranous; Axilla large, projecting, separated by not more than the length of an axilla, and each axilla with one seta; Forewing with submarginal vein having at least 5 setae (Plate 11.1g).

Species: In India, forty-eight species have been recorded from different scales of insects. Some of them are *Coccophagus acanthosceles* Waterston, *C. assamensis* Hayat, *C. burksi* Hayat, *C. candidus* Hayat, *C. dius* Hayat, *C. fumadus* Hayat, *C. indochraceus* Hayat, *C. nigricorpus* Shafee, *C. ochraceus* Howard, *C. parlobatae* Hayat, *C. redini* Girault, *C. silvestrii* Compere. *C. tschirchii* Mahdihassan, *C. zebratus* Howard, and *C. zeyai* Hayat. They are endoparasitoids of Coccoidea, mainly soft scales (Hemiptera: Coccidae), but a few species parasitise mealybugs (Pseudococcidae). Males (ecto- or endoparasitoids) are generally hyperparasitoids on other primary parasitoids, including females of their own species(Hayat, 1998; NBAIR, 2021).

11.5.8 Genus: *Encarsia* Foerster

Diagnosis: Female with antennae always 8-segmented, club when present 2- or 3-segmented. Male antenna 8- or 7-segmented; Mesoscutum with fewer than 20 setae, symmetrically arranged; axillae small, each axilla with one seta; Scutellum shaped more or less like a biconvex lens with a distinct central longitudinal area; Tarsal formula 5-4-5 or 5-5-5, i.e. tarsi of middle leg 4- or 5-segmented; Forewing with elongate marginal vein, submarginal vein having 1–2 setae, costal cell distinctly shorter than the marginal vein, parastigma short; Propodeum medially membranous; Hypopygium in female not prominent, usually not extending to apex of gaster (Plate 11.1h).

Species: Seventy-six species have been recorded from various parts of India and some of the important species are *Encarsia acaudaleyrodis* Hayat *E. albiscutellum* (Girault), *E. aurantii* (Howard), *E. azimi* Hayat, *E. bangalorensis* Hayat, *E. dialeurodis* Hayat, *E. dispersa* Polaszek, *E. flavoscutellum* Zehntner, *E. guadeloupae* Viggiani, *E. longicauda* Hayat, *Encarsia longifasciata* SubbaRao, *E. perniciosi* (Tower), *E. perplexa* Huang &Polaszek, *E. pseudococci* (Agarwal), *E. punicae* Hayat, and *E. pura* Hayat. They are endoparasitoids of diaspidids (Hemiptera: Diaspididae) and whiteflies (Hemiptera: Aleyrodidae). This genus includes some of the most successfully employed biocontrol agents in the world against whiteflies and diaspine scales. Some species are known to attack aphids. *Encarsia flavoscutellum* Zehntner is a successful biocontrol agent of the sugarcane woolly aphid, *Ceratovacuna lanigera* Zehntner in India (Hayat, 1998; NBAIR, 2021).

11.5.9 Genus: *Eretmocerus* Haldeman

Diagnosis: Body colour predominantly yellow; Female has antenna 5-segmented with 1121 formula, funicle segments short to anelliform, clava large, cylindrical or gradually widened distally or spindle-shaped, with sparse sensilla; Male antenna 3-segmented, formula 1101, clava long and curved with numerous sensillae; Forewing with marginal vein at most twice as long as stigma, parastigma well developed, stigmal vein long, with four sensilla of which one distinctly separated from the other three; Tarsal formula 4-4-4 (Plate 11.1i).

Species: They are well-known endoparasitoids of whiteflies, nearly cosmopolitan and known from all the major zoogeographical regions of the world. Nineteen species have been recorded from different parts of India and some of the important species are *Eretmocerus adustiscutum* Krishnan &Vasantharaj David, *E. bisetae* Hayat, *E. breviclavus* Subba Rao, *E. corni* Haldeman, *E. delhiensis* Mani, *E. dialeurolongae* Krishnan &Vasantharaj David, *E. flavus* Krishnan &Vasantharaj David, *E. sculpturatus* Hayat, *E. serius* Silvestri, *E. trialeurodis* Hayat(Hayat, 1998; NBAIR, 2021).

11.5.10 Genus: *Eriaphytis* Hayat

Diagnosis: Antenna is 7-segmented, formula 1132, with a distinct, large anellus; Axillae large, projecting forwards, each axilla with 2 setae; Scutellum large, hexagonal, with at least 10 setae; Forewing with marginal vein shorter than costal cell; submarginal vein with 5 setae, a seta present at the junction of parastigma and postmarginal vein; linea calva distinct; Tarsal formula 5-5-5.

Species: Two species recorded from India are *Eriaphytis chackoi* Subba Rao and *E. orientalis* Hayat. They are endoparasitic on scale insects belonging toCerococcidae and Coccidae (Hayat, 1998; NBAIR, 2021).

11.5.11 Genus: *Marlattiella* Howard

Diagnosis: Antenna with 4 segments, formula 1111; funicle segment small, quadrate or broader than long; clava long, slightly curved; Propodeum slightly longer than metanotum, with a triangular projection in middle of posterior margin, but without crenulae; Forewing as in *Aphytis*; costal cell distinctly shorter than marginal vein and with 2 setae; linea calva distinct, proximally bordered by several lines of setae; stigmal vein short, but with a distinct neck; Tarsal formula 5-5-5; Gaster as in Aphytis; hypopygium not extending to apex of gaster.

Species: The species *Marlattiella maculata* Hayat has been reported from Uttar Pradesh, India and is an endoparasitoid of diaspidids (Hayat, 1998; NBAIR, 2021).

11.5.12 Genus: *Paraphytis* Compere

Diagnosis: Their body is with distinctly mottled forewings and heavily pigmented/banded mesosoma, metasoma and legs; Antenna 5-segmented with 1121 formula, anelli absent, first two funicular segments reduced, clava elongate without specialized sensilla; Axillae slightly but distinctly advanced, with a single seta, scutellum with two pairs of setae; Ventral prepectus medially separated from mesepisternum; Posterior end of propodeum crenulate (Plate 11.1j).

Species: The important species recorded from India are *Paraphytis maculatus* (Shafee), *P. peculiaris* (Girault) and *P. transversus* (Huang). They are endoparasitic on soft scales. *Paraphytis* is externally very similar to *Marietta* in having distinctly mottled forewings and usually heavily pigmented mesosoma, metasoma and legs. However, *Paraphytis* has a strong constriction between the antennal clava and F3 and the forewing mottling pattern is formed by dark setae. *Marietta* has a less constricted connection of clava and F3 and mottled forewing formed by darkened areas of wing cuticle (Kim and Heraty, 2012).

11.5.13 Genus: *Proaphelinoides* Girault

Diagnosis: Their body is depressed/ flattened with ovipositor long and exserted at apex; with very long setae on body and wing veins; Antenna with 6 segments, formula 1122 or 1131; Pronotum dorsad long, medially divided by a suture; Forewing broad, with or without a bunch of dark setae behind proximal half of marginal vein; linea calva distinct or not clearly defined (Plate 11.1k).

Species: The species recorded in India are *Proaphelinoides anomalus* Hayat, *P. assamensis* Hayat, *P. australis* Girault, *P. chidambaramensis* Manickavasagam &Menakadevi. They are parasitic on diaspidid scales of the tribe Odonaspidini (Hemiptera: Diaspididae).

11.5.14 Genus: *Protaphelinus* Mackauer

Diagnosis: Antenna with 6 segments, formula 1131; Frontovertex broader than dorsal eye length; temples well-developed; Anterior pair of scutellar setae in line with posterior angles of axillae; Forewing with costal cell about as long as marginal vein; stigmal vein short, with a rounded stigma and 4 circular sensillae; submarginal vein with 5–6 setae; linea calva distinct; Tarsal formula 5-5-5; tarsal claws unequal in length; Gaster flat, ovate; hypopygium reaching apex of gaster and apically with long setae

Species: The species *Protaphelinus nikolskajae* (Yasnosh) has been recorded from Himachal Pradesh and Jammu & Kashmir, India. It is a specific parasitoid of gall-making aphids of the genus *Pemphigus* Hartig (Aphidoidea: Pemphigidae) on poplar trees, *Populus* spp. (Hayat, 1998).

11.5.15 Genus: *Pteroptrix* Westwood

Diagnosis: Antenna with 7 or 8 segments, antennal formula 1133, flagellum usually elongate, spindle-shaped, clava 3-segmented, usually longer than pedicel and funicle segments combined; Forewings with elongate marginal fringe, linea calva absent; Tarsal formula 4-4-4 (Plate 11.1l).

Species: Some of the important species recorded from different parts of India are *Pteroptrix albifemur* (Girault), *P. aster* Hayat, *P. chinensis* (Howard), *P. longiclavata* (Shafee, Siddiqui & Rizvi), *P.matpada* Hayat, *P. zoza* Hayat. They are endoparasitoids of diaspine scale insects (Hemiptera: Coccoidea: Diaspididae).

11.5.16 Genus: *Umairia* Hayat

Diagnosis: Head slightly flattened antero-posteriorly, subprognathous; malar space straight, and with a fine sulcus, Mandible rudimentary, lobe-like, tips of the two mandibles widely separated and hidden by ventral, rounded extension of clypeus. Antennal formula 1131; scape very slightly flattened, apically narrowed, and more than 3 as long as broad; pedicel longer than F1; funicle segments transverse; F1 with ventral side longer than dorsal side; F2 and F3 with dorsal sides longer than ventral sides; clava longer than funicle, and with apex slightly curved; Pronotum medially membranous, thus consisting of two plates; Mesoscutum with notaular lines present in about anterior half only; Forewing with costal cell slightly to distinctly longer than marginal vein; postmarginal vein absent; stigmal vein with a short neck, and stigma with 4 circular sensilla arranged in a curved line; Legs with tarsal formula 5-5-5; foretibial spur curved and bifid at apex; Gaster apically rounded; cercal plates situated near apex of gaster; ovipositor exserted; hypopygium extending to the level of cercal plates (Hayat, 2014) (Plate 11.1m).

Species: The species recorded from India are *Umairia laiba* Hayat and *Umairia zeera* Hayat and they are parasitic on eggs of *Papilio polytes* Linnaeus (Lepidoptera: Papilionidae) on curry leaves (*Murraya koenigii* (L.) (Rutaceae).

11.6 MASS PRODUCTION

The development of efficient mass-rearing methods of Aphelinid parasitoids is critical for the implementation of classical and augmentative biocontrol programmes. The mass production techniques of various important aphelinid parasitoids are discussed hereunder.

11.6.1 EncarsiaPerniciosi

Encarsia perniciosi is a monophagous, host specific, solitary ectoparasitoid of *Quadraspidiotus perniciosus*. It is believed to be a native of the Far East. It has been introduced into India from several places in USA, Russia and China. Most of these races are morphologically indistinguishable but differ in certain biological characteristics, such as fecundity, rate of development, presence of males and adaptability to extreme environmental conditions. The parasitoid has been established in all apple-growing San Jose scale endemic areas. Till recently the parasitoid along with other native natural enemies was quite effective but with the resurgence of apple scab disease, the orchardists have started spraying their orchards frequently not only with various fungicides but also sometimes by mixing insecticides. This has necessitated periodic colonisation of the parasitoid even in those areas where it had been well colonised.

Procedure: The pumpkin infested with scale, which is kept in a cage is exposed to *Encarsia perniciosi* females in a ratio of 20:1 for 24 hours when the scale insect is about 12 days old. The adult parasitoids are fed with honey streaks provided on butter paper. The female lays its eggs singly in the body cavities of the scales. The exposed pumpkins are either not disturbed from the original cage and the parasitoids are allowed to develop or one date of pumpkins are stacked in a rack and at least 2 days before the date of emergence, pumpkins which have been shifted to racks are placed in the cages for facilitating the easy collection of adult parasitoids. The pupation takes place inside the body of the parasitised scale, which gets mummified. The parasitoids emerge within 3 weeks of ovipositing by the female parasitoid. Each female is capable of parasitising up to 90 scales. The number of pumpkins, the frequency of infestation, etc. could be determined depending on need (NBAIR, 2021).

11.6.2 Aphelinus Mali

Aphelinus mali is a native of England that was imported and colonised for the first time in Shimla, Himachal Pradesh, India. Subsequently, it has spread to all the apple-growing areas where *Eriosoma lanigerum* is a serious problem.

Procedure: The parasitoid is produced on the potted plants of apples, which are first infested with *Eriosoma lanigerum* and then exposed to *Aphelinus mali*. The parasitoid has not been produced on a large scale but sufficient quantities are produced for inoculative releases (NBAIR, 2021).

11.6.3 Aphytis Melinus

Aphytis melinus is a common biparental ectoparasitoid of *Aonidiella aurantii* in India. The parasitoid is not only suppressing the pest population in India but some of its races have been exported to provide similar benefits to the citrus industry throughout the world.

Aphytis melinus is produced in a similar manner as many other *Aphytis* spp. production protocol. The pumpkins are infested with scale insects and when *Aonidiella aurantii* nymphs reach third instar stage, they are exposed to the adults of *Aphytismelinus* in the ratio of 40:1 for 24 hours. The parasitoids are then gently aspirated out and the pumpkin carrying *Aonidiella aurantii* parasitised by *Aphytis melinus* are either stacked in racks for convenience sake or each of them placed in a cage. The parasitoid completes its development in 10–13 days. The adults also feed on the body fluids of young nymphs of *A. aurantii* in addition to fine honey streaks provided on a butter paper.

11.6.4 Aphytis Proclia

It is an important indigenous ectoparasitoid of *Quadraspidiotus pernisiosus* in Jammu & Kashmir state. It has now spread to Himachal Pradesh and other apple or temperate tree-growing states.

Procedure: *Aphytis proclia* is also multiplied on San Jose scale as per the procedure used in *Encarsia perniciosi*. *Aphytis proclia* females are offered third instar scale insects (it can attack second instar nymphs, young adult female scales and male pupae) in the host parasitoid ratio of 40:1. The parasitoid completes development in 19 days. The adults live for about 12 days.

11.7 ROLE IN BIOLOGICAL CONTROL

Aphelinids have been employed successfully in the biocontrol of homopterous insects, such as aphids, whiteflies, armoured scale insects, and soft scales in their native environments. This family accounts for the majority of biocontrol programme accomplishments around the world. For instance, Noyes (1985) found that the Aphelinidae was ranked the highest in the ratio of the actual number of introductions to the number of cases in which full economic control of the pest was accomplished (5:1).

Aphelinidae is one of the most important Chalcidoidea families found associated with 335 species of scale insects. Among them, species under the genus *Aphytis* Howard have been recorded on 165 species of scale insects (García et al., 2016; Noyes, 2017). *Aphytis* species develop exclusively as primary ectoparasitoids in diaspidid scales. The ovipositor of the adult female is inserted through the scale cover, and the eggs are laid in the area between the body and the scale cover (Yu and Luck, 1988; Rosen and Rose, 1989). As natural enemies, several species of this genus have been employed successfully in biocontrol programs for economically important armoured scale insects (Rosen and DeBach, 1979; García et al., 2016; Noyes, 2017).

The genus *Coccophagus* Westwood comprises 200 described species. Most of them are parasitising the species of Coccoidea, mainly Coccidae. Some of these parasitoids have also been associated with species of Diaspididae, Eriococcidae, Kermesidae, Pseudococcidae, and Stictococcidae (Hayat, 1997; García et al., 2016; Noyes, 2017).

Encarsia Förster has over 400 species described worldwide, and is one of the most important genera for the biocontrol, parasitising species of Aleyrodidae and Diaspididae (Evans et al., 1995; Heraty et al., 2007; Myartseva and Evans, 2008). *Encarsia formosa*, specifically the greenhouse whitefly parasitoid is exploited for the management of greenhouse whiteflies. These tiny wasps are best used for preventing the establishment of the greenhouse whitefly (*Trialeurodes vaporariorum*). In addition to the greenhouse whitefly, *Encarsia formosa* can parasitise a few other whitefly species, including the sweet potato whitefly (*Bemisia tabaci*). With the exception of the adult, all stages of *Encarsia* occur within the whitefly host. Adults are tiny wasps that lay eggs inside 2-week-old whitefly scales (second and third whitefly larval stages). Each egg will kill one whitefly scale. Adults lay up to 200 eggs. The parasite then develops inside the whietfly scale, and midway through the development of *Encarsia* within the whitefly, the scale turns dark. It becomes black for the greenhouse whitefly, and transparent brown for the sweet potato whitefly.

Aphelinus mali is a parasitoid wasp that exploits the woolly apple aphid (*Eriosoma lanigerum*), a pest of apple trees. It is native to the North-Eastern United States but has been introduced to other parts of the world as a biological pest control agent. Introduced from its native region to other apple-growing areas, *A. mali* was one of the earliest examples of controlling pests biologically. It works best in conjunction with other generalist predators, such as lady beetles, lacewings, the larvae of hoverflies and the plant bug *Deraeocoris brevis*, and is negatively impacted by the use of chemical pesticides.

11.8 FIELD RELEASE ACROSS THE GLOBE

The chalcidoid parasite *Aphelinus mali* has been introduced into many countries in attempts to control *E. lanigerum*. It was originally native to the USA, but has become acclimatised in Europe and has now been introduced into apple-growing regions worldwide. Where aboveground infestations dominate, control of *E. lanigerum* has been very successful, but where root-feeding populations are important, the results have been less good. In laboratory studies performed by Mueller et al. (1992), A. mali parasitised all stages of *E. lanigerum*, but preferred third-instar nymphs, whereas rates of parasitism were inversely proportional to host colony size, with small colonies and long, thin colonies having a greater proportion of individuals parasitized. *A. mali* was first introduced into India during the 1930s, and 98% control was soon achieved in the Kullu Valley, Himachal Pradesh (Thakur and Dogra, 1980).

In field studies undertaken in New Zealand, Shaw et al. (1996) reported that parasitism of *E. lanigerum* by *A. mali* exceeded 80% by late April, and control was achieved without the need for specific aphicide sprays. High levels of parasitism by *A. mali* (80%) were also recorded at low population densities of *E. lanigerum* in Mexico (Tejada and Rumayor, 1986), whereas parasitism rates of over 50% were recorded during the summer in West Virginia, USA (Brown and Schmitt, 1994). *E. lanigerum* was controlled by *A. mali* in pesticide-free apple orchards in Israel (Oppenheim et al., 1997).

Biocontrol is mostly based on the chalcid wasp, *Encarsia formosa* has been widely utilised in glasshouses, especially since the advent of insecticide-resistant whiteflies (Osborne and Landa, 1992). *E. formosa* parasitises *T. vaporariorum* and the parasitoid is usually shipped as parasitised scales on strips. For cucumbers, a typical rate of release is 12,500 parasitoids per ha in areas where *T. vaporariorum* is a habitual pest, starting at the beginning of the season. In tomatoes, most growers introduce parasitised scales weekly at the rate of 5000/ha as soon as plants are available. Once the parasitoid is established, it is important to avoid removing from the glasshouse leaves with black scales from which adult wasps have not emerged. Empty scales can be spotted by holding leaves up to the light and examining them for emergent holes (Roermund et al., 1997).

The aphelinid-parasitoid, *Encarsia perniciosi* was reared, released and established to control the destructive pest of apple, sanjose scale, *Diaspidiotus perniciosus* in different parts of the world, especially in the USA, Europe and Russia (Rosen, 1990). It gives substantially good results only after a relatively long time, and only in the absence of toxic insecticides. The additional release of different predators, and conserving the local natural enemies can increase the efficiency of control.

Aphytis melinus and *Enacarsia citrina* have successfully controlled the *Aoindilella aurantii* infestation on papaya in Queensland, Australia. *Aphytis melinus* was augmentatively released and rates of parasitism were increased, although this parasitoid had to be continually re-released. *Encarsia citrina* occurred naturally, producing levels of up to 80% parasitism, and was therefore important in the biological control of *A. orientalis* (Elder and Smith, 1995).

11.9 SUCCESS STORIES

Aphelenid parasitoids have been employed for the successful management of various insect pests across the globe. Some of the remarkable success stories are given below.

Quadraspidiotus perniciosus is a polyphagous pest of Oriental origin. Its native home is believed to be somewhere in the general area of northern China, the Far East, and North Korea. It is a serious pest of apples in North-Western India. It also attacks other deciduous trees, poplars and willows. About 50 hosts have been recorded in India. The scale insect colonises all parts of the plant. Young tree bark when attacked shows characteristic deep red or purplish stain. The bark of old trees develops cracks and exudes a large amount of gummy substance. The fruits attacked may also develop reddish spots around the scale infestations. In case of severe infestation, the attacked tree may get defoliated. Such trees do not bear any fruits. The overwintering nymphs become active in mid-March; the males emerge in April. The life cycle is completed in 35–40 days. The scale insect passes through 4–5 generations before mid-October when hibernation starts.

Aphelinid parasitoid, *Encarsia perniciosus* strain from California was introduced in 1958 and Illinois, Chinese and Russian strains were introduced in 1960 for the biological suppression of San Jose scale. Russian strain gave 89% parasitism in Himachal Pradesh. In Kashmir, the Russian and Chinese strains appeared to be superior to Californian and Illinois strains. American (Illinois) and Chinese strains were also released in Kumaon hills of Uttar Pradesh; the population of *Q. perniciosus* was reduced by about 95% t. *E. perniciosi* has been established in many areas where the population of San Jose scale is low. But due to frequent sprays in apple orchards, mass multiplication and release of *E. perniciosi* as well as *Aphytis proclia* g at the beginning of the season is recommended (Singh, 2004).

Eriosoma lanigerum is a native of the Eastern United States. It was probably accidentally introduced to India from England as indicated from its record in the Shimla district of Himachal Pradesh, where nursery stocks were imported. Currently it has spread to all the apple-growing areas of the country. The aphids prefer to feed on the new bark at pruning and wound scars on the trunk and branches. The aphids feeding on the roots cause maximum direct injury, although large numbers may be found on the aerial parts of the tree. The aphid sucks sap from the bark of the trunk and the roots. It also infests fruit stalks and calyx end. The infested twigs shrivel and die. The nursery plants when attacked suffer the most. Points on the stems and roots where the aphids crowd together and feed develop gall-like swellings.

For the control of woolly aphid, exotic aphelinid parasitoid, *Aphelinus mali*, a native of North America, was introduced from UK at Saharanpur (Uttar Pradesh). The parasitoid failed to establish and provide effective control of woolly aphid in the Kumaon hills because of the intense activity of coccinellid beetle, *Coccinella septempunctata*, which fed indiscriminately on parasitised as well as unparasitised woolly aphids. *C. septempunctata,* however, provided satisfactory control of the pest. The parasitoid could control woolly aphid in Kullu valley (Himachal Pradesh) and has spread to Kashmir valley. *A. mali* was transferred to Coonoor (Tamil Nadu) from Kullu where it was established. The parasitoid was sent to Shillong(Meghalaya) from Coonoor in 1961 and Kullu in 1963 and establishment was achieved after periodic releases. At present, it has been established in all apple-growing areas of the country but is more effective in valleys rather than on mountain slopes. During the period of activity of other predators, the population of this exotic parasitoid is diminished. In general, *A. mali* has emerged as a key population regulatory agent of *E. lanigerum* in the areas of its successful establishment (Singh, 2004).

Spiralling whitefly, *Aleurodicus dispersus*, a native of Carribbean region and Central America was introduced into India in 1995. First reported from Kerala, it has spread to all the southern states causing serious damage to several plants. It attacks over 253 plant species belonging to 176 genera and 60 families. It causes serious damage to avocado, banana, cassava, guava, papaya and mango, besides several ornamental and avenue trees. Severe infestations cause yellowing, crinkling and curling, and premature fall of leaves. The nuisance created by the wind-borne, copious white waxy flocculent material and the flying adults are very revealing. The whitefly is suspected to be a possible vector of lethal yellow disease.

Exotic aphelinid parasitoids, *Encarsia guadeloupae* and *E.* sp. nr. *meritoria* (Origin: Carribbean region/Central America) collected from Minicoy Island of Lakshadweep and brought to the mainland have established well, causing a perceptible reduction in the pest population. Parasitism levels due to both parasitoids vary from 29% to 70% and exceed 90% during some parts of the year. Wherever parasitism was heavy, the pest population was substantially reduced subsequently. *E. guadeloupae* is performing better than *E.* sp. nr. *Meritoria* and has displaced it almost completely in Bangalore. It has been established in Kerala, Karnataka and several parts of Andhra Pradesh, where it was earlier absent (Singh, 2004).

11.10 CONCLUSIONS

The aphelinid parasitoids represent one of the most fascinating groups among the parasitic-Hymenoptera. Though they are moderate-sized in the Chalcidoidea, it comprises parasitoids of sternorrhynchous Homopteran, such as aphids, scale insects and whiteflies, and play a major role in biocontrol of various insect species belonging to these groups. They are highly diverse and distributed worldwide with more than 41 genera and 1000 species. They are parasitic on various stages of homopteran pests, viz. egg, nymphs and adults and some are parasitic on eggs of other insects; and some of them do exhibit hyper parasitism. It is an economically important family of Chalcidoidea and includes some of the most successful and widely used parasitoids in applied biocontrol, such as *Aphytis, Encarsia* and *Aphelinus*. India has provided many parasitoids that were introduced into other countries, notably the United States for biocontrol. Although no extensive work on the 'introduction' and release of foreign species in India was carried out, it is a well-known fact that success or failure in any biocontrol programme primarily depends on the correct identification of the host and its parasitoids. The important genera with their key taxonomic features, mass production techniques and potential in biocontrol have been highlighted in this chapter.

11.11 POINTS TO REMEMBER

- Aphelind parasitoids are very small to minute parasitic hymenopteran, commonly associated with the families of homoptera including aphids, scale insects, whiteflies, mealybugs, etc.
- Though the majority of aphelinids are primary ecto-or- endoparasitoids of stemorrhynchous Homoptera, a few are reported to develop on other hosts, such as eggs of Orthoptera and Lepidoptera and puparia of Diptera. However, some species exhibit hyperparasitism.
- They are distributed worldwide with more than 41 genera and more than 1000 described species and in India, a total of 269 species belonging to 24 genera have been documented.
- There are 16 important genera under Aphelinidae, which comprise parasitic species associated with homopteran pests damaging various crops in India.
- Various mass production techniques have been standardised for the important species of aphelinid parasitoids used for managing pests in crop ecosystems.
- It is also a compliment to this family that it is an economically important Chalcidoidea family, with parasitoids, like Aphytis, Encarsia, and *Aphelinus* being the most successful and commonly employed parasitoids in applied biocontrol.

REFERENCES

Agarwal, M.M. (1964). Studies of forms of Aphelinidae (Hymenoptera: Chalcidoidea) collected at aligarh (India). II. *Proc. Ind. Acad. Sciences (B)*, 60(5): 316.

Ahmad, M.E. & Singh, R. (1996). Records of aphid parasitoids from the North Bihar and associations with their hosts and food plants. *J. Advan. Zoology*, 17(1): 26–33.

Amin Masoodi, M. & Trali, A.R. (1987). New record of scale insects (Diaspididae:Homoptera) from and aphelinid parasite (Hymenoptera) from Kashmir, India. *Entomon*, 12(3): 227.

Andre, M.E. (1878). Memoire pour server a histoire de la TriozacentranthiVallot. *Ann. Soc. Ent., Franc*, 8(5): 77–86.

Annecke, D.P. & Insley, H.P. (1970). New and little-known species of *Azotus*Howard, *Ablerus*Howard and *Physcus*Howard (Hymenoptera, Aphelinidae) from Africa and Mauritius. *Bull Entomol. Res.*, 60(2): 244–246.

Antony, B., Palaniswami, M.S., Kirk, A.A. & Henneberry, T.J. (2006). Aphelinid parasitoids of *Bemisiatabaci* (Gennadius) (Homoptera: Aleyrodidae) in India. *Entomon*, 31(217): 224.

Ashmead, W.H. (1900). Description of a new genus in the Aphelinidae. *Canad. Entomol.*, 32: 349.

Azim, M.N. & Shafee, S.A. (1980). Indian species of the genus *Trichaporus*Foerster (Hymenoptera: Aphelinidae). *J. Bombay Nat. Hist. Soc.*, 76(2): 337.

Berlese, A., & Paoli, G. (1916). Un endofagoesoticoefficace contra il *Chrysomphalus dictyospermi. Morg. Redia*, 11: 305–307.

Bonnemaison, L. (1965). Observations ecologiques sur *Aphelinus mali* Haldeman parasite du puceronlanigere (*Eriosomala nigerum* Hausmann). *Ann. Soc. Emomol. Fr. (NS)*, 1: 143–176.

Begum. S., Anis, S.B., Farooqi, M.K., Rehmat, T. & Fatma, J. (2011). Aphelinid parasitoids (Hymenoptera; Aphelinidae) of whiteflies (Homoptera: Aleyrodidae) from India. *Biology and Medicine*, 3 (2): 222–231.

Brown, M.W. & Schmitt, J.J. (1994). Population dynamics of woolly apple aphid (Homoptera: Aphididae) in West Virginia apple orchards. *Environmental Entomology*, 23(5): 1182–1188.

Campbell, B., Heraty, J.M., Rasplus, J.Y., Chan, K., Steffen-Campbell, J. & Babcock, C. (2000). Molecular systematics of the Chalcidoidea using 28S-D2 rDNA. *The Hymenoptera: Evolution, Biodiversity and Biological Control*, ed. Austin, A.D. & Dowton, M., pp. 57–71. Melbourne: CSIRO Publishing.

Cameron, P. (1891). Hymenopterological notices. *Memoirs & Proceedings of the Manchester Literary and Philosophical Society*, 4(4): 183.

Cendana, S.M. (1937). Studies on the biology of Coccophagous (Hymenoptera), a genus parasitic on nondiaspidineCoccidae. *University of California Publications in Entomology*, 6: 337–399.

CIBC, (1976). Report of work carried out during 1975. Pp. 49. Commonwealth Institute of Biological Control, Commonweth Agricultural Bureaux, Slough, U.K.

Clausen, C.P. (1940). *Entomophagouslnsects*. New York: McGraw-Hill. 688 pp.

Compere, H. & Annecke, D.P. (1961). Descriptions of parasitic Hymenoptera and comments (Hymenoptera: Aphelinidae, Encyrtididae, Eulophidae). *J. Entomol. Soc. South. Africa*, 24: 17–71.

Compere, H. (1925). A new genus and species of Aphelinidae (Hymenoptera) from China. *Trans. Ameri. Entomol. Soc.*, 51: 129–134.

Compere, H. (1926). Descriptions of new coccid inhabiting chalcidoid parasites (Hymenoptera). *Univ. Califor. Publi. Entomo.*, 4(1): 2.

Compere, H. (1931). A revision of the species of *Coccophagus*, a genus of hymenopterous, coccid inhabiting parasites. *Proc. Unit. Sta. Nat. Muse.*, 78(7): 67.

Compere, H. (1933). The parasites of *Pseudococcus comstocki* Kuw. *Cana. Entomol.*, 65(11): 245.

Compere, H. (1955). A systematic study of the genus *Aphytis* Howard (Hymenoptera, Aphelinidae) with descriptions of new species. *Univ. Califor. Publi. Entomo.*, 10: 271–320.

Craw, A. (1891). Internal parasites discovered in the San Gabriel Valley; recommendations and notes. *Bull. Califo. State Boa. Horti.*, 57: 4.

Dalman, J.W. (1820). Forsok till Uppställningaf Insect-familjen Pteromalini, isynnerhet med afseen de på de i Sverige funne Arter. *Kungliga Svenska Vetenskapsakademiens Handlingar*, 41(1): 181.

DeBach, P. & Rosen, D. 1976. Twenty new species of *Aphytis* (Hymenoptera: Aphelinidae) with notes and new combinations. *Ann. Entom. Soc Ame.*, 69(3): 545.

DeBach, P. (1959). New species and strains of *Aphytis* (Hymenoptera, Eulophidae) parasitic on the california red scale, *Aonidiella aurantii* (Mask.), in the Oriental. *Ann. Entom. Soc Ame.*, 52(4): 363.

DeBach, P. (1960). The importance of taxonomy to biological control as illustrated by the cryptic history of *Aphytis holoxanthus* n.sp. (Hymenoptera: Aphelinidae), a parasite of *Chrysomphalus aonidum*, and *Aphytis coheni* n.sp., a parasite of *Aonidiella aurantii. Ann. Entom. Soc Ame.*, 53(6): 704. MANICK-AVASAGAM and MENAKADEVI: A checklist of Aphelinidae.....of India

Doutt, R.L. (1966). A taxonomic analysis of parasitic Hymenoptera reared from *Parlatoria oleae* (Colvée). *Hilgardia*, 37: 224.

Dozier, H.L. (1932). Miscellaneous notes and descriptions of Chalcidoid parasites (Hymenoptera). *Procee. Entomol. Soci. Washington*, 35: 85–100.

Dozier, H.L. (1937). Descriptions of miscellaneous chalcidoid parasites from Puerto Rico (Hymenoptera). *J. Agri. Univer. Puer. Rico*, 21(2): 128.

Easwaramoorthy, S., David, H. & Bai, K.S. (1996). Studies on *Botryoideclava bharatiya* Subba Rao, a parasite of sugarcane scale insect, *Melanaspis glomerata* (Green). *Entomon*, 21(1): 55–64.

Elder, R.J. & Smith, D. (1995). Mass rearing of *Aonidiella orientalis* (Newstead) (Hemiptera: Diaspididae) on butternut gramma. *Journal of the Australian Entomological Society*, 34(3): 253–254; 6 ref.

Erdos & Novicky, J. (1953). Additamentaadcognitionem faunae Chalcidoidarum in Hungaria et regionibus fi nitimis. V. 11. Thysanidae, 12. Aphelinidae, 17. Tetracampidae, 18. Elasmidae. *Folia Entomologica-Hungarica*, 6: 180.

Evans, G.A., Polaszek, A. & Bennett, F.D. (1995). The taxonomy of the *Encarsiaflavoscutellum* species-group (Hymenoptera: Aphelinidae) parasitoids of Hormaphididae (Homoptera: Aphidoidea). *Oriental Insects*, 29: 33–45.

Ferriere, C. (1931). New chalcidoid egg-parasites from South Asia. *Bull. Entom. Research*, 22(2): 286.

Fry, J.M. (1989). Natural enemy databank, 1987. *A catalogue of natural enemies of arthropods derived from records in the CIBC Natural Enemy Databank.* pp. 58–59. CAB International, Wallingford, oxford, UK.

García, M., Denno, B., Miller, D.R., Miller, G.L., Ben-Dov, Y. & Hardy, N.B. (2016). ScaleNet: A Literature-based model of scale insect biology and systematics. http://www.scalenet.info (last accessed 23 Apr 2017).

Gerling, D. (1966). Studies with whitefly parasites of southern California. I. *Encarsiapergandie*Ua Howard (Hymenoptera: Aphelinidae). *Can. Entomol.*, 98: 707–724.

Gibson, G.A.P. (1989) Phylogeny and classification of Eupelmidae, with a revision of the world genera of Calosotinae and Metapelmatinae (Hymenoptera: Chalcidoidea). *Memoirs of the Entomological Society of Canada*, 149, 1–121.

Girault, A.A. (1913). Some chalcidoid Hymenoptera from North Queensland. *Arch, Naturg*, 76: 70–90.

Girault, A.A. (1914). The third genus of the family Elasmidae (Hymenoptera). *The Cana. Entomol.*, 46: 285–286.

Girault, A.A. (1915). Australian Hymenoptera Chalcidoidea III. Second Supplement. *Memo. Queens. Muse.*, 3: 170–179.

Girault, A.A. (1916). New miscellaneous chalcidoid Hymenoptera with notes on described species. *Ann. Entom. Soc Ame.*, 9(3): 292.

Girault, A.A. (1917). Notes and descriptions of miscellaneous chalcid-flies (Hymenoptera). *Proc. Unit. Sta. Nat. Muse.*, 53(2213): 445.

Girault, A.A. (1923). Miscellaneous species of chalcid- flies from Australia (Hymenoptera, Chalcidoidea). *Insec. Insci. Menst.*, 9: 188.

Girault, A.A. (1926). New pests from Australia III. 2pp. private publication, Brisbane.

Girault, A.A. (1932). *New lower Hymenoptera from Australia and India* pp.2 Brisbane, private publication.
Girault, A.A. (1933). *Some beauties inhabitant not of commercial boudoirs but of natures bosom, notably new insects* pp. 4 Brisbane, private publication.Haldeman, S. S. (1851). Description of *Eriophilusmali* (Hymenoptera). *Pen. Farm Jour.*, 1: 130–131.
Haldeman, S. S. (1851)Description of Eriophilus mali (Hymenoptera).Pennsylvanian Farm Journal, 1(1):130–131.
Hayat, M. & Anis, S.B. (2004). Records of *Ablerusconnectens*Silvestri (Hymenoptera:Aphelinidae) from India. *Bionotes*, 6: 99.
Hayat, M. & Fatima, K. (1992). Taxonomic studies on *Aphelinus*(Hymenoptera: Aphelinidae) 5. Description of new species and further records of *A. gossypii* with a new synonymy. *Entomon*, 17(1/2): 103–106.
Hayat, M. & Khan, (2010). Additions to the Aphelinidae of India (Hymenoptera – Chalcidoidea): 1. On species of *Ablerus*Howard, *Coccobius*Ratzeburg, *Coccophagus*Westwood, *Pteroptrix* Westwood, and *Idoicoccobius*Hayatgen.nov. *Colemania*, 21: 5–6.
Hayat, M. & Khan, (2012). Two new species of *Coccobius*Ratzeburg (Hymenoptera:Aphelinidae) from India. *Biosystematica*, 6(1): 13–17.
Hayat, M. & Polaszek, A. (1992), *Encarsia aseta* new species (Hymenoptera; Aphelinidae) a parasitoid of *Dialeurolonga elongata* Dozier (Homoptera; Aleyrodidae) from India. *Orien. Inse.*, 26: 107–109.
Hayat, M. & Singh, S. (1989). Description of a new *Coccophagus of the lycimnial*-group (Hymenoptera: Aphelinidae) from the Khasi hills, with some other species records from India. *Colemania*, 5: 29–33.
Hayat, M. & Viggiani, G. (1984). A preliminary catalogue of the Oriental *Trichogrammatidea* (Hym: Chalcidoidea). *Boll. del Labo. di Entom. Agr.Flip. Silv. Port.*, 41: 27.
Hayat, M. & Zeya, S.B. (1993). Records and descriptions of some Indian Aphelinidae (Hymenoptera: Chalcidoidea). *Hexapoda*, 5(1): 58.
Hayat, M. (1971). Two new and known species of Aphelinidae (Hymenoptera: Chalcidoidea) from India. *Bull. Entom. Soc. India*, 12(2): 118–120.
Hayat, M. (1972a). The species of *Aphelinus*Dalman, 1820 (Hymenoptera: Aphelinidae) from India. *Entomophaga*, 17(1): 52–56.
Hayat, M. (1972b). The species of *Eretmocerus*Haldeman, 1850 (Hymenoptera: Aphelinidae) from India. *Entomophaga*, 17(1): 99–106.
Hayat, M. (1972c). A new aphelinid genus *Eriaphytis*(Hymenoptera:Chalcidoidea) reared from *Cerococus*spp. *Polskie Pismo Entomologiezne*, 423: 151–156.
Hayat, M. (1973). Two new Indian species of *Centrodora*Foerster, 1878 (Hymenoptera: Aphelinidae) parasitic in the eggs of membracids (Homoptera: Membracidae). *Entomophaga*, 18(1): 41- 44.
Hayat, M. (1974a). Three new species of Aphelinidae (Hymenoptera: Chalcidoidea) parasitic on *Aonidiellaorientalis*(Newst.) from India. *J. Bom. Nat. Hist.*, 71(1): 64–71.
Hayat, M. (1974b). On some Indian species of Aphelinidae, with a description of a new *Coccophagus* (Hymenoptera; Chalcidoidea). *J. Nat. Hist.*, 8 pp. 179–186.
Hayat, M. (1976). Two new species of Aphelinidae parasitic on *Acaudaleyrodesrhachipora*from India. *Entomophaga*, 21(2): 158–160.
Hayat, M. (1984). *Encarsiabennetti*sp.n. (Hymenoptera: Aphelinidae) from India, associated with the citrus black fl y *Aleurocanthuswoglumi*(Homoptera: Aleyrodidae). *Bull. Entom. Rese.*, 74(3): 399–401.
Hayat, M. (1985). Notes on some species of *Coccobius*and*Prophyscus*(Hymenoptera: Aphelinidae), with special reference to Girault and Howard types. *Orien. Inse*, 18: 296.
Hayat, M. (1986). Family Aphelinidae. (In: Subba Rao, B.R.; Hayat, M. (Eds) - *The Chalcidoidea (Insecta: Hymenoptera) of India and the adjacent countries. Part II.) Oriental Insects*, 20: 159.
Hayat, M. (1988). Description of a new species of *Centrodora*Foerster and notes on the type of *Eretmocerusdelhiensis*Mani (Hymenoptera: Chalcidoidea: Aphelinidae). *Entomon*, 13: 128–131.
Hayat, M. (1989). A revision of the species of *Encarsia*Foerster (Hymenoptera: Aphelinidae) from India and the adjacent countries. *Orien. Inse.*, 23: 47.
Hayat, M. (1990). Taxonomic studies on *Aphelinus*(Hymenoptera: Aphelinidae). 2. A new subgenus from India with comments on *Mesidia*and*Mesidiopsis*. *Orien. Inse.*, 24: 253–254.
Hayat, M. (1993). The *malthusi*-group of *Coccophagus*(Hymenoptera: Aphelinidae), with descriptions of three new species from India. *Orien. Inse.*, 27: 178–179.
Hayat, M. (1994). Taxonomic studies on *Aphelinus*(Hymenoptera: Aphelinidae). VI. Records of two known and descriptions of two new species from the Oriental region. *Entomon*, 19(1): 2.
Hayat, M. (1997). Aphelinidae. World Crop Pests [Key to genera of Aphelinidae associated with soft scale (Coccidae).]7:112–115
Hayat, M. (1998). Aphelinidae of India (Hymenoptera: Chalcidoidea): a taxonomic revision. Memoirs on Entomology. *International*, 13: 1–416.

Hayat, M. (2006). Description of a new species of *Encarsia*from India, with records of four other species (Hymenoptera: Aphelinidae). *Orien. Inse.*, 40: 292–293. MANICKAVASAGAM and MENAKADEVI: A checklist of Aphelinidae.....of India

Hayat, M. (2007). Description of a new species of *Coccophagus*(Hymenoptera: Aphelinidae), parasitoid of *Paratachardina lobata* (Homoptera:Kerridae) from India. *Orien. Inse.*, 41: 387–388.

Hayat, M. (2010). Additions to the Aphelinidae of India (Hymenoptera: Chalcidoidea)–2, Genus *Centrodora*-Foerster. *Orien. Inse.*, 44: 57.

Hayat, M. (2011a). On *Coccophagoides*Girault and *Diaspiniphagus*Silvestri (Hymenoptera: Chalcidoidea: Aphelinidae). *Zootaxa*, 3070: 19–21.

Hayat, M. (2011b). Additions to the Indian Aphelinidae (Hymenoptera: Chalcidoidea)- III: the genus *Encarsia*Foerster. *Orien. inse.*, 45(2-3): 202–274.

Hayat, M. (2012). Records and descriptions of some Aphelinidae (Hymenoptera: Chalcidoidea) from India. *Zootaxa*, 3521: 39–50.

Hayat, M. (2014). Eutrichosomellini (Hymenoptera; Chalcidoidea: Aphelinidae) from India, with description of two new genera. *Zootaxa*, 3821 (4): 425–442

Hempel, A. (1904). Notassobredoisinimigos da laranjeira. *Boletim da Agricultura, Sao Paulo*, 1(5): 19.

Heraty, J.M. (2016). Aphelinidae. Accessed at http://cache.ucr.edu/~heraty/Aphelinidae.html (Accessed June 2016).

Heraty, J.M. & Polaszek, A. (2000). Morphometric analysis and descriptions of selected species in the *Encarsiastrenua*group (Hymenoptera: Aphelinidae). *J. Hym. Rese.*, 9: 142–169.

Heraty, J., Woolley, J. & Polaszeck, A. (2007). *Catalogue of the Encarsia of the World.* University of California, Riverside, California, USA.

Howard, L.O.. (1910). Two new aphelinine parasites of scale insects. *Entomological News*, 21(4): 162.

Howard, L.O. (1881). (In Comstock, J.H. - Report of the entomologist for 1880.) Report of the parasites of Coccidae in the collections of the U.S. Department of Agriculture Part III. *Report. UnitedStates Department of Agriculture. Washington. (Entomology)*, 1880: 355.

Howard, L.O. (1885). Descriptions of North American Chalcididae from the collections of the U.S. Department of Agriculture and of Dr. C. V. Riley, with biological notes. (First paper). Together with a list of the described North American species of the family. *Bull. Unit Stat. Depar.Agricul. Bure. Entom.*, 5: 25.

Howard, L.O. (1894). The hymenopterous parasites of the California red scale. *Insect life*, 6: 231.

Howard, L.O. (1895). Two parasites of important scale-insects. *Insect Life*, 7: 5–8.

Howard, L.O. (1907). New genera and species of Aphelinidae with a revised table of genera. *United States Department of Agriculture Technical Series, Bureau of Entomology, United States Department of Agriculture*, 12(4): 69–88.

Howard, L.O. (1908). Upon the aphis feeding species of *Aphelinus*. *Entomological News*, 19(8): 367.

Howard, L.O. (1911). A note on the Indian enemies of *Aleyrodescitri*R. and H., with description of a new species of *Prospaltella*. *J. Econ. Entom.*, 4: 132.

Howard, L.O. (1914). Concerning some Aphelininae. *Proc. Entomol. Soc. Washi.*, 16: 85.

Huang, J. & Polaszek. (1998). A revision of Chinese species of *Encarsia*Foerster (Hymenoptera: Aphelinidae): parasitoids of whiteflies, scale insects and aphids (Hemiptera: Aleyrodidae, Diaspididae, Aphidoidea). *J. Nat. Hist.*, 32: 1825–1966.

Haung, J. (1994). *Systematic studies on Aphelinidae (Hymenoptera: Chalcidoidea) of China.* 348 pp. Chongqing Publishing House, Chongqing.

Husain, T. & Agarwal, M.M. (1982). Studies on some aphelinid (Hymenoptera: Chalcidoidea) parasites of hemipterous insects of India. *J. Bom. Nat. Hist. Soc.*, 79(1): 157–159.

Jadhav, R.B. & Varma, A. (2001). Natural enemies of sugarcane scale insect and their seasonal acitivity in Ahmednagar district of Maharashtra. *Ind. J. Sugar. Tech.*, 16(1): 123–126.

Janssen, M. (1961). Pigmentmodifikation und neuerFundort von *Aphelinusserniflavus*How. [Hymenoptera: Chalcidoidea]. *Beitr.Entomol.*, 11:671–678.

Jasnosh, V.A. (1979). Food specialization in the family Aphelinidae (Hymenoptera, Chalcidoidea). *Eruomol. Rev.*, 58: 61–67.

Khan, M.Y. & Shafee, S.A. (1976). Indian species of the genus *Azotus*(Hymenoptera, Aphelinidae). *Geobios, Jodhpur*, 3(5): 173.

Kim, J.W. & Heraty, J. (2012). A phylogenetic analysis of the genera of Aphelininae (Hymenoptera: Aphelinidae), with generic key and descriptions of new taxa. *Systematic Entomology*, 37: 497–549.

Koningberger, J.C. (1897). De dierkijkevijanden der koffiecutuur op Java, deel 1. *Mededelingenuit 'S Lands Plantentuin*, 20: 14–15.

Krishnan, B. & Vasantharaj David, B. (1996). Records and descriptions of some aphelinid parasitoids of Aleyrodidae (Homoptera: Insecta) from India. *Jai Research Foundation*, Valvada, Gujarat, pp. 21–22.

Lal, K.B. (1938). The fi rst record of the eulophid genus *Azotus*from India (Chalcidoidea: Hymenoptera). *Rec. Ind. Mus.*, 40: 2.

Le Barron, W. (1870). The chalcideous parasite of the apple-tree bark-louse (*Chalcis* (*Aphelinus*) *mytilaspidis*n, sp.). *Ame. Entomo. Bota*, 2(12): 360–362.

Lopez-Avila, A. (1987). Two new species of *Encarsia*Foerster (Hymenoptera: Aphelinidae) from Pakistan, associated with the cotton whitefly, *Bemisiatabaci*(Gennadius) (Hemiptera: Aleyrodidae). *Bull. Entomol. Rese.*, 77(3): 427–429. MANICKAVASAGAM and MENAKADEVI: A checklist of Aphelinidae.....of India

Mahdihassan, S. (1923). Classifi cation of lac insects from physiological stand point. *Journal of Scientifi c Association Maharajah's Coll, Vizianagaram*, 1: 47–99.

Mani, M.S. & Kurian, C. 1953. Descriptions and records of chalcids from India. *Ind. J. Entomo.*, 15(1): 1–22.

Mani, M.S. (1941). Studies on Indian parasitic Hymenoptera I. *Ind. J. Entomo.*, 3: 25–46.

Manickavasagam, S. & Menakadevi C. (2014). A checklist of Aphelinidae (Hymenoptera: Chalcidoidea) of India. *Rec. zool. Surv. India*, 114(1): 13-13.

Manickavasagam, S. & Menakadevi, C. 2012. A new species of *Eutrichosomella*Girault (Chalcidoidea: Aphelinidae) from India, with Additional distribution records of Chalcids. *Madras. Agric. J.*, 99(10-12): 877–881.

Masi, L. (1909). ContribuzioniallaconoscenzadeiChalcididiitaliani. *Bollett. Del. Labora. Di. Zool. Generale e Agrariadella R. ScuolaSuperiorre d Agricol. In Port.*, 4: 3–37.

Masi, L. (1911). ContribuzioniallaconoscenzadeiCalcididiItaliani (parte IV). *Bollettino del Laboratorio di ZoologiaGenerale e Agrariadella R. ScuolaSuperiored'Agricoltura, Portici*, 5: 140–171.

Menakadevi, C. & Manickavasagam, S. (2011). First report of *Encarsiacolima*Myartseva from India and other distributional records oaf aphelinidae (Hymenoptera: Chalcidoidea) from the states of Kerala and Tamil Nadu, India. *Madras Agric. J.*, 98(4-6): 182–186.

Menakadevi, C. & Manickavasagam, S. (2012). A new species of *Proaphelinoides*Girault (Hymenoptera: Aphelinidae) with a key to world species and additional records from Tamil Nadu, India. *Zootaxa*, 3193: 62–68.

Mercet, R.G. (1912a). Los enemigos de los parasites de las plantas. Trabajos del Museo de Ciencias Naturales, *Madrid*, 10: 30.

Mercet, R.G. (1912b). Un parasito dell 'pollroig'. *Boletin de la Real Sociedad Española de Historia Natural*, 12: 135.

Mercet, R.G. (1927). La fauna afelininaespañola. *Asoc. Esp. Progr. Cienc. Cádiz, Sección IV (Ciencas Nat.)* pp. 240.

Mercet, R.G. (1931). NotassobreAphelinidos (Hym. Chalc.) 4a nota. *Eos. Revista Española di Entomologia. Madrid*, 7: 395.

Motschulsky, V. de (1863). Essai d'un catalogue des insectes de l'Ile Ceylon (Suite). *Bulleten' MoskovskogoObshchestvaIspytateleyPrirody (OtdelBiologicheskiy)*, 36(3): 51.

Mueller, T.F., Blommers, L.H.M. & Mols, P.J.M., (1992). Woolly apple aphid (*Eriosoma lanigerum*Hausm., Hom., Aphididae) parasitism by *Aphelinusmali* Hal. (Hym., Aphelinidae) in relation to host stage and host colony size, shape and location. *Journal of Applied Entomology*, 114(2):143–154.

Munro, J.B., Heraty, J., Burks, R.A. Hawks D., Mottern J., Cruaud, A., Rasplus J.Y. and Jansta P. (2011) A molecular phylogeny of the Chalcidoidea (Hymenoptera). *PLoS ONE*, 6(11): 1–27.

Myartseva, S.N. (2005). Notes on the species of the genus *Encarsia*Foerster (Hymenoptera: Aphelinidae) introduced to Mexico for biological control of the blackfl y *Aleurocanthuswoglumi*Ashby (Homoptera: Aleyrodidae), with description of a new species. *ZoosystematicaRossica*, 14(1): 148–150.

Myartseva, S.N. & Evans, G.A. (2008). Genus *Encarsia*Förster of Mexico (Hymenoptera: Chalcidoidea: Aphelinidae). A revision, key and description of new species. Serie AvispasParasíticas de Plagas y otrosInsectos, Cd. Victoria, Mexico.

NBAIR (National Bureau of Agricultural Insect Resources). (2021). Indian general of aphelinids. https://databases.nbair.res.in//Aphelinidae/index.php (Last updated: July, 2020).

Neil, H.W., Sevenster-van der Lelie, L.A., Woets, J. & van Lenteren, J.C. (1976). The parasite-host relationship between *Encarsiaformosa* (Hymenoptera:Aphelinidae) and *Trialeurodesvaporariorum* (Homoptera:-Aleyrodidae). II. Selection of host stages for oviposition and feeding by the parasite. *Z. Angew. Entomol*, 81:372–376.

Noyes, J.S. (2003). Universal Chalcidoidea Database Website [WWW. Document]. URL www.nhm.ac.uk/entomology/chalcidoids/index.html [December, 2013-updated].

Noyes, J.S. (1985). Chalcidoids and biological control. *Chalcid Forum,* 5 (1): 5–10.

Noyes, J.S. (2012). An inordinate fondness of beetles, but seemingly even more fond of microhymenoptera Hamuli. *TheNewsletter of the International Society of Hymenopterists*, 3: 5–8.

Noyes, J.S. (2016). Universal Chalcidoidea Database. World Wide Web Electronic Publication. London, UK: Natural History Museum. Available online at http://www.nhm.ac.uk/our-science/data/chalcidoids/database/.

Noyes, J.S. (2017). Universal Chalcidoidea Database. http://www.nhm.ac.uk/chalcidoids (last accessed 24 Apr 2017).

Oppenheim, D., Palevsky, E., Horovitz, I., Shaltiel, L., Reuveny, H. & Aconis, O. (1997). The influence of poison-free pest management on the fauna of arthropod pests and their natural enemies in an apple orchard at Havat Matityahu, Israel, during the seasons of 1994-96. *Alon Hanotea*, 51(8): 346–356; 16 ref.

Osborne, L.S. & Landa, Z. (1992). Biological control of whiteflies with entomopathogenic fungi. *Florida Entomologist*, 75(4):456–471.

Polaszek, A. & Hayat, M. (1992). A revision of the genera *Dirphys*Howard and *Encarsiella*Hayat (Hymenoptera: Aphelinidae). *Syst. Entomo.*, 17: 193.

Polaszek, A. (1991) Egg parasitism in Aphelinidae (Hymenoptera: Chalcidoidea) with special reference to *Centrodora* and *Encarsia*. *Bulletin of Entomological Research*, 81, 97–106.

Polaszek, A.S., Manzari & D., Quicke. (2004). Morphological and molecular taxonomic analysis of the *Encarsiameritoria*species-complex (Hymenoptera, Aphelinidae), parasitoids of whiteflies (Hemiptera,-Aleyrodidae) of economic importance. *Zoologica Scripta*, 33: 403–421.

Raina, S.K., Rathod, M.K., Khurad, A.M., Mategaonkar, D., Adolkar, V.V. & Rai, M.M. (1996). Survey and identifi cation of citrus blackfl y and its parasitoids in the Vidarbha region of Maharashtra (India). *Conservation of hymenopteran parasitoids for management of the citrus blackfl y inoranges*, eds. Raina, S.K. & Overholt, W.A., pp. 18–21. Nagpur, Maharashtra, India: CSBR, Nagpur University.

Roermund, H.J.W., van; Lenteren, J.C. & van; Rabbinge, R. (1997). Biological control of greenhouse whitefly with the parasitoid*Encarsiaformosa* on tomato: an individual-based simulation approach. *Biological control: theory and applications in pest management (USA)*, 9(1): 25–47.

Rose, M. & Zolnerowich, G. (1997). *Eretmocerus* Haldeman (Hymenoptera: Aphelinidae) in the United States, with descriptions of new species attacking *Bemisiatabaci* complex (Homoptera: Aleyrodidae). Procee. *EntomolSoci. Washin.*, 99(1): 1–27.

Rosen, D. & De Bach, P. (1979). Species of *Aphytis*of the World (Hymenoptera: Aphelinidae), *Series Entomologica*, 17, Dr. W. Junk. KV. Publisher, The Hague.

Rosen, D. & De Bach, P. 1986. Three new species of *Aphytis*(Hymenoptera: Aphelinidae), parasites of *Pseudaulacaspis*spp. (Hymenoptera: Aphelinidae) in Indian and Australia. *Entomophaga*, 31: 139–151.

Rosen, D. & Rose, M. (1989). *Aphytis fi oriniae*sp. nov. (Hymenoptera: Aphelinidae), a parasite of Tea scale, *Fioriniatheae*Green, from India. *Orie. Inse.*, 23: 270–273.

Rosen, D. (1990). *World Crop Pests. 4B. Armoured Scale Insects: their biology, natural enemies and control.* Amsterdam, Netherlands: Elsevier Science Publishers, 688 pp.

Schmidt, S., Naumann, I.D. & De Barro, P.J. 2001. *Encarsia*species (Hymenoptera: Aphelinidae) of Australia and the Pacific islands attacking *Bemisiatabaci*and*Trieurodesvaporariorum* (Hemiptera: Aleyrodidae)-a Pictorial key and descriptions of four new species. *Bull. Entomol.Rese.*, 91: 377.

Shafee, S.A., Siddiqui, J.M. & Rizvi, S. (1988). Description of four new species of Aphelinidae (Hymenoptera) from India. *Ind. J. Sys. Ent.*, 5(1): 6–7.

Shafee, S.A. (1970). A new genus of Aphelinidae recorded from Ootacumund (India) (Hymenoptera). *Mushi*, 43: 145–146.

Shafee, S.A. (1972). Indian species of the genus *Coccophagus*Westwood (Hymenoptera: Aphelinidae). *Bull. Entomo. Soci. India*, 13(1): 25.

Shafee, S.A. (1973). Indian species of the genus *Prospaltella*Ashmead (Hymenoptera:Aphelinidae). *Entomophaga*, 18: 251–258.

Shafee, S.A. (1974). A new family of Chalcidoidea (Insecta: Hymenoptera) Euryischidaefam.n. *Current Science*, 43(23): 768. MANICKAVASAGAM and MENAKADEVI: A checklist of Aphelinidae.....of India

Shafee, S.A., Azim, M.N. & Khan, M.Y. (1988). Taxonomic notes on some genera of Aphelinidae (Hymenoptera: Chalcidoidea). *Ind. J. Syst. Entomol.*, 2(2): 27–30.

Shaw, P.W., Walker, J.T.S. & O'Callaghan, M. (1996). Biological control of woolly apple aphid by *Aphelinusmali* in an integrated fruit production programme in Nelson. Proceedings of the Forty Ninth New Zealand Plant Protection Conference, Nelson, New Zealand, 13-15 August, 1996. Rotorua, New Zealand: New Zealand Plant Protection Society, 59-63

Sivestri, F. (1909). Notizie e descrizioni preliminari di insetti parassti della lDiaspis pentagonal. Atti della Reale Accademia dei Lincei. Rendiconti. Roma. Classe Scienze Fiziche, Matematiche e Naturali.18(1):563-565.

Silvestri, F. (1926). Descrizione di tre specie di *Prospaltellae* di una di *Encarsia*(Hym. Chalcididae) parassite di *Aleurocanthus*(Aleyrodidae). *Eos. Revista Española di Entomologia. Madrid*, 2: 179.

Silvestri, F. (1927). Contribuzioneallaconoscenzadegli Aleyrodidae viventisu citrus in estremooriente e deiloroparassiti. *Bollettino del Laboratorio di ZoologiaGenerale e AgrariadellaR. IstitutoSuperiored'-Agricoltura, Portici*, 21: 28.

Singh, S. & Srinivasa, B. (2010). Description of a new species of the genus *Eutrichosomella*Girault (Hymenoptera: Chalcidoidea: Aphelinidae) from the Western Ghats, India. *Ann. Entomo*, 28(2): 1–5.

Singh, S.P. (2004). Some success stories in classical biocontrol of agricultural pests in India. *APPARI*. 80 p.

Stringer, D., Mantel S., Jennings, J. & Austin, A. (2012). Family Aphelinidae. [WWW. Document].URL:https:// www.researchgate.net/publication/262008500_Family_APHELINIDAE.[July, 2012-uploaded].

Subba Rao, B.R. (1980). *Botryoideclavabharatiya*, gen et spnov and a new species of *Eriaphytis*Hayat from India (Hymenoptera: Aphelinidae). *Orie. Inse.*, 14: 41–45.

Subba Rao, B.R. (1984). Descriptions of new species of Oriental Mymaridae and Aphelinidae (Hymenoptera: Chalcidoidea). *Proc. Ind. Acad. Scie., (Animal Sciences)*, 93: 261.

Tejada, M.L.O. & Rumayor, R.O.L. (1986). Phytosanitary study of apple cultivation (*Malus sylvestris* Mill) in the municipalities of Arteaga and Saltillo, Coahuila. Informe de Investigacion, Division de CienciasAgropecuarias y Maritimas, Instituto Technologico y de EstudiosSuperiores de Monterrey, Mexico, No. 19:25-26.

Thakur, J.N., Rawat, U.S. & Pawar, A.D. (1988). Survey for natural enemies of Pomegranate butterfly *Deudorixepijarbas*Moore in Himachal Pradesh, *J. Biol. control*, 2: 1–2.

Thakur, J.R. & Dogra, G.S. (1980). Woolly apple aphid, *Eriosomalanigerum*, research in India. *Tropical Pest Management*, 26(1): 8–12.

Thompson, W.R. (1955). *A catalogue of the parasites and predators of insect pests. Section 2. Host parasite catalogue, Part 3. Hosts of the Hymenoptera (Calliceratid to Evaniid)*. pp. 291 Commonwealth Agricultural Bureaux, The Commonwealth Institute of Biological Control, Ottawa, Ontario, Canada.

Timberlake, P.H. (1924). Descriptions of new chalcid-flies from Hawaii and Mexico (Hymenoptera). *Proc. Hawa. Entomol. Soci.*, 5: 408.

Tower, D.G. (1913). A new hymenopterous parasite on *Aspidiotusperniciosus*Comst. *Ann. Entomol. Soc. Amer.*, 6(1): 125.

Traboulsi, R. (1969). Contribution à l'étude des *Aphytis* Howard du Liban[Hymenoptera, Chalcidoidea, Aphelinidae]. *Ann. Soc. Entomol. Fr.n.s.* 5(1):5–72.

Tripathi, G.M. (1995). Records of parasite and predator complex of sugarcane woolly aphid, *Ceratovacunalangera*Zehnt. In Nagaland. *Indian Sugar*, 44(11): 839–841.

Viggiani, G. & Mazzone, P. (1978). Morfologia, biologia e utilizzazione di *Prospaltellalahorensis* How. (Hym. Aphelinidae), parassitaesoticointrodotto in Italia per la lotta biologic a al *Dialeurodescitri* (Ashm.). *Boll. Lab. Entomol. Agrar. F. Silvestri Portici*, 35: 99–161

Viggiani, G. & Mazzone. (1980). Le specie paleartiche di *Encarsia*delgruppo lutea Masi (Hym. Aphelinidae), con descrizione di due nuove specie. *Ricerchesugli* Hymenoptera Chalcidoidea. *LXVIII.-Boll. del Labor. di Zool. Gene. e Agra. della R. Istituto Supe. d' Agrico., Por.*, 37: 51–57.

Viggiani, G. (1981). Hyperparasitism and sex differentiation in the Aphelinidae. In The Role of Hyperparasitism in Biological Control: A Symposium, Div. Agric. Sci. Univ. Calif. Publ., ed.D. Rosen, 45. 410:19–26.

Viggiani, G. (1984). Bionomics of the Aphelinidae. *Annual Review of Entomology*: 29, 257–276.

Viggiani, G. (1985). Notes on a few Aphelinidae, with description of five new species of *Encarsia*Foerster (Hymenoptera, Chalcidoidea). *Boll. del Labo. di Entom. Agra. 'Fili. Silve.', Por.*, 42: 87.

Viggiani, G. (1987). New species of *Encarsia*Förster (Hymenoptera: Aphelinidae), parasitoids of whiteflies. *Boll. del Labo. di Entom. Agra. 'Fili. Silve.', Por.*, 44: 35–37.

Walker, F. (1839). *MonographiaChalciditum*. 1: 9 London.

Waterson, J. (1916). Notes on Coccid infesting Chalcidoidea. *Bull. Entomol. Rese., London*, 7: 137- 144.

Woolley, J.B. (1988) Phylogeny and classification of the Signiphoridae (Hymenoptera: Chalcidoidea). *Systematic Entomology*, 13,465–501.

Yasnosh, V.A. (1963). New species of the genus *Aphelinus*Dalm. (Hymenoptera, Chalcidoidea) in the fauna of the USSR. *Entomolo. Obozr.*, 42(1): 181.

Yasnosh, V.A. (1979). Food specialization in the family Aphelinidae (Hymenoptera, Chalcidoidea). *Entomological Review*, 58, 61–70.

Yu, D.S. & Luck, R.F. 1988. Temperature-dependent size and development of California red scale (Homoptera: Diaspididae) and its effect on host availability for the ectoparasitoid, *Aphytismelinus*DeBach (Hymenoptera: Aphelinidae). *Environmental Entomology*, 17: 154–161.

Zehntner, L. (1896). De Plantenluizen van het suikerroet op Java. *(I. AleurodesbergiSignoret.) Mededelingen van het Proefstation Oost-Java (Nieuwe Serie)*, 29: 11.

Zehntner, L. (1900). De Plantenluizen van het Suikerriet op Java. (X. *CeratovacunalanigeraZ*ehnt. de 'Witte Luis' der bladeren.). *Mededelingen van het ProefstationvoorSuikerriet in West-Java.'Kagok' TePekalongan*, 49: 12.

Zolnerowich, G. & Rose, M. (1998). *Eretmocerus*Haldeman (Hymenoptera: Aphelinidae) imported and released in the United States for control of (*Bemisiatabaci*complex) (Homoptera: Aleyrodidae). *Proceedings of the Entomological Society of Washington*, 100(2): 310–323.

12 Beneficial Diptera

Dhriti Banerjee, Atanu Naskar, and Jayita Sengupta
Diptera Section, Zoological Survey of India, New Alipore, Kolkata

CONTENTS

LEARNING OBJECTIVES

i. Basic knowledge of the insect order Diptera: morphology, classification, phylogeny, habits and habitat.
ii. Understanding the biological attributes of the order Diptera: Life history and development and feeding behaviour.
iii. Functional attributes of the different types of feeding patterns in different life cycle stages and the role played by the relevant functional groups in the ecosystems.
iv. Analysing the behavioural attributes of Carnivory: Parasitism and Predation in Diptera.
v. Analysing the beneficial role of Parasitoids, and Predators in the biocontrol of pests.
vi. Harnessing the beneficial parasitoids for rearing and mass production in both *in vivo* and *in vitro* for commercial release for Integrated Pest Management.

DOI: 10.1201/9781003354239-12

vii. Evaluating the efficacy of the beneficial dipteran parasitoids and predators as an alternative to chemical pesticides and understanding their usefulness in ecosystem restoration and sustainable agriculture.

12.1 INTRODUCTION

The order Diptera (true flies) is a group of two-winged insects deriving its name from the Greek word *diptera* (*di* meaning two and *pteron* meaning wings), coined 2000 years ago by the Greek philosopher Aristotle and espoused by Linnaeus in 1744. The hindwings of this group of insects have highly evolved into mechano-sensorial rotational halteres for balancing during flight and advanced 'aerobatics' (Dickinson and Michael, 1999). The Diptera, commonly known as flies, are a biologically and morphologically diverse group of insects, making up 10–15% of the known animal species. They are found all over the world being cosmopolitan in their distribution. They are generally considered a nuisance since they are vectors of dreaded diseases, like malaria, dengue, kala-azar, yellow fever and sleeping sickness. However, the myriad of ecosystem services, beneficial Diptera provide, underpin all dimensions of human well-being. Flies have a significant contribution to the dynamic ecological functions, like crop pollination, pest control, decomposition, and maintenance of wildlife species in agrarian and wild ecosystems. They are major detectors of cadavers and carcasses, providing useful information regarding the time of death, manner of death and localisation of the dead, playing an important role in forensics and criminology. They are also model organisms (*Drosophila)* in the field of genetics for the past century and are key players in the nascent field of evolutionary developmental biology in the 21st century. The role played by the beneficial Diptera in ecosystem service provision is essential to fully assess synergies between functional ecology, community ecology, and biodiversity conservation under current global changes (Noriega et al., 2018). The actual economics of the ecosystem services provided by the Diptera remains to be estimated, though a very traditional estimate by Losey and Vaughan (2006), shows that the ecological services provided by primarily native insects in the United States were more than $57 billion ($0.38 billion for dung burial, $3.07 billion for pollination, $4.49 billion for pest control of native herbivores, and $49.96 billion for recreation).

12.1.1 Phylogeny and Classification

The Diptera are known to be around, 240 million years ago, during the Middle Triassic and they became widespread during the Mid-Triassic and Late-Triassic. Dipterans are most closely related to the Mecoptera (scorpionflies) and the Siphonaptera (fleas) who were their presumed ancestors. Traditionally, the flies are divided into two major groups based on their antennal structure: the Lower Diptera (erstwhile 'Nematocera'), flies with long antennae like mosquitoes and the Brachycera, stout and fast-moving flies with short antennae. The Lower Diptera is composed of Tipulomorpha (crane flies), Culicomorpha(mosquitoes), Psychodomorpha (sandflies) and Bibinomorpha (march flies and gall midges). The Brachycera comprises Eremoneura (flies with three larval instars), Cyclorrapha (flies that pupate in a puparium), Schizophora (flies that escape from the puparium using an eversible frontal pouch: *Drosophila*), and Calyptratae (larger flies with wings having an enlarged basal lobe called calypter: houseflies) (Weigmann et al., 2011). The majority species of Brachycera, including the *Drosophila* and the houseflies belong to the clade Cyclorrapha, characterised by their adaptable larval stage(maggot) and their means of metamorphosis (puparium) (Weigmann et al., 2011).

12.1.2 Habits and Habitat

The Diptera occur almost in all places, from high mountainous elevations up to 5600 m in the Himalayas to the seas at depths of more than 1000 m and from hot deserts to the cold ice waters of Antarctica in the contrasting extreme. Adult dipterans can be broadly classified as terrestrial, aquatic and semi-aquatic. They are found in flowers and herbages, in dry to moist heaths, dunes, grasslands, forests, and uplands extending to the shoreline vegetation. The habitat of dipterans varies from fungoid growths, dead decaying organic matter, rotten tree trunks, decaying wood, animal remains, inside bird nests, fresh wounds on tree trunks, animal sheds of horses, sheep, goats and cattle, ants, termites and wasps nests, to human dwellings, like basements, bathrooms, roof thatches, sewers, drains, and stagnant water in potholes and ditches.

12.1.3 Ecosystem Services

A traditional estimation of the value of ecosystem services provided by insects is almost $60 billion a year in the United States, which is only a fraction of the value of all the services insects provide (Losey and Vaughan, 2006). Ecosystem services are services, outputs, conditions, or processes of natural systems that directly or indirectly benefit humans or enhance social welfare, either directly or as inputs into the production of other goods and services. However, this role is often assumed, with limited or no experimental quantification of its real value. Dipteran insects play a key role in the regulation and dynamics of our ecosystems through their ecological services. Pollination, biological control, food provisioning, and recycling organic matter are the most common and valued ecosystem services provided by flies. Based on the ecosystem services, *viz.* pollination, biological pest control and nutrient recycling in agroecosystems, the most commonly studied functional groups in Diptera are pollinators, predators, parasitoids and decomposers. In this chapter, the role of Diptera in bio-control is focussed upon. Contemporary research on Diptera has served as a model system for studies in the field of predation –parasitism- parasitoids, with regard to their optimal foraging behavior, adaptive radiations, sex allocation strategies and information processing (Lacey et al., 2001). Their ability to survive in different habitats is the chief reason behind their diversity and their life stages made the dipteran flies highly capable as potential biocontrol agents.

12.1.4 Life Cycle and Developmental Stages

Adults: Typical adult flies have mouthparts designed for piercing and sucking (mosquitoes, black flies and robber flies), or for lapping and sucking, and sponging type in the other groups. Their head is highly mobile, with a pair of large compound eyes. Thorax is short, and stout with a pair of highly manoeuvrable wings and legs with claws and pads on their feet that enable them to cling to smooth surfaces. The adult flies feed, mate and lay eggs. The flies have much quicker and more reproductive capacity than most insect groups. The torsion of the male genitals in flies allows the male to mate while remaining upright and hence gives them a higher reproductive advantage over most insects.

Flies mostly have short lives as adults and undergo complete metamorphosis, with four distinct life stages – egg, larva, pupa and adult. Eggs are laid on the larval food-source and the apodous larvae (which lack true limbs) develop in a protected environment, often inside their food source and the pupa is a tough capsule from which the adult mature fly emerges. The predatory beneficial Diptera are commonly the adult members of the family Asilidae, Dolichopodidae, Empididae, Lonchaidae and Syrphidae, who provide both ecological services of biological control and pollination some being both predators (as larvae) and pollinators(as adults).

Larva: In most cases, the larval stage is long compared to the adult flies. Usually, the larvae prefer to develop in sheltered environments, like moist places such as carrion, fruit, vegetable matter, and fungi, but in the case of parasitic species, they develop inside their hosts. They tend to have thin cuticles and become desiccated if exposed to the air. In the case of the endoparasitic, carrion and detritus-feeding larvae, the eyes and antennae of larvae are reduced or absent, and the abdomen also lacks appendages (Hoell et al., 1998), while free-living larvae generally have well-developed eyes and antennae. Diptera larvae lack true jointed appendages but some dipteran larvae, such as species of Simuliidae, Tabanidae and Vermilionidae, have prolegs adapted to hold onto a substrate in flowing water, host tissues or prey (Chapman, 1998). The majority of dipterans are oviparous and lay batches of eggs, but some species are ovoviviparous, where the larvae start their development inside the eggs before they hatch, or viviparous, where the larvae hatch and mature in the body of the mother before being externally deposited. These are found especially in groups that have larvae dependent on food sources that are short-lived or are accessible for brief periods (Meier et al., 1999). This is widespread in families, such as the Sarcophagidae. The Glossinidae, Hippoboscidae, Nycteribidae and Streblidae exhibit adenotrophic viviparity while some Chironomidae reproduce by parthenogenesis, and some gall midges have larvae that can produce eggs (paedogenesis) (Gillott, 2005; Mcmahon et al., 2016). Diptera larvae may be predaceous that prey upon the larvae of other insects or even other Diptera and they can be parasitoids. Diptera parasitoids are immature larvae that develop on or within an insect host and ultimately kill it. After feeding on host body fluids and organs, most parasitoids leave their hosts to pupate or emerge as adults. Both predator and parasitoid larvae are effective biocontrol agents of pest populations.

Pupa: Diptera pupae are either 'obtect' or 'coarctate'. In the Nematocera, the pupa is intermediate between the larval and adult form with visible future appendages, bear spines, respiratory features or locomotor paddles. The appendages are not visible in other groups, where the actual pupa is concealed within a puparium. When the adult insect is ready to emerge from this tough, desiccation-resistant capsule, it inflates a balloon-like structure on its head, and forces its way out (Resh et al., 2009). The pupae, however, are non-invasive and do not play any role in predation or parasitism.

12.1.5 Feeding Behaviour

The feeding behaviour and food habits of most dipterans are largely unknown. Based on their food habits, the fly communities are divided into four major groups: i.e. saprophagous/necrophagous, carnivorous, phytophagous and hematophagous. Broadly, phytophagous Diptera feeds on pollen, nectar and plant sap. Hence they are important crop pollinators and plant pests. Adult Brachycera, like Syrphidae, Conopidae, Bombyliidae, Nemestrinidae, and adult Nematocera, like Tipulidae, Bibionidae, are major plant feeders. The Agromyziodae are leaf miners, while Cecidomyiidae and some Tephritidae are leaf miners or gall formers. Many Diptera, like Sarcophagidae, Calliphoridae, Muscidae and Stratiomyidae are saprophagous detrivores. Adult Asilidae, Empididae, Scathophagidae, Dolichopodidae and Ephydridae are carnivorous predators and feed on other insects, including smaller Diptera. Adult Culicidae (mosquitoes), both male and female, feed on nectar and plant juices, but in many species the mouthparts of the females are adapted for piercing the skin of animal hosts and feed on blood as ectoparasites. Most females, likeTabanidae (horse flies), feed on mammal blood, but some species are known to feed on birds, amphibians or reptiles. Other haematophagous (blood-feeding) Diptera include Ceratopogonidae, Phlebotomidae, Hippoboscidae, some Muscidae and Rhagionidae which are obligate blood-feeders.

The Diptera larvae interestingly feed on a diverse array of nutrients, often widely different from those of adults. In most cases, they feed in leaf-litter, leaves, stems, roots, flower and seed heads of plants, moss, fungi, rotting wood, rotting fruit or other organic matter, such as slime, flowing sap, and rotting cacti, carrion, dung, detritus in mammal, bird or wasp nests, fine organic material including insect and microorganisms. However, several Diptera larvae are carnivorous, being predatory or parasitoid and feed on larvae of other insects, and sometimes on the larvae of other Diptera as well. The syrphid larvae, for instance, are flower-feeding saprotrophs, eating decaying plant or animal matter, or insectivorous, eating plant-sucking insects, like thrips and aphids, thus serve as a dual ecosystem service provider of crop pollination and biocontrol of pests. Parasitic larvae are common in Diptera. The larvae of all Oesteridae are obligate parasites of mammals. The Calliphoridae larvae cause myiasis in both humans and livestock. Tachinidae larvae are parasitic and parasitoids on other insects. Larval Pyrgotidae are endoparasites of adult scarab beetles while larval Conopidae are endoparasites of bees and wasps or of cockroaches and calyptrate Diptera. Larval Sciomyzidae are associated with terrestrial and freshwater snails, and slugs feed on snails as predators, parasitoids, or scavengers. Odiniidae larvae are predatory scavengers living in the tunnels of wood-boring larvae of Coleoptera, Lepidoptera, and other Diptera. *Oedoparena* larvae feed on barnacles. The hyper-metamorphic larvae of Bombyliidae and Acroceridaeare well-known predators in natural ecosystems.

12.2 CARNIVORY IN DIPTERA: PREDATORS, PARASITES AND PARASITOIDS

The feeding behaviour of Diptera includes three basic types of carnivory: predation, parasitoidism and parasitism (Vinson and Barbosa, 1987). All three comprise the third trophic level characterised by the consumption of organisms and other carnivores. Parasitism is characterised by multiple hosts that survive and are only partially consumed (Schmid-Hempel, 1998). Consumption of tissues, which does not result in the death of the host, has also been termed parasitism (Schoonhoven et al., 1998). The life histories of parasitoids are ecologically transitional between predators – which rapidly consume multiple and often unrelated individuals during a lifetime – and parasites, which feed through several generations on a near-infinite food resource from mostly single, generally surviving individuals. Parasitoids pursue an ecologically intermediate strategy of feeding generally on a predictable and finite food resource consisting of a single host that is eventually killed. Moreover, they are characterised by a life cycle that includes a free-living reproductive stage, whose habitat is separate from that of the developing larvae (Yeates and Greathead, 1997). Additional distinctions between predators and parasites help to define the role of parasitoids. The host-accommodation exhibited by parasites and parasitoids, in contrast to the rapid consumption of prey by predators, is a major theme in the evolutionary history of fly carnivory. The trophic diversity in dipterans is reflected in their feeding habits of both larvae and adults. The adults belong to the same trophic grouping but may adopt divergent feeding habits. They may be predaceous, parasitic or parasitiods. Again adults of parasitoids are free-living and may be predaceous. Biological control or the regulation of insects by other insects involves predators, parasites, or parasitoids. Predators consume multiple hosts, and parasites take their nutrition from a host but do not necessarily kill it. Fly parasitoids though often called as parasites, have an immature life stage that develops on or within a single insect host, ultimately killing the host, hence the value of parasitoids as natural enemies.

The research on beneficial Diptera as a potential biocontrol agent is comparatively young compared to other insect orders, even though 20% of their total species database, i.e. approximately 16,000 species, projects them as useful biocontrol agents (Feener and Brown, 1997). Currently more than 51 families are of carnivorous Diptera (Table 12.1) of which 30 dipteran larval families and 15 adult dipteran families are well-known as effective biocontrol agents. The major families of beneficial flies are Bombyliidae, Syrphidae, Cecidomyiidae, Sciomyzidae, Lonchaeidae, Dolichopodidae, Ephydridae, Conopidae, Agromyzidae, Phoridae, Sarcophagidae, Fannidae, Muscidae, Calliphoridae and Tachnidae. The research on predation-parasitism-parasitoidism is comparatively a new subject, and only a few families, i.e., Tachinidae, Sarcophagidae, Phoridae and Bombyliidae, have been taken on as subjects of interest for applied biocontrol that would also be of concern in augmenting crop production. The focus is usually Tachinidae centric, as it is the most investigated group in terms of the rearing and production techniques, both *in vivo* as well as *in vitro*. There are also other Dipteran groups though not that well documented but play major roles in the biocontrol of pests and can be agents of the future.

TABLE 12.1
Dipteran Families of Predators Parasitoids and Parasites

Predators		Parasitoids	Parasite
Larvae	Adults	Invertebrate	Vertebrate (Larvae/Adults)
Apioceridae	Asilidae	Acroceridae	Carnidae
Asilidae	Dolichopodidae	Bombyliidae	Oestridae
Athericidae	Empididae	Calliphoridae	Sarcophagidae
Cecidomyiidae	Hybotidae	Conopidae	Hippoboscidae
Ceratopogonidae	Muscidae	Cryptochetidae	Calliphoridae
Chamaemyiidae	Scathophagidae	Nemestrinidae	Piophilidae
Chironomidae		Phaeomyiidae	
Chloropidae		Phoridae	
Corthrellidae		Pipunculidae	
Culicidae		Pyrgotidae	
Dolichopodidae		Rhinophoridae	
Drosophilidae		Sarcophagidae	
Dryomyzidae		Sciomyzidae	
Empididae		Tachinidae	
Keroplatidae		Tachiniscidae	
Muscidae			
Mydidae			
Odiniidae			
Pallopteridae			
Pelecorhynchidae			
Rhagionidae			
Scenopinidae			
Sciomyzidae			
Stratiomyidae			
Syrphidae			
Tabanidae			
Therevidae			
Tipulidae			
Vermileonidae			
Xylomyidae			
Xylophagidae			

12.2.1 Evolution of Carnivory in Diptera

The intricate web of selection pressures, like environmental cues for the selection of hosts, interaction with hosts, reproductive behaviour of the adult parasites, host discrimination, the evolution of host specificity, patterns of host use, behavioural interaction with host, over long periods of time have allowed dipterans to evolve in terms of ecological and evolutionary contexts. This has further mediated the emergence of dipteran parasitoids as crucial natural biocontrol agents of ecosystems. Despite their diversity and ecological impact, little is known about the early evolution and ecology of parasitoid insects, especially parasitoid Diptera barring that most of these have radiated from ancestors with saprophagous lineages. A bizarre fly, *Zheniaxiai* (family Eremochaetidae), an endoparasitoid insect, from late Cretaceous Burmese amber (about 99 million years old) has a highly developed, hypodermic-like ovipositor and is the first fossil record of parasitism from Cretaceous amber.

Flies are one of the most ancient pollinators of flowering plants (Labandeira, 1998), and played an important role in the origin of co-evolutionary relationships with flowering plants and insects (Ssymank et al., 2008). Since angiosperms did not appear until 140 million years ago (around the Cretaceous), so nectar was never their primary source of nutrition. This allowed the flies to evolve and made them capable of occupying a whole range of trophic niches (Weigmann et al., 2011). Nevertheless, the angiosperms rose to dominance during the Cretaceous. The angiosperm radiations provided new food resources and habitats and had a profound effect on flies, beetles, and other insects (Wang et al., 2013). However, the interval from the middle early Cretaceous to early Late Cretaceous witnessed a significant transformation to the modern terrestrial world. Between this time (from 125 million years ago to 90 million years ago), there were significant shifts in the major ecological associations among plants, insects, and other organismic groups dominant on land (Labandeira, 2010). The phenomenon of larval parasitism may have evolved during this switch coupled with multiple interspecific insect interactions and close proximity to the host species in gruelling conditions.

12.2.2 Diptera Predators

The primary mode of carnivory which has made dipterans beneficial to our ecosystem is predation. Predatory Diptera are flies that capture and eat another insect (the prey). Predatory flies are usually larger than their prey or overwhelm their prey by attacking. Most predatory flies do not have any specific prey. Death to the victim is usually immediately after capture. Adults or young both may be predatory. In several groups, the young ones are predatory but the adults are not. Carnivory is widespread in Diptera since the larval stages range from predatory generalists taking a broad range of prey species to highly prey-specific parasitoids. The predaceous fly larvae usually kill multiple hosts, while parasitoids and parasites are single-host-specific. The predaceous flies can attack invertebrates as a primary food source but in some cases they may also exploit vertebrate prey/hosts, like frogs and salamanders. The predatory tactic is particularly manifested in Apioceridae, Athericidae, Cecidomyiidae, Chamaemyiidae, Chaoboridae, Dolichopodidae, Dryomyzidae, Empididae, Mydidae, Rhagionidae, Scenopinidae, Syrphidae, Tabanidae, Therevidae and Sciomyzidae, with the latter family involving malacophagy.

However, classical adult predation is displayed by one of the leading predators from the dipteran ranks, the family Asilidae (Joernand Rudd, 1982). An extremely efficient population controller of the insect world is represented by this group of robber flies (Ghahari et al., 2007). They have been reported preying on multiple arthropods including flies, beetles, grasshoppers, dragonflies, hymenopterans, and even spiders (Dennis et al., 2012) and indirectly maintain the balance in the trophic levels of the concerned ecosystem (Dennis and Lavigne, 2007). The carnivorous Asilidae can feed on adult insects, double its body size as well as their larvae and play an important role in controlling populations as primary predators. They are well known for their role in maintaining ecosystem balance. Besides the family Asilidae, the adults of the family Dolichopodidaeare an established group of predator flies. They prey on mostly nematoceran Diptera families, likeChironomidaeand Culicidae, which are a common nuisance to human well-being (Ulrich, 2004). Both Dolichopodidae and Lonchaeidae play an important role in controlling the population on bark beetles and are extremely important to the pine wood industry. They can feed voraciously on the eggs, larvae, and adults of the bark beetles, which wreak havoc on the conifers, thus saving millions of dollars of forest produce.

12.2.3 Diptera Parasitoids

The larval stages of flies always play a crucial role in controlling insect populations by preying and parasitizing them. This makes them classically beneficial in both wild and managed ecosystems. The most well-known of them are the larvae of the family Syrphidae (Chambers and Adams, 1986). The larvae of almost all the species are commercially available as aphid biocontrol agents in agricultural and horticultural crop fields (Sommaggio, 1999). Feeding on aphids of different food crops along different ecosystems has helped this group to emerge as natural biocontrol agents (Sommaggio and Burgio, 2014). Syrphidae have also been found to be parasitoids on *Brassica* pests, whereas the family Agromyzidae is a parasitoid on different species of leaf miners including potato and pea leaf miners.

The gall midges, i.e. Cecidomyiidae, are reportedly parasitoidic on the banana aphids and are being studied for its potential in biocontrol. The agricultural production of banana fruit is getting affected by BBTV virus, transmitted by banana aphid, *Pentalonia nigronervosa* (Coquerel) (Robson et al., 2007). Flies from family Cecidomyiidae use this aphid as an obligatory host and thereby emerge as effective biocontrol agents by acting as aphid parasitoids (Gagne et al., 2004). The five species, which mainly act as biocontrol agents, are *Endaphis perfida* (Kieffer), *Endaphis compitalis* (Mamaev), *Endaphis gregaria* (Gagne), *Endaphis aphidomyza* (Grove andKasyap) and *Endaphis maculans*. These koinobiont exoparasitoids mediate this host killing by penetrating banana exoskeleton (Waring and Price, 1989). They are also known for their ability to control coccoids, several species of predatory mites as well as parasitic nematodes on vegetables. The predatory gall midge, *Feltiella acarisuga* (Vallot), is a biocontrol agent for two-spotted spider mites on greenhouse vegetable crops (Gillespie et al., 2000).

The biting midges or Ceratopogonidae are commonly known for its disease-causing ability, but the larvae of the Ceratopogonidae feed on a wide range of food items including algae, diatoms, certain bacteria, protozoa, and small invertebrates, such as rotifers and nematodes (Reeves, 2010) and have established themselves as controlling agents in aquatic ecosystems. The larvae of the family Chamaemyiidae are predators of scale insects, mealybugs and aphids (Fréchette et al., 2008). The larvae of the family Chironomidae have a great effect on controlling the transmission of onchocerciasis, or river blindness disease. This disease is transmitted by repeated bites by blackflies of the genus *Simulium*. The larvae of the family Chironomidae feed upon the larvae of the black flies, by this way Chironomidae larvae have a great effect on controlling the deadly disease of onchocerciasis. Besides this disease-controlling efficiency, as most of the larvae are benthic in nature, they are able to control aquatic ecosystems, acting as natural predators.

The dipteran larvae are also able to maintain the ecosystem balance by controlling the invasive plant species. The larvae of the species *Platycephala planifrons* of the family Chloropidae are generally used to control the invasive plant species of *Phragmites australis* to maintain the ecosystem stability of the native plant species (Häfliger et al., 2005). Mosquitoes are known worldwide for their disease-causing capabilities, but the largest non-biting mosquito genus *Toxorhynchites* is completely harmless to man, rather the larvae are predaceous on other mosquitoes and aquatic organisms. This predaceous habit of *Toxorhynchites* has established the genus as larval predators of the deadly *Aedes aegyptii* and a potent biocontrol agent under demanding conditions. The family Drosophilidae displays marked parasitism. The larvae of the subfamily Steganinae of the family Drosophilidae feed upon the lachrymal secretions of living mammals and while feeding these lachryphagous species parasitizes the spirurid nematode, *Thelazia callipaeda*, which parasitizes the eyes of domestic and wild carnivores and also the humans (Mácaand Otranto, 2014). The larvae of the family Mydidae are known to be effective soil-dwelling predators. They have well-developed mouthparts to prey on other soil-dwelling insect larvae, especially on harmful coleopteran larvae and mosquito larvae. Even the small group of dipteran family Odiniidae, having only 58 species recorded worldwide, has an impact on ecosystem services by acting as a biocontrol agent. The family Ephydridae is found to be accountable for the biocontrol of submersed aquatic weed and is a natural controlling agent in aquatic ecosystems. The family Sciomyzidae is larval parasitoid of snails, slugs and fingernail clams and larvae of one particular Sciomyzidae even prey on eggs of aquatic and semiaquatic snails. The larvae of the family Dryomyzidae are also known to be parasitoids feeding on barnacles.

The vast majority of the family Bombyliidae are well-known parasitoids. Their recorded host range spans over numerous insect groups including spiders, bees and wasps, grasshoppers, locusts, reduviid bugs and lepidopterous larvae (Muniz-Reyes et al., 2011; Boesi et al., 2009).

Topical research on Diptera as a biocontrol agent is restricted mostly to the taxonomic reports but its detailed suitability in the perspective of applied entomology is still not well documented. Tachinidae represents the largest and most important group of non-hymenopteran parasitoids and is notable because of the diverse oviposition strategies that allow them to parasitize hosts in varying environments (Dindo et al., 2019). These flies have been used as biocontrol agents from the beginning of the century and these parasitoids have been used for inoculative, innundative and augmentative releases in the field most effectively.

They play a major role in regulating the populations of herbivore insects, primarily, larval lepidopterans, which represent 70% of their known host species. There are several studies conducted from the Nearctic and Neotropical regions and few from the Palearctic, Australian and Oceanian regions for successful biocontrol programme, both in the case of classical and augmentative biocontrol. The biology, behavior and ecology of the species as well as proper rearing techniques and shipment procedures (Dindo and Grenier, 2014) have been studied in these areas, but data are deficient from the Oriental region (Dindo et al., 2019).

The first report of the use of Tachinidae flies in the biocontrol programme was reported from the Northern United States to control the gypsy moth, *Lymantria dispar* (L.) and the brown tail moth, *Nygmia phaerrhoea* Don. The parasitoid tachinids are able to exploit a wide range of plant-feeding insects. The insect orders which are utilised by tachinids as their hosts are mostly the larvae of the family Lepidoptera, Heteroptera, Dermaptera, Phasmatodea, Orthoptera, Coleoptera and larval Hymenoptera (Cerretti and Tschorsnig, 2010). About 70% of the known host species of tachinids are Lepidoptera, and among them, more than 50% belong to four families (Noctuidae, Geometridae, Tortricidae, and Pyralidae)of economical importance of Lepidoptera (Grenier and Liljestrhöm, 1991). Two species of Tachinidae, i.e. *Blepharipa pratensis* Meigen and *Compsilura concinnata* Meigen were successfully introduced from Europe to the Northern Unites states (Dindo and Grenier, 2014). In 1993, the species *Paratheresia menezesi* Townsend was introduced for the biocontrol of the coleopteran pest species, *Rhynchophorus palmarum* (L.) in the agricultural field (Moura et al., 2006). The tachinid species, *Strongygaster triangulifera* (Loew) was used for controlling the ladybird beetle, *Harmonia axyridis* (Pallas), in Western North America (Katsoyannos and Aliniazee, 1998). Augmentative releases of *L. diatraeae* to control *D. Saccharalis* and other lepidopterous sugarcane borers have been (and are currently) widely and successfully carried out in Cuba, which is the country of origin of the bio-controlling tachinids (Nicholls et al., 2002; Montes, 2008). In India, tachinid flies have been used for mostly classical and occasionally augmentative control programme in control of one of the most serious agricultural pests, *Helicoverpa armigera* (Huebner) (Lepidoptera: Noctuidae), which causes substantial losses to legume, fibre, cereal, oilseed and vegetable crop. Romeis and Shanower (1996) reported seven tachinids as natural parasitoids of this lepidopteran pest species, i.e. *Carcelia kockiana* Townsend, *Compsilura concinnata* Meigen, *Exoris tabombycis* (Louis), *Exorista japonica* (Townsend), *Exoris taxanthaspis* (Wiedemann), *Peribea orbuta* (Wiedemann), and *Sisyropa firmosa* Mesnil.

The families Phoridae and Sarcophagidae have displayed their potential for biocontrol too. Sarcophagid flies of the genus *Blaesoxipha* are known as natural enemies of grasshoppers, but their potential as biocontrol agents is not well documented (Mason and Erlandson, 1994; Danyk et al., 2005). The sarcophagid fly, i.e. *Seniorwhitea reciproca* (Walker), is also reported to be effective as a biocontrol agent in India. The family Phoridae is hyperparasitic on parasitic wasps, and can invade the red and black fire ants, which cause huge crop damages, take control over their brain and ultimately decapitate them, thus saving huge agricultural losses.

12.2.4 Biology of Diptera Parasitoids

The application of the parasitoid groups of dipteran flies as a biocontrol agent depends upon the successful study of the entire biology of the applied species, especially host range, oviposition strategies and host-parasite interaction. To locate hosts, dipteran parasitoids use environmental cues and the nature of these cues is often subtle and complex, reflecting the intense selection pressure faced by parasitoids in searching for hosts (Feener and Brown, 1997). A series of reviews were conducted from the different geographical locations worldwide for mass production. The first group-specific review of the parasitoid biology of Sarcophagidae was conducted by Pape (1987), and later a generalised review of dipteran parasitoid biology was initiated by Feenerand Brown (1997) mostly focusing on the family Phoridae. A detailed study of the biology of Bombyliidae was conducted by Yeates and Greathead (1997) and of Tachinidae by Dindo (2011).

Feener and Brown (1997) first gave a lucid description of the mechanism evolved by this group of parasitoid insects for searching the host and it was of strategic interest to the scientific community. According to them, the host searching behaviour is largely categorised into three broad partially overlapping mechanisms. First mechanism involves the signals originating from the microhabitat or food plants of potential hosts; second, signals indirectly associated with the activity of hosts in the microhabitat; and finally signals arising directly from the hosts themselves. The effective use of these three mechanisms simultaneously makes them effective parasitoid groups for commercial use. The host ranges are also an important phenomenon in parasitoid biology. This maximum range of utilisation of host species goes a long way to improving the success rate of a biocontrol agent.

The oviposition strategies adopted by the parasitoid flies are always important for successful and continuous oviposition in *in vivo* culture of parasitoids. In some families (e.g. Acroceridae, Bombyliidae and Rhinophoridae), first instar larvae are planidiform when hatched and move actively to seek hosts (Schlinger, 1981; Yeates, 1997), but most of the dipteran parasitoids lay eggs directly on the host species. Reproductive strategies, such as indirect oviposition and ovo-larviparity have been adopted in tachinids (Wood, 1987). Indirect oviposition appears to have been adopted for hosts that are nocturnal or are concealed in leaves or stems of plants or in the ground, although in some species (e.g. *Belvosia*) an indirect ovipositional strategy is reported to attack frequently exposed hosts (Stireman and Singer, 2003). The ovo-larviparous habit is considered effective in reducing the mortality of the eggs and early stages and is found also in Calliphoridae and Sarcophagidae (Rognes, 1997; Pape, 1998a). Sarcophagid flies are both ovoviviparous and viviparous in nature and display either direct or indirect oviposition (or larviposition) strategies.

With the lack of a strong piercing ovipositor, dipteran parasitoids cannot directly affect the host immune system, like the hymenopterans. Many dipteran larvae escape the host immune response and turn the host response to their advantage by forming respiratory funnels, which are sclerotised folders around the hind part of their body. Funnels may be primary, built in the host integument by first instars as soon as they enter, or secondary, formed by late-first or early-second instars in the host integument or tracheae. These have evolved resourceful strategies to avoid both host encapsulation and effective parasitoidism.

12.3 REARING, MASS PRODUCTION, AND FIELD APPLICATION OF DIPTERA PARASITOIDS

Dipterans can be reared using both in-*vivo* and *in-vitro* techniques. Adults are first reared in natural conditions and then harnessed for egglaying. The eggs are laid in masses and larval rearing of dipteran parasitoids can be done both in-*vivo* and *in-vitro*. The larvae are harvested at the suitable stage (larval instars) of their development for application and then released in the field. The adults are also harvested from mature pupal stages and reared and made suitable for maximum oviposition and successful egg laying in the successive stages. The adults can be transported using proper mechanisms to prevent desiccation and death, to distant places where they can be cultivated and harvested for their larvae and subsequently deployed in fields. The successful mass production of the fly parasitoids as biocontrol agents depends primarily on factors, like food (natural and artificial), oviposition strategies, time period of application and abiotic factors, like temperature relative-humidity and photoperiod in both techniques.

12.3.1 *In-vivo* Rearing Technique

In-vivo technique basically depends upon the interaction between the host and fly parasitoid in natural environments. Successful *in-vivo* rearing involves the proper adoption of the following strategies.

Food of the parasitoid and the host: Under natural circumstances, the rearing and mass production of entomophagous insects is dependent upon the host insect and the food of that host insect (it may be plant or other natural food resources) more than the parasitoid itself. For better production as well as the quality of the parasitoids, the food of the host species is of primary importance. Artificial diets are usually used for replacing plants, or other natural food to grow the insect hosts when parasitoids are required to be produced on a mass scale. Rossi and Fowler (2003) conducted experiments in Brazil where the tachinid parasitoid species, *L. minense* and *P. claripalpis* were reared on their natural host *Diatraea* sp. The natural host was reared on an artificial diet to avoid the use of sugarcane, thus doing away with the natural food and introducing an artificial diet for the host species for the first time. This in turn made possible the large-scale production of hosts as well as the mass rearing of the parasitoids.

Oviposition strategies: Oviposition strategies and mechanisms of host selection in parasitoids may be crucial for the success of parasitoid production. For example, most of the tachinid species prefer an indirect mode for oviposition strategies, where the females, which are usually ovoviviparous, deposit their eggs close to a host. The tachinid species *Archytas marmoratus* wait for a host to pass by to deposit eggs. If the indirect oviposition strategies can be modified artificially, then modified indirect oviposition strategies enhance the effect of parasitoids on the host. For instance, the tachinid fly, *Pseudogonia rufifrons* oviposits micro-type eggs on foliage, and natural hosts (noctuid lepidopterans) become infested by eating 'contaminated' leaves. In captivity, oviposition was obtained on wax substrates simulating leaves, and high parasitization percentages were obtained leading to an increase in parasitoid development manifold in *G. mellonella* larvae after they had fed on these artificial 'leaves'.(Mellini andCoulibaly, 1991).

Time period of application: Another most important factor, which is decisive in the application of parasitoid flies as biocontrol agents, is the time period when the parasitoids are applied and released in the field. Depending upon the larval stages, the degree of effect on host species is determinant. For instance, most of the tachinid larvae produce a better result if they are applied in the third instar stage rather than first instar stage (Mellini and Coulibaly, 1991) and this needs to be kept in mind during the field applications.

Abiotic factors: Lastly, for the successful production of Diptera parasitoids, abiotic factors, like temperature, relative humidity (RH) and day length are crucial. Vogt et al. (2003) proved that in a rearing system for the phorid fly, *Pseudacteon curvatus,* parasitized host ants if held for 13 days at 27°C and 60% humidity, and then moved to well-ventilated boxes ('attack boxes') placed at 28°C, 80% RH, and a 12:12 photoperiod (light:day), emergence occurred in 10 days. It was further proved that of all the factors, high relative humidity was most vital for fly parasitization and survival.

12.3.2 *In-vitro* Rearing Technique

In vitro rearing is an effective tool to produce entomophagous insects as biocontrol agents at a reasonable cost in the absence of hosts or prey. For that purpose, the availability of an artificial and economically desirable medium is of utmost importance. However, for dipteran parasitoids, it is more appropriate to speak of *in vitro* 'rearing' on a relatively small scale instead of actual 'production', as high adult yields have so far been obtained in only a few species on artificial media. Meager information is available on this practice and only for Tachinidae and Sarcophagidae. Proper evaluation of the nutritional requirements of the insect is the first step to succeed in *in vitro* rearing.

Preparation of food media: The food media mostly depends upon the nutritional need of parasitoids rather than the host species. The preparation of media is species-specific and the proper knowledge is essential for the preparation of artificial media during rearing. The basic qualitative requirements for dipteran parasitoids are very similar to those of free-living insects, except for the specific requirements of aromatic amino acids and surfactants. The main nutritional components required for Diptera rearing in artificial media are nitrogen, lipids, carbohydrates and miscellaneous micronutrients.

Nitrogen: Dipteran parasitoids are carnivorous insects and they exhibit rapid growth during larval stages (Grenier, 1980), especially when they have access to atmospheric oxygen through respiratory funnels. Thus, a protein-rich medium with a well-balanced composition of proteins, peptides, or free amino acids is essential to avoid the loss of time and energy in conversion metabolism. In Tachinidae, aromatic amino acids are required, especially at the end of larval development for cuticle tanning (Bonnot et al., 1976). The supply of aromatic amino acids is complicated due to the fact that some free amino acids, like phenylalanine, may be toxic at high concentrations and also have low solubility (Grenier et al., 1994; Grenier, 2012). The best way to deliver aromatic amino acids is within peptides or proteins rich in tyrosine and/or phenylalanine since β-alanyl-tyrosine, a tyrosine-rich dipeptide, is present in mature tachinid larvae, for example in *P. caudate* (Delobel and Bonnot, 1976). Casein, serum albumine, ovalbumine, lactalbumine, soybean extract, and yeast are commonly used as crude proteins or hydrolysates.

Lipids: The energy source lipids may be absorbed with slight modifications and they significantly improve the survival and fecundity of dipteran parasitoids. Apart from nutritional values, polyunsaturated fatty acids are necessary for development and reproduction (Delobel and Pageaux, 1981). Phospholipids, the key component in the parasitoid cell membranes and the free fatty acids or triglycerides are usually supplied and incorporated within aqueous base formulations with the help of surfactants (House, 1961), but Thompson (1977) has reported some toxicity.

Carbohydrates: A very specific qualitative supply of carbohydrates is suggested for the rearing process. Trehalose, the most common non-reducing disaccharide, plays an important role in the general metabolism (Thompson, 2003). Monosaccharides (e.g. glucose) are replaced with polysaccharides (e.g. glycogen) to maintain the osmotic potential at an acceptable level. Sucrose may act as a feeding stimulant (Cohen, 2004) in all parasitoid cultures.

Miscellaneous: Other nutrients, such as organic acids, vitamins, nucleic acids, and minerals, are required in addition to the three basic types of nutrients (proteins, lipids, and carbohydrates). These requirements are generally similar among insects, but they are poorly investigated in dipteran parasitoids.

Though several researchers have tried to rear dipteran flies in a controlled environment, success was only seen for Tachinidae and Sarcophagidae families. The Sarcophagid flies were reared to adult with high adult yields in media based on raw materials (salmon and fresh liver) (House and Traer, 1949; Coppel et al., 1959) and on chemically defined media containing amino acids, RNA, dextrose, inorganic salts, B vitamins, choline, inositol, and agar as a gelling agent (House, 1954). *Pseudogonia rufifrons* was reared from first-instar larva to adult (with very low yields) on a sub-natural medium and on a medium containing bovine serum as the main ingredient, combined with trehalose, chicken egg yolk (Mellini et al., 1994). Complete *in vitro* development on an insect-free artificial medium has been obtained only for tachinids that do not display developmental synchrony with their hosts. The rearing was performed on an artificial medium gelled with agarose and containing organic acids, inorganic salts, amino acids, B and fat-soluble vitamins, gelatin, protein hydrolysates, glycogen, adenosine triphosphate, lecithin, corn oil, and cholesterol and was found to be successful (Grenier et al., 1978).

12.3.3 Adult Rearing and Management

The most crucial part of adult rearing apart from providing proper nutrition is space management to increase the reproductive capability of flies. For better management of parasitoid rearing, it could be helpful to know the age of the adults *vis a vis* their reproductive physiology. Dipteran parasitoid adults are generally good flyers. The space available (cage volume) and adult density (number of individuals per cage) are two of the parameters to be considered for survival and mating success, and this is often a crucial phase for life cycle completion. Usually, mating was best achieved when the adult sex ratio was male-biased (Odell and Godwin, 1979; Zhang et al., 2003) or when newly emerged females were kept with older males (Ho et al., 2011). According to Wäckers (2003), nectar might also contribute to egg maturation and other physiological processes despite the low content in amino acids, proteins, and lipids. In nature, some tachinids are able to feed on pollen (Campadelli, 1977). In captivity, adult tachinids are often provided with absorbent cotton soaked in honey- or sugar-water solutions and/or sugar cubes (Sourakov and Mitchell, 2002). Research is needed on the varied food sources and their comparative efficacy in optimising the performance of adult parasitoids. Distilled water may be supplied to dipteran parasitoid adults by spraying over the cages (Quednau, 1993) or through moistened absorbent cotton in drinking troughs. A special automatic collector was designed by Grenier and Ogier (1978) for controlling the emergence of tachinid flies, allowing an accurate determination of egg maturation dynamics (Grenier et al., 1982). Oviposition or larviposition substrates (hosts for direct strategists or artificial devices for indirect strategists) have to be periodically exposed to the adults inside cages. The approach also allows the evaluation of the weight and fecundity of emerging adults because adult weight correlates with fecundity.

12.3.4 Storage and Transportation of Reared Flies

Quality control is very important during the storage and shipment processes of entomophagous insects. Each stage responds differently to short- and long-term storage at low temperature and their ability to tolerate cold storage depends on the nutrient origin (artificial vs. natural food). Storage of parasitoids is usually done by placing them, often at immature stages, at temperatures between 4 and 15 °C for short periods (van Lenteren and Tommasini, 2003). *Sturmiopsis inferens* (Townsend) was stored in the pupal stage for 30 days at 15 °C without any adverse effects on fly emergence, female mating, and male longevity, whereas female fertility was slightly reduced (Easwaramoorthy et al., 2000). When the medium with eggs was restored at the standard rearing temperature of 26°C, the eggs hatched and the parasitoids developed to the adult stage; however, fewer adults were obtained compared to controls, especially from eggs stored at 15°C. Studies on shipment procedures for dipteran parasitoids are very scant. In recent years, a number of developments have occurred in insulation, energy storage, and humidity control that provide the opportunity for improving the conditions during the transportation of insects. New vacuum panel insulation is up to 10 times more efficient than conventional expanded polystyrene, providing longer temperature control in smaller containers during shipping. Semipermeable membrane sachets of saturated salt solutions can maintain required humidity conditions inside sealed containers for extended periods.

12.4 THE FUTURE OF DIPTERANS AS BIOCONTROL AGENTS

Production of dipteran parasitoids is deeply related to the utilisation of these beneficial insects in applied biocontrol. From the above literature, it is evident that there is only meager information available on rearing and management methods. The main hindrances are: (1) the programmes appeared to be expensive, mainly due to rearing costs –among other things, the bigger size of tachinids, compared to hymenopteran parasitoids, is unfavourable for mass production, as more food and space are necessary to multiply them, (2) the general image of flies was (and still is) unpopular, which may lead to a preference for other programs using other kinds of insects, and (3) some biological characteristics may be considered as negative for inundative biocontrol (e.g. adult flies are good fliers and may disperse too much from release points). Increasing the use of dipteran parasitoids in applied biocontrol requires the improvement of their rearing techniques for mass production. Upcoming research on these issues may deal with species that are recognised as promising biocontrol agents but are not yet exploited due to difficulties encountered in their rearing. Future studies should focus on the possibility of better exploiting dipteran parasitoids (in particular, but not exclusively, tachinids) in biocontrol strategies.

12.5 CONCLUSIONS

The biocontrol method for agricultural pest control has been considered among the most encouraging technologies for sustainable agriculture. It reduces the reliance on synthetic pesticides, minimises the negative impact on the environment and improves workers' safety, while at the same time maintaining the economic viability of crop production. Yet the rate of biocontrol adoption as compared to conventional pesticide-based methods is not profound, as the structure of the biocontrol industry is defined by diversity and the absence of big international corporations (Ehlers, 2011). The rate of adoption of biocontrol in Integrated Pest Management has still remained negligible as compared to the use of conventional pesticides. Even though the major international agencies, such as the UN Food and Agriculture Organization, the World Bank and others have continuously promoted 'farmer-driven ecologically based pest control' (IAASTD, 2009), the actual national implementations of these global policy recommendations are weak (Karel, 2004). Yet the world's agricultural systems need these methods more than ever.

In this context, the market for the use of biocontrol agents for pest control has evolved, though slowly but at a constant aggregate rate and this includes a long-range use of insects including the Diptera. Unfortunately, in spite of serving as effective predators and parasitoids, this group of flies has received much less attention in comparison to other insect groups as effective pest control agents. Proper strategic documentation is needed, which defines the promotion and implementation of dipteran flies as potential bioagents in the Integrated Pest Management programme. Diptera research over the past decade has shown that flies have always served as surrogates for pest control in the natural conservation action plan. It is high time to acknowledge and conserve flies. It should be prioritised as a pragmatic and responsible approach of environmental biologists on a global scale. The flies have profound and well-identified influences on many ecosystem services, like crop pollination, resource improvisation, and biocontrol and it is crucial to integrate these ecosystem services into strategies for achieving our Sustainable Development Goals (SDGs) (Figures 12.1–12.5 and Tables 12.2).

(a) (b)

(c)

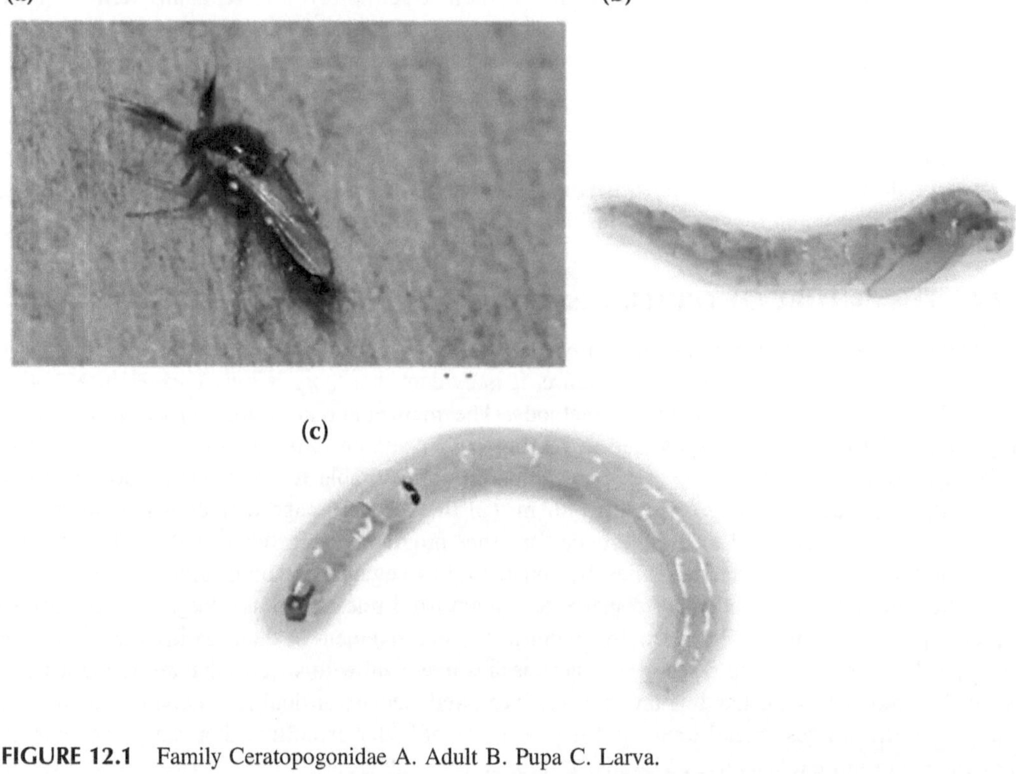

FIGURE 12.1 Family Ceratopogonidae A. Adult B. Pupa C. Larva.

FIGURE 12.2 Family Chironomidae: A. Adult B. Pupa C. Larva.

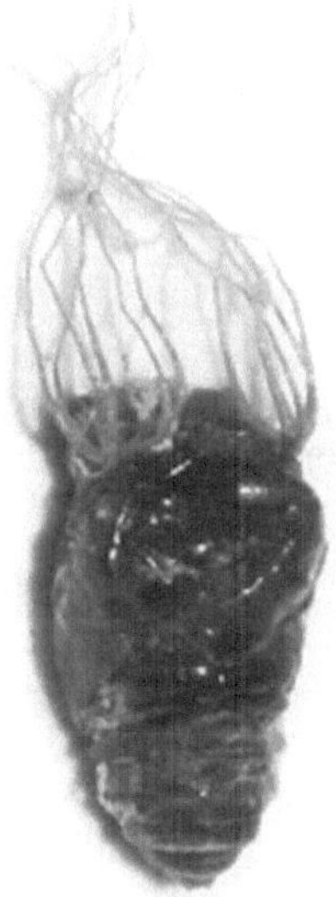

FIGURE 12.3 Larva of Family Simulidae.

FIGURE 12.4 Adult of the Family Asilidae capturing the prey.

12.6 POINTS TO REMEMBER

i. Dipterans, commonly known as flies, contribute significantly to vital ecological functions, such as crop pollination, pest control, decomposition of organic matter, and maintenance of wildlife species in agrarian and wild ecosystems.

ii. Adult dipterans can be broadly classified as terrestrial, aquatic and semiaquatic, and found in a wide variety of habitats.

iii. This versatility allows them to exploit all trophic levels. Based on their food habits, the fly communities are divided into four major groups: i.e. saprophagous/necrophagous, carnivorous, phytophagous and hematophagous.

iv. Flies have short lives as adults and undergo complete metamorphosis, with four distinct life stages – egg, larva, pupa and adult.

v. Typical adults are usually free living, have mouthparts designed for piercing and sucking (mosquitoes, black flies and robber flies), or sponging for lapping and sucking, while larvae are free-living, parasitic or parasitoids. Larvae in the latter case are apodous, the eyes and antennae of larvae are reduced or absent, and the abdomen lacks appendages.

vi. Diptera larvae may be predaceous and prey upon larvae of other insects or even other Diptera and they can be parasitoids. Diptera parasitoids are immature larvae that develop on or within an insect host and ultimately kill it. After feeding on host body fluids and organs, most parasitoids leave their hosts to pupate or emerge as adults. These make them efficient biocontrol agents.

vii. Currently, more than 30 dipteran larval families and 15 adult dipteran families are well-known as effective biocontrol agents. 20% of their total species database, i.e. approximately 16,000 species, project them as beneficial biocontrol agents.

FIGURE 12.5 Adult A. Family Tachinidae B. Family Dolichopodidae C. Family Syrphidae D. Family Sarcophagidae E. Family Bombylidae.

viii. Adults of the family Asilidae, Dolichopodidae, Empididae and Lonchaidae are well-known predators. Larval predators and parasitoids are extensively found in families, like Tachnidae, Syrphidae, Bombyliidae, Sciomyzidae, Cecidomyiidae, Dolichopodidae, Ephydridae, Conopidae, Apioceridae, Athericidae, Chamaemyiidae, Chaoboridae, Agromyzidae, Phoridae, Sarcophagidae, Fannidae, Muscidae and Calliphoridae. The Syrphidae, Bombyliidae, Mydidae, Rhagionidae, Scenopinidae, Tabanidae and Therevidae serve the dual purpose of pollination as adults and biocontrol agents in their larval stages.

TABLE 12.2
Dipteran Parasitoid Species Used in the Biological Control Programme

Family	Parasitoid Species	Host
Tachinidae	*Compsilura concinnata*	*Nygmia phaerrhoea*
	Paratheres iamenezesi	*Rhynchophorus palmarum*
	Strongygaster triangulifera	*Harmonia axyridis*
	Lixophaga diatraeae	*Diatraea saccharalis*
	Carcelia kockiana	*Helicoverpa armigera*
	Sisyropa firmosa	
	Compsilura concinnata	
	Exorista bombycis	
	Exorista japonica	
	Exorista xanthaspis	
	Peribea orbuta	
	Billaea claripalpis	*Diatraea* ssp.
	Lydella minense	
Sarcophagidae	*Seniorwhite areciproca*	*Helicoverpa armigera*
	Blaesoxipha sp.	Grasshoppers
Dryomyzidae	*Oedoparena* sp.	barnacles
Culicidae	*Toxorhynchites* sp.	larval *Aedes aegyptii,* aquatic organisms
Cecidimyiidae	*Endaphis perfida*	*Pentalonia nigronervosa*
	Endaphis compitali	
	Endaphis gregari	
	Endaphis aphidomyz	
	Endaphis maculans	
	Feltiella acarisuga.	two-spotted spider mites

ix. The effective recognition of three host searching signals simultaneously makes parasitoid groups effective biocontrol agents for commercial use: (i) the signals originating from the microhabitat or food plants of potential hosts, (ii) signals indirectly associated with the activity of hosts in the microhabitat and (iii) the signals arising directly from the hosts themselves.
x. The application of the parasitoid groups of dipteran flies as a biocontrol agent depends upon the successful study of the entire biology of the applied species, especially host range, oviposition strategies and host-parasite interaction.
xi. The most successful commercially utilised parasitoid flies belong to the Dipteran families of Tachinidae, Syrphidae and Sarcophagidae. The Tachinid flies happen to be the most reared group for mass production both through *in vitro* and *in vivo* techniques.
xii. The successful mass production of fly parasitoids as biocontrol agents depends primarily on factors, like food (natural and artificial), oviposition strategies, time period of application and abiotic factors, like temperature, relative humidity and photoperiod.
xiii. The standard artificial medium for successful larval rearing of parasitoids is gelled agarose containing organic acids, inorganic salts, amino acids, B and fat-soluble vitamins, gelatin, protein hydrolysates, glycogen, adenosine triphosphate, lecithin, corn oil and cholesterol.
xiv. Not only the proper rearing of larval parasitoids but also the adults are important along with proper storage during transportation for viable commercial use.

xv. The dipteran flies have always served as surrogates for pest control in the natural conservation action plan. Recognising their role and conserving the beneficial Diptera should be prioritised as it is crucial to integrate this ecosystem service into the strategies for achieving our Sustainable Development Goals (SDGs).

REFERENCES

Boesi, R., Polidori, C. & Andrietti F. (2009). Searching for the Right Target: Oviposition and Feeding Behavior in *Bombylius* Bee Flies (Diptera: Bombyliidae). *Zoological Studies*, 48(2): 141–150.

Bonnot, G., Delobel, B. & Grenier, S. (1976). Composition corporelleenacides amines du parasitoidc *Phryxecaudata* (Diptera) au cours de sacroissancelarvaire. *Journal of Insect Physiology*, 22(4): 505–514.

Campadelli, G. (1977). Rearing techniques for Diptera Tachnidae with particular regard to the adults [insects parasite of other insects]. *Bollettinodell'Instituto di Entomologiadell'UniversitadegliStudi di Bologna (Italy)*, 33: 215–240.

Cerretti, P. & Tschorsnig, H.P. (2010). Annotated host catalogue for the Tachinidae (Diptera) of Italy. *Stutt. Beit. Natur.*, 3: 305–340.

Chambers, R.J. & Adams, T.H.L. (1986). Quantification of the impact of hoverflies (Diptera: Syrphidae) on cereal aphids in winter wheat: an analysis of field populations. *Journal of Applied Ecology*, 23: 895–904.

Chapman, R.F. (1998). *The Insects; Structure & Function*. Cambridge: Cambridge University Press. pp. 1–770.

Cohen, A.C. (2004). *Insect Diets: Science and Technology*. CRC Press, Boca Raton, USA, pp. 1–473.

Coppel, H.C., House, H.L. & Maw, M.G. (1959). Studies on Dipterous Parasites of the Spruce Budworm, *Choristoneura fumiferana* (Clem.) (Lepidoptera: Tortricidae): Vii. *Agria affinis* (Fall.) (Diptera: Sarcophagidae). *Canadian Journal of Zoology*, 37(6): 817–830.

Danyk, T., Mackauer, M. & Johnson, D.L., (2005). The influence of host suitability and host range of grasshopper species utilized by *Blaesoxiphaatlantis* (Diptera: Sarcophagidae) in the field. *Bull. Entomol. Res.*, 95: 571–578.

Delobel, B. & Bonnot, G. (1976). Delobel, B. & Bonnot, G. (1976). Présence de beta alanyltyrosine chez le Diptère Tachinaire *Phryxe caudata* Rond. Importance relative des acides aminés libres, peptidiques et protéiques. *Ann.Zool. Ecol. Anim.*, 8: 493–497.

Delobel, B. & Pageaux, J.F. (1981). Influence de l'alimentation sur la composition en acides gras totaux des Diptères Tachinaires. *Ent. Exp. Appl.*, 29: 281–288.

Dennis, D.S. & Lavigne, R.J. (2007). Hymenoptera as prey of robber flies (Diptera: Asilidae) with new prey records. *Journal of the Entomological Research Society*, 9(3). 23–42.

Dennis, D.S., Lavigne, R.J. & Dennis, J.G. (2012) Spiders (Araneae) as Prey of Robber Flies (Diptera: Asilidae). *Journal of Entomological Research Society*, 14(1): 65–76.

Dickinson, M.H. (1999). "Haltere–mediated equilibrium reflexes of the fruit fly, *Drosophila melanogaster*". *Philosophical Transactions of the Royal Society of London B: Biological Sciences*, 354 (1385): 903–916.

Dindo, M.L. (2011). Tachinid parasitoids: are they to be considered as koinobionts. *BioControl*, 56(3): 249–255.

Dindo, M.L. & Grenier, S. (2014). Production of Dipteran Parasitoids. In: *Mass Production of Beneficial Organisms*, ed. Morales-Ramos, J.A. et al., pp. 101–143. Academic Press.

Dindo, M.L., Rezaei, M. & De Clercq, P. (2019). Improvements in the rearing of the tachinid parasitoid *Exoris talarvarum* (Diptera: Tachinidae): influence of adult food on female longevity and reproduction capacity. *Journal of Insect Science*, 19(2): 1–6.

Easwaramoorthy, S., Kurup, N.K., Santhalakshmi, G. & Shanmugasundaram, M. (2000). Effect of low temperature storage on the viability of puparia of *Sturmiopsis inferens* Townsend (Diptera: Tachnidae) a larval parasitoid of sugarcane moth borers. *Journal of Biological Control*, 14(1): 63–65.

Ehlers, R.U. (2011). *Regulation of biological control agents*. New York: Springer Science, p. 1–412.

Feener Jr, D.H., Brown, B.V. (1997) Diptera as parasitoids; *Ann. Rev. Entomol.*, 42:73–97.

Feener Jr, D.H. & Brown, B.V. (1997) Diptera as parasitoids. *Annual review of Entomology*, 42(1): 73–97.

Fréchette, B., Larouche, F. & Lucas, E. (2008) *Leucopis annulipes* larvae (Diptera: Chamaemyiidae) use a furtive predation strategy within aphid colonies. *European Journal of Entomology*, 105: 399–403.

Gagn´e, R.J., Sosa, A. & Cordo, H. (2004). A new Neotropical species of *Clinodiplosis* (Diptera: Cecidomyiidae) injurious to alligator weed, *Alternanthera philoxeroides* (Amaranthaceae). *Proceedings of the Entomological Society of Washington*, 106: 305–311.

Ghahari, H., Lavigne, R.J. & Geller-Grimm, F. (2007). Bibliography of Asilidae (Insecta: Diptera) 1996-2006. *Far Eastern Entomologist*, 176: 1–39.

Gillespie, D.R., Opit, G. & Roitberg, B. (2000) Effects of Temperature and Relative Humidity on Development, Reproduction, and Predation in *Feltiella acarisuga* (Vallot) (Diptera: Cecidomyiidae). *Biological Control*, 17 (2): 132–138.

Gillott, C. (2005). *Entomology* (3 ed.). Springer. pp. 614–615.

Grenier, S. & Liljestrhöm, G. (1991) Préférencesparasitairesetparticularitésbiologiques des Tachinaires (Diptera Tachinidae). *Bull. Soc. Linn. Lyon*, 60: 128–141.

Grenier, S. & Ogier, P. (1978). Description d'un appareil collecteur automatique pour contrôlerl'émergence de mouches. *Ann. Zool. Ecol. Anim.* 10: 139–145.

Grenier, S., Bonnot, B. Delobel & P. Laviolette. (1978). Developpmenten milieu artificiel du parasitoid *Lixophaga diatraeae* (Towns.) (Diptera, Tachinidae). *Obtention de l'imago a partir de l'oeuf. C. R. Acad. Sci., Paris, Ser. D.* 287: 535–538.

Grenier, S. (1980). Endoparasitic development and pond larval growth of *Lixophaga diatraeae* (Diptera: Tachinidae) in a substitution host: *Galleria mellonella* (Lepidoptera: Pyralidae) [Sugarcane pests, biological control]. *Entomophaga*, 25: 17–26.

Grenier, S., Barthelemy, S. & Bonnot, G. (1982). Dynamique de la maturation des œufs chez le parasitoïde *Lixophaga diatraeae* (Diptera, Tachinidae) élevédans un hôte de substitution *Galleria mellonella* (Lepidoptera, Pyralidae). *Reproduction Nutrition Développement*, 22(3): 523–535.

Grenier, S., Greany, P. & Cohen, A.C. (1994). Potential for mass release of insect parasitoids and predators through development of artificial culture techniques, In: *Pest management in the subtropics: biological control – a Florida perspective*, Eds. Rosen, D., Bennett, F.D., Capinera J.L., pp. 181–205. Andover, UK: Intercept Pub.

Grenier, S. (2012). Artificial rearing of entomophagous insects, with emphasis on nutrition and parasitoids-general outlines from personal experience. *Karaelmas Fen veMühendislikDergisi*, 2(2): 1–12.

Häfliger, P., Schwarzlaender, M. & Blossey, B. (2005) Biology of *Platycephala planifrons* (Diptera: Chloropidae) and its potential effectiveness as biological control agent for invasive *Phragmites australis* in North America. *Biological Control*, 34: 302–311.

Ho, G.T., Ichiki, R.T. & Nakamura, S. (2011). Reproductive biology of the microtype tachinid fly, *Zenillia dolosa* (Meigen)(Diptera: Tachinidae). *Entomological science*, 14(2), 210–215.

Hoell, H.V., Doyen, J.T. & Purcell, A.H. (1998). *Introduction to Insect Biology and Diversity* (2nd ed.). Oxford University Press. pp. 493–499.

House, H.L. (1954). Nutritional studies with *Pseudosarcophaga affinis* (Fall.), a dipterous parasite of the spruce bud-worm, *Choristoneura fumiferana* (Clem.) in a chemically defined medium and aseptic-culture technique. *Canadian Journal of Zoology*, 32(5): 331–341.

House, H.L. & Traer, M.G. (1949). An artificial food for rearing *Pseudosarcophaga affinis* (Fall.), a parasite of the spruce budworm, *Choristoneura fumiferana* (Clem.). *Ann. Rept. Entomol. Sot. Ontario,* 79:50–53.

House, H.L. (1961). Insect nutrition. *Annual Review of Entomology*, 6(1): 13–26.

IAASTD (2009). Agriculture at a crossroads. *The global report of the International Assessment of Agricultural Knowledge, Science, and Technology.* Washington DC: Island Press.

Joern, A. & Rudd, N.T. (1982). Impact of predation by the robber fly *Proctacanthus milbertii* (Diptera: Asilidae) on grasshopper (Orthoptera: Acrididae) populations. *Oecologia*, 55: 42–46.

Karel, B. (2004). *The Persistence of pesticide dependence: A review of World Bank projects and their compliance with the World Bank's pest management policy, 1999-2003.* San Francisco: PANNA. Kelowna Land and Orchard Research and Education, pp. 1–34.

Katsoyannos, P. & Aliniazee, M.T. (1998). First record of *Strongygaster triangulifera* (Loew)(Diptera: Tachinidae) as a parasitoid of *Harmonia axyridis* (Pallas)(Coleoptera: Coccinellidae) in western North America. *The Canadian Entomologist*, 130(6): 905–906.

Labandeira, C.C. (1998). How old is the flower and the fly? *Science* 280: 57–59.

Labandeira, C.C. (2010). The pollination of mid Mesozoic seed plants and the early history of long-proboscid insects. *Ann. Mo. Bot. Gard.*, 97: 469–513.

Lacey, L.A., Frutos, R., Kaya, H.K. & Vail, P. (2001) Insect pathogens as biological control agents: do they have a future. *Biological control*, 21(3):230–248.

Losey, J.E. & Mace Vaughan, M. (2006). The Economic Value of Ecological Services Provided by Insects. *BioScience*, 56(4): 311–323.

Máca, J. & Otranto, D. (2014). Drosophilidae feeding on animals and the inherent mystery of their parasitism. *Parasites Vectors*, 7 (516): 1–8.

Mason, P.G. & Erlandson, M.A. (1994). Invitation paper (C. P. Alexander Fund): the potential of biological control for management of grasshoppers (Orthoptera: Acrididae) in Canada. *Can. Entomol.*, 126: 1459–1491.

McMahon, D.P. & Hayward, A. (2016). Why grow up? A perspective on insect strategies to avoid metamorphosis. *Ecological Entomology*, 41(5): 505–515.

Meier, R., Kotrba, M., Ferrar, P., (1999). Ovoviviparity and viviparity in the Diptera. *Biological Reviews*, 74 (3): 199–258.

Mellini, E. & Campadelli, G. & Dindo, M. (1994). Possibileimpiego di plasma bovinonell'allestimento di dieteartificiali per le larve del parassitoide, *Pseudogonia rufifrons*Wied. M.A.F.-Convegno Lotta Biologica. *Acireale*, 1991: 145–150.

Mellini, E. & Coulibaly, A.K. (1991). Un decennio di sperimenta zioncsulsistcmaospitc parassita *Galleria mellonella* L.—*Pseudogonia rufifrons*Wied.:sintesideirisultati. *Boll. Ist. Ent. "G. Grandi" Univ. Bologna*, 45: 191–249.

Montes, M. (2008). Producciónmasiva y aplicación de biorreguladoresnativosen Cuba. *Fitosanidad*, 12: 241–242.

Moura, J.I.L., Toma, R., Sgrillo, R.B. & Delabie, J.H. (2006) Natural efficiency of parasitism by *Billaea rhynchophorae* (Blanchard) (Diptera: Tachinidae) for the control of *Rhynchophorus palmarum* (L.) (Coleoptera: Curculionidae). *Neotropical Entomology*, 35(2): 273–274.

Muniz-Reyes, E., Lomell-Flores, J.R. & Sanchez-Escudero, J. (2011) Parasitoides nativos de *Rhagoletis pomonella* Walsh (Diptera: Tephritidae) entejocote *Crataegus* spp. enelcentro de Mexico. *Acta Zool. Mex.*, 27: 425–440.

Nicholls, I.C., Pérez, N., Vasquez, L. & Altieri, M.A. (2002). The development and status of biologically based integrated pest management in Cuba. *Int. Pest Manage. Rev.*, 7: 1–16.

Noriega, J.A. (2018). Research trends in ecosystem services provided by insects. *Basic and Applied Ecology*, 26: 8–23.

Odell, T.M. & Godwin, P.A. (1979). Laboratory techniques for rearing of *Blepharipa pratensis*, a tachinid parasite of gypsy moth. *Annals of the Entomological Society of America*, 72(5): 632–635.

Pape, T. (1987). *The Sarcophagidae (Diptera) of Fennoscandia and Denmark. Fauna Entomologica Scandinavica*, Volumemac_mac 19, Leiden, the Netherlands, pp. 1–203.

Pape, T. (1998a) Family Sarcophagidae. In: *Manual of Palaearctic Diptera) Volume- 3*, ed. L. Papp & B. Darvas, pp. 649–678. Budapest: Science Herald

Quednau, F.W. (1993). Reproductive biology and laboratory rearing of *Ceranthias amarensis* (Villeneuve) (Diptera: Tachinidae), a parasitoid of the gypsy moth, *Lymantria dispar* (L.). *The Canadian Entomologist*, 125(4): 749–759.

Reeves, W.K. (2010). Behavior of larval *Culicoides sonorensis* (Diptera: Ceratopogonidae) in response to an invertebrate predator, *Hydra littoralis* (Anthomedusae: Hydridae). *Entomological news*, 121(3): 298–301.

Resh, V.H. & Cardé, R.T. (2009). *Encyclopedia of Insects*. Academic Press.

Robson, J.D., Wright, M.G. & Almeida, R.P. (2007) Effect of imidacloprid foliar treatment and banana leaf age on *Pentalonia nigronervosa* (Hemiptera, Aphididae) survival. *New Zealand Journal of Crop and Horticultural Science*, 35(4): 415–422.

Rognes, K. (1997) The Calliphoridae (Blowflies) (Diptera: Oestroidea) are not a monophyletic group. *Cladistics*, 13: 27–66.

Romeis, J. & Shanower, T.G. (1996). Arthropod natural enemies of *Helicoverpa armigera* (Hübner) (Lepidoptera: Noctuidae) in India. *Biocontrol Science and Technology*, 6(4): 481–508.

Rossi, M.N. & Fowler, H.G. (2003). The sugarcane borer *Diatraea saccharalis* (Fabr.)(Lep., Crambidae) and its parasitoids: a synchrony approach to spatial and temporal dynamics. *Journal of Applied Entomology*, 127(4), 200–208.

Schlinger, E.I. (1981). Acroceridae. In: *Manual of Nearctic Diptera (Monograph 27)*, ed. J.F. McAlpine, et al., pp. 575–584. Ottawa: Research Branch, Agriculture Canada.

Schmid-Hempel. (1998). *Parasites in Social Insects*. Princeton University Press, 2019, pp. 1–392.

Schoonhoven, L.M., Jermy T. & van Loon J.J.A. (1998). *Insect–Plant Biology: From Physiology to Evolution*. Chapman & Hall, London, p. 421.

Sommaggio, D. (1999). Syrphidae: can they be used as environmental bioindicators. *Agriculture, Ecosystems & Environment*, 74(1–3): 343–356.

Sommaggio, D. & Burgio, G. (2014). The use of Syrphidae as functional bioindicator to compare vineyards with different managements. *Bulletin of Insectology*, 67(1): 147–156.

Sourakov, A. & Mitchell, E.R. (2002). Laboratory biology of *Chetogena scutellaris* (Diptera: Tachinidae), a parasitoid of Noctuidae, reared on fall armyworm and cabbage looper. *Florida Entomologist*, 85(2): 341–343.

Ssymank, A., Kearns, C.A., Pape, T. & Thompson, F.C. (2008). Pollinating flies (Diptera): a major contribution to plant diversity and agricultural production. *Biodiversity*, 9: 86–89.

Stireman, J.O. & Singer, M.S. (2003). What determines host range in parasitoids? An analysis of a tachinid parasitoid community. *Oecologia*, 135: 629–638.

Thompson, S.N. (1977). Lipid nutrition during larval development of the parasitic wasp, *Exeristes. J. Insect Physiol.*, 23: 279–283.

Thompson, S.N. (2003). Trehalose—the insect 'blood'sugar. *Advances in insect physiology*, 31(Supplement C): 205–285.

Ulrich, H. (2004). Predation by adult Dolichopodidae (Diptera): a review of literature with an annotated prey–predator list. *StudiaDipterologica*, 11: 369–403.

Van Lenteren, J.C. & Tommasini, M.G. (2003). Mass production, storage, shipment and release of natural enemies. *Quality control and production of biological control agents: theory and testing procedures* 14:181–189.

Vinson, S.B. & Barbosa, P. (1987). Interrelationships of nutritional ecology of parasitoids. In *Nutritional ecology of Insects, Mites, and Spiders and Related Invertebrates*, ed. F. Slansky & J.G. Rodriguez, pp. 673–695. New York: John Wiley & Sons. 880 p.

Vogt, J.T., Porter, S.D., Nordlund, D.A. & Smith, R. (2003). A modified rearing system for production of *Pseudacteon curvatus* (Diptera: Phoridae), a parasitoid of imported fire ants. *Biological Control*, 28(3): 346–353.

Wäckers, F.L. (2003). The parasitoids' needs for sweets: sugar in mass rearing and biological control. In: *Quality Control and Production of Biological Control Agents—Theory and Testing Procedures*, Ed. J.C. van Lenteren, pp. 59–72. Oxon, UK: CABI Publishing.

Wang, B., Zhang, H. & Jarzembowski, E. (2013). Early Cretaceous angiosperms and beetle evolution. *Front. Plant Sci.*, 4(360): 1–6.

Waring, G.L. & Price, P.W. (1989). Parasitoid pressure and the radiation of a gall forming group (Cecidomyiidae: *Asphondylia* spp.) on creosote bush (*Larrea tridentata*). *Oecologia*, 79(3): 293–299.

Wiegmann, B.M., Trautwein, M.D., Winkler, I.S., Bar, N.B., Kim, J.W., Lambkin, C., Bertone, M.A., Cassel, B.K., Baylessa, K.M., Heimberge, A.M., Wheeler, B.M., Petersone, K.J., Pape, T., Sinclair, B.J., Skevington, J.H., Blagoderov, V., Caravas, J., Kutty, S.N., Schmidt-Ott, U., Kampmeier, G.E., Thompson, F.C., Grimaldi, D.A., Beckenbach, A.T., Courtney, G.W., Friedrich, M., Meier, R. & Yeates, D.K. (2011). Episodic radiations in the fly tree of life. *PNAS*, 108(14): 5690–5695.

Wood, D.M. (1987). Tachinidae. In: *Manual of the Nearctic Diptera*, Ed.J.F.M. McAlpine, pp. 1193–1269, Volumemac_mac 2, Ottawa: Agriculture Canada, Academic Press.

Yeates, D.K. & Greathead, D. (1997). The evolutionary pattern of host use in the Bombyliidae (Diptera): a diverse family of parasitoid flies. *Biol. J. Linn. Soc.*, 60: 149–185.

Yeates, D.K. (1997). The evolutionary pattern of host use in the Bombyliidae (Diptera): a diverse family of parasitoid flies. *Biological Journal of the Linnean Society*, 60: 149–185.

Zhang, F., Toepfer, S. & Kuhlmann, U. (2003). Basic biology and small-scale rearing of *Celatoria compressa* (Diptera: Tachinidae), a parasitoid of *Diabrotica virgifera* (Coleoptera: Chrysomelidae). *Bulletin of Entomological Research*, 93(6): 569–575.

Index

Note: Locators in *italics* represent figures and **bold** indicate tables in the text.